Quantum Spin Glasses, Annealing and Computation

Quantum annealing is a new-generation tool of information technology, which helps in solving combinatorial optimization problems with high precision, based on the concepts of quantum statistical physics.

This book focuses on the recent developments in quantum spin glasses, quantum annealing and quantum computations. It offers a detailed discussion on quantum statistical physics of spin glasses and its application in solving combinatorial optimization problems. Separate chapters on simulated annealing, quantum dynamics and classical spin models are provided for enhanced understanding. Notes on adiabatic quantum computers and quenching dynamics make it apt for the readers. This text will be useful for the students of quantum computation, quantum information, statistical physics and computer science.

Shu Tanaka is Assistant Professor at Waseda Institute for Advanced Study, Waseda University, Tokyo, Japan. He is also PRESTO Researcher, JST. He has worked in various fields of the theory of statistical physics, materials science, and quantum information theory. He has studied both of fundamental and application sides of quantum annealing. He has also edited three books of Kinki University Series on Quantum Computing. He got the 9th Award for the Encouragement of Young Physicists from the Physical Society of Japan (2015).

Ryo Tamura is Researcher at International Center for Materials Nanoarchitectonics (MANA) in National Institute for Materials Science (NIMS), Ibaraki, Japan. He has worked in the field of materials science, in particular, his research interests focus on magnetic materials including random spin systems and materials informatics.

Bikas K. Chakrabarti is Professor of Theoretical Condensed Matter Physics at Saha Institute of Nuclear Physics (SINP), Kolkata, India. His research interests include physics of fracture, quantum glasses and the interdisciplinary sciences of optimisation, brain modelling, and econophysics. Professor Chakrabarti is a recipient of the S. S. Bhatnagar Award (1997). He has also received the Outstanding Referee Award of the American Physical Society (2010). He is fellow of Indian National Science Academy & Indian Academy of Sciences.

Quantum Spin Glasses, Annealing and Computation

Shu Tanaka
Ryo Tamura
Bikas K. Chakrabarti

CAMBRIDGE
UNIVERSITY PRESS

CAMBRIDGE
UNIVERSITY PRESS

Shaftesbury Road, Cambridge CB2 8EA, United Kingdom

One Liberty Plaza, 20th Floor, New York, NY 10006, USA

477 Williamstown Road, Port Melbourne, VIC 3207, Australia

314–321, 3rd Floor, Plot 3, Splendor Forum, Jasola District Centre,
New Delhi – 110025, India

103 Penang Road, #05–06/07, Visioncrest Commercial, Singapore 238467

Cambridge University Press is part of the University of Cambridge.

It furthers the University's mission by disseminating knowledge in the pursuit of education, learning and research at the highest international levels of excellence.

www.cambridge.org
Information on this title: www.cambridge.org/9781107113190

First Published 2017
Reprint 2018, 2019, 2020, 2022, 2025

Printed in India by Thomson Press India Ltd.

A catalogue record for this publication is available from the British Library

ISBN 978-1-107-11319-0 Hardback

Additional resources for this publication at www.cambridge.org/9781107113190

This book is dedicated to the memory of Prof. Jun-ichi Inoue

The book contains three important notes contributed by Eliahu Cohen, Uma Divakaran, Sudip Mukherjee, Atanu Rajak and Boaz Tamir.

Contents

Figures xiii
Tables xxxi
Preface xxxiii

1 Introduction

References 3

Part One Quantum Spin Glass, Annealing and Computation

2 Classical Spin Models: From Ferromagnetic Spin Systems to Spin Glasses

2.1 Ising Model 8
2.1.1 Phase transition and critical phenomena 8
2.1.2 Mean-field approximation and infinite-range model 10
2.1.3 Ising model on finite dimensional lattices 15
2.2 Basic Concept of Classical Spin Glass 22
2.2.1 Frustration and randomness 22
2.2.2 Spin glass order parameter 25
2.2.3 Replica method 26
2.3 Sherrington–Kirkpatrick Model 26
2.3.1 Free energy of the Sherrington–Kirkpatrick model 27
2.3.2 Replica symmetric solution 32
2.3.3 Replica symmetry breaking 36
2.4 Edwards–Anderson Model on the Finite Dimensional Lattices 42
2.A Several Spotlights on Classical Models 45
2.A.1 Potts model 46
2.A.2 XY model 48
2.A.3 Heisenberg model 50

References 51

3 Simulated Annealing

3.1 Relation Between Random Ising Models and Combinatorial Optimization Problems 60
3.2 Representative Combinatorial Optimization Problems 64
3.2.1 Computational complexity theory 64
3.2.2 Classification of combinatorial optimization problems 66
3.2.3 Examples of combinatorial optimization problems 67
3.3 Conventional Computation Methods Including Simulated Annealing 73
3.3.1 Monte Carlo methods 73
3.3.2 Mean-field methods 77
3.3.3 Preceding theoretical studies on simulated annealing 79
3.4 Convergence Theorem for Simulated Annealing 84
3.4.1 Inhomogeneous Markov chain 84
3.4.2 Ergodicity 85
3.4.3 Proof of the convergence theorem 86
References 97

4 Quantum Spin Glass

4.1 Ferromagnetic Ising Model in a Transverse Field 98
4.1.1 Semiclassical analysis and a simple mean-field approximation of the ferromagnetic Ising model in a transverse field 100
4.1.2 Phase transition in the finite-dimensional ferromagnetic Ising models in a transverse field 103
4.1.3 Hushimi-Temperely–Curie–Weiss model in a transverse field 110
4.2 Introduction of Quantum Spin Glass 115
4.3 Sherrington-Kirkpatrick Model in a Transverse Field 117
4.3.1 Mean-field calculations 118
4.3.2 Simulation results 129
4.4 Edwards–Anderson Model in a Transverse Field 137
4.5 Existence of Replica Symmetric Spin Glass Phase in the Random Quantum Ising Model 142
References 146

5 Quantum Dynamics

5.1 Landau–Zener Transition 151
5.2 Kibble–Zurek Mechanism 158
5.2.1 Basic concept of the Kibble–Zurek mechanism 159
5.2.2 Kibble–Zurek mechanism in the Ising chain with random ferromagnetic interactions 162
References 167

6 Quantum Annealing

6.1 General Concept of Quantum Annealing 170
6.2 Quantum Annealing by the Schrödinger Equation 174
6.3 Quantum Annealing by the Path-integral Monte Carlo Method 180
6.4 Quantum Annealing by the Green's Function Monte Carlo Method 185
6.5 Quantum Annealing by the Density Matrix Renormalization Group 191
6.6 Quantum Annealing by the Mean-field Approximation 195
6.7 Quantum Field Response to Magnetic Materials 201
6.8 Quantum Adiabatic Evolution 203
6.8.1 Adiabatic quantum computation 204
6.8.2 Quantum adiabatic theorem 205
6.8.3 Quantum adiabatic approximation 215
6.9 Convergence Theorem for Quantum Annealing 222
6.9.1 Convergence theorem for quantum annealing by the Schrödinger equation 222
6.9.2 Correspondence between classical and quantum systems 226
6.9.3 Convergence theorem for quantum annealing by the path-integral Monte Carlo method 231
6.9.4 Convergence theorem for quantum annealing by the Green's function Monte Carlo method 242
References 248

Part Two Additional Notes

7 Notes on Adiabatic Quantum Computers

7.1 Introduction 255
7.2 The Circuit Model 256
7.3 Adiabatic Computation 259
7.3.1 Simulated annealing and adiabatic computation 266
7.3.2 Different paths from an initial to a final eigenstate 267
7.3.3 Imaginary time and simulations 268
7.3.4 Complexity and universality 268
7.3.5 Additional methods 270
7.4 Quantum Annealing 270
7.4.1 Relation between simulated annealing, quantum annealing, and adiabatic computation 272
7.5 D-Wave Computer 274
7.5.1 Is the D-Wave a quantum computer? The future of quantum annealers 276
7.5.2 Universality 276
7.5.3 Coherence time of the SQUIDs 277

7.5.4 Scalability 278
7.5.5 Speed up 278
7.5.6 "Quantumness" 279
7.5.7 Does the computer exhibit entanglement? 281
7.5.8 Open questions and possible future routes 281

References 282

8 Quantum Information and Quenching Dynamics

8.1 Quantum Information Theoretic Measures: Indicator of a QCP 290
8.1.1 Quantum fidelity 290
8.1.2 Loschmidt echo 292
8.1.3 Entanglement 295
8.2 Dynamics in Quantum Manybody Systems 296
8.3 Fidelity Susceptibility and Loschmidt Echo for Generic Paths 296
8.3.1 Model and phase diagram 297
8.3.2 Fidelity susceptibility 299
8.3.3 Loschmidt echo 304
8.4 Loschmidt Echo: Effect of Gapless Phase 307
8.4.1 Model, phase diagram and anisotropic quantum critical point (AQCP) 307
8.4.2 Loschmidt echo for the Kitaev model 310
8.5 Effect of Double Local Quenches on Loschmidt Echo and Entanglement Entropy 313
8.5.1 Model 314
8.5.2 Semiclassical theory of quasiparticles 315
8.5.3 Loschmidt echo and entanglement entropy for a critical chain 316
8.5.4 Entanglement entropy: Ferromagnetic phase 321
8.5.5 QP^1 and QP^2: A comparative study 323
8.6 Topology in Condensed Matter Systems 324
8.6.1 Majorana fermions and condensed matter systems 326
8.6.2 Kitaev chain: Model and topological phases 327
8.7 One-dimensional Generalized Cluster Model 333
8.8 Sudden Quench: Dynamics of a Majorana Edge State 335
8.8.1 Quenching schemes and results 337
8.9 Slow Quench: Dynamics of a Majorana Edge State 340
8.9.1 Model, phase diagram and energy spectrum 341
8.9.2 Quenching dynamics of a Majorana edge state 343
8.10 Summary and Conclusion 347

References 350

9 A Brief Historical Note on the Studies of Quantum Spin Glasses, Annealing and Computation

9.1 Introduction 359
9.2 A Short History of the Development 360
9.3 Review of Some Recent Discussion Papers and Arxiv Preprints 367
9.4 Applications of QA 374
9.5 Summary and Discussions 375
References 379

Index 383

Figures

2.1 Temperature dependence of the magnetization per spin m of the Ising model with homogeneous ferromagnetic interactions by the mean-field approximation. We display the result of $h \to +0$ for $h = 0$. 12

2.2 Lattice structures of the infinite-range model given by Eq. (2.25) for $N = 4$ (left) and $N = 8$ (right). 14

2.3 Lattice structures of (a) chain (one-dimensional lattice), (b) square lattice (two-dimensional lattice), and (c) cubic lattice (three-dimensional lattice). 16

2.4 Temperature dependence of the exact solution for magnetization per spin of the Ising model on a chain under finite magnetic fields. 18

2.5 Temperature dependence of the exact solution of magnetization per spin of the Ising model on a square lattice when the magnetic field is zero. The dotted curve is the result from the mean-field approximation. The transition temperature from the exact solution is lower than that from the mean-field approximation. 19

2.6 Monte Carlo simulation results of the Ising model on a cubic lattice. The linear dimension is L. (a) Temperature dependence of the Binder ratio g of the magnetization. At $T_c/J \simeq 4.51$, the Binder ratios cross at the same point. The curves between points are a guide to the eye. (b) Finite-size scaling results of the Binder ratio of the magnetization using Eq. (2.59), where $T_c/J = 4.51$ and $\nu = 0.627$. 21

2.7 (a) Ground state spin configuration of the Ising model with homogeneous ferromagnetic interactions on a square lattice. The magnetic interaction between nearest-neighbor spins indicated by the solid line is the ferromagnetic interaction ($J_{ij} = +J$). (b) Ground state spin configuration of the Ising model with homogeneous antiferromagnetic interactions on a square lattice. The magnetic interaction between nearest-neighbor spins indicated by the dotted line is the antiferromagnetic interaction ($J_{ij} = -J$). (c) Ground state spin configuration of the $\pm J$ model. $J_{ij} = \pm J$ are randomly distributed. The plaquette marked by the star indicates the frustration plaquette and an energy loss exists at interactions indicated by the crosses. The energy is the same if the spins surrounded by the dashed squares are flipped. 23

2.8 Phase diagram of the SK model under the replica symmetric approximation. Depending on the temperature T/J and the mean value of interactions J_0/J, paramagnetic phase, ferromagnetic phase, and spin glass phase appear. 35

2.9 Phase diagram of the SK model including the replica symmetry breaking. In the mixed phase, the ferromagnetic order and the spin glass order coexist. The dashed line indicates the AT line. Below the AT line, the replica symmetric solution becomes unstable. 37

2.10 (a) Schematic of $q(x)$ defined in $1 \le x \le n$, that is, Eq. (2.131) is satisfied. (b) Schematic of $q(x)$ defined in $0 \le x \le 1$, that is, Eq. (2.133) is satisfied. (c) Schematic of $q(x)$ in the limit of $M \to \infty$. $q(x)$ becomes a continuous increasing function defined in $0 \le x \le 1$. 39

2.11 Schematic of the many-valley structure of the free energy landscape in the spin glass phase. 41

2.12 Temperature dependence of the correlation length ξ_L obtained by Monte Carlo simulations for the Edwards–Anderson model on a square lattice. The probability distribution of magnetic interactions is the Gaussian distribution with zero mean and standard deviation unity. The slope of this figure is the critical exponent ν (from Katzgraber et al. 2004). 43

2.13 Results obtained by Monte Carlo simulations of the Edwards–Anderson model on a cubic lattice. The probability distribution of magnetic interactions is the Gaussian distribution with zero mean and standard deviation unity. (a) Inverse temperature, $\beta = T^{-1}$, dependence of the Binder ratio of the spin glass order parameter. (b) Finite-size scaling result of the Binder ratio of the spin glass order parameter when $T_c = 0.931$ and $\nu = 2.67$ are used (from Katzgraber et al. 2006). 45

2.14 Schematic of the boundary between first-order and second-order phase transitions depending on the value of q and the spatial dimension d in the Potts model (Wu, 1982). 46

3.1 Typical snapshots of the ferromagnetic Ising model on a square lattice whose linear dimension is 128; these were obtained by Monte Carlo simulations. (Upper panel) Snapshots for rapid cooling case. (Lower panel) Snapshots for slow cooling case. 62

3.2 Schematics of the computational complexity for (a) P = NP and (b) P $\neq$ NP. 65

3.3 (a) Position of cities of a traveling salesman problem for $N = 6$. (b) A non-best solution. The lines denote the selected path. (c) The best solution where the length of path is the minimum value. 69

3.4 (a) Position of vertices of a vertex coloring problem for $N = 6$. (b) The answer for the three-coloring problem ($k = 3$). (c) The "best" solution for the two-coloring problem ($k = 2$). This solution does not satisfy the constraint: two adjacent vertices do not share the same color. 72

3.5 Two-level systems considered by Huse and Fisher (1986). 80

3.6 Energy landscape of the model considered by Shinomoto and Kabashima (1991). 83

4.1 (Left panel) A schematic of classical phase transition. The horizontal and vertical axes represent the temperature and the order parameter, respectively. (Right panel) A schematic of quantum phase transition. The horizontal and vertical axes represent the quantum field such as the transverse field and the order parameter, respectively. 99

4.2 $\Gamma/J(0)$-dependence of $\langle\hat{\sigma}^z\rangle$ (solid line) and $\langle\hat{\sigma}^x\rangle$ (dotted line). When $\Gamma \geq J(0)$, quantum paramagnetic state ($\langle\hat{\sigma}^x\rangle = 1$) appears. 102

4.3 Phase diagram of the ferromagnetic Ising model in a transverse field at the mean-field level. 103

4.4 (Left panel) Phase diagram of the ferromagnetic quantum Ising model in one dimension. The bold line indicates the ferromagnetic phase and the rest is the paramagnetic phase. (Right panel) Schematic phase diagram of the ferromagnetic quantum Ising model in $d = 2, 3$ dimensions. The bold curve indicates the boundary between the ferromagnetic and paramagnetic phases. 104

4.5 Schematic representations of the Suzuki–Trotter decomposition. 106

4.6 Specific heat C and the internal energy E as a function of temperature for a transverse Ising chain at $\Gamma/J = 1/2$. The diamond, solid, and open circles represent the quantum Monte Carlo data for $P = 4, 8$, and 12, respectively. The solid curves are exact solutions (from Nagai et al., 1986). 108

4.7 Specific heat C and the longitudinal susceptibility χ for the square lattice whose lattice size is 24×24 at $\Gamma/J = 1$. The data was obtained by quantum Monte Carlo simulations for $P = 24$. The solid curves are a guide to the eye (from Nagai et al., 1986). 109

4.8 Phase boundaries for the transverse Ising models on square ($d = 2$) and cubic ($d = 3$) lattices. The solid and broken curves represent results obtained by the series expansion (from Nagai et al., 1986). 110

4.9 Phase diagram of the Husimi–Temperely–Curie–Weiss model in a transverse field Γ for $h \to +0$. 115

4.10 Transverse field, Γ, dependence of the longitudinal magnetization m_z of the Husimi–Temperely–Curie–Weiss model in a transverse field Γ when $T = 0$. The quantum phase transition occurs at $\Gamma_c/zJ = 1$. 116

4.11 Schematic phase diagram of the Ising model with random interactions in a transverse field Γ. O_{cl}, the spin glass phase, is characterized by the classical spin glass order. In the region indicated by O_q, there are possibilities of the existence of the replica symmetric spin glass phase due to the quantum fluctuation effect. 117

4.12 Phase diagram of the SK model in a transverse field under the single-spin approximation. 119

4.13 Phase diagram of the SK model in a transverse field under the replica symmetric approximation and static approximation. 129

4.14 (Left panel) Phase diagram of the SK model in a transverse field obtained by the replica symmetry breaking solution without using static approximation (from Goldschmidt and Lai, 1990). (Right panel) Phase diagram of the SK model in a transverse field obtained by the replica symmetric solution without using static approximation. "Perturbative" denotes the naive Landau expansion result which is the same as in Fig. 4.12. m stands for the number of modes expressing the dynamical nature; $m = 0$ is the result of static approximation. "Quantum" denotes the result of the improved Landau expansion in a quantum regime where the quantum fluctuation is dominant. The quantum regime is shown by the shaded portion (from Takahashi, 2007). 130

4.15 Monte Carlo simulation results of the SK model in a transverse field in a high-temperature region for $J = 1$. In these cases, the classical fluctuation is dominant. (a) Temperature, T, dependence of the Binder ratio g for $\Gamma = 0$. (b) Scaling plot of the Binder ratio g when $x_T = 0.31$ for $\Gamma = 0$. (c) Transverse field, Γ, dependence of g for $T = 0.65$. (d) Scaling plot of g when $x_\Gamma = 0.31$ and $T = 0.65$. (e) Γ-dependence of g for $T = 0.6$. (f) Scaling plot of g when $x_\Gamma = 0.31$ and $T = 0.6$ (from Mukherjee et al., 2015). 133

4.16 Monte Carlo simulation results of the SK model in a transverse field in a low-temperature region for $J = 1$. In these cases, the quantum fluctuation is dominant. (a) Transverse field, Γ, dependence of the Binder ratio g for $T = 0.30$. (b) Scaling plot of the Binder ratio g when $x_\Gamma = 0.50$ for $T = 0.30$. (c) Γ-dependence of g for $T = 0.25$. (d) Scaling plot of g when $x_\Gamma = 0.50$ for $T = 0.25$ (from Mukherjee et al., 2015). 134

4.17 Exact diagonalization results of the SK model in a transverse field at zero temperature for $J = 1$. (a) Transverse field, Γ, dependence of the Binder ratio

g. (b) Scaling plot of the Binder ratio g when $x_T = 0.51$ and $\Gamma_c = 1.62$ (from Mukherjee et al., 2015). 136

4.18 Schematic phase diagram of the SK model in a transverse field obtained by simulation results. The solid and dotted curves indicate the phase boundary dominated by the classical fluctuation and that by the quantum fluctuation, respectively. At $T/J \simeq 0.45$, crossover occurs. 137

4.19 L_τ-dependence of g for various K for the effective classical model corresponding to the EA model in a transverse field on the square lattice. The diamond, plus, square, cross, and triangular points indicate results for $L = 4$, 6, 8, 12, and 16, respectively. At $K^{-1} = 3.30$ (center figure), the maximum values at the peak become about the same value for various L. This result was obtained by Monte Carlo simulations (from Rieger and Young, 1994). 139

4.20 (a) Scaling plot of the Binder ratio g when the value of z is 1.5 for the effective classical model corresponding to the EA model in a transverse field on the square lattice. (b) First derivative of Binder ratio with respect to K at K_c : $g'(K = K_c)$ depending on L. The slope of the fitting line is $1/\nu = 1.0$. This result was obtained by Monte Carlo simulations (from Rieger and Young, 1994). 140

4.21 Scaling plot of the Binder ratio g when $K^{-1} = T = 4.3$ and $z = 1.3$ for the effective classical model corresponding to the EA model in a transverse field on the cubic lattice. This result was obtained by Monte Carlo simulations (from Guo et al., 1994). 141

4.22 K^{-1}-dependence of the Binder ratio g for the effective classical model corresponding to the EA model in a transverse field on the cubic lattice. Inset is the scaling plot of the Binder ratio g when $K^{-1} = T = 4.3$ and $1/\nu = 1.3$. This result was obtained by Monte Carlo simulations (from Guo et al., 1994). 142

4.23 Overlap distribution function $P_N(q)$ which is indicated by $W(q)$ in this figure as a function of q for various lattice sizes N at $\Gamma/J = 0.15$ and $T/J = 0.4$, which is located in the spin glass phase of the SK model in a transverse field. The dotted curve indicates the upper profiles of $P_N(q)$ (from Ray et al., 1989). 144

4.24 Overlap distribution function $P_N(q)$ which is indicated by $P(q)$ in this figure as a function of q for various lattice sizes N at $\Gamma/J = 0.4$ and $T/J = 0.4$, which is located in the spin glass phase of the SK model in a transverse field. The plus, cross, diamond, square, and circle points indicate results for $N = 20$, 40, 60, 80, and 100, respectively (from Lai and Goldschmidt, 1990). 145

5.1 Landau–Zener transition of a system whose Hamiltonian is given by $\mathscr{H} = -h(t)\sigma^z - \Gamma\sigma^x$, where $h(t) = -t/\tau$. (a) The instantaneous eigenenergies of the system. (b) Γ-dependence of the expectation value of σ^z as a function of $h(t)$ with $\tau = 100$. Here, the initial state is set to the ground state for $h = -5$. (c) τ-dependence of the expectation value of σ^z as a function of $h(t)$ with $\Gamma = 0.1$. Here, the initial state is set to the ground state for $h = -5$. 158

5.2 (a) Schematics of the temperature dependence of correlation length $\xi(g(t))$ in simulated annealing where the annealing speed τ_Q^{-1} ($\tau_{Q1} < \tau_{Q2} < \tau_{Q3}$). $T_i := T(t_s)$ for $i = 1,2,3$ denotes the temperature at which the system cannot reach the equilibrium state under simulated annealing whose annealing speed is τ_{Q_i}. The stopped correlation length is written as $\xi_i := \xi(g(t_s))$ for $i = 1,2,3$. Here, $\xi_1 < \xi_2 < \xi_3$. (b) Schematics of the temperature dependence of the density of the domain wall $n(t)$ in simulated annealing where the annealing speed τ_Q^{-1} is different. The stopped density of the domain wall is written as $n_i := n(t_s)$ for $i = 1,2,3$. Here, $n_1 > n_2 > n_3$. 161

6.1 Schematic of simulated annealing and quantum annealing. The purpose of these methods is to obtain the best solution at the position indicated by the star ($T = 0, \Gamma = 0$). 172

6.2 Schematic of the concept of quantum annealing to search for the ferromagnetic ground state. 172

6.3 Ground state spin configuration of the Ising model with homogeneous ferromagnetic interactions for $N = 8$. The magnetic field is applied for all spins, i.e., $h_i = 0.1J$ for all i to avoid the trivial degeneracy when the transverse field is absent. 177

6.4 Schematic of the schedules of annealing methods defined by Eqs. (6.21), (6.22), and (6.23). 178

6.5 Time dependence of $P_{QA}(t)$, $P_{QA}^{st}(\Gamma(t))$, $P_{SA}(t)$, and $P_{SA}^{st}(T(t))$ for the Ising model with homogeneous ferromagnetic interactions. (a) The schedule of annealing is $\Gamma(t)/J = T(t)/J = 3/\ln(t+1)$. (b) The schedule of annealing is $\Gamma(t)/J = T(t)/J = 3/\sqrt{t}$ (from Kadowaki and Nishimori, 1998). 179

6.6 Time dependence of $P_{QA}(t)$, $P_{QA}^{st}(\Gamma(t))$, $P_{SA}(t)$, and $P_{SA}^{st}(T(t))$ of an Ising model with random interactions. The schedule of annealing is $\Gamma(t)/J = T(t)/J = 3/\sqrt{t}$ (from Kadowaki and Nishimori, 1998). 180

6.7 Monte Carlo step, τ, dependence of residual energy $\varepsilon_{res}(\tau)$ obtained by simulated annealing (classical annealing: CA) and quantum annealing for various temperature T and the Trotter number P when $PT = 1$ (from Martoňák et al., 2002). 182

6.8 Monte Carlo steps, τ, dependence of the excess length after annealing ε_{exc} obtained by simulated annealing (SA) and quantum annealing (QA). In order to compare the results by the simulation time, the result of QA multiplied by P is also shown. The dotted line indicates the value of $\varepsilon_{exc} \simeq 1.57$ obtained by the Lin–Kernigham algorithm which is a standard local-search algorithm (from Martoňák et al., 2004). 183

6.9 (Left panel) Total annealing time (Monte Carlo step) dependence of the residual energy after simulated annealing (classical annealing, CA) and quantum annealing (QA). The dotted line indicates the Gardner energy for the 3-SAT problem, which is a lower bound of the energy of the metastable states. Inset is the Trotter number, P, dependence of the residual energy for the 3-SAT problem (Battaglia et al., 2004). (Right panel) Schedule of the temperature and the transverse field $J_\Gamma = \ln\coth(\beta\Gamma(t)/P)/2\beta$ in field cycling. (from Battaglia et al., 2005). 184

6.10 Monte Carlo step, τ, dependence of the residual energy obtained by quantum annealing based on the Green's function Monte Carlo method (GFMC) and the path-integral Monte Carlo method (PIMC-QA), and simulated annealing (classical annealing: CA) for the Ising model with random interactions. The results of PIMC-QA and CA are from Fig. 6.7. Two types of simulations for GFMC were performed using different trial wave function. When the trial wave function is constant, results of upper diamonds are obtained (GFMC $\psi_T = 1$). On the other hand, when the trial wave function is the Boltzmann type, results of lower diamonds are obtained (GFMC-QA). The time unit in GFMC is a step that a single-spin flip event is tried, while in PIMC-QA and CA, time unit is a step that the update of all spins is tried. Thus, in order to compare these results when the time unit is a single-spin flip, the results of PIMC-QA and CA multiplied by N are also shown (from Stella and Santoro, 2007). 190

6.11 Schematic of division of blocks in DMRG. 192

6.12 Residual energy, E_{res}, dependence of τ when $J = 1.6$ obtained by the density matrix renormalization group for the random field Ising model. The dotted line is the guide to the eyes for logarithmic scaling (from Suzuki and Okada, 2007). 195

6.13 Residual energy depending on the annealing time τ when (a) $J = 2.0$ and (b) $J = 0.6$ for the random field Ising model on the square lattice by mean-field calculation based on the Bethe approximation. The obtained results of quantum annealing by transverse field defined by Eq. (6.83) (TF) and by transverse field and ferromagnetic interaction defined by Eq. (6.117) (FI) and simulated annealing (thermal) are compared. These results are averaged out over eighty configurations of random fields (from Suzuki et al., 2007). 201

6.14 Phase diagram of the magnetic materials $\mathrm{LiHo}_x\mathrm{Y}_{1-x}\mathrm{F}_4$. The horizontal axis corresponds to the concentration x of Ho^{3+} ions, and the vertical axis is the temperature. PM, FM, and SG denote the paramagnetic phase, ferromagnetic phase, and spin glass phase, respectively. The arrow shows the location of the decoupled cluster glass in which essentially isolated clusters with zero net moments appears at zero temperature. The coupling between these clusters increases with temperature; decoupled cluster glass does not appear at a finite temperature (from Reich et al., 1990). 202

6.15 Schematic phase diagram of the temperature T versus the magnetic field H_t corresponding to Γ of $LiHo_xY_{1-x}F_4$ for $x = 0.56$. The arrows indicate the quantum annealing (QA) process and the thermal annealing (simulated annealing: SA) process toward the point at $H_t = 7.2$ kOe under $T = 0.03$ K in the spin glass phase, respectively. 203

6.16 Time evolution of the coefficients of $\mathscr{H}_0$ and $\mathscr{H}_q$ in the adiabatic quantum computation. 204

7.1 Quantum annealing vs thermal annealing in a graph of energy as a function of configuration space. 256

7.2 The set of amplitudes following the first step of the Grover iteration in the 2-qubit case. 257

7.3 The set of amplitudes following the second step of the Grover iteration in the 2-qubit case. 258

7.4 8 qubit cell. 275

7.5 The vertex 0 in G is mapped into two such vertices to satisfy the physical requirements of C_n, that is, no more than 4 couplings to each physical qubit. 275

8.1 The phase diagram of the one-dimensional three-spin interacting TFIM with different paths along which the MCP (point A) is approached. Three different critical lines are shown satisfying the relations $h = 1$, $h = J_x - 1$ and $h = -J_x - 1$, whereas, Paths I, II, III and IV are denoted by lines with arrows. A shaded region also has been shown in the figure where quasicritical points exist (from Rajak and Divakaran, 2014). 298

8.2 The plot shows χ_F as a function of h at $J_x = 2, J_3 = -1$. Here the system size is $L = 100$. The first peak that occurs at $h = -3$ is an Ising critical point which shows linear scaling with L, and the second peak occuring at $h = 1$ corresponds to the MCP where L^5 scaling is found. The inset figure shows that the FS oscillates close to the MCP, indicating the presence of quasicritical points there (from Rajak and Divakaran, 2014). 301

8.3 The scaling of χ_F when the MCP is approaced along four different paths. The plots shown in (a), (b) and (c) are for Paths I, II and III resulting in $\chi_F \propto L^5$ that shows the effect of quasicritical points; whereas it is linear in L for Path IV as shown in (d) due to the absence of quasicritical points along this path or line (from Rajak and Divakaran, 2014). 302

8.4 Variation of the LE as a function of h. The sharp dips in the LE are seen at the Ising critical points ($h = -2, 0$) as well as at anisotropic critical point at $h = 1$ (from Rajak and Divakaran, 2014). 305

8.5 Variation of LE as a function of scaled time t/L^{α} to highlight the scaling of time period with system size L, with α being the scaling exponent ($T \propto L^{\alpha}$) as obtained in the text for various types of critical points, i.e., (a) the anisotropic critical point where $T \propto L$, (b) multicritical point where $T \propto L^2$ and (c) quasicritical point with $T \propto L^3$. In (d), we present the almost linear variation of $\ln \mathscr{L}$ with t^2 for small times with $L = 100$ at the various critical points confirming the general small time behavior given by $\mathscr{L} \sim e^{-\Gamma t^2}$ (from Rajak and Divakaran, 2014). 306

8.6 The figure shows a schematic representation of the Kitaev model on a honeycomb lattice with the interactions denoted by J_1, J_2 and J_3; $\vec{M}_1 = \frac{\sqrt{3}}{2}\hat{i} - \frac{3}{2}\hat{j}$ and $\vec{M}_2 = \frac{\sqrt{3}}{2}\hat{i} + \frac{3}{2}\hat{j}$ are the spanning vectors of the lattice (from Sharma and Rajak, 2012). 307

8.7 The phase diagram of the Kitaev model in which all the points satisfy the relation $J_1 + J_2 + J_3 = 4$. The gapless phase of the model is given by the inner equilateral triangle satisfying the relations $J_1 \leq J_2 + J_3$, $J_2 \leq J_3 + J_1$ and $J_3 \leq J_1 + J_2$. The study of LE is carried out along Paths I, II and III where J_3 is varied (from Sharma and Rajak, 2012). 309

8.8 The plots show variation of LE with the parameter J_3 (J_3 is changed Path I). The LE shows a sharp dip at $J_3 = 2 - \delta$ (point A) and following a small but finite revival throughout the gapless phase, it further decays at point B, $J_3 = 0$ (see Fig. (8.7)). Here, $N_x = N_y = 200$, $\delta = 0.01$ and $t = 10$. Inset (a) shows LE as a function of J_3 when the change in J_3 is carried out along Path II ($J_1 = J_2 + 1$) with $N_x = N_y = 200$, $\delta = 0.01$ and $t = 10$ exhibiting a sharp dip at point P ($J_3 = 2 - \delta$) and again there is a sharp increase at point Q ($J_3 = 1 - \delta$). Inset (b) shows the dip in LE at $J_3 = 2 - \delta$ when J_3 is changed along the line $J_1 + J_3 = 4$ (Path III). For this case, $N = 400$, $\delta = 0.01$ and $t = 10$ (from Sharma and Rajak, 2012). 311

8.9 The LE as a function of time t at the AQCP A (see Fig. 8.7) with $J_1 = J_2 = 1$, $\delta = 0.01$ and $J_3 = 2 - \delta$, considering $N_x (= 100)$ as fixed and N_y as changing confirms the scaling relations discussed in the text. The inset figure shows that the time period of the collapse and revival does not depend on N_x (from Sharma and Rajak, 2012). 312

8.10 The variation of LE with time for different values of δ when J-quench and h-quench are performed at $L_1 = L/2$ and $L' = L/3$ respectively. The h-quenching here is made by changing its value from 1 to $1 + \delta$ at L'. In the case of only J-quenching (i.e., $\delta = 0$ case), the LE marks a peak at $t_3 = L/v_{\max} = 150$ and $T = 2t_3$. Here, $v_{\max} = 2$ and $L = 300$. By applying both the local quenches simultaneously at time $t = 0$, we find a small peak at $t' = 50$ and a relatively stronger peak at $t'' = 200$. It can also be noted that there are small fluctuations close to $t_1 = 100$ which is more clear for the $\delta = 1$ curve (from Rajak and Divakaran, 2016). 317

8.11 Main: The time evolution of the LE for two local quenches having $L_1 = 100, L_A = 100$ and $L = 300$ and different L'. The first peak in the LE is seen at time $t' = 2(L_1 - L')/v_{\max}$; $t' = 25$ and 50 for $L' = L/4$ and $L/6$ respectively. The other time scales are given by $t_3 = 100$, $t_4 = 200$ and t''. It can be noted that $t'' = 225$ and 250 for $L' = 75$ and 50, respectively. Inset: The LE as a function of time t for only h-quench at $L/3$ with $L = 300$. Time scales $t_1 = 100$ and $t_2 = 200$ both are observed in this case. It is easy to find that the magnitude of decay in this case is very small (from Rajak and Divakaran, 2016). 318

8.12 The plot shows time evolution of the entanglement entropy having $L_1 = L_A = 100$ and different values of L'. For $L' = L/4$, the departure from the only J-quench case is seen at $t'/2 = 12.5$, whereas at $t' = 25$, QP_L^2 get in the system B after reflection at L' in which QP_R^2, the other partner, is already present. The EE keeps on decreasing till QP_R^2 enters system A at time $t = L_2 = 200$. A dip is also seen at $t = t'' = 225$ where QP_L^2 and QP_R^2 interchanges their systems. In a similar fashion, we can make an argument for the time evolution of EE when $L' = 3L/4$. The deviation from the case of J-quench alone starts at $t = 62.5$. There is a sharp decay in EE at $t = 100$ when QP_L^2 enters system B. On the other hand, QP_R^2 goes into system A at $t = 125$ that results in an increase in the EE since its other partner remains in system B. The normal rise at $t = 200$ which is clearly present for the single quenching (J-quenching) case can also be found here. This may be an artefact of getting a transmitted component of the QP wavepacket at L' which indicates that the reflection of the QPs at L' is not perfect, as already discussed in Sec. 8.5.4. We can also observe the time scales $t'' = 225$ and $T = 300$ (from Rajak and Divakaran, 2016). 320

8.13 (a) Variation of the EE as a function of time following single and double quenches when the total spin chain is in the ferromagnetic region at $h = 0.5$. Here, we consider $L_1 = 100$, $L_A = 100$ and $L = 300$ and different L'. For $L' = L/4$, the EE deviates from the only J- quenching case, starting at $t = t'/2$ with $t' = 50$ and $v_{\max} = 2h = 1$. For time $t > 50$, both the quasiparticles QP_L^2 and QP_R^2 exist in subsystem B that leads to the decrease in the EE continuing up to $t_4 = 400$. Again, for $L' = L/5$, we find $t' = 80$ and observe the expected behavior. The explanation of the absence of t'' is given in the text. The plot (b) is the same with $h = 0.99$. The observed time scales are given by $t' = 25.25$, $t'/2$, $t_3(\sim 101)$ $t_4(\sim 202)$ and $t''(\sim 227)$ calculated with $v_{\max} = 1.98$ (from Rajak and Divakaran, 2016). 322

8.14 The plot shows f_q as a function of q for a critical chain with $L_1 = 100$, $L' = 60$ and $L = 300$. The dashed line represents the case of the h-quench alone which is clearly smaller than the J-quench alone (the dotted line) by an order of magnitude. The solid line is for the double quench, which almost overlaps with the J-quenching case. A similar type of behavior can also be observed for a spin chain residing in the ferromagnetic region (from Rajak and Divakaran, 2016). 324

8.15 Phase diagram of the p-wave superconducting model in one dimension (see Eq. (8.52)). In this diagram, Phases I and II are topologically non-trivial, whereas, Phase III is topologically trivial. Sudden quenching in this system is performed along Paths A and B (see Sec. 8.8). 329

8.16 Schematic representation of the Majorana chain with the Hamiltonian as defined in Eq. (8.57), considered a virtual ladder. It may be thought that the middle of each vertical bond here represents a fermionic site j which supports two Majorana fermions a_j and b_j represented by black and gray filled circles respectively with their intra-site interaction as μ. In this representation, other two couplings are symbolized by diagonal links. 330

8.17 Schematic representation of the Hamiltonian in Eq. (8.57) with different special conditions as defined in the text. (a) $\mu \neq 0$, $w = \Delta = 0$, (b) $\mu = 0$, $w = \Delta$ and (c) $\mu = 0$, $w = -\Delta$. In (a), the pair of Majoranas at each site is connected to each other by the strength μ resulting in a topologically trivial phase. (b) In this regime, there are two isolated Majorana modes a_1 and b_N at the left end and right end respectively. (c) This regime also represents a topologically non-trivial phase with two isolated Majorana modes b_1 and a_N. 331

8.18 Two Majorana edge states are localized at two ends of a 100-site open Majorana chain in Phase I ($\Delta = 0.1$ and $\mu = 0.0$) with an exponential decay into the bulk of the chain. j labels the Majorana sites $1, 2, \ldots 200$ (from Rajak and Dutta, 2014). 332

8.19 (a)–(c) Schematic representation of three terms in Eq. (8.63) using the language of Majorana fermions where (a), (b), and (c) describe the first, second, and third terms respectively. The unpaired Majorana modes enclosed in dotted lines exist at two ends of the chain. (d) The plot shows the phase diagram of the generalized cluster model of Eq. (8.63) at zero temperature for $J^{XZX} > 0$. The thick solid curves are the phase boundaries where the energy gap vanishes at a particular k. The different phases like ferromagnetic, antiferromagnetic, paramagnetic, cluster and dual cluster are represented by F, AF, P, C and C* respectively. The superscripts in F and AF denote the direction of the order. (e) These phases are determined by string order parameters. The string order parameters O_{XZX}, O_{XX} and O_{YY}

have been calculated along the thick dotted line $J^{YZY}/J^{XZX} = 0.5$ (see Fig. 8.19(d)) using the iTEBD (from Ohta et al., 2015). 335

8.20 The diagram for the energy levels of the Majorana chain of system size $N = 100$ as a function of $\xi = \Delta/w$ with $\mu = 0$ and $w = 1$ considering (a) open boundary condition and (b) periodic boundary condition, respectively. One can observe that two zero energy Majorana edge modes exist in case (a) but not in case (b). It is mentioned that the energy spectrum is scaled by a constant factor $1/4$ in Eq. (8.65) compared to Eq. (8.54) (from Rajak and Dutta, 2014). 337

8.21 Majorana survival probability of a Majorana edge state for various quench protocols along Path A and probability of Majorana at certain time t after quench. (a) There is a sudden decay in the MSP due to a quenching from Phase I ($\Delta = 0.1$) to Phase II ($\Delta = -0.1$) and after that, it remains minimum without any revival. (b) On the other hand, following a quench from Phase I ($\Delta = 0.1$) to the QCP ($\Delta = 0.0$), MSP exhibits nearly perfect oscillation with a time period scaling linearly with the system size N. (c) Probability of the Majorana end mode at time $t = 100$ for quenching at the QCP with the Majorana site j. At this time, the probability of the Majorana mode is maximum, close to the center of the chain (from Rajak and Dutta, 2014). 338

8.22 A sudden quench is performed within Phase I from ($\Delta = 0.2$ and $\mu = 0.0$) to ($\Delta = 0.1$ and $\mu = 0.0$) and the MSP becomes (a) a rapidly oscillating function of time t having nearly 0.8 mean value and (b) probability of the Majorana end mode at $t = 100$, with the Majorana sites j following the quench (from Rajak and Dutta, 2014). 340

8.23 Phase diagram of the system with Hamiltonian in Eq. (8.52) for various phases of complex hopping term (a) $\phi = 0$, (b) $\phi = \pi/4$, (c) $\phi = 2\pi/5$ and (d) $\phi = \pi/2$. In this plot, I and II represent two different topological phases, whereas III is the topologically trivial phase. The path $\mu = 0$ is represented by the vertical arrow along which quenching is performed (from Rajak et al., 2014). 341

8.24 The energy levels of the Hamiltonian in Eq. (8.52) as a function of parameter $\xi = \frac{\Delta}{w_0}$ (where $w_0 = 1$) for (a) periodic and (b) open boundary conditions with $N = 100$ and $\phi = \pi/10$. The inverted energy levels inside the gapless phase of the system is indicated by the red color (from Rajak et al., 2014). 343

8.25 (a) Variation of the probability of defect (P_{def}) with τ for different values of ϕ exhibiting dip at different values of $\tau \geq \tau_c$ where the value of τ_c increases as ϕ decreases. (b) The probability of Majorana (P_m) exhibits a peak precisely at those values of τ where P_{def} shows dips. Here, $N = 100$ (from Rajak et al., 2014). 345

8.26 (a) P_{def}, P_{neg} and P_m as a function of τ with $\phi = \pi/10$ exhibit that all of them add up to unity for any τ. (b) The plot shows that $\ln(\tau_c)$ varies almost linearly with ln(sin ϕ) having slope ($= -0.9$) nearly equal to -1 (from Rajak et al., 2014). 346

8.27 (a) Variation of P_{m} as a function of τ for different system sizes N with $\phi = \pi/5$. P_{m} exhibits peak at different values of $\tau \geq \tau_c$. (b) The figure shows a plot between $\ln \tau_c$ and $\ln N$ for ϕ with slope ($= 0.94$) nearly equal to 1 suggesting $\tau_c \sim N$ (from Rajak et al., 2014). 347

8.28 Variation of P_{def} as a function of τ determining overlaps between the time-evolved Majorana state at final time and different number of positive energy states (near zero energy) at the final parameter values. Let us consider the case with $\phi = \pi/10$ and $N = 100$, where the number of inverted energy levels in both positive and negative sides of the zero energy is given by 18 (see Ref. (Rajak et al., 2014)). The plot indicates that for $\tau \leq \tau_c$ ($\Delta t < \Delta t_{\text{th}}$), the initial edge Majorana mixes with all the bulk energy levels. For $\tau \geq \tau_c$, on the other hand, the P_{def} determined by using only the inverted positive bulk energy levels nearly coincides with the exact P_{def} procured considering all positive energy levels. This clearly verifies that quenched Majorana state interacts only with the inverted energy levels for a passage time $\Delta t \geq \Delta t_{\text{th}}$ (from Rajak et al., 2014). 348

9.1 Title and abstract from the first published paper proposing the idea that quantum tunneling across the free energy barriers in the Sherrington–Kirkpatrick spin glass model can lead to an efficient way of searching for its ground state(s). It was argued that "quantum tunneling between the classical 'trap' "states", separated by infinite (but narrow) barriers in the free energy surface, is possible, as quantum tunneling probability is proportional to the barrier area which is finite." They suggested that any amount of transverse field would lead to the collapse of the overlap distribution to a delta function. It may be noted that computationally hard problems can often be mapped into such long-range spin glass models; the advantage of quantum tunneling in such quantum spin glass models has lead ultimately to the development of the quantum annealer. A related reference is Chakrabarti (1981). (Permission to use title and abstract from the paper is given by American Physical Society) 360

9.2 Title and abstract from the first paper reporting on the experimental studies of an Ising spin glass sample under the influence of a tunable transverse field. The observed nature of the tunneling induced phase diagram compared well with that predicted by Ray et al. (1989). (Permission to use title and abstract from the paper is given by American Physical Society) 361

9.3 Title and abstract of the first published paper demonstrating the search of the ground state of a Lennard–Jones system using ‘quantum annealing’ (term appearing for the first time in paper title). (Permission to use title and abstract from the paper is given by Elsevier) 361

9.4 Title and abstract from a paper demonstrating the clear advantage of quantum annealing in Ising models with frustrating interactions. It was found that in case of quantum annealing not only there was a higher probability of the initial state to converge in the ground state of such models but also the convergence time was comparatively less than that of simulated annealing. A related reference is Chakrabarti et al. (1996). (Permission to use title and abstract from the paper is given by American Physical Society) 362

9.5 Title and abstract of the first published experimental paper reporting on the advantage of quantum annealing in finding the ground state of disordered magnets. Such experimental observations have put quantum annealing on firm physical ground. Some related references are Wu et al. (1991), Wu et al. (1993), Bitko et al. (1996), Kadowaki and Nishimori (1998). (Permission to use title and abstract from the paper is given by The American Association for the Advancement of Science) 363

9.6 Title and abstract of a paper reporting on the zero temperature quantum adiabatic algorithm for NP-hard problems. A related reference here is Kadowaki and Nishimori (1998). (Permission to use title and abstract from the paper is given by The American Association for the Advancement of Science) 364

9.7 Title and abstract of a paper on the application of quantum annealing in estimating the remaining fraction of undesired solutions in some optimization searches in Ising spin glasses. They also indicated the relative fastness of quantum annealing with respect to simulated annealing. Some related references here are Wu et al. (1993), Finnila et al. (1994), Kadowaki and Nishimori (1998), Brooke et al. (1999), Brooke et al. (2001) and Farhi et al. (2001). (Permission to use title and abstract from the paper is given by The American Association for the Advancement of Science) 365

9.8 Two important early reviews helped the subsequent development of quantum annealing significantly. Title and abstract for the first review on adiabatic quantum computation and annealing, proclaiming that the idea of quantum tunneling through the infinitely high energy barriers in long-range frustrated spin glasses was introduced in Ray et al. (1989). Other related references here are Finnila et al. (1994), Chakrabarti et al. (1996), Kadowaki and Nishimori (1998), Brooke et al. (1999) and Das et al. (2005). (Permission to use title and abstract from the paper is given by Institute of Physics) 366

9.9 Title and abstract from the second (early) review on quantum annealing and quantum computation. Authors summarized that following the indication of Ray et al. (1989), both theoretical and experimental successful studies on QA made this technique extremely useful in solving hard optimization problems, in binary to analog quantum computers. Some related references here are Finnila et al. (1994), Kadowaki and Nishimori (1998), Brooke et al. (1999), Farhi et al. (2001) and Santoro et al. (2002). (Permission to use title and abstract from the paper is given by American Physical Society) 367

9.10 Title and abstract from the review discussing the detailed mathematical structure and theorems related to QA. Some related references here are Finnila et al. (1994), Kadowaki and Nishimori (1998), Santoro et al. (2002), Santoro and Tosatti (2006), Das et al. (2005), Das and Chakrabarti (2005) and Das and Chakrabarti (2008). (Permission to use title and abstract from the paper is given by American Institute of Physics) 368

9.11 Title and abstract from a paper extending and clarifying the quantum annealing method used by Brooke et al. (1999) for quantum glasses. Authors argued that at low enough temperatures, and "in cases where barriers to relaxation are tall and narrow, quantum mechanics can enhance the ability to traverse the free energy surface (Ray et al. 1989)". Other related references cited are Wu et al. (1991), Wu et al. (1993). (Permission to use title and abstract from the paper is given by American Physical Society) 369

9.12 Title and abstract of the first paper by the D-Wave group giving the basic architecture of their quantum annealing processor. Some related references here are Finnila et al. (1994), Kadowaki and Nishimori (1998), Brooke et al. (1999) and Farhi et al. (2001). (Permission to use title and abstract from the paper is given by Nature Publishing Group) 370

9.13 Title, abstract and some excerpts from a paper by scientists from the University of Southern California, University of California, ETH Zurich and Microsoft Research. They argued in favour of the quantum nature of the D-Wave quantum annealer by comparing its performance with that of a classical one. In the introduction, the authors had commented "The phenomena of quantum tunneling suggests that it can be more efficient to explore state space quantum mechanically in a quantum annealer (Ray et al. 1989, Finnila et al. 1994, Kadowaki and Nishimori 1998)". Some other related references here are Brooke et al. (1999), Farhi et al. (2001) and Johnson et al. (2011). (Permission to use title and abstract from the paper is given by Nature Publishing Group) 371

9.14 Title and abstract from a paper where authors critically investigate the claim of the probable quantum speed-up over classical annealing. Their investigations suggest that the previously observed (Santoro et al., 2002; Martoňák et al., 2002) advantage of QA compared to SA in searching for the ground state of 2D spin glasses was due to the discretization of time in the quantum Monte Carlo algorithm. They have not found any advantage in continuous time limit. Other related cited papers are Ray et al. (1989), Finnila et al. (1994), Kadowaki and Nishimori (1998), Farhi et al. (2001) and Das and Chakrabarti (2008). (Permission to use title and abstract from the paper is given by The American Association for the Advancement of Science) 372

9.15 Title and abstract from a paper discussing the out of equilibrium quantum Monte Carlo study of spin glasses on 3-regular graphs. Authors observe the inability of quantum algorithm regarding fastness relative to SA in bringing down the system to glassy state. Some related references here are Ray et al. (1989), Finnila et al. (1994), Kadowaki and Nishimori (1998), Brooke et al. (1999), Farhi et al. (2001), Santoro et al. (2002), Das and Chakrabarti (2008) and Boixo et al. (2014). (Permission to use title and abstract from the paper is given by American Physical Society) 373

9.16 Title and abstract from a paper where authors scrutinize the role of decoherence in quantum computation as well as quantum annealing. They indicate that decoherence does not affect adiabatic quantum computation. The related references here are Ray et al. (1989), Kadowaki and Nishimori (1998), Farhi et al. (2001), Das and Chakrabarti (2008), Boixo et al. (2013) and Boixo et al. (2014). (Permission to use title and abstract from the paper is given by American Physical Society) 374

9.17 Title and abstract from a paper reporting on an interesting implementation of singular-value decomposition by QA in cases of big data and image analysis. The related references are Ray et al. (1989), Finnila et al. (1994), Kadowaki and Nishimori (1998), Farhi et al. (2001), Das and Chakrabarti (2005), Das and Chakrabarti (2008) and Suzuki et al. (2013). (Permission to use title and abstract from the paper is given by American Physical Society) 375

9.18 Title and abstract from the paper suggesting a generalized method for the reduction of the dimension of the Hilbert space of quantum systems. The related references are Ray et al. (1989), Finnila et al. (1994), Kadowaki and Nishimori (1998), Farhi et al. (2001) and Perdomo-Ortiz et al. (2012). (Permission to use title and abstract from the paper is given by American Physical Society) 376

9.19 Title and abstract from the paper reporting on experimental evidence of the success of QA using quantum tunneling across free energy barriers. The related references are Ray et al. (1989), Finnila et al. (1994), Kadowaki and

Nishimori (1998), Brooke et al. (1999), Santoro et al. (2002), Johnson et al. (2011) and Boixo et al. (2013). 377

9.20 Title and abstract from a paper reporting on the failure of the quantum annealer in context of speed enhancement with respect to SA. Some related references here are Finnila et al. (1994), Kadowaki and Nishimori (1998), Brooke et al. (1999), Santoro et al. (2002), Das and Chakrabarti (2008), Boixo et al. (2014) and Mukherjee and Chakrabarti (2015). (Permission to use title and abstract from the paper is given by American Physical Society) 378

9.21 Title, abstract and excerpts from the first major paper supporting the claim of D-Wave quantum computer used in searching for low energy conformations of the lattice protein model, reported by the Harvard University group. The authors commented in the introductory section "Harnessing quantum-mechanical effects to speed up the solving of classical optimization problems is at the heart of quantum annealing algorithms (Finnila et al., 1994; Kadowaki and Nishimori, 1998; Farhi et al., 2001; Santoro and Tosatti, 2006; Das and Chakrabarti, 2008; Ray et al., 1989)". Some other relevant references here are Amara et al. (1993), Brooke et al. (1999), Farhi et al. (2001) and Johnson et al. (2011). (Permission to use title and abstract from the paper is given by Nature Publishing Group) 379

Tables

2.1 Critical exponents of the Ising model obtained by the mean-field approximation. 13

2.2 Critical exponents of the Ising model on two-dimensional lattices. 20

2.3 Critical exponents of the Ising model on three-dimensional lattices. 20

2.4 Critical exponents of the Potts model with homogeneous ferromagnetic interactions depending on the value of q and the spatial dimension (Wu, 1982). Note that critical exponents in the two-state Potts model are equivalent to that in the Ising model (see Table 2.2 and Table 2.3). 47

2.5 Critical exponents of the XY model with homogeneous ferromagnetic interactions on three-dimensional lattices obtained by Monte Carlo simulations (Campostrini et al. 2001). 49

2.6 Critical exponents of the Heisenberg model with homogeneous ferromagnetic interactions on three-dimensional lattices obtained by Monte Carlo simulations (Holm and Janke, 1993). 50

Preface

This book intends to introduce the readers to the developments in the researches on phase transition of quantum spin glasses, their dynamics near the phase boundary and applications in the efforts to solve multi-variable optimization problems by using quantum annealing. In view of the recent successes of both theoretical and experimental studies, major efforts were undertaken in employing these ideas in developing some prototype quantum computers (e.g., by D-wave Systems Inc.) and these have led to a revolution in quantum technologies.

The ideas developed in solving the dynamics of frustrated random systems like spin glasses, in particular of the celebrated Sherrington–Kirkpatrick model (SK model) (1975), have led to the understanding of the intrinsic nature of the problems involved in searches for the least cost solutions in multi-variable optimization problems. In particular, the pioneering idea of simulated or classical annealing technique by Kirkpatrick, Gelatt, and Vecchi (1983) had already led to major breakthroughs. It has also led to some crucial concepts regarding how the hardness of such optimization problems come about through their mapping to the ruggedness of the cost function landscape of the SK model. The idea that quantum fluctuations in the SK model can lead to some escape routes by tunneling through such macroscopically tall but thin barriers (Ray, Chakrabarti, and Chakrabarti, 1989) those which are difficult to scale using classical fluctuations, have led to some important clues. With this and some more developments, the quantum annealing technique was finally launched through a landmark paper by Kadowaki and Nishimori in 1998. Since then, as mentioned earlier, a revolution has taken place through a surge of outstanding papers both in theory and in technological applications, leading finally to the birth of this new age of quantum technologies.

This book intends to present and review these developments in a step-by-step manner, mainly from the point of view of theoretical statistical physicists. We hope the book will also be useful to physicists in general and to computer scientists as well. As one can easily see, the subject is growing at a tremendous rate today, and many more materials will soon be needed to supplement our knowledge on quantum annealing. We believe, however, the materials discussed in the book will prove indispensable for young researches and Ph. D course students who are eager to get into this exciting field of research! Indeed, we have added three notes by other experts (B. Tamir and E. Cohen, A. Rajak and U. Divakaran, and S. Mukherjee) to provide some complementary ideas and historical accounts for the benefit of the readers.

This book is dedicated to the loving memory of Professor Jun-ichi Inoue with whom we shared many ideas and developed part of the studies described here. Indeed, we had an early plan to write this book together with him. His untimely death has robbed us of that opportunity. We still hope, he would be delighted to see this book and its contents.

We are grateful to the Cambridge University Press, in particular to M. Choudhary, R. Dey, and D. Majumdar, for their immense patience and constant encouragements. We do hope the book will be useful and enjoyable to the readers.

Shu Tanaka, Tokyo
Ryo Tamura, Tsukuba
Bikas K. Chakrabarti, Kolkata
August 2016

CHAPTER 1 Introduction

Quantum annealing is a new-generation quantum information technology, which is expected to rapidly solve combinatorial optimization problems with high precision. It is based on quantum statistical physics. The realization of a quantum annealing machine (e.g., D-Wave developed by D-Wave Systems Inc.; Johnson et al., 2011) is a big step in opening the door to new development of quantum information technology as well as a historically significant event in the history of quantum information processing. Nowadays, not only the academic side but also industry have high expectations in the field of quantum information processing.

The target of quantum annealing is to solve combinatorial optimization problems. Combinatorial optimization problems are problems which seek to find the best answer from a vast number of candidates. The most typical example of a combinatorial optimization problem is the traveling salesman problem which is to find the path along which we can visit all locations efficiently. More precisely, the traveling salesman problem is to find the path that minimizes costs such as distance and travel expenses under the condition that all locations are visited once. A naive method is to make a list of all candidates of paths and to find the best solution that minimizes the cost from this list. If we use the naive method, the number of candidates increases exponentially with the number of locations. This situation is called the combinatorial explosion. Thus, the naive method is not efficient in practice. Of course, other combinatorial optimization problems have the same problem.

Combinatorial optimization problems can be expressed by random Ising models. We can map the cost function of a combinatorial optimization problem to the Hamiltonian of a random Ising spin system. Thus, the knowledge of random Ising spin systems is useful to consider combinatorial optimization problems. In order to consider cooperative phenomena in random magnets, Edwards and Anderson considered the Ising model with random interactions, called the Edwards–Anderson model or the EA model (Edwards and Anderson, 1975). The EA model is a prototype with which we consider random spin systems. After the proposition of the EA model, Sherrington and Kirkpatrick proposed the infinite-range interaction version of the EA model, called the Sherrington–Kirkpatrick model or the SK model (Sherrington and Kirkpatrick, 1975). The interactions in the SK model is infinite ranged and the probability distribution of the value of interactions is the Gaussian distribution. Depending on the value of the average and the standard deviation of interactions, spin glass phase appears.

In information science and engineering, particular algorithms to solve each combinatorial optimization problem have been proposed. Since these methods are useful in practice, they are often used in each industrial area. However, a particular algorithm proposed for a combinatorial optimization problem is not necessarily suitable for other combinatorial optimization problems. Thus, if we encounter new types of combinatorial optimization problems, we cannot handle them by using particular algorithms. In statistical physics which is an established discipline in physics, on the other hand, generic algorithms have been developed exhaustively. Simulated annealing, or classical annealing proposed by Kirkpatrick, Gelatt, and Vecchi is a famous algorithm inspired by the concept of statistical physics (Kirkpatrick et al., 1983). Simulated annealing is an algorithm that uses the temperature effect, i.e., the thermal fluctuation effect. The convergence of simulated annealing is guaranteed by the Geman–Geman theorem (Geman and Geman, 1984). Simulated annealing has been widely used for solving combinatorial optimization problems, because of the ease of implementation.

Through the analysis of the EA model and the SK model, the thermal fluctuation effect of random spin systems has been considered. In physics, there is another important fluctuation effect – the quantum fluctuation effect. Chakrabarti first investigated critical behavior of the EA model with a transverse field (Chakrabarti, 1981). After that, phase transition nature and dynamical properties of the random Ising spin models under the transverse field have been studied exhaustively. Ray, Chakrabarti, and Chakrabarti considered dynamical properties of the SK model under the transverse field (Ray et al., 1989). They pointed out that quantum fluctuation lead to some escape routes by tunneling through macroscopically tall but thin barriers which are difficult to scale using classical (thermal) fluctuations. This idea is the basis of quantum annealing.

Quantum annealing is a cousin of simulated annealing. In quantum annealing, we introduce a quantum effect to the combinatorial optimization problem. By decreasing the quantum effect, we can obtain the best solution of the combinatorial optimization problem. This scheme corresponds to simulated annealing. In simulated annealing, we introduce a thermal effect to the combinatorial optimization problem. By decreasing the temperature, we can obtain the best solution of the combinatorial optimization problem. The milestone in the use of quantum annealing is the study done by Kadowaki and Nishimori (Kadowaki and Nishimori, 1998). Their paper is based on the statistical physics of random Ising models. In preceding studies, the quantum effect for random Ising models plays the same role as the thermal one.

As mentioned earlier, quantum annealing is a hot topic in science since quantum annealing is expected to be a powerful and generic algorithm to solve combinatorial optimization problems. Thus, the purpose of this book is to provide an overview of quantum annealing by reviewing famous studies. In addition, related topics of quantum annealing and the background of quantum annealing – quantum statistical physics and information science are also discussed.

The organization of the book is as follows. Part 1 has six chapters. Chapter 1, this chapter, is devoted to introducing the topic of the book. The rest is as follows. In Chapter 2, the physical properties of classical spin models, especially, random Ising models are reviewed. This is a fundamental knowledge needed to study quantum annealing. In Chapter 3, simulated annealing is explained. Simulated annealing is a generic algorithm inspired by statistical physics and developed to find the best solution of combinatorial optimization problems. Simulated annealing has been used in broad areas of science and industry. Quantum annealing which is the main topic of the book is based on simulated annealing. Thus, the knowledge of simulated annealing is useful to study quantum annealing. In Chapter 4, we briefly review the physical properties of quantum random Ising models and compare them with the classical random Ising models explained in Chapter 2. Since the instantaneous Hamiltonian of quantum annealing is described by quantum random Ising models, the knowledge learned in the chapter is used to consider the performance of quantum annealing. In Chapter 5, two typical quantum dynamic properties are shown. One is the Landau–Zener transition and the other is the Kibble–Zurek mechanism. Both of them are used for analysis of the performance of quantum annealing. In Chapter 6, we review quantum annealing that is the main topic of the book from the theoretical to the experimental aspects. In this chapter, we briefly show famous studies on quantum annealing. Part 2 has three chapters written by guest authors. In Chapter 7, adiabatic quantum computers are reviewed by Boaz Tamir and Eliahu Cohen. In this chapter, the architecture of D-Wave, that is a quantum annealing hardware, is also explained. In Chapter 8, the relation between quantum information and quenching dynamics is reviewed by Atanu Rajak and Uma Divakaran. In this chapter, the recent development of theoretical studies on quenching dynamics of a number of models including topological systems is explained. In Chapter 9, the brief history of quantum annealing is presented by Sudip Mukherjee.

References

Chakrabarti, B. K. 1981. 'Critical behavior of the Ising spin-glass models in a transverse field'. *Phys. Rev. B* 24: 4062.

Edwards, S. F., and P. W. Anderson. 1975. 'Theory of spin glasses'. *J. Phys.F: Metal Phys.*, 5, 965.

Geman, S., and D. Geman. 1984. 'Stochastic relaxation, Gibbs distributions, and the Bayesian restoration of images'. *IEEE Trans. Patt. Anal. Machine Intel.* 6: 721.

Johnson, M. W., M. H. S. Amin, S. Gildert, T. Lanting, F. Hamze, N. Dickson, R. Harris, A. J. Berkley, J. Johansson, P. Bunyk, E. M. Chapple, C. Enderud, J. P. Hilton, K. Karimi, E. Ladizinsky, N. Ladizinsky, T. Oh, I. Perminov, C. Rich, M. C. Thom, E.Tolkacheva, C. J. S. Truncik, S. Uchaikin, J. Wang, B. Wilson, and G. Rose 2011. 'Quantum annealing with manufactured spins'. *Nature* 473: 194.

Kadowaki, T., and H. Nishimori.1998. 'Quantum annealing in the transverse Ising model'. *Phys. Rev. E* 58: 5355.

Kirkpatrick, S., C. D. Gelatt Jr., and M. P. Vecchi. 1983. 'Optimization by simulated annealing'. *Science* 220: 671.

Ray, P., B. K. Chakrabarti, and A. Chakrabarti. 1989. 'Sherrington–Kirkpatrick model in a transverse field: Absence of replica symmetry breaking due to quantum fluctuations'. *Phys. Rev. B* 39:11828.

Sherrington, D., and S. Kirkpatrick. 1975. 'Solvable model of a spin-glass'. *Phys. Rev. Lett.* 35: 1792.

PART ONE

QUANTUM SPIN GLASS, ANNEALING AND COMPUTATION

CHAPTER

2 Classical Spin Models: From Ferromagnetic Spin Systems to Spin Glasses

In this chapter, we explain the properties of classical spin models and some analytic methods for them. As will be described in Chapter 3, combinatorial optimization problems can be expressed by random Ising models which are a kind of classical spin models. The most fundamental classical spin system is the Ising model where the spin takes either $+1$ (up-spin state) or -1 (down-spin state). The Ising model was introduced to understand the phase transition in magnetic systems and the cooperative phenomena in bistable systems about a century ago. In Section 2.1, the phase transition nature of the Ising model with *homogeneous* ferromagnetic interactions will be reviewed. In this model, two degenerate ground states exist: in one ground state, all spins take $+1$, whereas, in the other ground state, all spins take -1. When the external field is absent, the Ising model exhibits the second-order phase transition with two-fold symmetry breaking at a finite temperature, called the Curie temperature. On the other hand, if magnetic interactions include *inhomogeneity*, the properties are entirely changed from the model with homogeneous ferromagnetic interactions, and the spin glass appears depending on the degree of inhomogeneity in magnetic interactions. In Section 2.2, the basic concept of classical spin glass will be described, and we will introduce the Edwards–Anderson model and $\pm J$ model. In both models, the spin glass phase appears. In order to study the spin glass nature analytically, some results obtained in the Sherrington–Kirkpatrick model will be shown in Section 2.3. The Sherrington–Kirkpatrick model is a random Ising model with infinite-ranged two-body interactions. In Section 2.4, the results obtained by the numerical simulations in the Edwards–Anderson model in finite dimensions will be introduced. Finally, in Section 2.A, a spin glass in other classical models: Heisenberg model, XY model, and Potts model will be described.

2.1 Ising Model

The Ising model (Ising, 1925) consists of the Ising spins which take either $\sigma_i^z = +1$ (up-spin state) or $\sigma_i^z = -1$ (down-spin state). Here, σ_i^z denotes the Ising spin at site i on the given lattice. The Hamiltonian of the Ising model with N spins is defined as

$$\mathcal{H} = -\sum_{\langle i,j \rangle} J_{ij}\sigma_i^z\sigma_j^z - \sum_{i=1}^{N} h_i\sigma_i^z, \quad (\sigma_i^z = \pm 1), \tag{2.1}$$

where J_{ij} is the magnetic interaction between the ith spin and the jth spin and h_i is the magnetic field at the ith spin. The first sum is taken over all interacting spin pairs on the given lattice. Note that the interaction between the ith and the jth spins is ferromagnetic for $J_{ij} > 0$ and antiferromagnetic for $J_{ij} < 0$.

The Ising model was introduced to understand the phase transition nature in magnetic materials and the cooperative phenomena in bistable systems. If the magnetic interactions are homogeneous and ferromagnetic, the Ising model in two or more higher dimensions exhibits the phase transition between a disordered phase and an ordered phase (order–disorder transition) under zero magnetic field. In the disordered phase, spins are arranged randomly, whereas in the ordered phase, spins are arranged uniformly. In this section, we show some physical properties of the Ising model with homogeneous ferromagnetic interactions.

2.1.1 Phase transition and critical phenomena

Order of the phase transition

Before we consider the physical properties of the Ising model, a general framework of the phase transition is shown. Depending on the anomaly in the free energy at the phase transition point, phase transition can be categorized into two types, that is, first-order phase transition and second-order phase transition. The Helmholtz free energy of the magnetic systems $F(T,H)$ is a function of the temperature T and magnetic field H. The total differential of the free energy is represented as

$$dF(T,H) = -MdH - SdT, \tag{2.2}$$

where M is the magnetization and S is the magnetic entropy. At the first-order phase transition point, the first derivative of $F(T,H)$ is discontinuous. In other words, the magnetization and the magnetic entropy become discontinuous at the first-order phase transition point, since the magnetization M and the magnetic entropy S are given by the first derivative of $F(T,H)$ as

$$M := -\left(\frac{dF(T,H)}{dH}\right)_T, \tag{2.3}$$

$$S := -\left(\frac{dF(T,H)}{dT}\right)_H. \tag{2.4}$$

The first-order phase transition is called the discontinuous phase transition. On the other hand, at the second-order phase transition point called the critical point, the first derivative of $F(T,H)$ has no singularities but the second derivative of $F(T,H)$ is discontinuous or diverges. That is, the magnetic susceptibility and the specific heat become discontinuous or diverge at the second-order phase transition point, since the magnetic susceptibility χ and the specific heat C are given by the second derivative of $F(T,H)$ as

$$\chi := -\left(\frac{d^2F(T,H)}{dH^2}\right)_T, \tag{2.5}$$

$$C := -T\left(\frac{d^2F(T,H)}{dT^2}\right)_H. \tag{2.6}$$

Note that at the second-order phase transition point, spontaneous symmetry breaking occurs and long-range order appears. The second-order phase transition is called the continuous phase transition.

Critical phenomena

Near the second-order phase transition point, a power-law behavior of physical quantities is observed due to the divergence of fluctuation – this behavior which is called critical phenomena. The power-law behavior of physical quantities is expressed by the critical exponents $\{\alpha,\beta,\gamma,\delta,\eta,\nu\}$. Let us consider the case where a second-order phase transition occurs at finite temperature T_{c} under zero magnetic field. The most fundamental example of this situation is the homogeneous ferromagnetic Ising model as described before. The behaviors of the physical quantities C, χ, and M around the second-order phase transition point are characterized by

$$C \propto \left|\frac{T-T_{\mathrm{c}}}{T_{\mathrm{c}}}\right|^{-\alpha} \quad (T > T_{\mathrm{c}}), \quad C \propto \left|\frac{T-T_{\mathrm{c}}}{T_{\mathrm{c}}}\right|^{-\alpha'} \quad (T < T_{\mathrm{c}}), \tag{2.7}$$

$$\chi \propto \left|\frac{T-T_{\mathrm{c}}}{T_{\mathrm{c}}}\right|^{-\gamma} \quad (T > T_{\mathrm{c}}), \quad \chi \propto \left|\frac{T-T_{\mathrm{c}}}{T_{\mathrm{c}}}\right|^{-\gamma'} \quad (T < T_{\mathrm{c}}), \tag{2.8}$$

$$M \propto \left|\frac{T-T_{\mathrm{c}}}{T_{\mathrm{c}}}\right|^{\beta} \quad (T < T_{\mathrm{c}}), \quad \chi \propto |H|^{1/\delta} \quad (T = T_{\mathrm{c}}). \tag{2.9}$$

Furthermore, the behavior of the correlation function between two spins whose distance is given by r, $G(r) := \langle \sigma_i^z \sigma_{i+r}^z \rangle$, near the second-order phase transition point is represented as

$$G(r) \propto r^{-d+2-\eta} \quad (T = T_c), \tag{2.10}$$

where d is the spatial dimension. In addition, the correlation length ξ is expressed by

$$\exp(-r/\xi) \propto G(r), \tag{2.11}$$

and the behavior of ξ is

$$\xi \propto \left| \frac{T - T_c}{T_c} \right|^{-\nu} \quad (T > T_c), \quad \xi \propto \left| \frac{T - T_c}{T_c} \right|^{-\nu'} \quad (T < T_c). \tag{2.12}$$

These critical exponents are not independent, and the following relations, called the scaling relations, are satisfied:

$$\nu d = 2 - \alpha = 2\beta + \gamma = \beta(\delta + 1) = \gamma \frac{\delta + 1}{\delta - 1}, \tag{2.13}$$

$$2 - \eta = \frac{\gamma}{\nu} = d \frac{\delta - 1}{\delta + 1}. \tag{2.14}$$

Thus, in principle, from only two critical exponents and the spatial dimension, other critical exponents can be obtained. The concept of universality is efficient since second-order phase transitions can be categorized by a combination of critical exponents which depend on only the spatial dimension and the symmetry which breaks at the phase transition point. In other words, we do not need to take care of the details of the model we have considered as long as we focus on the physical properties near the second-order phase transition point. As will be shown in Section 5.2, the performance of annealing in systems with second-order phase transition is also characterized by critical exponents.

2.1.2 Mean-field approximation and infinite-range model

In order to thoroughly understand the finite-temperature properties of the Ising model, we should solve the many-body problem in principle. In practice, however, it is difficult to solve the many-body problem except for some exactly solvable models (Onsager, 1944; Baxter, 1982). Thus, by introducing some approximation, we can treat the problem qualitatively. The obtained results, based on an appropriate approximation, help us to understand the physical properties of the model we have considered. The most famous approximation is the mean-field theory in which the spin variable is replaced by the average of the spin variable. The mean-field approximation reduces the original many-body problem to an effective single-body problem which can be analyzed by simple calculations.

Mean-field approximation of the Ising model with homogeneous ferromagnetic interactions

Here, we demonstrate the mean-field approximation of the Ising model with homogeneous ferromagnetic interactions to study its finite-temperature properties. The Hamiltonian of the model is defined by

$$\mathscr{H} = -J\sum_{\langle i,j\rangle} \sigma_i^z\sigma_j^z - h\sum_{i=1}^{N}\sigma_i^z, \quad (\sigma_i^z = \pm 1), \tag{2.15}$$

and the partition function is written as

$$\begin{aligned} Z &= \mathrm{Tr}\exp\left(-\beta\mathscr{H}\right) \\ &= \mathrm{Tr}\exp\left[\sum_{i=1}^{N}\beta\sigma_i^z\left(J\sum_{j\in\partial i}\sigma_j^z + h\right)\right], \end{aligned} \tag{2.16}$$

where ∂i denotes the set of sites which are neighboring on site i, and β is the inverse temperature defined by $\beta := 1/k_{\mathrm{B}}T$. k_{B} is the Boltzmann constant and set to unity herein. Here, Tr is the summation over all spin configurations. In the mean-field approximation, we replace σ_j^z with the average value of magnetization per spin $m = \sum_{i=1}^{N}\sigma_i^z/N$ and obtain

$$Z_{\mathrm{MF}} = \mathrm{Tr}\exp\left[\sum_{i=1}^{N}\beta\sigma_i^z\left(zJm + h\right)\right], \tag{2.17}$$

where z is the number of interacting sites for each site, say,

$$z = \sum_{j\in\partial i} 1. \tag{2.18}$$

z is called the coordination number. Here we assume that the number of neighboring sites do not depend on the site. We can obtain the single-body problem by using the approximation. Then, we can perform the trace, and Z_{MF} is calculated as

$$\begin{aligned} Z_{\mathrm{MF}} &= \left\{\sum_{\sigma_i^z=\pm 1}\exp\left[\beta\sigma_i^z\left(zJm + h\right)\right]\right\}^N \\ &= \left[2\cosh\beta\left(zJm + h\right)\right]^N. \end{aligned} \tag{2.19}$$

By using the partition function, the magnetization per spin is obtained as

$$m = \frac{1}{N}\frac{\partial}{\partial(\beta h)}\ln Z_{\mathrm{MF}}$$

$$= \tanh\left[\beta(zJm + h)\right]. \tag{2.20}$$

This is the self-consistent equation of the magnetization. The behavior of the magnetization is obtained by solving this equation of state. Figure 2.1 shows the temperature dependences of the magnetization of the Ising model for various magnetic fields h. When the magnetic field is absent $h = 0$, at $k_{\mathrm{B}}T_{\mathrm{c}}/zJ = 1$, the magnetization becomes a finite value from zero, as decreasing with the temperature. The magnetization is not discontinuous at $k_{\mathrm{B}}T_{\mathrm{c}}/zJ = 1$, and thus the phase transition under $h = 0$ is of the second-order. In this case, two ground states are degenerated, that is, the all-up spin state ($m = 1$) and the all-down spin state ($m = -1$). Thus, at the second-order phase transition point, the twofold symmetry (Z_2 symmetry) breaks spontaneously. On the other hand, under finite magnetic fields, the model does not exhibit the phase transition, because the magnetic field always breaks the twofold symmetry and spontaneous symmetry breaking does not occur.

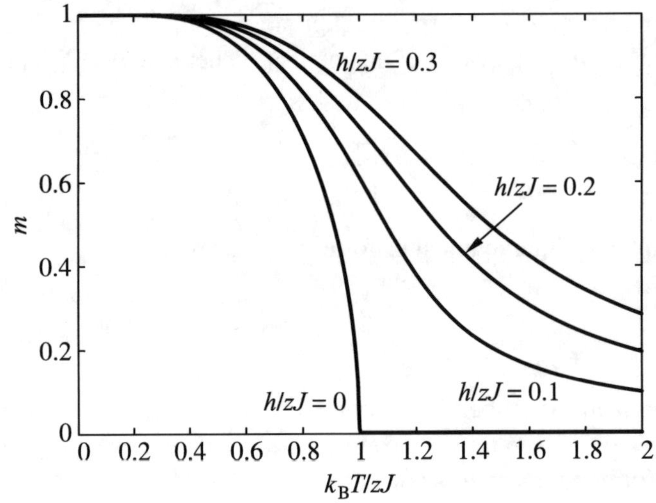

Fig. 2.1 Temperature dependence of the magnetization per spin m of the Ising model with homogeneous ferromagnetic interactions by the mean-field approximation. We display the result of $h \to +0$ for $h = 0$.

From the behavior of the magnetization, the critical exponent of β can be determined. We expand Eq. (2.20) with respect to m, and obtain

$$m \simeq \beta zJm - \frac{1}{3}(\beta zJ)^3 m^3. \tag{2.21}$$

Thus, the solution of magnetization m except for $m = 0$ is

$$m = \frac{T}{T_c}\sqrt{3\frac{T_c - T}{T_c}}, \tag{2.22}$$

and we obtain $\beta = 1/2$. Furthermore, the magnetic susceptibility is defined as $\chi = \partial m/\partial h$, and thus we expand Eq. (2.20) with respect to m and h:

$$m \simeq \beta zJm + \beta h. \tag{2.23}$$

Therefore, the magnetic susceptibility is

$$\chi = \frac{m}{h} = \frac{1}{T - T_c}, \tag{2.24}$$

and we obtain $\gamma = 1$. From β and γ, α and δ can be determined by the scaling relation shown in Eq. (2.13). Note that the correlation function between two spins cannot be obtained from this calculation because the local fluctuation effect is not included. By considering the local fluctuation of spins, critical exponents η and ν in the mean-field approximation are obtained. The critical exponents of the Ising model obtained by the mean-field approximation are summarized in Table 2.1. Notice that the critical exponents shown here are the same as the mean-field approximation of any model, e.g., XY model and Heisenberg model.

Table 2.1 Critical exponents of the Ising model obtained by the mean-field approximation.

Critical exponents	α	β	γ	δ	η	ν
Mean-field approximation	0	1/2	1	3	1/2	0

Infinite-range model (Husimi–Temperely–Curie–Weiss model)

In fact, there is a model so that the solution of the mean-field theory is the exact solution. The Hamiltonian of the model is defined by

$$\mathcal{H} = -\frac{zJ}{N}\sum_{1\le i<j\le N} \sigma_i^z\sigma_j^z, \quad (\sigma_i^z = \pm 1). \tag{2.25}$$

This model is called the infinite-range model or the Husimi–Temperely–Curie–Weiss model. Lattice structures of the model given by Eq. (2.25) are shown in Fig. 2.2.

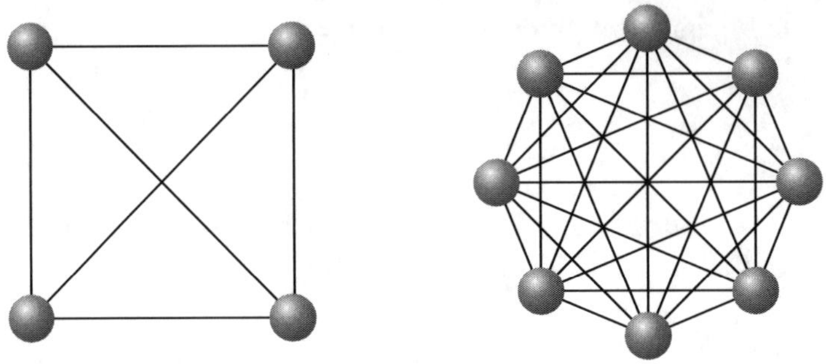

Fig. 2.2 Lattice structures of the infinite-range model given by Eq. (2.25) for $N = 4$ (left) and $N = 8$ (right).

In this model, the ferromagnetic interaction exists between all spin pairs. The partition function of the infinite-range model is written as

$$Z_{\rm IR} = {\rm Tr}\exp\left(\beta\frac{zJ}{N}\sum_{1\le i<j\le N}\sigma_i^z\sigma_j^z\right) \tag{2.26}$$

$$= {\rm Tr}\exp\left\{\beta\frac{zJ}{2N}\left[\left(\sum_{i=1}^{N}\sigma_i^z\right)^2 - N\right]\right\} \tag{2.27}$$

$$\simeq {\rm Tr}\exp\left[\beta\frac{zJ}{2N}\left(\sum_{i=1}^{N}\sigma_i^z\right)^2\right], \tag{2.28}$$

where the last approximation is valid for the thermodynamic limit, say, $N \to \infty$. By using the following identity:

$$e^{bx^2} = \int_{-\infty}^{\infty}\frac{da}{\sqrt{2\pi}}e^{-a^2+2a\sqrt{b}x}, \tag{2.29}$$

$Z_{\rm IR}$ can be rewritten as

$$Z_{\rm IR} = {\rm Tr}\int_{-\infty}^{\infty}\frac{da}{\sqrt{2\pi}}\exp\left[-a^2 + 2a\sqrt{\frac{\beta zJ}{2N}}\sum_{i=1}^{N}\sigma_i^z\right] \tag{2.30}$$

$$= \int_{-\infty}^{\infty}\frac{da}{\sqrt{2\pi}}e^{-a^2}\left[2\cosh\left(2a\sqrt{\frac{\beta zJ}{2N}}\right)\right]^N \tag{2.31}$$

$$= \int_{-\infty}^{\infty} \frac{da}{\sqrt{2\pi}} \mathrm{e}^{-a^2} \exp\left\{N \ln\left[2\cosh\left(2a\sqrt{\frac{\beta zJ}{2N}}\right)\right]\right\}. \tag{2.32}$$

We put $a = \sqrt{\beta zJN/2}m$, and obtain

$$Z_{\mathrm{IR}} = \int_{-\infty}^{\infty} \sqrt{\frac{\beta zJN}{4\pi}} dm \exp\left[-N\left(\frac{\beta zJ}{2}m^2 - \ln\{2\cosh(\beta zJm)\}\right)\right]. \tag{2.33}$$

Since the power of the exponential is proportional to N, we can apply the saddle-point method in the limit of $N \to \infty$. As a result,

$$Z_{\mathrm{IR}} \simeq \exp\left[-N\left\{\frac{\beta zJ}{2}m^2 - \ln[2\cosh(\beta zJm)]\right\}\right], \tag{2.34}$$

is obtained and m is calculated by the following equation:

$$\frac{\partial}{\partial m}\left\{\frac{\beta zJ}{2}m^2 - \ln[2\cosh(\beta zJm)]\right\} = 0 \tag{2.35}$$

$$\Rightarrow\ m = \tanh(\beta zJm). \tag{2.36}$$

This self-consistent equation is the same as the mean-field approximation, that is, Eq. (2.20) for $h = 0$. Thus, the solution of the infinite-range model in the thermodynamic limit ($N \to \infty$) is equivalent to that from the mean-field approximation. Notice that this fact is satisfied when $h \neq 0$ definitely. In this subsection, we demonstrated the mean-field approximation for the Ising model and showed the equivalence of the mean-field approximation and the infinite-range model. This means that the mean-field approximation can be regarded as the theory for infinite dimensional systems. Next, we review the homogeneous ferromagnetic Ising model in finite dimensions.

2.1.3 Ising model on finite dimensional lattices

In this subsection, we introduce the finite-temperature properties of the Ising model with homogeneous ferromagnetic interactions on finite dimensional lattices. In the one-dimensional case, that is, the Ising model on a chain, a phase transition does not occur at finite temperatures, which is rigorously proved. On the other hand, the second-order phase transition with the twofold symmetry breaking occurs in Ising models on two- and three-dimensional lattices when the magnetic field is absent. Lattice structures of chain (one-dimensional lattice), square lattice (two-dimensional lattice), and cubic lattice (three-dimensional lattice) are shown in Fig. 2.3. Notice that d-dimensional Ising models ($d \geq 2$) exhibit a second-order phase transition.

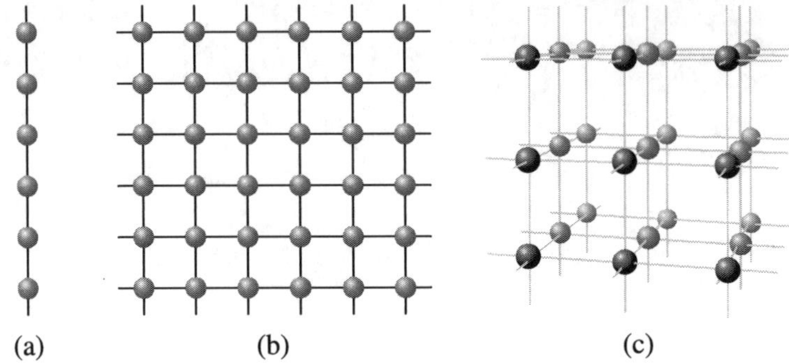

Fig. 2.3 Lattice structures of (a) chain (one-dimensional lattice), (b) square lattice (two-dimensional lattice), and (c) cubic lattice (three-dimensional lattice).

One-dimensional case

The Ising model on a chain can be exactly solved by the transfer matrix. The Hamiltonian of the Ising model on a chain is written as

$$\mathscr{H} = -J\sum_{i=1}^{N} \sigma_i^z \sigma_{i+1}^z - h\sum_{i=1}^{N} \sigma_i^z, \quad (\sigma_i^z = \pm 1), \tag{2.37}$$

where $J > 0$ and N is the number of spins. Here, we impose the periodic boundary condition: $\sigma_{N+1}^z = \sigma_1^z$. The partition function is written as

$$Z = \sum_{\sigma_1^z=\pm 1} \cdots \sum_{\sigma_N^z=\pm 1} \exp\left[\beta J \sum_{i=1}^{N} \sigma_i^z \sigma_{i+1}^z\right] \exp\left[\beta h \sum_{i=1}^{N} \sigma_i^z\right]. \tag{2.38}$$

Furthermore, the partition function can be rewritten as

$$Z = \sum_{\sigma_1^z=\pm 1} \cdots \sum_{\sigma_N^z=\pm 1} \sum_{\sigma_1^{z\prime}=\pm 1} \cdots \sum_{\sigma_N^{z\prime}=\pm 1} \exp(\beta J \sigma_1^{z\prime} \sigma_2^z) \exp(\beta h \sigma_2^z) \delta_{\sigma_2^z \sigma_2^{z\prime}} \times$$

$$\cdots \times \exp(\beta J \sigma_N^{z\prime} \sigma_1^z) \exp(\beta h \sigma_1^z) \delta_{\sigma_1^z \sigma_1^{z\prime}}, \tag{2.39}$$

where δ_{ij} is the Kronecker delta, i.e., $\delta_{ij} = 1$ when $i = j$ and $\delta_{ij} = 0$ when $i \neq j$. Here, we introduce the following matrices where elements are $\exp(\beta J \sigma_i^z \sigma_{i+1}^z)$ and $\exp(\beta h \sigma_i^z) \delta_{\sigma_i^z \sigma_{i+1}^z}$, respectively:

$$V_1 := \begin{pmatrix} e^{\beta J} & e^{-\beta J} \\ e^{-\beta J} & e^{\beta J} \end{pmatrix}, \tag{2.40}$$

$$V_2 := \begin{pmatrix} e^{\beta h} & 0 \\ 0 & e^{-\beta h} \end{pmatrix}, \tag{2.41}$$

and the partition function can be represented by

$$Z = \mathrm{Tr}\,(V_1 V_2)^N = \mathrm{Tr}\left(V_2^{\frac{1}{2}} V_1 V_2^{\frac{1}{2}}\right)^N =: \mathrm{Tr}\,V. \tag{2.42}$$

This symmetric matrix V is called the transfer matrix and given by

$$V := \begin{pmatrix} e^{\beta(J+h)} & e^{-\beta J} \\ e^{-\beta J} & e^{\beta(J-h)} \end{pmatrix}. \tag{2.43}$$

Let λ_+ and λ_- ($\lambda_+ > \lambda_-$) be the eigenvalues of V. Since V is a symmetric matrix, λ_+ and λ_- are real-valued, definitely. By using the eigenvalues, the partition function can be represented as

$$Z = \mathrm{Tr}\begin{pmatrix} \lambda_+ & 0 \\ 0 & \lambda_- \end{pmatrix}^N \tag{2.44}$$

$$= \lambda_+^N + \lambda_-^N \tag{2.45}$$

$$\underset{N\to\infty}{\to} \lambda_+^N. \tag{2.46}$$

Since λ_+ is expressed by

$$\lambda_+ = e^{\beta J}\cosh\beta h + \sqrt{e^{2\beta J}\sinh^2\beta h + e^{-2\beta J}}, \tag{2.47}$$

we obtain

$$Z \underset{N\to\infty}{\to} \left[e^{\beta J}\cosh\beta h + \sqrt{e^{2\beta J}\sinh^2\beta h + e^{-2\beta J}}\right]^N. \tag{2.48}$$

Then, the free-energy per spin in the thermodynamic limit ($N \to \infty$) is obtained as

$$f = -k_{\mathrm{B}}T\ln\left[e^{\beta J}\cosh\beta h + \sqrt{e^{2\beta J}\sinh^2\beta h + e^{-2\beta J}}\right]. \tag{2.49}$$

From the free energy, the behaviors of almost all physical quantities can be calculated. For example, the magnetization per spin is obtained as

$$m = \frac{1}{N}\frac{\partial}{\partial(\beta h)}\ln Z \tag{2.50}$$

$$= \frac{\sinh \beta h}{\sqrt{\sinh^2 \beta h + e^{-4\beta J}}}. \tag{2.51}$$

From this, we can understand that the magnetization at finite temperatures is zero when the magnetic field is zero ($h = 0$). Figure 2.4 shows the temperature dependence of the magnetization per spin under finite magnetic fields. No anomaly is observed in Fig. 2.4, and thus the Ising model in one dimension does not exhibit a phase transition at finite temperatures.

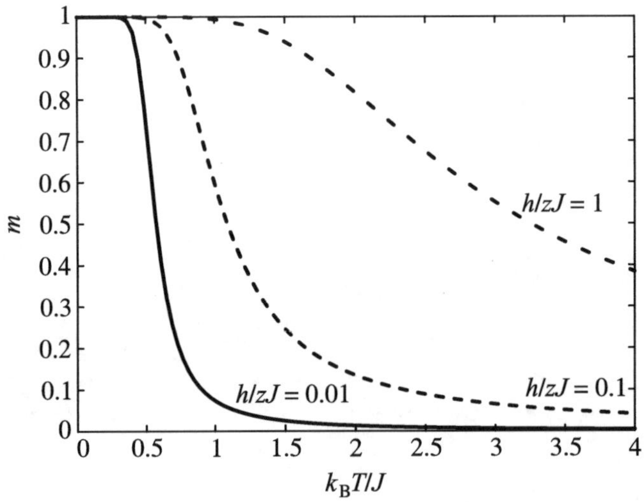

Fig. 2.4 Temperature dependence of the exact solution for magnetization per spin of the Ising model on a chain under finite magnetic fields.

Two-dimensional case

The Ising model on a two-dimensional square lattice under zero magnetic field can be exactly solved. Onsager first obtained the exact solution of this model in 1944 (Onsager, 1944). After the celebrating work, many studies have been done (Kaufman, 1949; Nambu, 1950; Kac and Ward, 1952; Potts and Ward, 1955; Schultz et al. 1964; Mcoy and Wu, 1973). According to these studies, this model under zero magnetic field exhibited the second-order phase transition at a finite temperature. In fact, the second-order phase transition temperature and critical phenomena were determined in these studies. However, till now, there is no exact solution of this model under finite magnetic fields. Here, we review the obtained results for the Ising model on a square lattice under zero magnetic field.

The Hamiltonian of the Ising model on a square lattice is defined as

$$\mathcal{H} = -J \sum_{\langle i,j \rangle} \sigma_i^z \sigma_j^z, \quad (\sigma_i^z = \pm 1), \tag{2.52}$$

where the summation runs over the nearest-neighbor spin pairs on the square lattice. The Helmholtz free energy per spin is exactly obtained as follows:

$$f = -T\left\{\ln(2\cosh 2\beta J) + \frac{1}{2\pi}\int_0^{\pi} \ln\left[\frac{1}{2}\left(1+\sqrt{1-k^2\cos^2\omega}\right)\right] d\omega\right\}, \tag{2.53}$$

where

$$k := \frac{2\sinh 2\beta J}{\cosh^2 2\beta J}. \tag{2.54}$$

From the free energy, the behaviors of almost all physical quantities can be calculated. For example, the magnetization per spin is obtained as

$$m = \begin{cases} \left[1 - \dfrac{1}{\sinh^4 2\beta J}\right]^{\frac{1}{8}} & (T < T_c) \\ 0 & (T > T_c) \end{cases}. \tag{2.55}$$

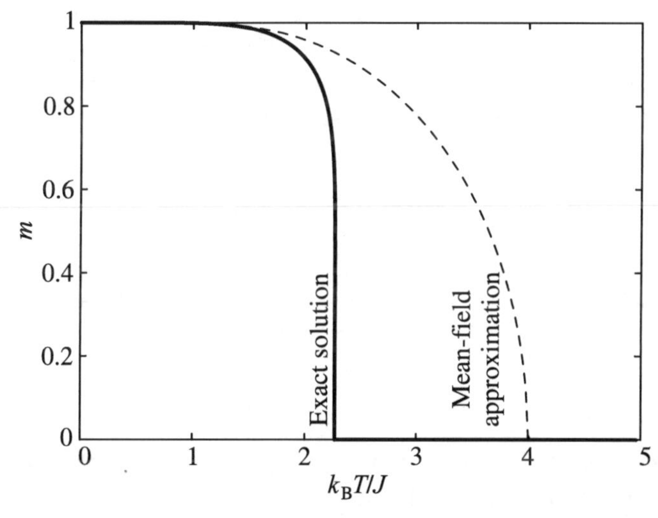

Fig. 2.5 Temperature dependence of the exact solution of magnetization per spin of the Ising model on a square lattice when the magnetic field is zero. The dotted curve is the result from the mean-field approximation. The transition temperature from the exact solution is lower than that from the mean-field approximation.

Figure 2.5 shows the temperature dependence of the magnetization per spin. At $k_B T_c/J = 2/\sinh^{-1} 1 \simeq 2.26918$, the magnetization becomes a finite value from zero, decreasing with temperature. Note that the second-order phase transition temperature obtained by the exact method is lower than that obtained by the mean-field approximation (see Fig. 2.5). This is because the mean-field approximation does not include the local fluctuation effect. The magnetization is not discontinuous at the phase transition temperature, and thus the phase transition is of the second order. The critical exponents of the Ising model on two-dimensional lattices were also exactly determined, and these are summarized in Table 2.2.

Table 2.2 Critical exponents of the Ising model on two-dimensional lattices.

Critical exponents	α	β	γ	δ	η	ν
2D Ising model	0	1/8	7/4	15	1/4	1

Three-dimensional case

There is no exact solution of the Ising model in three dimensions. The model has been exhaustively studied using renormalization groups and ε-expansion, and numerical simulations (Brezin et al. 1974; Landau, 1976; Le Guillou and Zinn-Justin, 1980; Barber et al. 1985; Parisi and Rapuano, 1985; Hoogland et al. 1985; Bhanot et al. 1986; Le Guillou and Zinn-Justin, 1987), and it has been determined that this model exhibits second-order phase transition at a finite temperature and no magnetic field. The critical exponents of the Ising model on the three-dimensional lattices are summarized in Table 2.3.

Table 2.3 Critical exponents of the Ising model on three-dimensional lattices.

Critical exponents	α	β	γ	δ	η	ν
3D Ising model	0.119	0.321	1.239	10.57	0.024	0.627

Here, we introduce the method to obtain the critical exponents of this model by Monte Carlo simulations which will be introduced in Section 3.3.1. The critical phenomena only occur in the thermodynamic limit ($N \to \infty$). To investigate the critical phenomena through the calculations for finite lattice systems, the finite-size scaling which is introduced by the renormalization group is useful (Privman, 1990). For example, the magnetization per spin m and susceptibility per spin χ in a system where the linear dimension is L are scaled by the following relations:

$$\langle m \rangle \sim L^{\beta/\nu} \Phi \left[L^{1/\nu}(T - T_c) \right], \tag{2.56}$$

$$\chi = \frac{L^d}{k_B T} \left(\langle m^2 \rangle - \langle m \rangle^2 \right) \sim L^{\gamma/\nu} \Psi \left[L^{1/\nu}(T - T_c) \right], \tag{2.57}$$

where $\langle\cdot\rangle$ indicates the thermal average. $\Phi(\cdot)$ and $\Psi(\cdot)$ are called the scaling functions. The critical exponents and transition temperature for physical quantities which are calculated for some lattice sizes can be estimated by these scaling laws. Furthermore, to estimate the critical exponents and transition temperature more easily, the Binder ratio was introduced by Binder (Binder, 1981). The Binder ratio is defined by

$$g := \frac{1}{2}\left[3 - \frac{\langle m^4\rangle}{\langle m^2\rangle^2}\right], \tag{2.58}$$

and the scaling law for the Binder ratio is given as

$$g \sim g\left[L^{1/\nu}(T - T_{\mathrm{c}})\right]. \tag{2.59}$$

Thus, at $T = T_{\mathrm{c}}$, the Binder ratios calculated for some lattice sizes coincide. By using the estimated T_{c}, the critical exponent ν can be easily obtained by the scaling law. For example, Fig. 2.6 (a) shows the temperature dependence of the Binder ratio g of the Ising model on a simple cubic lattice, calculated by Monte Carlo simulations (explained in Section 3.3.1). At $T_{\mathrm{c}}/J \simeq 4.51$, the Binder ratios cross. Figure 2.6 (b) shows the finite-size scaling results found by using Eq. (2.59) for $\nu = 0.627$. The Binder ratios for $L = 8, 16, 24$, and 32 collapse with the same curve.

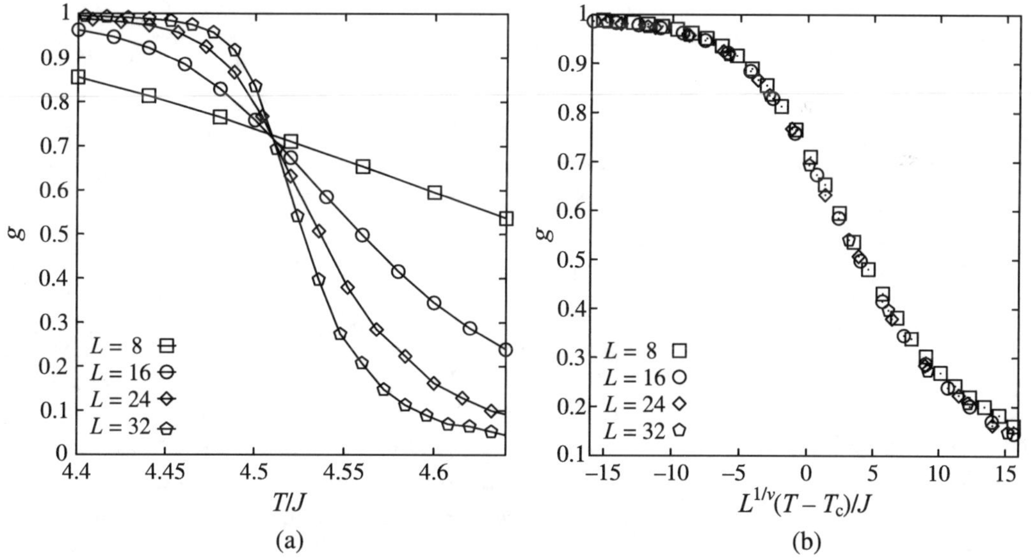

Fig. 2.6 Monte Carlo simulation results of the Ising model on a cubic lattice. The linear dimension is L. (a) Temperature dependence of the Binder ratio g of the magnetization. At $T_{\mathrm{c}}/J \simeq 4.51$, the Binder ratios cross at the same point. The curves between points are a guide to the eye. (b) Finite-size scaling results of the Binder ratio of the magnetization using Eq. (2.59), where $T_{\mathrm{c}}/J = 4.51$ and $\nu = 0.627$.

2.2 Basic Concept of Classical Spin Glass

As will be shown in Chapter 3, the Ising model with random magnetic interactions is used to express combinatorial optimization problems. The model was introduced to explain the physical properties of random spin systems such as spin glass materials (Chowdhury, 1986; Mézard et al. 1986; Fischer and Hertz, 1993; Young, 1997; Nishimori, 2001; Dominicis and Giardina, 2006; Bolthausen and Bovier, 2007). Spins in spin glass materials are not aligned in a regularly ordered pattern such as ferromagnetic and antiferromagnetic configurations in the low-temperature phase. At the freezing temperature T_c, spins are frozen in the disordered spin configuration, and the magnetic susceptibility exhibits a sharp cusp (Cannella and Mydosh, 1972). Thus, it has been suggested that phase transition occurs at T_c in spin glass materials. In order to study a spin glass, some theoretical models where a spin glass phase exists were proposed.

As one of the models where a spin glass phase exists, Edwards and Anderson introduced the following model (Edwards and Anderson, 1975):

$$\mathcal{H} = -\sum_{\langle i,j \rangle} J_{ij}\sigma_i^z\sigma_j^z, \quad (\sigma_i^z = \pm 1), \tag{2.60}$$

where the probability distribution of J_{ij} is given as

$$P(J_{ij}) = \left(\frac{N}{2\pi J^2}\right)^{\frac{1}{2}} \exp\left[-\frac{N}{2J^2}\left(J_{ij} - \frac{J_0}{N}\right)^2\right]. \tag{2.61}$$

This model is called the Edwards–Anderson model. In the model, magnetic interactions $\{J_{ij}\}$ are randomly distributed by a Gaussian distribution with mean J_0/N and standard deviation $J/\sqrt{N}$. On the other hand, when the Hamiltonian is given by Eq. (2.60), where $J_{ij} = \pm J$ are randomly distributed for each magnetic interaction, the model is called the $\pm J$ model. The probability distribution of the $\pm J$ model is given by

$$P(J_{ij}) = p\delta(J_{ij} - J) + (1-p)\delta(J_{ij} + J), \tag{2.62}$$

where $\delta(\cdot)$ is the Dirac delta function. As long as we focus on the finite-temperature phase transition, there is no essential difference between the Edwards–Anderson model and the $\pm J$ model, although low-temperature behaviors are different in these models. By using these models, the properties of a spin glass have been exhaustively investigated by theoretical works.

2.2.1 Frustration and randomness

Frustration appears in the Ising model with random interactions. This causes changes in the phase transition nature and magnetic ordered state. Here, we explain the frustration effect in the case of the $\pm J$ model.

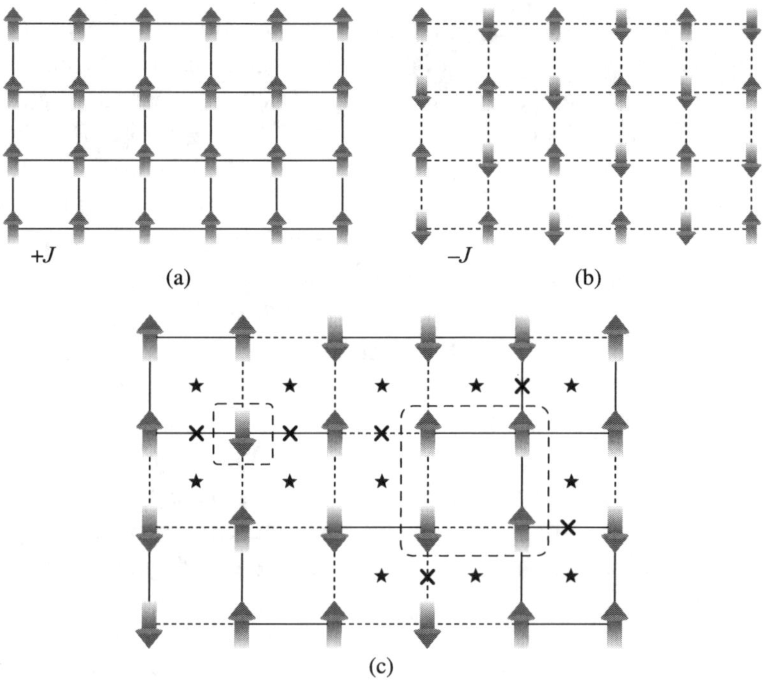

Fig. 2.7 (a) Ground state spin configuration of the Ising model with homogeneous ferromagnetic interactions on a square lattice. The magnetic interaction between nearest-neighbor spins indicated by the solid line is the ferromagnetic interaction ($J_{ij} = +J$). (b) Ground state spin configuration of the Ising model with homogeneous antiferromagnetic interactions on a square lattice. The magnetic interaction between nearest-neighbor spins indicated by the dotted line is the antiferromagnetic interaction ($J_{ij} = -J$). (c) Ground state spin configuration of the $\pm J$ model. $J_{ij} = \pm J$ are randomly distributed. The plaquette marked by the star indicates the frustration plaquette and an energy loss exists at interactions indicated by the crosses. The energy is the same if the spins surrounded by the dashed squares are flipped.

At first, we consider the ground state spin configurations in the Ising model with homogeneous interactions on a square lattice. When all magnetic interactions are ferromagnetic ($J_{ij} = +J$) in Eq. (2.60), in the ground state, all spins are aligned in the same direction, which is shown in Fig. 2.7 (a). Even if all spins are flipped, the energy is not changed, which means that the ground state is twofold degenerate. Next, let us consider the case that all magnetic interactions are antiferromagnetic ($J_{ij} = -J$). Figure 2.7 (b) shows one of the two ground state spin configurations. Similar to the ferromagnetic system, the ground state is twofold degenerate. Notice that the ground state of the Ising model with homogeneous antiferromagnetic interactions is not necessarily twofold degenerate (it depends on the lattices) whereas the homogeneous ferromagnetic

Ising model has twofold degenerate ground states on any lattices. If the lattice structure is based on triangles such as the triangular lattice and Kagomé lattice, the antiferromagnetic Ising model has the frustration effect known as geometrical frustration, and as a result, macroscopically degenerated ground states appear (Wannier, 1950; Husimi and Syôzi, 1950; Syozi, 1950; Houtappel, 1950; Kanô and Naya, 1953; Wannier, 1973; Ramirez et al. 1999; Bramwell and Gingras, 2001; Udagawa et al. 2002). Geometrical frustration causes interesting properties such as unconventional excitations and exotic dynamic behavior (Kawamura and Miyashita, 1984a,b; Liebmann, 1986; Katsura et al. 2005; Tanaka et al. 2006; Onoda and Nagaosa, 2006; Castelnovo et al. 2007; Tanaka and Miyashita, 2007; Miyashita et al. 2007; Tamura and Kawashima, 2008; Tanaka and Miyashita, 2009; Tamura and Kawashima, 2011; Seki et al. 2012; Nagaosa and Tokura, 2013; Diep, 2013; Nisoli et al. 2013; Tamura et al. 2013; Gingras and McClarty, 2014; Tokura et al. 2014). Here, since the square lattice is a bipartite lattice, the ground state is twofold degenerate, which is also satisfied in the homogeneous antiferromagnetic Ising model on bipartite lattices. In the homogeneous ferromagnetic and antiferromagnetic Ising models on bipartite lattices, the second-order phase transition with twofold symmetry breaking occurs at a finite temperature.

However, if randomness of magnetic interactions exists, frustration is induced by the competition among interactions between spins, and the properties of the ground state are drastically changed. We consider the ground-state spin configuration in the model where $J_{ij} = \pm J$ are randomly distributed for each nearest-neighbor interaction. Figure 2.7 (c) is an example of random magnetic interactions ($J_{ij} = \pm J$). Without loss of generality, we assume that the upper left spin is the up-spin state. We try to decide each spin direction such that the magnetic interaction energies have the lowest value. However, we notice that there is no spin configuration so that all magnetic interaction energies have the lowest value at the plaquette marked by the star. This situation is called frustration and the plaquette having frustration is called the frustration plaquette. In this case, the spin configuration of Fig. 2.7 (c) is one of the ground states. Since energy loss exists at interactions indicated by crosses, the ground-state energy of the system including frustration is higher than that of models with homogeneous ferromagnetic and antiferromagnetic interactions such as shown in Figs 2.7 (a) and (b). Furthermore, in systems including the randomness of magnetic interactions, the frustration plaquettes are randomly distributed on the lattices. From this effect, the number of states which have nearly the same energy increases in the low-temperature phase. In other words, condensation of states in the density of states occurs, which causes unusual static and dynamic properties of random spin systems (Néel, 1949; Matsubara, 1974; Aharony, 1975; Matsubara, 1975; Guy, 1978; Fishman and Aharony, 1979; Kinzel, 1979; Oguchi and Ueno, 1979; Ferré et al. 1981; Katsumata et al. 1982; Oseroff et al. 1982; Grassberger and Procac-cia, 1982; Chamberlin et al. 1984; Ngai et al. 1984; Chamberlin, 1985; Ogielski, 1985; Uemura et al. 1985; Mézard et al. 1986; Fisher and Huse, 1988; Takano and Miyashita, 1989; Le och et al. 1992; Fischer and Hertz, 1993; Kawasaki and Miyashita,

1993; Shirakura and Matsub-ara, 1993; Dotsenko, 1995; Matsubara et al. 1996; Nielsen et al. 1996; Hukushima et al. 1997; Young, 1997; Jonsson et al. 1999; Dupuis et al., 2001; Nishimori, 2001; Bekhechi and Southern, 2004; Ladieu et al. 2004; Jönsson and Takayama, 2005; Beath and Ryan, 2006; Dominicis and Gi-ardina, 2006; Arai et al. 2007; Bolthausen and Bovier, 2007; Takayama and Hukushima, 2007; Vincent, 2007; Tamura et al. 2011; Jesi, 2016). For example, if the spins indicated by the dashed squares are flipped in Fig. 2.7 (c), the energy does not change, which means that the state is also the ground state. As shown here, many ground states degenerate in the $\pm J$ model. Properties of the ground state of the Ising model with random interactions is quite different from that in the models with homogeneous ferromagnetic or antiferromagnetic interactions. Thus, it is expected that the phase transition nature in the Ising model with random interactions is different from that in models with homogeneous interactions.

2.2.2 Spin glass order parameter

Although the Ising model with random interactions has no spin glass phase in one or two dimensions at finite temperatures, the model has a spin glass phase in three or higher dimensions at finite temperatures. In the spin glass phase, the spin configuration is disordered and each spin is frozen. Thus, magnetization which is the order parameter of the ferromagnetic model becomes zero in the spin glass phase. Moreover, the ground states cannot be characterized by any wave vectors. More precisely, the generalized magnetization with $\mathbf{q}$ should be zero. Generalized magnetization is defined by

$$m(\mathbf{q}) := \frac{1}{N}\sum_{i=1}^{N} \sigma_i^z e^{i\mathbf{q}\cdot\mathbf{r}_i}, \tag{2.63}$$

where $\mathbf{r}_i$ is the position of site i. In order to detect the spin glass phase, Edwards and Anderson introduced the spin glass order parameter by focusing on the spin freezing (Edwards and Anderson, 1975; Parisi, 1983). The spin glass order parameter is given by

$$q_{\mathrm{EA}} = \lim_{t\to\infty} \langle \sigma_i^z(0)\sigma_i^z(t)\rangle, \tag{2.64}$$

where t is the time, $\sigma_i^z(t) = \pm 1$ is the spin state at the time t, and $\langle\cdot\rangle$ indicates the thermal average. To calculate the spin glass order parameter is difficult since the definition of the order parameter includes the time t. Therefore, $\sigma_i^z(t)$ in the limit as $t\to\infty$ is replaced by the thermal average of spin $\langle\sigma_i^z\rangle$, and the spin glass order parameter is written by

$$q_{\mathrm{EA}} = \langle\sigma_i^z\rangle^2. \tag{2.65}$$

By using this order parameter instead of Eq. (2.64), the spin glass phase can be detected.

2.2.3 Replica method

In general, thermal properties do not depend on the detail of the distribution of magnetic interactions $\{J_{ij}\}$. Thus, we define the free energy of a system with random interactions by taking the average over the configuration of $\{J_{ij}\}$ as follows:

$$\overline{f} = -\frac{1}{N} k_{\mathrm{B}} T \overline{\ln Z}, \tag{2.66}$$

where the bar indicates the configurational average. More precisely, $\overline{\mathscr{O}}$ is represented as

$$\overline{\mathscr{O}} = \int_{-\infty}^{\infty} \prod_{\langle i,j \rangle} dJ_{ij} P(J_{ij}) \mathscr{O}, \tag{2.67}$$

where $\mathscr{O}$ is a physical quantity. However, the calculation of $\overline{\ln Z}$ is involved. To calculate $\overline{\ln Z}$ relatively easily, Edwards and Anderson introduced the following identity (Edwards and Anderson, 1975):

$$\overline{\ln Z} = \lim_{n \to 0} \frac{\overline{Z^n} - 1}{n}. \tag{2.68}$$

This calculation method is called the replica method. The replica method has been used for analysis of random spin systems and random electric systems. Here, for the case of Eq. (2.60), $\overline{Z^n}$ is written as

$$\overline{Z^n} = \overline{\mathrm{Tr} \exp\left(\beta \sum_{\alpha=1}^{n} \sum_{\langle i,j \rangle} J_{ij} \sigma_{i\alpha}^z \sigma_{j\alpha}^z \right)}, \tag{2.69}$$

where α is the index of replica which is replicated by using the same $\{J_{ij}\}$. Thus, by using the replica method, the configurational average can be calculated before the calculation of trace, and estimation of $\overline{Z^n}$ is easier than that of $\overline{\ln Z}$, which is the merit of the replica method.

2.3 Sherrington–Kirkpatrick Model

The Hamiltonian of the Sherrington–Kirkpatrick model (SK model) (Sherrington and Kirkpatrick, 1975; Kirkpatrick and Sherrington, 1978) is written as follows:

$$\mathscr{H} = - \sum_{1 \le i < j \le N} J_{ij} \sigma_i^z \sigma_j^z - h \sum_{i=1}^{N} \sigma_i^z, \quad (\sigma_i^z = \pm 1), \tag{2.70}$$

where the number of spin is N. Here, the distribution of J_{ij} is given as

$$P(J_{ij}) = \left(\frac{N}{2\pi J^2} \right)^{\frac{1}{2}} \exp\left[-\frac{N}{2J^2} \left(J_{ij} - \frac{J_0}{N} \right)^2 \right]. \tag{2.71}$$

That is, the SK model is the infinite-range Edwards–Anderson model whose probability distribution of interactions is a Gaussian distribution. The model is the most fundamental model which exhibits a spin glass.

2.3.1 Free energy of the Sherrington–Kirkpatrick model

Here, we calculate the free energy of the SK model. The configurational average of the partition function defined by Eq. (2.69) based on the replica method for the SK model is written as

$$\overline{Z^n} = \int_{-\infty}^{\infty} \prod_{1\leq i<j\leq N} dJ_{ij} P(J_{ij})$$

$$\times \mathrm{Tr} \exp\left(\beta \sum_{1\leq i<j\leq N} J_{ij} \sum_{\alpha=1}^{n} \sigma_{i\alpha}^z \sigma_{j\alpha}^z + \beta h \sum_{i=1}^{N}\sum_{\alpha=1}^{n} \sigma_{i\alpha}^z\right), \tag{2.72}$$

where α is the index of replica, and here we consider the case where the number of replica is n. By using the definition of $P(J_{ij})$ (Eq. (2.71)), we take the integral of the right-hand side of Eq. (2.72) with respect to J_{ij}, and obtain,

$$\overline{Z^n} = \mathrm{Tr}\exp\left[\frac{1}{N}\sum_{1\leq i<j\leq N}\left(\frac{1}{2}\beta^2 J^2 \sum_{1\leq\alpha,\beta\leq n} \sigma_{i\alpha}^z\sigma_{j\alpha}^z\sigma_{i\beta}^z\sigma_{j\beta}^z\right.\right.$$

$$\left.\left. + \beta J_0 \sum_{\alpha=1}^{n} \sigma_{i\alpha}^z\sigma_{j\alpha}^z\right) + \beta h \sum_{i=1}^{N}\sum_{\alpha=1}^{n}\sigma_{i\alpha}^z\right]. \tag{2.73}$$

When N is large enough, the argument of the right-hand side of Eq. (2.73) is rewritten as

$$\frac{1}{N}\sum_{1\leq i<j\leq N}\left(\frac{1}{2}\beta^2 J^2 \sum_{1\leq\alpha,\beta\leq n} \sigma_{i\alpha}^z\sigma_{j\alpha}^z\sigma_{i\beta}^z\sigma_{j\beta}^z + \beta J_0 \sum_{\alpha=1}^{n}\sigma_{i\alpha}^z\sigma_{j\alpha}^z\right)$$

$$= \frac{1}{N}\frac{1}{2}\sum_{1\leq i,j\leq N}\left[\frac{1}{2}\beta^2 J^2\left(2\sum_{1\leq\alpha<\beta\leq n}\sigma_{i\alpha}^z\sigma_{j\alpha}^z\sigma_{i\beta}^z\sigma_{j\beta}^z\right.\right.$$

$$\left.\left. + \sum_{\alpha=1}^{n}\sigma_{i\alpha}^z\sigma_{j\alpha}^z\sigma_{i\alpha}^z\sigma_{j\alpha}^z\right) + \beta J_0\sum_{\alpha=1}^{n}\sigma_{i\alpha}^z\sigma_{j\alpha}^z\right]$$

$$= \frac{\beta^2 J^2}{2N}\sum_{1\leq\alpha<\beta\leq n}\left(\sum_{i=1}^{N}\sigma_{i\alpha}^z\sigma_{i\beta}^z\right)^2 + \frac{1}{4}\beta^2 J^2 nN + \frac{\beta J_0}{2N}\sum_{\alpha=1}^{n}\left(\sum_{i=1}^{N}\sigma_{i\alpha}^z\right)^2. \tag{2.74}$$

In Eq. (2.74), we used the fact that $\sigma^z_{i\alpha}\sigma^z_{j\alpha}\sigma^z_{i\alpha}\sigma^z_{j\alpha} = 1$. Then, the partition function is written as follows:

$$\overline{Z^n} = e^{\beta^2 J^2 nN/4} \mathrm{Tr} \exp\left[\frac{\beta^2 J^2}{2N}\sum_{1\le\alpha<\beta\le n}\left(\sum_{i=1}^N \sigma^z_{i\alpha}\sigma^z_{i\beta}\right)^2\right.$$

$$\left. + \frac{\beta J_0}{2N}\sum_{\alpha=1}^n\left(\sum_{i=1}^N \sigma^z_{i\alpha}\right)^2 + \beta h \sum_{i=1}^N\sum_{\alpha=1}^n \sigma^z_{i\alpha}\right]. \tag{2.75}$$

Next, we use the Hubbard–Stratonovich transformation:

$$\exp\left(\frac{\lambda a^2}{2}\right) = \left(\frac{\lambda}{2\pi}\right)^{\frac{1}{2}}\int_{-\infty}^{\infty} dx \exp\left(-\frac{\lambda x^2}{2} + a\lambda x\right), \tag{2.76}$$

and obtain

$$\exp\left[\frac{\beta^2 J^2}{2N}\left(\sum_{i=1}^N \sigma^z_{i\alpha}\sigma^z_{i\beta}\right)^2\right]$$

$$= \left(\frac{N\beta^2 J^2}{2\pi}\right)^{\frac{1}{2}}\int_{-\infty}^{\infty} dq_{\alpha\beta} \times \exp\left[-\frac{N\beta^2 J^2}{2} q^2_{\alpha\beta} + \beta^2 J^2 q_{\alpha\beta}\left(\sum_{i=1}^N \sigma^z_{i\alpha}\sigma^z_{i\beta}\right)\right], \tag{2.77}$$

$$\exp\left[\frac{\beta J_0}{2N}\left(\sum_{i=1}^N \sigma^z_{i\alpha}\right)^2\right]$$

$$= \left(\frac{N\beta J_0}{2\pi}\right)^{\frac{1}{2}}\int_{-\infty}^{\infty} dm_\alpha \times \exp\left[-\frac{N\beta J_0}{2} m^2_\alpha + \beta J_0 m_\alpha\left(\sum_{i=1}^N \sigma^z_{i\alpha}\right)\right]. \tag{2.78}$$

By using these relations, the partition function is written as follows:

$$\overline{Z^n} = e^{\beta^2 J^2 nN/4}\int_{-\infty}^{\infty}\prod_{1\le\alpha<\beta\le n} dq_{\alpha\beta}\int_{-\infty}^{\infty}\prod_{\alpha=1}^n dm_\alpha$$

$$\times \exp\left(-\frac{N\beta^2 J^2}{2}\sum_{1\le\alpha<\beta\le n} q^2_{\alpha\beta} - \frac{N\beta J_0}{2}\sum_{\alpha=1}^n m^2_\alpha\right)$$

$$\times \mathrm{Tr}\exp\left[\beta^2 J^2 \sum_{1\le\alpha<\beta\le n} q_{\alpha\beta}\left(\sum_{i=1}^{N}\sigma_{i\alpha}^z\sigma_{i\beta}^z\right)\right.$$

$$\left.+\beta\sum_{\alpha=1}^{n}(J_0 m_\alpha + h)\left(\sum_{i=1}^{N}\sigma_{i\alpha}^z\right)\right] \tag{2.79}$$

$$= \mathrm{e}^{\beta^2 J^2 nN/4}\int_{-\infty}^{\infty}\prod_{1\le\alpha<\beta\le n} dq_{\alpha\beta}\int_{-\infty}^{\infty}\prod_{\alpha=1}^{n} dm_\alpha$$

$$\times\exp\left(-\frac{N\beta^2J^2}{2}\sum_{1\le\alpha<\beta\le n}q_{\alpha\beta}^2 - \frac{N\beta J_0}{2}\sum_{\alpha=1}^{n}m_\alpha^2\right)$$

$$\times\left\{\mathrm{Tr}\exp\left[\beta^2J^2\sum_{1\le\alpha<\beta\le n}q_{\alpha\beta}\sigma_\alpha^z\sigma_\beta^z + \beta\sum_{\alpha=1}^{n}(J_0 m_\alpha + h)\sigma_\alpha^z\right]\right\}^N. \tag{2.80}$$

Here, L is defined by

$$L := \beta^2J^2\sum_{1\le\alpha<\beta\le n}q_{\alpha\beta}\sigma_\alpha^z\sigma_\beta^z + \beta\sum_{\alpha=1}^{n}(J_0 m_\alpha + h)\sigma_\alpha^z. \tag{2.81}$$

By using L, the argument of the trace of the exponential in the right-hand side of Eq. (2.80) is expressed as

$$\exp\left[\beta^2J^2\sum_{1\le\alpha<\beta\le n}q_{\alpha\beta}\sigma_\alpha^z\sigma_\beta^z + \beta\sum_{\alpha=1}^{n}(J_0 m_\alpha + h)\sigma_\alpha^z\right] = \exp\left(\ln\mathrm{Tr}\,\mathrm{e}^L\right). \tag{2.82}$$

Then, we obtain

$$\overline{Z^n} = \mathrm{e}^{\beta^2 J^2 nN/4}\int_{-\infty}^{\infty}\prod_{1\le\alpha<\beta\le n} dq_{\alpha\beta}\int_{-\infty}^{\infty}\prod_{\alpha=1}^{n} dm_\alpha$$

$$\times\exp\left(-\frac{N\beta^2J^2}{2}\sum_{1\le\alpha<\beta\le n}q_{\alpha\beta}^2 - \frac{N\beta J_0}{2}\sum_{\alpha=1}^{n}m_\alpha^2 + N\ln\mathrm{Tr}\,\mathrm{e}^L + \frac{1}{4}\beta^2J^2nN\right). \tag{2.83}$$

Since the power of the exponential is proportional to N, we can take these integrals to the right-hand side of Eq. (2.83) with respect to $q_{\alpha\beta}$ and m_α by the saddle-point method. We obtain

$$\overline{Z^n} \simeq \exp\left(-\frac{N\beta^2 J^2}{2}\sum_{1\leq\alpha<\beta\leq n} q_{\alpha\beta}^2 - \frac{N\beta J_0}{2}\sum_{\alpha=1}^{n} m_\alpha^2 + N\ln \mathrm{Tr}\,\mathrm{e}^L + \frac{1}{4}\beta^2 J^2 nN\right) \quad (2.84)$$

$$\simeq 1 + Nn\left(-\frac{\beta^2 J^2}{2n}\sum_{1\leq\alpha<\beta\leq n} q_{\alpha\beta}^2 - \frac{\beta J_0}{2n}\sum_{\alpha=1}^{n} m_\alpha^2 + \frac{1}{n}\ln \mathrm{Tr}\,\mathrm{e}^L + \frac{1}{4}\beta^2 J^2\right). \quad (2.85)$$

Therefore, the free energy per spin f can be calculated by the replica method explained in Section 2.2.3:

$$-\beta f = \frac{1}{N}\lim_{n\to 0}\frac{\overline{Z^n}-1}{n} \quad (2.86)$$

$$= \lim_{n\to 0}\left(-\frac{\beta^2 J^2}{2n}\sum_{1\leq\alpha<\beta\leq n} q_{\alpha\beta}^2 - \frac{\beta J_0}{2n}\sum_{\alpha=1}^{n} m_\alpha^2 + \frac{1}{n}\ln \mathrm{Tr}\,\mathrm{e}^L + \frac{1}{4}\beta^2 J^2\right), \quad (2.87)$$

where $q_{\alpha\beta}$ and m_α are decided by the saddle-point conditions. The saddle-point conditions for $q_{\alpha\beta}$ and m_α are given as follows:

$$\frac{\partial(-\beta f)}{\partial q_{\alpha\beta}} = -\frac{\beta^2 J^2}{n} q_{\alpha\beta} + \frac{1}{n}\frac{\mathrm{Tr}\,\beta^2 J^2 \sigma_\alpha^z \sigma_\beta^z \mathrm{e}^L}{\mathrm{Tr}\,\mathrm{e}^L} = 0, \quad (2.88)$$

$$\frac{\partial(-\beta f)}{\partial m_\alpha} = -\frac{\beta J_0}{n} m_\alpha + \frac{1}{n}\frac{\mathrm{Tr}\,\beta J_0 \sigma_\alpha^z \mathrm{e}^L}{\mathrm{Tr}\,\mathrm{e}^L} = 0, \quad (2.89)$$

and we obtain

$$q_{\alpha\beta} = \frac{\mathrm{Tr}\,\sigma_\alpha^z \sigma_\beta^z \mathrm{e}^L}{\mathrm{Tr}\,\mathrm{e}^L} = \langle \sigma_\alpha^z \sigma_\beta^z \rangle_L, \quad (2.90)$$

$$m_\alpha = \frac{\mathrm{Tr}\,\sigma_\alpha^z \mathrm{e}^L}{\mathrm{Tr}\,\mathrm{e}^L} = \langle \sigma_\alpha^z \rangle_L, \quad (2.91)$$

where $\langle\cdot\rangle_L$ is the thermal average by L. Here, m_α and $q_{\alpha\beta}$ mean the order parameters of the ferromagnetic and spin glass phases, respectively. By the same formula deformation from Eq. (2.72) to Eq. (2.79), the following relation is obtained:

$$\overline{\mathrm{Tr}\,\sigma^z_{i\alpha}\sigma^z_{i\beta}\mathrm{e}^{-\beta\sum_\gamma \mathscr{H}_\gamma}} = \mathrm{e}^{\beta^2 J^2 nN/4}\int_{-\infty}^{\infty}\prod_{1\le\alpha<\beta\le n} dq_{\alpha\beta}\int_{-\infty}^{\infty}\prod_{\alpha=1}^{n} dm_\alpha$$

$$\times\exp\left(-\frac{N\beta^2J^2}{2}\sum_{1\le\alpha<\beta\le n} q^2_{\alpha\beta} - \frac{N\beta J_0}{2}\sum_{\alpha=1}^{n} m^2_\alpha\right)$$

$$\times\left(\mathrm{Tr}\,\sigma^z_{i\alpha}\sigma^z_{i\beta}\mathrm{e}^L\right)\left(\mathrm{Tr}\,\mathrm{e}^L\right)^{N-1}, \tag{2.92}$$

where $\mathscr{H}_\gamma$ is the Hamiltonian of the replica γ, that is,

$$\mathscr{H}_\gamma = -\sum_{1\le i<j\le N} J_{ij}\sigma^z_{i\gamma}\sigma^z_{j\gamma} - h\sum_{i=1}^{N}\sigma^z_{i\gamma}. \tag{2.93}$$

By using the saddle-point method and taking the limit of $n\to 0$, we obtain

$$\lim_{n\to 0}\overline{\mathrm{Tr}\,\sigma^z_{i\alpha}\sigma^z_{i\beta}\mathrm{e}^{-\beta\sum_\gamma \mathscr{H}_\gamma}} = \mathrm{Tr}\,\sigma^z_\alpha\sigma^z_\beta\mathrm{e}^L. \tag{2.94}$$

Here, we used the relation given by $\mathrm{Tr}\,\mathrm{e}^L\to 1$ in the limit of $n\to 0$. Thus, in the limit $n\to 0$, the following equation is satisfied:

$$q_{\alpha\beta} = \frac{\mathrm{Tr}\,\sigma^z_\alpha\sigma^z_\beta\mathrm{e}^L}{\mathrm{Tr}\,\mathrm{e}^L} = \frac{\overline{\mathrm{Tr}\,\sigma^z_{i\alpha}\sigma^z_{i\beta}\mathrm{e}^{-\beta\sum_\gamma \mathscr{H}_\gamma}}}{\mathrm{Tr}\,\mathrm{e}^{-\beta\sum_\gamma \mathscr{H}_\gamma}} \tag{2.95}$$

$$= \overline{\frac{\mathrm{Tr}\,\sigma^z_{i\alpha}\mathrm{e}^{-\beta\mathscr{H}_\alpha}}{\mathrm{Tr}\,\mathrm{e}^{-\beta\mathscr{H}_\alpha}}\cdot\frac{\mathrm{Tr}\,\sigma^z_{i\beta}\mathrm{e}^{-\beta\mathscr{H}_\beta}}{\mathrm{Tr}\,\mathrm{e}^{-\beta\mathscr{H}_\beta}}} \tag{2.96}$$

$$= \overline{\langle\sigma^z_{i\alpha}\rangle\cdot\langle\sigma^z_{i\beta}\rangle} \tag{2.97}$$

$$= \overline{\langle\sigma^z_i\rangle^2} \tag{2.98}$$

$$= \overline{q_{\mathrm{EA}}}. \tag{2.99}$$

Here, we also used $\mathrm{Tr}\,\mathrm{e}^L\to 1$ and $\mathrm{e}^{-\beta\sum_\gamma \mathscr{H}_\gamma}\to 1$ in the limit of $n\to 0$. Thus, we confirm that $q_{\alpha\beta}$ is the spin glass order parameter which is defined by Eq. (2.65). From the same calculations, we can confirm that m_α is the magnetization which is the order parameter of the ferromagnetic phase, that is,

$$m_\alpha = \overline{\langle\sigma^z_i\rangle} = \overline{m}. \tag{2.100}$$

Thus, depending on the values of m_α and $q_{\alpha\beta}$, phases are characterized as follows:

1. Paramagnetic phase $(m_\alpha = 0,\ q_{\alpha\beta} = 0)$,
2. Ferromagnetic phase $(m_\alpha > 0,\ q_{\alpha\beta} = 0)$,
3. Spin glass phase $(m_\alpha > 0,\ q_{\alpha\beta} > 0)$.

2.3.2 Replica symmetric solution

Free energy of SK model for the replica symmetric approximation

In this subsection, we consider the replica symmetric approximation: $m_\alpha = m$, and $q_{\alpha\beta} = q$. Under the approximation, the dependence of replica vanishes. In this case, the free energy per spin is written as follows:

$$-\beta f = \frac{1}{N}\lim_{n\to 0}\frac{\overline{Z^n}-1}{n} \tag{2.101}$$

$$= \lim_{n\to 0}\left[-\frac{\beta^2 J^2}{4n}n(n-1)q^2 - \frac{\beta J_0}{2n}nm^2 + \frac{1}{n}\ln \mathrm{Tr}\,\mathrm{e}^{L} + \frac{1}{4}\beta^2 J^2\right], \tag{2.102}$$

where

$$L = \beta^2 J^2 q \sum_{1\le\alpha<\beta\le n} \sigma_\alpha^z \sigma_\beta^z + \beta(J_0 m + h)\sum_{\alpha=1}^{n}\sigma_\alpha^z. \tag{2.103}$$

By using the Hubbard–Stratonovich transformation, we can obtain the following relation:

$$\exp\left(\beta^2 J^2 q \sum_{1\le\alpha<\beta\le n}\sigma_\alpha^z\sigma_\beta^z\right)$$

$$= \exp\left[\frac{\beta^2 J^2}{2}q\left(\sum_{\alpha=1}^{n}\sigma_\alpha^z\right)^2\right]\exp\left[-\frac{\beta^2 J^2}{2}q\sum_{\alpha=1}^{n}(\sigma_\alpha^z)^2\right] \tag{2.104}$$

$$= \sqrt{\frac{\beta^2 J^2 q}{2\pi}}\int_{-\infty}^{\infty} dx \exp\left[-\frac{\beta^2 J^2 q}{2}\left(x^2 - 2x\sum_{\alpha=1}^{n}\sigma_\alpha^z\right)\right]$$

$$\times \exp\left(-\frac{\beta^2 J^2}{2}nq\right). \tag{2.105}$$

By using Eq. (2.105), we can rewrite

$$\ln \mathrm{Tr}\, \mathrm{e}^{L}$$

$$= \ln \mathrm{Tr} \sqrt{\frac{\beta^2 J^2 q}{2\pi}} \int_{-\infty}^{\infty} dx$$

$$\times \exp\left[-\frac{\beta^2 J^2 q}{2} x^2 + \beta(\beta J^2 q x + J_0 m + h) \sum_{\alpha=1}^{n} \sigma_\alpha^z - \frac{\beta^2 J^2}{2} n q \right]. \tag{2.106}$$

By using a new variable r defined by

$$\beta J \sqrt{q} x =: r, \tag{2.107}$$

we obtain,

$$\ln \mathrm{Tr}\, \mathrm{e}^{L}$$

$$= \ln \frac{1}{\sqrt{2\pi}} \int_{-\infty}^{\infty} dr \mathrm{e}^{-\frac{r^2}{2}} \mathrm{Tr} \exp\left[\beta h^z(r) \sum_{\alpha=1}^{n} \sigma_\alpha^z - \frac{\beta^2 J^2}{2} n q \right] \tag{2.108}$$

$$= \ln \frac{1}{\sqrt{2\pi}} \int_{-\infty}^{\infty} dr \mathrm{e}^{-\frac{r^2}{2}} \left[2\cosh \beta h^z(r) \right]^n \mathrm{e}^{-\frac{\beta^2 J^2}{2} n q} \tag{2.109}$$

$$= \ln \frac{1}{\sqrt{2\pi}} \int_{-\infty}^{\infty} dr \mathrm{e}^{-\frac{r^2}{2}} \exp\left[n \ln\left(2\cosh \beta h^z(r)\right) - \frac{\beta^2 J^2}{2} n q \right] \tag{2.110}$$

$$\simeq \ln \left[1 + n \frac{1}{\sqrt{2\pi}} \int_{-\infty}^{\infty} dr \mathrm{e}^{-\frac{r^2}{2}} \ln\left(2\cosh \beta h^z(r)\right) - \frac{\beta^2 J^2}{2} n q \right] \tag{2.111}$$

$$\simeq n \frac{1}{\sqrt{2\pi}} \int_{-\infty}^{\infty} dr \mathrm{e}^{-\frac{r^2}{2}} \ln\left(2\cosh \beta h^z(r)\right) - \frac{\beta^2 J^2}{2} n q. \tag{2.112}$$

In these equations, we assume that n is very small, and put

$$h^z(r) := J\sqrt{q} r + J_0 m + h. \tag{2.113}$$

Therefore, the free energy per spin can be expressed as:

$$-\beta f = -\frac{\beta^2 J^2}{4}(1-q)^2 - \frac{\beta J_0}{2} m^2 + \frac{1}{\sqrt{2\pi}} \int_{-\infty}^{\infty} dr \mathrm{e}^{-\frac{r^2}{2}} \ln\left[2\cosh \beta h^z(r)\right]. \tag{2.114}$$

The saddle-point condition of m which is the order parameter of the ferromagnetic phase is

$$m = \frac{1}{\sqrt{2\pi}} \int_{-\infty}^{\infty} dr e^{-\frac{r^2}{2}} \tanh \beta h^z(r). \tag{2.115}$$

On the other hand, the saddle-point condition of q which is the order parameter of the spin glass phase becomes

$$-\frac{\beta^2 J^2}{2}(1-q) + \frac{1}{2\pi} \int_{-\infty}^{\infty} dr e^{-\frac{r^2}{2}} \frac{\beta J r}{2} q^{-\frac{1}{2}} \tanh \beta h^z(r) = 0, \tag{2.116}$$

and we perform the partial integration in the second term, and obtain

$$q = 1 - \frac{1}{\sqrt{2\pi}} \int_{-\infty}^{\infty} dr e^{-\frac{r^2}{2}} \operatorname{sech}^2 \beta h^z(r) \tag{2.117}$$

$$= \frac{1}{\sqrt{2\pi}} \int_{-\infty}^{\infty} dr e^{-\frac{r^2}{2}} \tanh^2 \beta h^z(r). \tag{2.118}$$

Phase diagram of the SK model under the replica symmetric approximation

We consider the phase diagram of the SK model under the replica symmetric approximation. First, we study the phase diagram at the limit of $h = 0$ and $J_0 = 0$, i.e., when the magnetic field does not exist and the interactions are distributed around zero. Then, $h^z(r)$ becomes

$$h^z(r) = J\sqrt{q} r, \tag{2.119}$$

and $\tanh \beta h^z(r)$ is an odd function. Therefore, the order parameter of the ferromagnetic phase obtained by Eq. (2.115) is always $m = 0$, which means that the ferromagnetic phase does not appear. From this fact, the free energy per spin in the limit of $h = 0$ and $J_0 = 0$ is written as:

$$-\beta f = \frac{\beta^2 J^2}{4}(1-q)^2 + \frac{1}{\sqrt{2\pi}} \int_{-\infty}^{\infty} dr e^{-\frac{r^2}{2}} \ln\left[2\cosh \beta h^z(r)\right]. \tag{2.120}$$

The point at which q becomes nonzero is the phase boundary between the paramagnetic phase and the spin glass phase. In order to obtain this phase boundary, we consider the case for small q. We expand the free energy with respect to q, the first and second derivatives of free energy are calculated as follows:

$$\frac{\partial}{\partial q}(-\beta f) = -\frac{\beta^2 J^2}{2}(1-q) + \frac{1}{2\pi} \int_{-\infty}^{\infty} dr e^{-\frac{r^2}{2}} \frac{\beta J r}{2} q^{-\frac{1}{2}} \tanh \beta h^z(r) \tag{2.121}$$

$$= -\frac{\beta^2 J^2}{2}(1-q) + \frac{\beta^2 J^2}{2} \frac{1}{\sqrt{2\pi}} \int_{-\infty}^{\infty} dr e^{-\frac{r^2}{2}} \operatorname{sech}^2 \beta h^z(r), \tag{2.122}$$

$$\frac{\partial^2}{\partial q^2}(-\beta f) = \frac{\beta^2 J^2}{2} - \frac{\beta^2 J^2}{2}\frac{1}{2\pi}\int_{-\infty}^{\infty} dr e^{-\frac{r^2}{2}} \times \frac{\beta J r}{2} q^{-\frac{1}{2}} 2\tanh\beta h^z(r)\mathrm{sech}^2\beta h^z(r) \quad (2.123)$$

$$= \frac{\beta^2 J^2}{2} - \frac{\beta^4 J^4}{2}\frac{1}{2\pi}\int_{-\infty}^{\infty} dr e^{-\frac{r^2}{2}} \times \left[\mathrm{sech}^4\beta h^z(r) - \tanh^2\beta h^z(r)\mathrm{sech}^2\beta h^z(r)\right]. \quad (2.124)$$

Then, the free energy is expressed as

$$-\beta f \simeq \frac{\beta^2 J^2}{4} + \ln 2 + \frac{\beta^2 J^2}{4}\left(1 - \beta^2 J^2\right) q^2 + \mathcal{O}(q^3). \quad (2.125)$$

According to the Landau theory, the point at which the coefficient of q^2 becomes zero is the critical point. Therefore, the spin glass transition temperature for $h = 0$ and $J_0 = 0$ is calculated as:

$$T_c = \frac{J}{k_B}. \quad (2.126)$$

Furthermore, this result can be obtained by expanding Eq. (2.116) with respect to q, that is,

$$q \simeq \beta^2 J^2 q. \quad (2.127)$$

By solving Eqs. (2.115) and (2.118) numerically, we can obtain the phase diagram for $h = 0$ and $J_0 \neq 0$. Figure 2.8 shows the phase diagram of the SK model under the replica symmetric approximation. When the value of J_0 is small, the spin glass phase appears at low temperatures in the SK model.

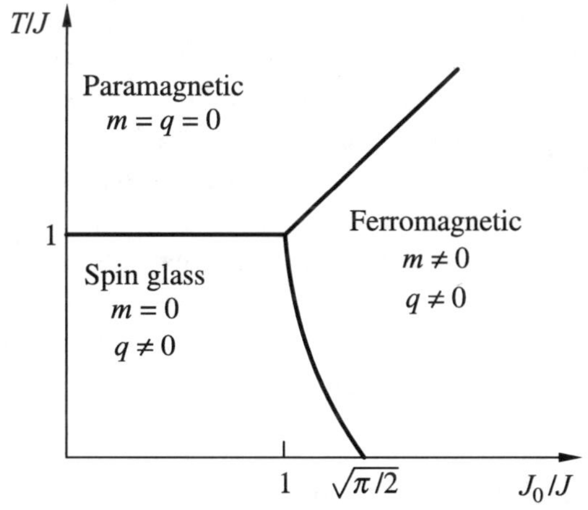

Fig. 2.8 Phase diagram of the SK model under the replica symmetric approximation. Depending on the temperature T/J and the mean value of interactions J_0/J, paramagnetic phase, ferromagnetic phase, and spin glass phase appear.

Problems in the replica symmetric solution

We obtained the phase diagram of the SK model under the replica symmetric approximation. However, the replica symmetric approximation also gives unphysical results. For example, the following problems exist.

(i) In the spin glass phase, the finite spin glass order parameter ($q > 0$) maximizes the free energy, while zero spin glass order parameter ($q = 0$) maximizes the free energy in the paramagnetic phase.

In Eq. (2.125), the coefficient of q^2 becomes positive when $T < T_{\mathrm{c}}$, i.e., in the spin glass phase. Thus, finite spin glass order parameter ($q > 0$) maximizes the free energy. Furthermore, the coefficient of q^2 becomes negative when $T > T_{\mathrm{c}}$. Thus, zero spin glass order parameter $q = 0$, i.e., in the paramagnetic phase, maximizes the free energy. This fact is opposite to general cases.

(ii) The entropy becomes negative at low temperatures.

The magnetic entropy S calculated by Eq. (2.120) becomes $S = -N/(2\pi)$ when $T = 0$. In Ising spin systems, the number of states should be a discrete value. Thus, the magnetic entropy should have a positive value at finite temperatures, and $S \to 0$ at $T = 0$ in the limit of $N \to \infty$ for systems in which the ground states are not macroscopically degenerated. This behavior of magnetic entropy cannot be obtained in the replica symmetric solution which will be explained in the next subsection.

2.3.3 Replica symmetry breaking

Instability of the replica symmetric solution

It is known that the replica symmetric solution is unstable at low temperatures. The de Almeida–Thouless (AT) line (de Almeida and Thouless, 1978) is the boundary between the condition when the replica symmetric solution is stable and the condition when the replica symmetric solution is unstable. Thus, below the AT line, we should take care of the replica symmetry breaking. The phase diagram of the SK model including the replica symmetry breaking is shown in Fig. 2.9. A significant point of the phase diagram is the existence of the mixed phase in which the ferromagnetic order and the spin glass order simultaneously exist. The AT line is the boundary between the ferromagnetic phase and the mixed phase. Furthermore, the replica symmetry is broken in the mixed phase and the spin glass phase. On the other hand, in the ferromagnetic phase and the paramagnetic phase, the replica symmetric solution exists.

The Parisi solution

The most important solution for replica symmetry breaking in the SK model is the Parisi solution (Parisi, 1979, 1980a,b) which is now considered as the exact solution of the SK model. In the Parisi solution, the replica symmetry is hierarchically broken.

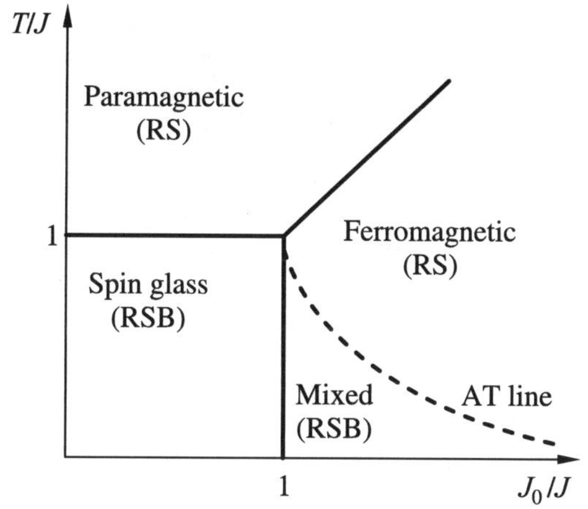

Fig. 2.9 Phase diagram of the SK model including the replica symmetry breaking. In the mixed phase, the ferromagnetic order and the spin glass order coexist. The dashed line indicates the AT line. Below the AT line, the replica symmetric solution becomes unstable.

In general, the spin glass order parameter $q_{\alpha\beta}$ is expressed as an $n \times n$ symmetric matrix, where n is the number of replicas. Here, we consider the case of $n = 12$ as an example. In the replica symmetric solution, the diagonal elements are zero whereas the off-diagonal elements are the same value q_0; thus we obtain

$$q_{\alpha\beta} = \begin{pmatrix} 0 & & & & & & & & & & & \\ & 0 & & & & & & & & & & \\ & & 0 & & & & & q_0 & & & & \\ & & & 0 & & & & & & & & \\ & & & & 0 & & & & & & & \\ & & & & & 0 & & & & & & \\ & & & & & & 0 & & & & & \\ & & & & & & & 0 & & & & \\ & & & & & & & & 0 & & & \\ & & q_0 & & & & & & & 0 & & \\ & & & & & & & & & & 0 & \\ & & & & & & & & & & & 0 \end{pmatrix}. \tag{2.128}$$

Next, we consider the first step of the hierarchical replica symmetry breaking. Let $m_1 (< n)$ be a positive integer; divide the $n \times n$ symmetric matrix into some blocks which are $m_1 \times m_1$ symmetric matrices. Furthermore, put all the elements in the off-diagonal blocks to q_0 and all the off-diagonal elements in the diagonal blocks to q_1, respectively. This is the spin glass order parameter $q_{\alpha\beta}$ of the first step of the hierarchical replica symmetry breaking (one-step RSB). For example, when $n = 12$ and $m_1 = 6$, the spin glass order parameter $q_{\alpha\beta}$ is as follows:

$$
q_{\alpha\beta} = \left(\begin{array}{cccccc|cccccc}
0 & & & & & & & & & & & \\
 & 0 & & q_1 & & & & & & & & \\
 & & 0 & & & & & & & q_0 & & \\
 & & & 0 & & & & & & & & \\
 & q_1 & & & 0 & & & & & & & \\
 & & & & & 0 & & & & & & \\
\hline
 & & & & & & 0 & & & & & \\
 & & & & & & & 0 & & q_1 & & \\
 & & & & & & & & 0 & & & \\
 & & q_0 & & & & & & & 0 & & \\
 & & & & & & & q_1 & & & 0 & \\
 & & & & & & & & & & & 0
\end{array}\right). \tag{2.129}
$$

Subsequently, we consider the second step of the hierarchical replica symmetry breaking. Let $m_2(< m_1)$ be a positive integer; divide the diagonal blocks having $m_1 \times m_1$ elements into some sub-blocks which are $m_2 \times m_2$ symmetric matrices. Furthermore, put all the elements in the off-diagonal sub-blocks to q_1 and all the off-diagonal elements in the diagonal sub-blocks to q_2, respectively. This is the spin glass order parameter $q_{\alpha\beta}$ of the second step of the hierarchical rcplica symmetry breaking. For example, when $n = 12$, $m_1 = 6$, and $m_2 = 3$, the spin glass order parameter $q_{\alpha\beta}$ is as follows:

$$
q_{\alpha\beta} = \left(\begin{array}{ccc|ccc|ccc|ccc}
0 & q_2 & q_2 & & & & & & & & & \\
q_2 & 0 & q_2 & & q_1 & & & & & & & \\
q_2 & q_2 & 0 & & & & & & & q_0 & & \\
\hline
 & & & 0 & q_2 & q_2 & & & & & & \\
 & q_1 & & q_2 & 0 & q_2 & & & & & & \\
 & & & q_2 & q_2 & 0 & & & & & & \\
\hline
 & & & & & & 0 & q_2 & q_2 & & & \\
 & & & & & & q_2 & 0 & q_2 & & q_1 & \\
 & & & & & & q_2 & q_2 & 0 & & & \\
\hline
 & & q_0 & & & & & & & 0 & q_2 & q_2 \\
 & & & & & & & q_1 & & q_2 & 0 & q_2 \\
 & & & & & & & & & q_2 & q_2 & 0
\end{array}\right). \tag{2.130}
$$

By repeating the aforementioned operation, we obtain the spin glass order parameter $q_{\alpha\beta}$ with hierarchical breaking of the replica symmetry. In the Parisi solution, this spin glass order parameter $q_{\alpha\beta}$ with the replica symmetry breaking is assumed.

Next, we consider the free energy of the Parisi solution. Here, let M be the number of divisions of $q_{\alpha\beta}$, that is, the number of steps of replica symmetry breaking. The following inequality is satisfied:

$$
n \geq m_1 \geq m_2 \geq \cdots \geq m_M \geq 1, \tag{2.131}
$$

where m_μ is the dimension of the μth sub-block. By using $q_i (i = 0, 1, \cdots, M-1)$, the step function is defined by

$$q(x) = q_i, \quad (m_{i+1} < x \leq m_i), \tag{2.132}$$

where $m_0 = n$. In order to take the limit of $n \to 0$, the function $q(x)$ defined in $1 \leq x \leq n$ connects the function $q(x)$ defined in $0 \leq x \leq 1$ by reversing the inequality of Eq. (2.131) as follows:

$$0 \leq m_1 \leq m_2 \leq \cdots \leq m_M \leq 1. \tag{2.133}$$

Furthermore, by considering the limit of $M \to \infty$, $q(x)$ becomes a continuous increasing function defined in $0 \leq x \leq 1$. The schematics of this operation are shown in Fig. 2.10. From these operations, the free energy of the Parisi solution can be obtained as

$$f_{\text{Parisi}} = \max\{f[q(x)]\}, \tag{2.134}$$

$$f[q(x)] = -\frac{\beta^2 J^2}{4}\left[1 + \int_0^1 q(x)^2 dx - 2q(1)\right]$$
$$- \frac{1}{\sqrt{2\pi q(0)}} \int_{-\infty}^{\infty} dy e^{-\frac{y^2}{2q(0)}} g(0, y + h/J). \tag{2.135}$$

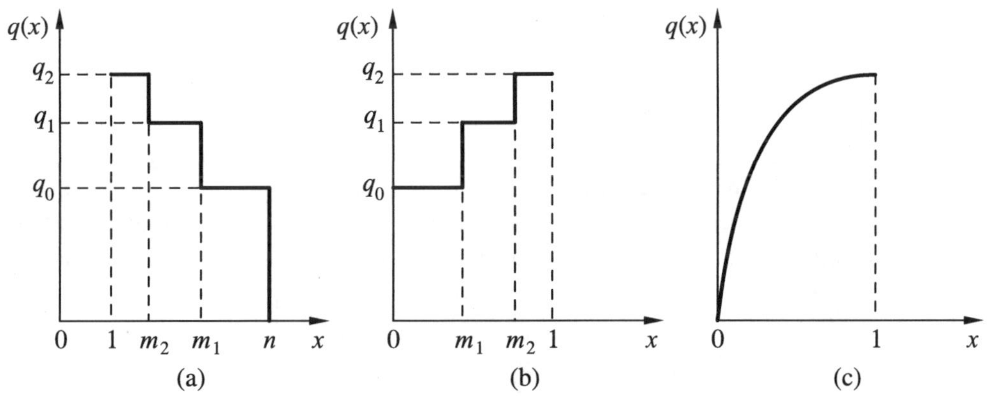

Fig. 2.10 (a) Schematic of $q(x)$ defined in $1 \leq x \leq n$, that is, Eq. (2.131) is satisfied. (b) Schematic of $q(x)$ defined in $0 \leq x \leq 1$, that is, Eq. (2.133) is satisfied. (c) Schematic of $q(x)$ in the limit of $M \to \infty$. $q(x)$ becomes a continuous increasing function defined in $0 \leq x \leq 1$.

In this case, $g(x,y)$ is the solution of the partial differential equation called the Parisi equation:

$$\frac{\partial g(x,y)}{\partial x} = -\frac{1}{2}\frac{dq(x)}{dx}\left\{\frac{\partial^2 g(x,y)}{\partial y^2} + x\left(\frac{\partial g(x,y)}{\partial y}\right)^2\right\}, \tag{2.136}$$

and the initial condition of this equation is put as

$$g(1,y) = \ln\left[2\cosh(\beta J y)\right]. \tag{2.137}$$

In the solution derived from f_{Parisi}, the negative magnetic entropy never appears, and a physically consistent solution is obtained. This solution is considered as the exact solution of the SK model. The phase diagram shown in Fig. 2.9 is obtained by the Parisi solution.

Thouless–Anderson–Palmer equation

So far, the property of the SK model was explained by the replica method. We now introduce research which does not use the replica method. The Thouless–Anderson–Palmer equation (TAP equation) (Thouless et al. 1977) is the equation of state for the spin variables in the SK model. In this method, the configurational average of randomness is postponed as will be shown. In the TAP equation, when one set of magnetic interactions $\{J_{ij}\}$ is given, the free energy is represented as:

$$\begin{aligned} F_{\text{TAP}}/N &= -\frac{1}{2}\sum_{1\le i\ne j\le N} J_{ij}m_i m_j - \sum_{i=1}^{N} h_i m_i \\ &\quad -\frac{\beta}{4}\sum_{1\le i\ne j\le N} J_{ij}^2(1-m_i^2)(1-m_j^2) \\ &\quad +T\sum_{i=1}^{N}\left(\frac{1+m_i}{2}\ln\frac{1+m_i}{2} + \frac{1-m_i}{2}\ln\frac{1-m_i}{2}\right), \end{aligned} \tag{2.138}$$

where $m_i := \langle s_i \rangle$ is the thermal average of spin at site i, and N is the number of spins. The first and second terms in the free energy are the internal energies from magnetic interactions and magnetic fields, respectively. The third term is the Onsager reaction term, and the fourth term indicates the magnetic entropy. The reason why the Onsager reaction term exists is explained as follows. The thermal average of spin m_i at site i generates the internal field at site j. The internal field originated by m_i affects m_j and as a result, the value of m_j is changed. Furthermore, the change of m_j affects m_i. Then, the change of m_i affects m_i itself through the change of m_j. The Onsager reaction term can cancel out the energy contribution of this rebound. The order of the Onsager reaction term is $\mathcal{O}(1/N)$. Since $J_{ij}^2 \sim \mathcal{O}(1/N)$ in the SK model, the Onsager reaction term plays an important role. On the other hand, since $J_{ij} \sim \mathcal{O}(J/N)$ in the ferromagnetic infinite-range model as explained in Section 2.1.2, the Onsager reaction term can be ignored against the internal energy term in the thermodynamic limit ($N \to \infty$). Recently, the perturbative interpretation of the TAP free energy was studied by (Yasuda et al. 2016).

By the variation of m_i of Eq. (2.138), the following equation of state for m_i is obtained:

$$m_i = \tanh\beta \left[\sum_{j=1}^{N} J_{ij} m_j + h_i - \beta \sum_{i=1}^{N} J_{ij}^2 (1 - m_j^2) m_i \right]. \tag{2.139}$$

The simultaneous equations for $\{m_i\}_{i=1,\cdots,N}$ are together called the TAP equation. By solving the TAP equation, we can understand the freezing direction of each spin when one set of magnetic interactions $\{J_{ij}\}$ is given. Although it is difficult to solve this equation, the number of solutions of the TAP equation can be estimated analytically. It is known that in the spin glass phase, enormous solutions exist in the TAP equation with the number of solutions being proportional to $\exp(\alpha N)$. The coefficient α is $\alpha \simeq 0.2$ at zero temperature, and its value of decreases with increase in temperature. This indicates that there are enormous types of spin freezing in the spin glass phase. However, almost all solutions of the TAP equation correspond to the local minima of Eq. (2.138); it is considered that the number of solutions which truly minimize Eq. (2.138) is proportional to the power of N. Furthermore, from the analysis of the TAP equation, the solutions having low free energy are separated from each other by an energy barrier whose height is proportional to $\mathcal{O}(N^a)$ $(0 < a < 0.5)$. This result indicates that the free energy landscape is the many-valley structure shown in Fig. 2.11 in the spin glass phase. In the many-valley structure, if the system falls in one of the valleys, escaping from the valley is impossible. In other words, the spin glass phase shows non-ergodicity. Thus, the phase transition to the spin glass phase is that from the ergodic phase at high temperatures to the non-ergodic phase at low temperatures. Furthermore, the characteristic free energy landscape in the spin glass phase indicates that it is extremely difficult to find the exact ground state in the SK model.

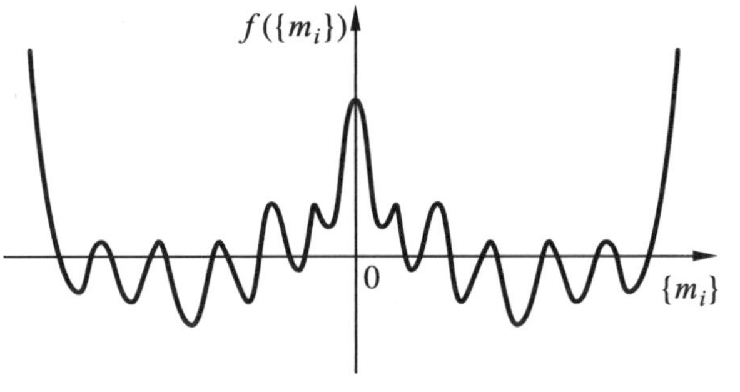

Fig. 2.11 Schematic of the many-valley structure of the free energy landscape in the spin glass phase.

2.4 Edwards–Anderson Model on Finite Dimensional Lattices

In Section 2.3, the results of the SK model were introduced. In the SK model, the spin glass phase appears at low temperatures. The SK model is the infinite-range version of the Edwards–Anderson model (EA model), where interactions between all spins exist. In other words, the SK model corresponds to the EA model in infinite dimensions. In this section, we introduce physical properties of the EA model in finite dimensions.

One-dimensional case

The ground state of the EA model on a one-dimensional lattice (chain) can be obtained easily–it can be constructed by putting spins from left to right, one by one. Furthermore, when the magnetic field is absent, by applying the local Gauge transformation, the model becomes equivalent to the Ising chain with ferromagnetic interactions explained in Section 2.1.3. Thus, in the EA model on a chain, the phase transition does not occur, and of course, the spin glass phase does not appear regardless of the characteristic values of the probability distribution of magnetic interactions such as the mean and standard deviation.

Two-dimensional case

In the two-dimensional EA model, the finite-temperature phase transition does not occur regardless of the characteristic values of the probability distribution of magnetic interactions. However, since the ground state becomes the spin glass state depending on the probability distribution of random interactions, the phase transition from paramagnetic phase to spin glass phase occurs at $T = 0$. The critical phenomena of the zero-temperature phase transition when $J_0 = 0$ were investigated by Monte Carlo simulations (Katzgraber et al. 2004).

In Monte Carlo simulations, the correlation length of spins is calculated by

$$\xi_L = \frac{1}{2\sin(|\mathbf{k}_{\text{min}}|/2)}\left[\frac{\chi(\mathbf{0})}{\chi(\mathbf{k}_{\text{min}})} - 1\right]^{\frac{1}{2}}, \tag{2.140}$$

where $\mathbf{k}_{\text{min}} = (2\pi/L, 0, 0)$ is the smallest nonzero wave vector, and $\chi(\mathbf{k})$ is defined by

$$\chi(\mathbf{k}) = \frac{1}{N}\sum_{i=1}^{N}\sum_{j=1}^{N}\overline{\langle\sigma_i^z\sigma_j^z\rangle^2 e^{i\mathbf{k}\cdot(\mathbf{r}_i - \mathbf{r}_j)}}, \tag{2.141}$$

where the bar indicates the configurational average and $\mathbf{r}_i$ is the lattice position of site i. This is the framework of the Ornstein–Zernike formula (Southern and Young, 1993). Here, L is the linear dimension of lattices. By using a scaling function $K[\cdot]$, the correlation length of spins ξ_L is scaled by

$$\xi_L \sim LK\left[L^{1/\nu}T\right], \tag{2.142}$$

where we used $T_c = 0$. Thus, the following power law is obtained:

$$\xi_L \sim T^{-\nu}, \tag{2.143}$$

and the critical exponent ν can be obtained by the temperature dependence of ξ_L. Katzgraber et al. studied the zero-temperature phase transition of the EA model on a square lattice (Katzgraber et al., 2004). They used the model defined by Eqs. (2.60) and (2.61), i.e., the EA model with random interactions whose probability distribution is the Gaussian distribution. Figure 2.12 shows the temperature dependence of ξ_L which was calculated by Monte Carlo simulations in the EA model on a square lattice. This result means that $\nu = 3.45$ of the zero-temperature phase transition. On the other hand, the critical exponent ν of the zero-temperature phase transition has also been obtained by other simulation methods, e.g., a cluster Monte Carlo simulation ($\nu = 2.0$, Liang, 1992) and a numerical transfer matrix calculation ($\nu = 2.08$, Kawashima et al. 1992). (Rieger et al. 1996) reviews ν in the EA model on two-dimensional lattices in detail.

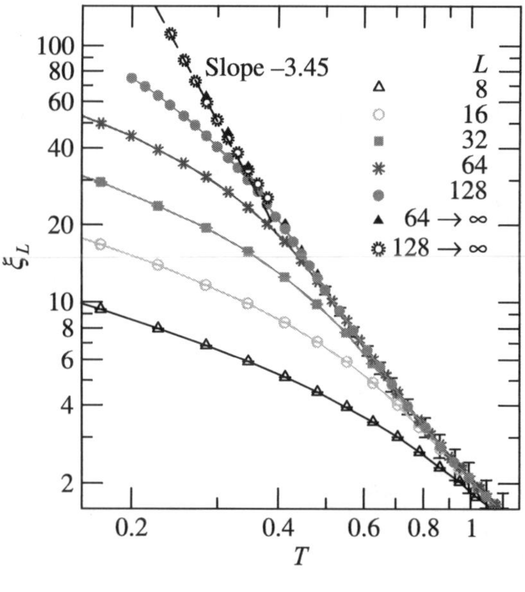

Fig. 2.12 Temperature dependence of the correlation length ξ_L obtained by Monte Carlo simulations for the Edwards–Anderson model on a square lattice. The probability distribution of magnetic interactions is the Gaussian distribution with zero mean and standard deviation unity. The slope of this figure is the critical exponent ν (from Katzgraber et al. 2004).

The boundary J_{0c} between the spin glass phase and the ferromagnetic phase at zero temperature has been investigated – it corresponds to the multicritical point. By using the duality technique, Nishimori and coworkers found $J_{0c} = 1.021770$ (Nishimori and Nemoto, 2002; Maillard et al. 2003). Furthermore, Ohzeki determined the magnitude of

J_{0c} more carefully by a systematic technique using duality as 1.021564 (Ohzeki, 2009a,b). Notice that the research by Ohzeki et al. (2008) is a related work of the multicritical point.

Three-dimensional case

In the EA model on three-dimensional lattices, the finite-temperature phase transition from the paramagnetic phase to the spin glass phase occurs depending on the probability distribution of random interactions. The critical phenomena of a finite-temperature spin glass transition were investigated by Monte Carlo simulations (Kawashima and Young, 1996; Katzgraber et al. 2006).

Katzgraber et al. studied the EA model on a cubic lattice (Katzgraber et al. 2006). In the study, they used the model defined by Eqs. (2.60) and (2.61), i.e., the probability distribution of magnetic interactions is a Gaussian distribution. Figure 2.13 (a) shows the temperature dependence of the Binder ratio g of the spin glass order parameter obtained by Monte Carlo simulations in the EA model on a cubic lattice. The Binder ratio is defined by

$$g = \frac{1}{2}\left[3 - \frac{\overline{\langle q_{\alpha\beta}^4\rangle}}{\overline{\langle q_{\alpha\beta}^2\rangle^2}}\right], \tag{2.144}$$

$q_{\alpha\beta}$ is the order parameter of thc spin glass phase and is defined by

$$q_{\alpha\beta} = \frac{1}{N}\sum_{i=1}^{N} \sigma_{i\alpha}^z \sigma_{i\beta}^z, \tag{2.145}$$

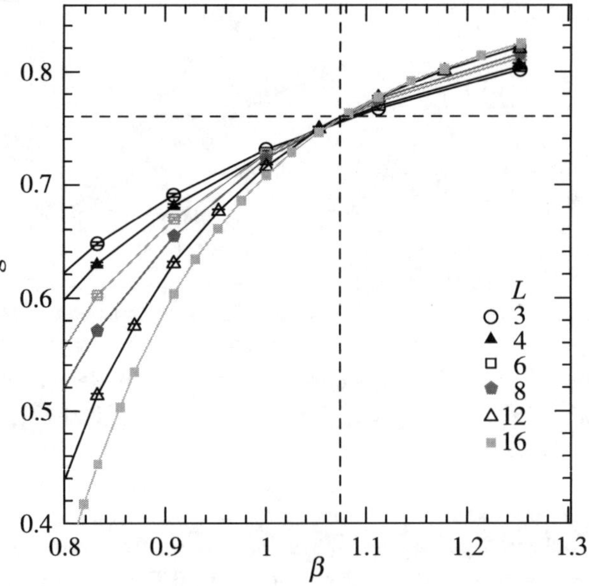

(a)

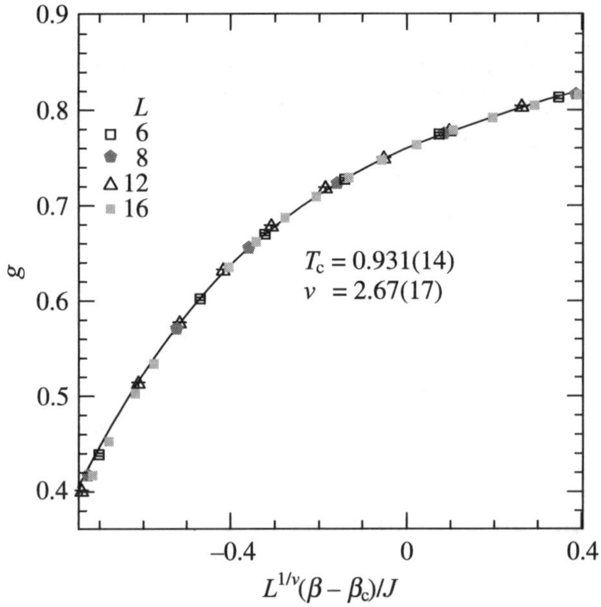

(b)

Fig. 2.13 Results obtained by Monte Carlo simulations of the Edwards–Anderson model on a cubic lattice. The probability distribution of magnetic interactions is the Gaussian distribution with zero mean and standard deviation unity. (a) Inverse temperature, $\beta = T^{-1}$, dependence of the Binder ratio of the spin glass order parameter. (b) Finite-size scaling result of the Binder ratio of the spin glass order parameter when $T_c = 0.931$ and $\nu = 2.67$ are used (from Katzgraber et al. 2006).

where α and β are the replica indices. At $T_c = \beta_c^{-1} = 0.9131$, where k_B is set to unity, the Binder ratios for all lattice sizes cross. Furthermore, the finite-size scaling result is shown in Fig. 2.13 (b) (see Eq. (2.59)), and this result means that $\nu = 2.67$ of the finite-temperature spin glass transition.

2.A Several Spotlights on Classical Models

In this chapter, we explained some properties of the Ising model. As mentioned earlier, the Ising model was introduced to analyze microscopically, phase transitions in magnetic systems; it has also been used to consider cooperative phenomena caused by interactions between elements. However, the Ising model is not a unique model to study phase transition and cooperative phenomena. Besides the Ising model, there are some important classical models which exhibit various interesting properties. In this appendix, we introduce some models which are generalizations of the Ising model and explain the properties of the respective models.

2.A.1 Potts model

Spins in the Ising model have two values ± 1 as explained earlier. Thus, it is convenient to describe a microscopic behavior of systems with bistability. The Potts model has been used to study physical properties of systems with multistability such as the absorption phenomena on a surface (Domany et al. 1977; Domany and Schick, 1979). The Hamiltonian of the q-state Potts model is defined by

$$\mathcal{H} = -\sum_{\langle i,j \rangle} J_{ij}\delta_{s_i s_j}, \quad (s_i = 1, 2, \cdots, q). \tag{2.146}$$

Thus, all spins take q states, and the interaction between spins is represented not by the product between spins but Kronecker's delta. Since the two-state Potts model ($q = 2$) is equivalent to the Ising model, the q-state Potts model is a straightforward generalization of the Ising model.

Homogeneous ferromagnetic interaction case

When all magnetic interactions are homogeneous ferromagnetic interactions, that is, $J_{ij} = J > 0$ for all interactions, all spins have the same state in the ground state. If this is the case, the ground state has a q-fold symmetry, and the phase transition with spontaneous breaking of the q-fold symmetry occurs at a finite temperature in two or higher dimensions. The phase transition nature of the q-state ferromagnetic Potts model depends on the value of q and the spatial dimension. In the model on two-dimensional lattices, the second-order phase transition occurs at a finite temperature when $q \leq 4$, while the first-order phase transition occurs when $q > 4$. On the other hand, in the model on three-dimensional lattices, the second-order phase transition occurs when $q = 2$, while the first-order phase transition occurs when $q \geq 3$ at a finite temperature. The boundary between first-order and second-order phase transitions depends on the value of q and the spatial dimension d (see Fig. 2.14). Table 2.4 summarizes the critical exponents of the second-order phase transition in the Potts model.

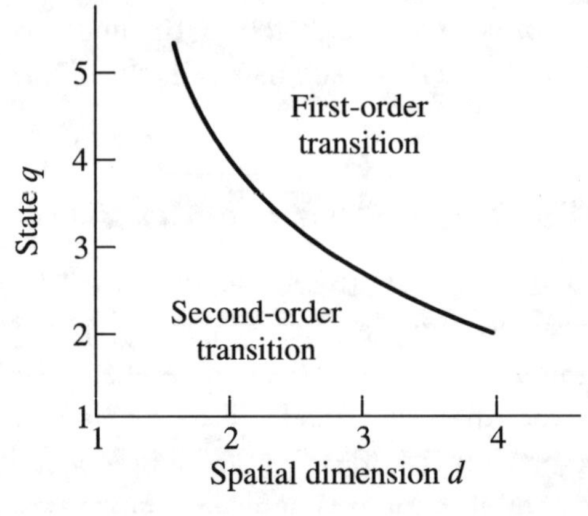

Fig. 2.14 Schematic of the boundary between first-order and second-order phase transitions depending on the value of q and the spatial dimension d in the Potts model (Wu, 1982).

Table 2.4 Critical exponents of the Potts model with homogeneous ferromagnetic interactions depending on the value of q and the spatial dimension (Wu, 1982). Note that critical exponents in the two-state Potts model are equivalent to that in the Ising model (see Table 2.2 and Table 2.3).

Critical exponents	α	β	γ	δ	η	ν
2D 2-state Potts model	0	1/8	7/4	15	1	1/4
2D 3-state Potts model	1/3	1/9	13/9	14	5/6	4/15
2D 4-state Potts model	2/3	1/12	7/6	15	2/3	1/2
3D 2-state Potts model	0.119	0.321	1.239	10.57	0.024	0.627

Invisible state

The invisible state was introduced as redundant states in the Potts model by Tamura, Tanaka, and Kawashima (Tamura et al. 2010; Tanaka et al., 2011; Tamura et al. 2012; Tamura and Tanaka, 2014). The invisible state does not contribute to the internal energy, but changes the structure of excited states without changing the ground state of the Potts model. Thus, the discrete symmetry which breaks at the phase transition point does not change when invisible states are added in the Potts model.

The Hamiltonian of the Potts model with invisible states is as follows:

$$\mathscr{H} = J \sum_{\langle i,j \rangle} \delta_{s_i,s_j} \sum_{\alpha=1}^{q} \delta_{s_i,\alpha}, \tag{2.147}$$

$$s_i = 1, \cdots, q, q+1, \cdots, q+r, \tag{2.148}$$

where r is the number of invisible states. For large r, the authors found that the first-order phase transition occurs at a finite temperature in the two-dimensional $q = 2,3,4$ Potts model by Monte Carlo simulations (Tamura et al. 2010). The ferromagnetic q-state Potts model for $q = 2,3,4$ exhibits the second-order phase transition when the invisible state is absent, and thus by introducing the invisible states, the order of the phase transition in the Potts model can be controlled, which is consistent with mean-field analysis. This model was firstly introduced to explain discrepancies between theoretical prediction and experimental discussion of the phase transition in frustrated magnetic materials (Tamura and Kawashima, 2008, 2011). After that, from the view point of fundamental studies of phase transition, the nature of the phase transition in the Potts model with invisible states has been exhaustively investigated by analytical techniques and numerical simulations (Tanaka and Tamura, 2011; van Enter et al. 2011, 2012; Johnston and Ranasinghe, 2013; Ananikian et al. 2013; Krasnytska et al. 2016). On top of that, the Potts model with invisible states has been used as a starting point in the analysis of the first-order phase transitiontransition (Komura and Okabe, 2010; Mori, 2012; Tamura and Tanaka, 2013; Tamura et al. 2014).

Random interaction case

When the probability distribution of the interactions $\{J_{ij}\}$ in the Potts model defined by Eq. (2.146) is given by Eq. (2.61) (the Gaussian distribution) or Eq. (2.62) ($\pm J$), the spin glass phase could appear depending on q and the spatial dimension. The finite-temperature spin glass phase transition does not occur in one or two dimensions as well as the Ising model. As explained in Section 2.4, the Ising model [two-state Potts model ($q = 2$)] with random interactions in three dimensions could exhibit a finite-temperature spin glass phase transition characterized by the EA order parameter. However, Brangian et al. (2002a, 2003) reported that the spin glass phase transition detected by the EA order parameter does not occur at finite temperatures in the $q = 10$ state Potts model with random interactions in three dimensions. Phase transition of the infinite-range q-state Potts model with random interactions depends on q. Since the two-state Potts model ($q = 2$) is equivalent to the Ising model, the spin glass phase transition occurs at a finite temperature as discussed in Section 2.3. On the contrary, two successive phase transitions occur at finite temperatures in the infinite-range q-state Potts model with random interactions for $q > 4$ (Kirkpatrick and Wolynes, 1987; Kirkpatrick and Thirumalai, 1988; Brangian et al. 2001, 2002b; Brangiana et al. 2002). One is the dynamical phase transition and the other is the phase transition characterized by the EA order parameter (Castellani and Cavagna, 2005; Binder and Kob, 2011).

Recently, a finite-dimensional random Potts model which exhibits a spin glass transition at a finite temperature was reported (Takahashi and Hukushima, 2015). In the model, a discontinuous jump of the spin glass order parameter appears but the latent heat is not discontinuous at the spin glass transition temperature. In addition, at the spin glass transition temperature, the distribution of the order parameter is the bimodal structure. These facts indicate that the model is a candidate for showing the random first-order phase transition in finite dimensions.

2.A.2 XY model

Spin variables in the Ising model and the Potts model have discrete values. In the Ising model, spins take $\sigma_i^z = \pm 1$ whereas in the q-state Potts model, spins take $s_i = 1, \cdots, q$. Here we consider a generalization of the Ising model called the XY model. The Hamiltonian of the XY model is defined by

$$\mathcal{H} = -\sum_{\langle i,j \rangle} J_{ij} \mathbf{s}_i \cdot \mathbf{s}_j, \tag{2.149}$$

where $\mathbf{s}_i$ denotes a two-dimensional real-valued vector given by $\mathbf{s}_i = (s_i^x, s_i^y)$ and $|\mathbf{s}_i| = 1$. Since the spin is defined as a two-component classical vector, it is called the classical XY spin. Notice that the discretized version of the XY model is referred to as the q-state clock model, where q is an integer. The two-state clock model is equivalent to the Ising model. Thus, the XY model can be regarded as a generalization of the Ising model.

Homogeneous ferromagnetic interaction case

Suppose that all magnetic interactions are homogeneous ferromagnetic interactions, that is, $J_{ij} = J > 0$ for all interactions. The directions of all spins become the same in the ground state. Thus, the ground state has a U(1) symmetry.

The XY model in one dimension does not exhibit the finite-temperature phase transition as well as the Ising model or the Potts model. When the spatial dimension is one or two, the long-range order of continuous spins at finite temperatures is forbidden by the Mermin–Wagner theorem (Mermin and Wagner, 1966). Thus, the XY model in two dimensions exhibits a topological phase transition at a finite temperature, which is not caused by the long-range order of bare spins. This topological phase transition is called the Kosterlitz–Thouless (KT) transition (Kosterlitz and Thouless, 1973). The KT transition occurs by the dissociation of the Z vortex which corresponds to the point defect of the XY model with homogeneous ferromagnetic interactions. Furthermore, in the XY model in three or higher dimensions, the phase transition with spontaneous breaking of the U(1) symmetry occurs at a finite temperature. Table 2.5 summarizes the critical exponents of the XY model on three-dimensional lattices.

Table 2.5 Critical exponents of the XY model with homogeneous ferromagnetic interactions on three-dimensional lattices obtained by Monte Carlo simulations (Campostrini et al. 2001).

Critical exponents	α	β	γ	δ	η	ν
3D XY model	-0.0146	0.3485	1.3177	4.780	0.0380	0.67155

Random interaction case

When the probability distribution of the interactions $\{J_{ij}\}$ in the XY model defined by Eq. (2.149) is given by Eq. (2.61) (the Gaussian distribution) or Eq. (2.62) ($\pm J$), the spin glass phase could appear depending on the spatial dimension. The infinite-range XY model with random interactions exhibits the finite-temperature spin glass transition. Thus, in higher spatial dimensions, the spin glass phase appears in the XY model as well as the Ising model. On the other hand, finite-temperature properties of the XY model with random interactions on three and four-dimensional lattices are not clear yet. In these models, the following two conflicting results have been reported.

1. Chiral-glass order appears at a temperature higher than the spin glass order (Kawamura and Tanemura, 1991; Maucourt and Grempel, 1998; Kawamura, 1992b, 1995; Kawamura and Li, 2001; Kawamura, 2011; Obuchi and Kawamura, 2013). The chiral-glass order appears with the long-range order of chirality. The chirality is defined as

$$\chi_i = \mathbf{s}_i \cdot (\mathbf{s}_{i+\delta} \times \mathbf{s}_{i+\delta'}), \tag{2.150}$$

where δ and δ' are displacements due to two distinct unit lattice vectors. Thus, the phase transition from the paramagnetic phase to the chiral glass phase is a novel phase transition.

2. Chiral-glass order and spin glass order appear at the same temperature (Granato, 2001; Lee and Young, 2003; Granato, 2004a,b; Yamamoto et al. 2004; Pixley and Young, 2008). In this case, the phase transition at a finite temperature occurs caused by the long-range order of bare spins, that is, this phase transition is the same type as the spin glass transition observed in the Ising model.

2.A.3 Heisenberg model

Two-component real-valued vectors express spin variables in the XY model as described in the previous subsection. Here we consider a model called the Heisenberg model whose spin variables are given by three-component real-valued vectors. The Hamiltonian of the Heisenberg model is defined by

$$\mathscr{H} = -\sum_{\langle i,j\rangle} J_{ij} \mathbf{s}_i \cdot \mathbf{s}_j, \tag{2.151}$$

where $\mathbf{s}_i = (s_i^x, s_i^y, s_i^z)$ and $|\mathbf{s}_i| = 1$; the spin is called the classical Heisenberg spin.

Homogeneous ferromagnetic interaction case

When all the magnetic interactions are homogeneous ferromagnetic interactions, that is, $J_{ij} = J > 0$ for all interactions, the directions of all spins become the same in the ground state. Thus, the ground state has an S_2 symmetry.

In the Heisenberg model in one or two dimensions, the finite-temperature phase transition does not occur. As stated earlier, when the spatial dimension is one or two, the long-range order of continuous spins at finite temperatures is forbidden by the Mermin–Wagner theorem (Mermin and Wagner, 1966) as well as the XY model. Thus, the finite-temperature phase transition caused by the spontaneous breaking of continuous symmetry does not occur. On the other hand, when the spatial dimension is more than two, the phase transition with spontaneous breaking of the S_2 symmetry occurs at a finite temperature. Table 2.6 summarizes the critical exponents of the Heisenberg model on three-dimensional lattices.

Table 2.6 Critical exponents of the Heisenberg model with homogeneous ferromagnetic interactions on three-dimensional lattices obtained by Monte Carlo simulations (Holm and Janke, 1993).

Critical exponents	α	β	γ	δ	η	ν
3D Heisenberg model	-0.112	0.362	1.389	4.834	0.027	0.704

Random interaction case

When the probability distribution of interactions $\{J_{ij}\}$ in the Heisenberg model defined by Eq. (2.151) is given by Eq. (2.61) (the Gaussian distribution) or Eq. (2.62) ($\pm J$), the spin glass phase could appear depending on the spatial dimension. The infinite-range Heisenberg model with random interactions exhibits the finite-temperature spin glass transition (Gabay and Toulouse, 1981; Cragg et al. 1982). Thus, in higher spatial dimensions, the spin glass phase appears in the Heisenberg model as well as the Ising model. Furthermore, when the spatial dimension is more than four, the existence of finite-temperature spin glass transition is confirmed (Imagawa and Kawamura, 2003). On the other hand, finite-temperature properties of the Heisenberg model with random interactions on three- and four-dimensional lattices are not clear yet. Besides the XY model, two conflicting results shown in the case of the XY spin systems have been reported in the Heisenberg model with random interactions (Kawamura, 1992a, 1998; Matsubara et al., 2000; Hukushima and Kawamura, 2000; Endoh et al. 2001; Matsubara et al. 2001; Kawamura, 2001; Kawamura and Imagawa, 2001; Imagawa and Kawamura, 2002, 2003; Lee and Young, 2003; Imagawa and Kawamura, 2004a,b; Hukushima and Kawamura, 2005; Viet and Kawamura, 2009a,b). In very recent research, a dynamic correlation-length scaling analysis suggests that the chiral-glass order and the spin glass order appear at the same temperature with a common critical exponent ν (Nakamura, 2016).

References

Aharony, A. 1975. 'Tetracritical points in mixed magnetic crystals'. *Phys. Rev. Lett.* 34: 590.

Ananikian N., N. Sh. Izmailyan, D. A. Johnston, R. Kenna, and R. P. K. C. M. Ranasinghe,. 2013. 'Potts models with invisible states on general Bethe lattices'. *J. Phys. A: Math. Theory.* 46: 385002.

Arai, R., K. Komatsu, and T. Sato. 2007. 'Aging behavior of spin glasses under bond and temperature perturbations from laser illumination' . *Phys. Rev. B* 75: 144424.

Barber, M. N., R. B. Pearson, D. Toussaint, and J. L. Richardson. 1985. 'Finite-size scaling in the three-dimensional Ising model'. *Phys. Rev. B* 32: 1720.

Baxter, R. J. 1982. *Exactly Solved Models in Statistical Mechanics*. Academic Press.

Beath, A. D., and D. H. Ryan. 2006. 'Phase diagrams of site-frustrated Heisenberg models on simple cubic, bcc, and fcc lattices'. *Phys. Rev. B* 73: 14445.

Bekhechi, S., and B. W. Southern. 2004. 'Chiral mixed phase in disordered 3d Heisenberg models'. *Phys. Rev. B* 70: 020405(R).

Bhanot, G., D. Duke, and R. Salvador. 1986. 'Finite-size scaling and the three-dimensional Ising model'. *Phys. Rev. B* 33: 7841(R).

Binder, K. 1981. 'Critical properties from Monte Carlo coarse graining and renormalization'. *Phys. Rev. Lett.* 47: 693.

Binder, K., and W. Kob. 2011. *Glassy Materials and Disordered Solids*. World Scientific Press.

Bolthausen, E., and A. Bovier. 2007. *Spin Glasses*. Springer.

Bramwell, S. T., and M. J. P. Gingras. 2001. 'Spin ice state in frustrated magnetic pyrochlore materials'. *Science* 294: 1495.

Brangian, C., W. Kob, and K. Binder. 2001. 'Finite-size scaling at the dynamical transition of the mean-field 10-state Potts glass'. *Europhys. Lett.* 53: 756.

Brangian, C., W. Kob, and K. Binder. 2002a. 'Evidence against a glass transition in the 10-state short-range Potts glass'. *Europhys. Lett.* 59: 546.

Brangian, C., W. Kob, and K. Binder. 2002b. 'Statics and dynamics of the ten-state mean-field Potts glass model: A Monte Carlo study'. *J. Phys. A: Math. Gen.* 35: 191.

Brangian, C., W. Kob, and K. Binder. 2003. 'Statics and dynamics of the ten-state nearest-neighbour Potts glass on the simple-cubic lattice'. *J. Phys. A: Math. Gen.* 36: 10847.

Brangiana, C., K. Kob, and K. Binder. 2002. 'The high temperature dynamics of a mean- field Potts glass'. *Phil. Mag. B* 82: 663.

Brezin, E., J. C. Le Guillou, and J. Zinn-Justin, 1974. 'Universal ratios of critical amplitudes near four dimensions'. *Phys. Lett. A* 47: 285.

Campostrini, M., M. Hasenbusch, A. Pelissetto, P. Rossi, and E. Vicari. 2001. 'Critical behavior of the three-dimensional XY universality class'. *Phys. Rev. B* 63: 214503.

Cannella, V., and J. A. Mydosh. 1972. 'Magnetic ordering in gold-iron alloys'. *Phys. Rev. B* 6: 4220.

Castellani, T., and A. Cavagna. 2005. 'Spin-glass theory for pedestrians'. *J. Stat. Mech.* P05012.

Castelnovo, C., R. Moessner, and S. L. Sondhi. 2007. 'Magnetic monopoles in spin ice'. *Nature* 451: 42.

Chamberlin, R. V. 1985. 'Time decay of the remanent magnetization in spin glasses'. *J. Appl. Phys.* 57: 3377.

Chamberlin, R. V., G. Mozurkewich, and R. Orbach. 1984. 'Time decay of the remanent magnetization in spin-glasses'. *Phys. Rev. Lett.* 52: 867.

Chowdhury, D. 1986. *Spin Glasses and Other Frustrated Systems*. World Scientific Press.

Cragg, D. M., D. Sherrington, and M. Gabay. 1982. 'Instabilities of an m-vector spin-glass in a field'. *Phys. Rev. Lett.* 49: 158.

de Almeida, J. R. L., and D. J. Thouless. 1978. 'Stability of the Sherrington–Kirkpatrick solution of a spin glass model'. *J. Phys. A: Math. Gen.* 11: 983.

Diep, H. T., ed. 2013. *Frustrated Spin Systems*. World Scientific Press.

Domany, E., and M. Schick. 1979. 'Classification of continuous order–disorder transitions in adsorbed monolayers. II'. *Phys. Rev. B* 20: 3828.

Domany, E., M. Schick, and J. S. Walker. 1977. 'Classification of order–disorder transitions in common adsorbed systems: Realization of the four-state Potts model'. *Phys. Rev. Lett.* 38: 1148.

Dominicis, C. D., and I. Giardina. 2006. *Random Fields and Spin Glasses: A field Theory Approach*. Cambridge University Press.

Dotsenko, V. 1995. *An Introduction to the Theory of Spin Glasses and Neural Networks*. World Scientific Press.

Dupuis, V., E. Vincent, J. P. Bouchaud, J. Hammann, A. Ito, and H. A. Katori. 2001. 'Aging, rejuvenation, and memory effects in Ising and Heisenberg spin glasses'. *Phys. Rev. B* 64: 174204.

Edwards, S. F., and P. W. Anderson. 1975. 'Theory of spin glasses'. *J. Phys. F: Metal Phys.* 5: 965.

Endoh, S., F. Matsubara, and T. Shirakura. 2001. 'Stiffness of the Heisenberg spin-glass model at zero- and finite-temperatures in three dimensions'. *J. Phys. Soc. Jpn.* 70: 1543.

Ferré, J., J. Rajchenbach, and H. Maletta. 1981. 'Faraday rotation measurements of time dependent magnetic phenomena in insulating spin glasses'. *J. Appl. Phys.* 52: 1697.

Ferrenberg, A. M., and D. P. Landau. 1991. 'Critical behavior of the three-dimensional Ising model: A high-resolution Monte Carlo study'. *Phys. Rev. B* 44: 5081.

Fischer, K. H., and J. A. Hertz. 1993. *Spin Glasses*. Cambridge University Press.

Fisher, D. S., and D. A. Huse. 1988. 'Nonequilibrium dynamics of spin glasses'. *Phys. Rev. B* 38: 373.

Fishman, S., and A. Aharony. 1979. 'Phase diagrams and multicritical points in randomly mixed magnets. II. Ferromagnet-antiferromagnet alloys'. *Phys. Rev. B* 19: 3776.

Gabay, M., and G. Toulouse. 1981. 'Coexistence of spin-glass and ferromagnetic orderings'. *Phys. Rev. Lett.* 47: 201.

Gingras, M. J. P., and P. A. McClarty, 2014. 'Quantum spin ice: A search for gapless quantum spin liquids in pyrochlore magnets'. *Rep. Prog. Phys.* 77: 056501.

Granato, E. 2001. 'Phase-coherence transition in granular super conductors with π junctions'. *J. Magn. Magn. Mater.* 364.

Granato, E. 2004a. 'Critical behavior and driven Monte Carlo dynamics of the XY spin glass in the phase representation'. *Phys. Rev. B* 69: 144203.

Granato, E. 2004b. 'Resistive transition in π-junction superconductors'. *Phys. Rev. B* 69: 012503.

Grassberger, P., and I. Procaccia. 1982. 'The long time properties of diffusion in a medium with static traps'. *J. Chem. Phys.* 77: 6281.

Guy, C. N. 1978. 'Spin glasses in low DC fields. II. Magnetic viscosity'. *J. Phys. F: Metal Phys.* 8: 1309.

Holm, C., and W. Janke. 1993. 'Critical exponents of the classical three-dimensional Heisenberg model: A single-cluster Monte Carlo study'. *Phys. Rev. B* 48: 936.

Hoogland, A., A. Compagner, and H. W. J. Blöte,. 1985. 'Smooth finite-size behaviour of the three-dimensional Ising model'. *Physica A* 132: 593.

Houtappel, R. M. F. 1950. 'Order–disorder in hexagonal lattices'. *Physica*, 16: 425.

Hukushima, K., and H. Kawamura. 2000. 'Chiral-glass transition and replica symmetry breaking of a three-dimensional Heisenberg spin glass'. *Phys. Rev. E* 61: R1008(R).

Hukushima, K., and Kawamura, H. 2005. Monte Carlo simulations of the phase transition of the three-dimensional isotropic Heisenberg spin glass. *Phys. Rev. B*, 72, 144416.

Hukushima, K., Y. Nonomura, Y. Ozeki, and H. Takayama. 1997. 'Phase diagram of three-dimensional site-random Ising model'. *J. Phys. Soc. Jpn.* 66: 215.

Husimi, K., and I. Syôzi. 1950. 'The statistics of honeycomb and triangular lattice. I'. *Prog. Theory. Phys.* 5: 177.

Imagawa, D., and H. Kawamura. 2002. 'Monte Carlo studies of the ordering of the three-dimensional isotropic Heisenberg spin glass in magnetic fields'. *J. Phys. Soc. Jpn.* 71: 127.

Imagawa, D., and H. Kawamura. 2003. 'Ordering of the Heisenberg spin glass in high dimensions'. *Phys. Rev. B* 67: 224412.

Imagawa, D., and H. Kawamura. 2004a. 'Monte Carlo study of the ordering of the weakly anisotropic Heisenberg spin glass in magnetic fields'. *Phys. Rev. B* 70: 144412.

Imagawa, D., and H. Kawamura. 2004b. 'Replica symmetry breaking transition of the weakly anisotropic Heisenberg spin glass in magnetic fields'. *Phys. Rev. Lett.* 92: 077204.

Ising, E. 1925. 'Beitrag zur theorie des ferromagnetismus'. *Z. Physik* 31: 253.

Jesi, M. B. 2016. *Spin Glasses: Criticality and Energy Landscapes*. Springer.

Johnston, D. A., and R. P. K. C. M. Ranasinghe. 2013. 'Potts models with (17) invisible states on thin graphs'. *J. Phys. A: Math. Theory.* 46: 225001.

Jönsson, P. E., and H. Takayama. 2005. '"Glassy dynamics" in Ising spin glasses: Experiment and simulation'. *J. Phys. Soc. Jpn.* 74: 1131.

Jonsson, T., K. Jonason, and P. Nordblad. 1999. 'Relaxation of the field-cooled magnetization of an Ising spin glass'. *Phys. Rev. B* 59: 9402.

Kac, M., and J. C. Ward. 1952. 'A combinatorial solution of the two-dimensional Ising model'. *Phys. Rev.* 88: 1332.

Kanô, K., and S. Naya. 1953. 'Antiferromagnetism. The kagomé Ising net'. *Prog. Theory. Phys.* 10: 158.

Katsumata, K., T. Nire, and M. Tanimoto. 1982. 'Thermoremanent magnetization in $Rb_2Mn_{(1-x)}Cr_xC_14$ spin-glass'. *Solid State Commun.* 43: 711.

Katsura, H., N. Nagaosa, and A. V. Balatsky. 2005. 'Spin current and magnetoelectric effect in noncollinear magnets'. *Phys. Rev. Lett.* 95: 057205.

Katzgraber, H. G., L. W. Lee, and A. P. Young. 2004. 'Correlation length of the two-dimensional Ising spin glass with Gaussian interactions'. *Phys. Rev. B* 70: 014417.

Katzgraber, H. G., M. Körner, and A. P. Young. 2006. 'Universality in three-dimensional Ising spin glasses: A Monte Carlo study'. *Phys. Rev. B*, 73, 224432.

Kaufman, B. 1949. 'Crystal statistics. II. Partition function evaluated by spinor analysis'. *Phys. Rev.* 76: 1232.

Kawamura, H. 1992a. 'Chiral ordering in Heisenberg spin glasses in two and three dimensions'. *Phys. Rev. Lett.* 68: 3785.

Kawamura, H. 1992b. 'Monte Carlo evidence of finite-temperature chiral ordering in a three-dimensional XY spin glass'. *J. Phys. Soc. Jpn.* 61: 3062.

Kawamura, H. 1995. 'Numerical studies of chiral ordering in three-dimensional XY spin glasses'. *Phys. Rev. B* 51: 12398.

Kawamura, H. 1998. 'Dynamical simulation of spin-glass and chiral-glass orderings in three-dimensional Heisenberg spin glasses'. *Phys. Rev. Lett.* 80: 5421.

Kawamura, H. 2001. 'Spin and chirality orderings of frustrated magnets – Stacked-triangular antiferromagnets and spin glasses'. *Can. J. Phys.* 79: 1447.

Kawamura, H. 2011. 'The ordering of XY spin glasses'. *J. Phys. Condens. Matt.* 23: 164210.

Kawamura, H., and D. Imagawa. 2001. 'Ordering of the three-dimensional Heisenberg spin glass in magnetic fields'. *Phys. Rev. Lett.* 87: 207203.

Kawamura, H., and M. S. Li. 2001. 'Nature of the ordering in the three-dimensional XY spin glass'. *Phys. Rev. Lett.* 87: 187204.

Kawamura, H., and S. Miyashita. 1984a. 'Phase transition of the two-dimensional Heisenberg antiferromagnet on the triangular lattice'. *J. Phys. Soc. Jpn.* 53: 9.

Kawamura, H., and S. Miyashita. 1984b. 'Phase transition of the two-dimensional Heisenberg antiferromagnet on the triangular lattice'. *J. Phys. Soc. Jpn.* 53: 4138.

Kawamura, H., and M. Tanemura. 1991. 'Chiral ordering of XY spin glasses in two and three dimensions-domain-wall renormalization-group studies. *J. Phys. Soc. Jpn.* 60: 608.

Kawasaki, T., and S. Miyashita. 1993. 'Anomalous size dependence of relaxation time in the fuzzy spin model: Mechanism of slow relaxation'. *Prog. Theory. Phys.* 89: 985.

Kawashima, N., and A. P. Young. 1996. 'Phase transition in the three-dimensional $\pm$ J Ising spin glass'. *Phys. Rev. B* 53: R484(R).

Kawashima, N., N. Hatano, and M. Suzuki. 1992. 'Critical behaviour of the two-dimensional EA model with a Gaussian bond distribution'. *J. Phys. A: Math. Gen.* 25: 4985.

Kinzel, W. 1979. 'Remanent magnetization in spin-glasses'. *Phys. Rev. B* 19: 4595.

Kirkpatrick, S., and D. Sherrington. 1978. 'Infinite-ranged models of spin-glasses'. *Phys. Rev. B* 17: 4384.

Kirkpatrick, T. R., and D. Thirumalai. 1988. 'Mean-field soft-spin Potts glass model: Statics and dynamics'. *Phys. Rev. B* 37: 5342.

Kirkpatrick, T. R., and P. G. Wolynes. 1987. 'Stable and metastable states in mean-field Potts and structural glasses'. *Phys. Rev. B* 36: 8552.

Komura, Y., and Y. Okabe. 2010. 'Phase transition of a two-dimensional generalized XY model'. *J. Phys. A: Math. Theory.* 44: 015002.

Kosterlitz, J. M., and D. J. Thouless. 1973. 'Ordering, metastability and phase transitions in two-dimensional systems'. *J. Phys. C: Solid State Phys.* 6: 1181.

Krasnytska, M., P. Sarkanych, B. Berche, Y. Holovatch, and R. Kenna. 2016. 'Marginal dimensions of the Potts model with invisible states'. *J. Phys. A: Math. Theor.* 49: 255001.

Ladieu, F., F. Bert, V. Dupuis, E. Vincent, and J. Hammann. 2004. 'The relative influences of disorder and of frustration on the glassy dynamics in magnetic systems'. *J. Phys. Condens. Matt.* 16: S735.

Landau, D. P. 1976. 'Finite-size behavior of the simple-cubic Ising lattice'. *Phys. Rev. B* 14: 255.

Le Guillou, J. C., and J. ZinnJustin. 1980. 'Critical exponents from field theory'. *Phys. Rev. B* 21: 3976.

Le Guillou, J. C., and J. ZinnJustin. 1987. 'Accurate critical exponents for Ising like systems in non-integer dimensions'. *J. Phys. France* 48: 19.

Lee, L. W., and A. P. Young. 2003. 'Single spin and chiral glass transition in vector spin glasses in three dimensions'. *Phys. Rev. Lett.* 90: 227203.

Leoch, F., J. Hammann, M. Ocio, and E. Vincent. 1992. 'Can aging phenomena discriminate between the droplet model and a hierarchical description in spin glasses?' *Europhys. Lett.* 18: 647.

Liang, S. 1992. 'Application of cluster algorithms to spin glasses'. *Phys. Rev. Lett.* 69: 2145.

Liebmann, R. 1986. *Statistical Mechanics of Periodic Frustrated Ising Systems.* Springer.

Maillard, J.-M., K. Nemoto, and H. Nishimori. 2003. 'Symmetry, complexity and multicritical point of the two-dimensional spin glass'. *J. Phys. A: Math. Gen.* 36: 9799.

Matsubara, F. 1974. 'Theory of the random magnetic mixture. I Ising system'. *Prog. Theory. Phys.* 52: 1124.

Matsubara, F. 1975. 'Theory of random magnetic mixture. II Classical Heisenberg model at T=0'. *Prog. Theory. Phys.* 53: 1603.

Matsubara, F., T. Tamiya, and T. Shirakura. 1996. 'Mixed phase of a longitudinal ferromagnetism and a transverse antiferromagnetism in a site-random Heisenberg model'. *Phys. Rev. Lett.* 77: 378.

Matsubara, F., S. Endoh, and T. Shirakura. 2000. 'Stiffness of the ground-state of a Heisenberg spin-glass model'. *J. Phys. Soc. Jpn.* 69: 1927.

Matsubara, F., T. Shirakura, and S. Endoh. 2001. Spin and chirality autocorrelation functions of a Heisenberg spin glass model. *Phys. Rev. B* 64: 092412.

Maucourt, J., and D. R. Grempel. 1998. 'Lower critical dimension of the XY spin-glass model'. *Phys. Rev. Lett.* 80: 770.

Mccoy, B. M., and T. T. Wu. 1973. *The two-dimensional Ising Model.* Harvard University Press.

Mermin, N. D., and H. Wagner. 1966. 'Absence of ferromagnetism or antiferromagnetism in one- or two-dimensional isotropic Heisenberg models'. *Phys. Rev. Lett.* 17: 1133.

Mézard, M., G. Parisi, and M. Virasoro. 1986. *Spin Glass Theory and Beyond.* World Scientific Press.

Miyashita, S., S. Tanaka, and M. Hirano. 2007. 'Nonmonotonic relaxation in systems with reentrant-type interaction'. *J. Phys. Soc. Jpn.* 76: 083001.

Mori, T. 2012. 'Microcanonical analysis of exactness of the mean-field theory in long-range interacting systems'. *J. Stat. Phys.* 147: 1020.

Nagaosa, N., and Y. Tokura. 2013. 'Topological properties and dynamics of magnetic skyrmions'. *Nature Nanotech.* 8: 899.

Nakamura, T. 2016. 'From measurements to inferences of physical quantities in numerical simulations'. *Phys. Rev. E* 93: 011301(R).

Nambu, Y. 1950. 'A note on the eigenvalue problem in crystal statistics'. *Prog. Theory. Phys.* 5: 1.

Néel, L. 1949. 'Théorie du trainage magnetique des ferromagnétiques au grains fins avec applications aux terres cuites'. *Ann. Geophys.* 5: 99.

Ngai, K. L., A. K. Rajagopal, and C. Y. Huang. 1984. 'Relaxations in spin glasses: Similarities and differences from ordinary glasses'. *J. Appl. Phys.* 55: 1714.

Nielsen, M., D. H. Ryan, H. Guo, and M. Zuckermann. 1996. 'Magnetic ordering in the three-dimensional site-disordered Heisenberg model'. *Phys. Rev. B* 53: 343.

Nishimori, H. 2001. *Statistical Physics of Spin Glasses and Information Processing: An Introduction*. Oxford University Press.

Nishimori, H., and K. Nemoto. 2002. 'Duality and multicritical point of two-dimensional spin glasses'. *J. Phys. Soc. Jpn.* 71: 1198.

Nisoli, C., R. Moessner, and P. Schiffer. 2013. 'Artificial spin ice: Designing and imaging magnetic frustration'. *Rev. Mod. Phys.* 85: 1473.

Obuchi, T., and H. Kawamura. 2013. 'Monte Carlo simulations of the three-dimensional XY spin glass focusing on chiral and spin order'. *Phys. Rev. B* 87: 174438.

Ogielski, A. T. 1985. 'Dynamics of three-dimensional Ising spin glasses in thermal equilibrium'. *Phys. Rev. B* 32: 7384.

Oguchi, T., and Y. Ueno. 1979. 'Statistical theory of the random ordered phase in the site model'. *J. Phys. Soc. Jpn.* 46: 729.

Ohzeki, M. 2009a. 'Accuracy thresholds of topological color codes on the hexagonal and square-octagonal lattices'. *Phys. Rev. E* 80: 011141.

Ohzeki, M. 2009b. 'Locations of multicritical points for spin glasses on regular lattices'. *Phys. Rev. E* 79: 021129.

Ohzeki, M., H. Nishimori, and A. N. Berker. 2008. 'Multicritical points for spin-glass models on hierarchical lattices'. *Phys. Rev. E* 77: 061116.

Onoda, M., and N. Nagaosa. 2006. 'Disorder-induced delocalization in an incommensurate potential'. *Phys. Rev. B* 74: 121101(R).

Onsager, L. 1944. 'Crystal statistics. I. A two-dimensional model with an order-disorder transition'. *Phys. Rev.* 65: 117.

Oseroff, S., M. Mesa, M. Tovar, and R. Arce. 1982. 'Time and field dependence of the magnetization in AuFe, AgMn, and ThGd spin glasses'. *J. Appl. Phys.* 53: 2208.

Parisi, G. 1979. 'Infinite number of order parameters for spin-glasses'. *Phys. Rev. Lett.* 43: 1754.

Parisi, G. 1980a. 'Magnetic properties of spin glasses in a new mean field theory'. *J. Phys. A: Math. Gen.* 13: 1887.

Parisi, G. 1980b. 'The order parameter for spin glasses: A function on the interval 0-1'. *J. Phys. A: Math. Gen.* 13: 1101.

Parisi, G. 1983. 'Order parameter for spin-glasses'. *Phys. Rev. Lett.* 50: 1946.

Parisi, G., and F. Rapuano. 1985. 'Effects of the random number generator on computer simulations'. *Phys. Lett. B* 157: 301.

Pixley, J. H., and A. P. Young. 2008. 'Large-scale Monte Carlo simulations of the three-dimensional XY spin glass'. *Phys. Rev. B* 78: 014419.

Potts, R. B., and J. C. Ward. 1955. 'The combinatrial method and the two-dimensional Ising model'. *Prog. Theory. Phys.* 13: 38.

Privman, V., ed. 1990. *Finite Size Scaling and Numerical Simulation of Statistical Systems*. World Scientific Press.

Ramirez, A. P., A. Hayashi, R. J. Cava, R. Siddharthan, and B. S. Shastry. 1999. 'Zero-point entropy in "spin ice". *Nature* 399: 333.

Rieger, H., L. Santen, U. Blasum, M. Diehl, M. Jünger, and G. Rinaldi. 1996. 'The critical exponents of the two-dimensional Ising spin glass revisited: Exact ground-state calculations and Monte Carlo simulations'. *J. Phys. A: Math. Gen.* 29: 3939.

Schultz, T. D., D. C. Mattis, and E. H. Lieb. 1964. 'Two-dimensional Ising model as a soluble problem of many fermions'. *Rev. Mod. Phys.* 36: 856.

Seki, S., X. Z. Yu, S. Ishiwata, and Y. Tokura. 2012. 'Observation of skyrmions in a multiferroic material'. *Science* 336: 198.

Sherrington, D., and S. Kirkpatrick. 1975. 'Solvable model of a spin-glass'. *Phys. Rev. Lett.* 35: 1792.

Shirakura, T., and F. Matsubara. 1993. 'Phase transition of three-dimensional site-random Ising model with short-range interactions'. *J. Phys. Soc. Jpn.* 62: 3007.

Southern, B. W., and A. P. Young. 1993. 'Spin stiffness in frustrated antiferromagnets. *Phys. Rev. B* 48: 13170(R).

Syozi, I. 1950. 'The statistics of honeycomb and triangular lattice. II'. *Prog. Theory. Phys.* 5: 341.

Takahashi, T., and K. Hukushima. 2015. 'Evidence of a one-step replica symmetry breaking in a three-dimensional Potts glass model'. *Phys. Rev. E* 91: 020102(R).

Takano, H., and S. Miyashita. 1989. 'Relaxation of the spin autocorrelation function in the kinetic Ising model with bond dilution'. *J. Phys. Soc. Jpn.* 58: 3871.

Takayama, H., and K. Hukushima. 2007. 'Computational experiment on glassy dynamic nature of field-cooled magnetization in Ising spin-glass model'. *J. Phys. Soc. Jpn.* 76: 013702.

Tamura, R., and N. Kawashima. 2008. 'First-order transition to incommensurate phase with broken lattice rotation symmetry in frustrated Heisenberg model'. *J. Phys. Soc. Jpn.* 77: 103002.

Tamura, R., and N. Kawashima. 2011. 'First-order phase transition with breaking of lattice rotation symmetry in continuous-spin model on triangular lattice'. *J. Phys. Soc. Jpn.* 80: 074008.

Tamura, R., and Tanaka, S. 2013. Interlayer-interaction dependence of latent heat in the Heisenberg model on a stacked triangular lattice with competing interactions. *Phys. Rev. E*, 88, 052138.

Tamura, R., and S. Tanaka, 2014. 'A method to change phase transition nature: Toward annealing method. *Kinki University Series on Quantum Computing Volume 9: 135.*

Tamura, R., S. Tanaka, and N. Kawashima. 2010. 'Phase transition in Potts model with invisible states'. *Prog. Theory. Phys.* 124: 381.

Tamura, R., N. Kawashima, T. Yamamoto, C. Tassel, and H. Kageyama. 2011. 'Random fan-out state induced by site-random interlayer couplings'. *Phys. Rev. B* 84: 214408.

Tamura, R., S. Tanaka, and N. Kawashima. 2012. 'A method to control order of phase transition: Invisible states in discrete spin models'. *Kinki University Series on Quantum Computing 7:217.*

Tamura, R., S. Tanaka, and N. Kawashima. 2013. 'Second-order phase transition in the Heisenberg model on a triangular lattice with competing interactions'. *Phys. Rev. B* 87: 214401.

Tamura, R., S. Tanaka, and N. Kawashima. 2014. 'Phase transitions with discrete symmetry breaking in antiferromagnetic Heisenberg models on a triangular lattice'. *JPS Conf. Proc.* 1: 012125.

Tanaka, S., and S. Miyashita. 2007. 'Slow relaxation of spin structure in exotic ferromagnetic phase of Ising-like Heisenberg kagomé antiferromagnets'. *J. Phys. Soc. Jpn.* 76: 103001.

Tanaka, S., and S. Miyashita. 2009. 'Mechanism of slow relaxation due to screening effect in a frustrated system'. *J. Phys. Soc. Jpn.* 78: 084002.

Tanaka, S., and R. Tamura. 2011. 'Dynamical properties of Potts model with invisible states'. *J. Phys.: Conf. Ser.* 320: 012025.

Tanaka, S., H. Katsura, and N. Nagaosa. 2006. 'Electron localization or delocalization in incommensurate helical magnets'. *Phys. Rev. Lett.* 97: 116404.

Tanaka, S., R. Tamura, and Kawashima, N. 2011. 'Phase transition of generalized ferromagnetic Potts model Effect of invisible states' –. *J. Phys.: Conf. Ser.* 297: 012022.

Thouless, D. J., P. W. Anderson, and R. G. Palmer. 1977. 'Solution of "Solvable model of a spin glass'". *Philosophical Magazine* 35: 593.

Tokura, Y., S. Seki, and N. Nagaosa. 2014. 'Multiferroics of spin origin'. *Rep. Prog. Phys.* 77: 076501.

Udagawa, M., M. Ogata, and Z. Hiroi. 2002. 'Exact result of ground-state entropy for Ising pyrochlore magnets under a magnetic field along [111] axis'. *J. Phys. Soc. Jpn.* 71: 2365.

Uemura, Y. J., T. Yamazaki, D. R. Harshman, M. Senba, and E. J. Ansaldo. 1985. 'Muon-spin relaxation in AuFe and CuMn spin glasses'. *Phys. Rev. B* 31: 546.

van Enter, A. C. D., G. Iacobelli, and S. Taati. 2011. 'First-order transition in Potts models with "invisible"' states rigorous proofs. *Prog. Theory. Phys.* 126: 983.

van Enter, A. C. D., G. Iacobelli, and S. Taati. 2012. 'Potts model with invisible colors: Random-cluster representation and Pirogov-Sinai analysis'. *Rev. Math. Phys.* 24: 1250004.

Viet, D. X., and H. Kawamura. 2009a. 'Monte Carlo studies of chiral and spin ordering of the three-dimensional Heisenberg spin glass'. *Phys. Rev. B* 80: 064418.

Viet, D. X., and H. Kawamura. 2009b. 'Numerical evidence of spin-chirality decoupling in the three-dimensional Heisenberg spin glass model'. *Phys. Rev. Lett.* 102: 027202.

Vincent, E. 2007. 'Aging, rejuvenation and memory: The example of spin glasses'. *Ageing and the Glass Transition. 7.*

Wannier, G. H. 1950. 'Antiferromagnetism. The triangular Ising net'. *Phys. Rev.*: 79: 357.

Wannier, G. H. 1973. 'Antiferromagnetism. The triangular Ising net'. *Phys. Rev. B* 7: 5017.

Wu, F. Y. 1982. 'The Potts model'. *Rev. Mod. Phys.* 54: 235.

Yamamoto, T., T. Sugashima, and T. Nakamura. 2004. 'Nonvanishing spin-glass transition temperature of the $\pm J$ XY model in three dimensions'. *Phys. Rev. B* 70: 184417.

Yasuda, M., C. Takahashi, and K. Tanaka. 2016. 'Perturbative interpretation of adaptive Thouless-Anderson-Palmer free energy'. *J. Phys. Soc. Jpn.* 85: 075001.

Young, A. P., ed. 1997. *Spin Glasses and Random Fields*. World Scientific Press.

CHAPTER

3 Simulated Annealing

In the first half of this chapter, we describe how to express combinatorial optimization problems in the language of statistical physics, whereas in the later half, a generic algorithm, called simulated annealing, to solve combinatorial optimization problems is explained.

As will be seen in Section 3.1, combinatorial optimization problems are represented by the random Ising model explained in Section 2.2. The best solution to combinatorial optimization problems corresponds to the ground state of the corresponding random Ising model. As the number of spins increases, the total number of states increases exponentially. Thus, brute-force algorithms fail to obtain the ground state. Moreover, random Ising models do not have any translational symmetries, in other words, no appropriate wave vector exists due to the randomness as explained in Section 2.2.2. Thus, to obtain the ground state of random Ising models is difficult. In Section 3.2, we will introduce representative combinatorial optimization problems and demonstrate a mapping from the combinatorial optimization problems to the Hamiltonian of the random Ising model or a generalization of the random Ising model.

Once we map combinatorial optimization problems to the random Ising model or its generalization, we can obtain the ground state of the corresponding model. To obtain the ground state of random Ising models, a generic algorithm – simulated annealing – was proposed (Kirkpatrick et al., 1983). As will be described in Section 3.3, simulated annealing is an algorithm inspired by statistical physics. In Section 3.4, the convergence theorem of simulated annealing proved by Geman and Geman will be reviewed (Geman and Geman, 1984). Owing to the convergence theorem, simulated annealing is a mathematically guaranteed algorithm to obtain the ground state of random Ising models.

3.1 Relation Between Random Ising Models and Combinatorial Optimization Problems

Combinatorial optimization problems exist in various fields such as natural science, social science, and industry and are about finding the minimum/maximum solution of cost/objective functions. Cost/objective functions measure some kind of "goodness" and depend on detailed information of systems. Since cost functions are of opposite sign to objective functions in the mathematical sense, we consider only cost functions. As will be shown later, cost functions correspond to the Hamiltonian of random Ising models or its generalization.

When we use a naive method in which all candidate solutions are listed and we choose the best solution, the computation time increases exponentially or much more as the number of objects increases. Number partitioning problem which will be explained in Section 3.2.3 is an example of a combinatorial optimization problem and belongs to Non-deterministic Polynomial complete (NP-complete) which will be defined in Section 3.2.1. If the best solution to combinatorial optimization problems was found, it could be mapped to other problems which belong to NP-complete such as graph coloring problems with polynomial time. Since the origin of difficulty in respective combinatorial optimization problems is not known at the present stage, there is no trigger to develop exact efficient methods. As mentioned earlier, however, in various fields, it is important to solve combinatorial optimization problems. Thus, we should develop efficient algorithms to solve combinatorial optimization problems.

To treat combinatorial optimization problems, heuristic methods whose computational cost is polynomial time were developed. Most heuristic methods can be categorized into two types: divide-and-conquer methods (Karp, 1977) and iterative improvement methods (Dunham et al., 1963; Lin, 1975). If an algorithm is specific for an individual combinatorial optimization problem in NP-complete, the algorithm cannot be applied to other combinatorial optimization problems in NP-complete.

In physics, efficient and generic algorithms to solve combinatorial optimization problems have been developed. One of them is based on the concept of statistical physics. Statistical physics is an established discipline in physics meant to treat emergent phenomena in materials where a vast number of elements such as atoms and molecules exist and interact with each other. In physics, it is an important problem to understand what happens in materials in the low-temperature limit; whether the substances are still in the liquid state or become solids. If the material we considered becomes solid, is it a crystal or glass? – This is a central issue in statistical physics since unusual events happens at low temperature, which will be explained later.

Here we consider the infinite-range ferromagnetic Ising model, i.e., the Husimi–Temperly–Curie–Weiss model considered in Section 2.1.2. The Hamiltonian of this model is given by

$$\mathcal{H} = -\frac{zJ}{N}\sum_{1\le i<j\le N}\sigma_i^z\sigma_j^z, \qquad (\sigma_i^z = \pm 1), \tag{3.1}$$

where the summation takes over all pairs of spins. The schematics of this model is shown in Fig. 2.2.

According to statistical physics, in the equilibrium state, the probability of the state with internal energy E is proportional to the Boltzmann weight, $\exp(-E/T)$, where the Boltzmann constant k_{B} is set to unity. Thus, all states could appear with equal probability at $T \to \infty$. In this model, the magnetization per spin m can characterize the state as well as the internal energy, where the magnetization per spin m is defined by

$$m = \frac{1}{N}\sum_{i=1}^{N}\sigma_i^z. \tag{3.2}$$

Using m, the Hamiltonian can be rewritten as

$$\mathcal{H} = -\frac{zJ}{N}\sum_{1\le i<j\le N}\sigma_i^z\sigma_j^z \tag{3.3}$$

$$= -\frac{zJ}{2N}\sum_{1\le i\ne j\le N}\sigma_i^z\sigma_j^z \tag{3.4}$$

$$= -\frac{zJ}{2N}\left[\left(\sum_{i=1}^{N}\sigma_i^z\right)^2 - N\right] \tag{3.5}$$

$$= -\frac{z}{2}JNm^2 + \frac{z}{2}J \simeq -\frac{z}{2}JNm^2, \tag{3.6}$$

where the last approximation is valid for $N \to \infty$. By the definition, the magnetization is respectively represented by the numbers of $+1$ and -1 spins, N_+ and N_-:

$$m = \frac{1}{N}(N_+ - N_-) = \frac{1}{N}(2N_+ - N). \tag{3.7}$$

The number of configurations for a given m is given by

$$\binom{N}{N_+} = \frac{N!}{N_+!(N-N_+)!}. \tag{3.8}$$

Let us first consider properties at high temperatures. The probability of the number of $+1$ spins, N_+, in random configurations is obtained by Eq. (3.8). This distribution is the Gaussian distribution of which the center is $m = 0$, i.e., $N_+ = N/2$. Because all spin configurations can be realized with equal probabilities in the limit of $T \to \infty$, zero-magnetization states ($m = 0$) tend to appear at high temperatures, which is the entropy effect. The statistical weight of the states with $m = \pm 1$ is $\exp(-N/2)$ times smaller than that of the states with $m = 0$.

Next we consider the case at low temperatures. As mentioned earlier, the probability of states is proportional to the Boltzmann weight, and the low-energy states tend to realize at low temperatures. However, we cannot obtain the stable state by only decreasing the temperature. If the solidity temperature (or phase-transition temperature) exists, dislocation appears unless the temperature decreases slow enough near the solidity temperature.

To display the phenomenon, we consider the ferromagnetic Ising model on a square lattice with periodic boundary condition:

$$\mathscr{H} = -J \sum_{\langle i,j \rangle} \sigma_i^z \sigma_j^z, \qquad (\sigma_i^z = \pm 1), \tag{3.9}$$

where $\langle i, j \rangle$ denotes the pairs of the nearest neighbor spins on the square lattice and $J > 0$. This model is the simplest model which exhibits a second-order phase transition at a finite temperature, and was explained in Section 2.1.3. The phase-transition temperature from the paramagnetic phase to the ferromagnetic phase in the model is $T_c/J = 2/\ln\left(1+\sqrt{2}\right)$ ($\simeq$ 2.269). In the ground state, all spins are directed at the same direction, i.e., either $\sigma_i^z = +1$ or $\sigma_i^z = -1$ for all i. Figure 3.1 shows typical snapshots of the model obtained by Monte Carlo simulations which will be introduced in Section 3.3.1. When the temperature decreases fast, domain walls which are the boundaries of $+1$ and -1 spins are generated. On the other hand, when the temperature decreases slow enough, the equilibrium states appear at each temperature. Thus, to obtain the stable state, we should decrease the temperature slow enough so that dislocations do not appear.

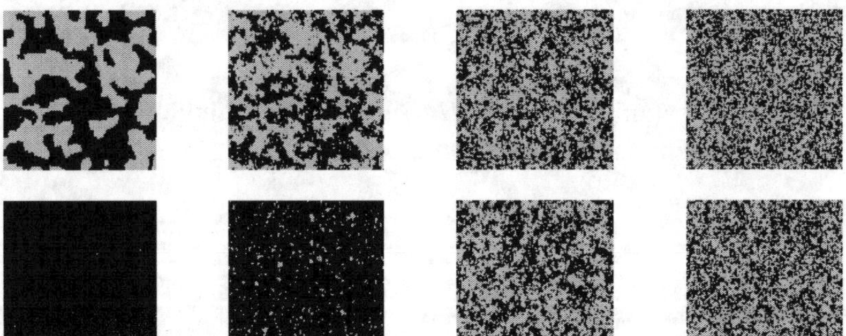

$T/J = 1$ $T/J = 2$ $T/J = 3$ $T/J = 4$

Fig. 3.1 Typical snapshots of the ferromagnetic Ising model on a square lattice whose linear dimension is 128; these were obtained by Monte Carlo simulations. (Upper panel) Snapshots for rapid cooling case. (Lower panel) Snapshots for slow cooling case.

To solve combinatorial optimization problems, iterative improvement methods are often used. In these methods, the cost function of the combinatorial optimization problem we considered is regarded as the internal energy of the corresponding random Ising model. However, only transitions from a state to another state in which the internal energy decreases are permitted in the method. The treatment corresponds to the quench from high temperature to zero temperature situation, in other words, infinitely fast cooling. In general, the method obtains metastable solutions as expected. The Metropolis method is a generalization of the iterative improvement which is often applied in Monte Carlo simulations (Metropolis et al., 1953). The method allows the transitions from a state to another state even when the internal energy increases according to a well-defined probability. The Metropolis method is often used to obtain the equilibrium state of systems at finite temperatures. The precise explanation of the Metropolis method will be given in Section 3.3.1; here, we give an intuitive explanation. Consider the Ising spin system given by Eq. (3.9). Suppose that a random configuration is prepared to the initial state. In each step, we try to update the state by flipping a spin which is randomly chosen one by one. Let ΔE be the energy difference between the present state and the state where the chosen spin is flipped. If ΔE is negative, the state is updated definitely, which is the same as the iterative improvement. The difference between the iterative method and the Metropolis method exists when ΔE is positive. If ΔE is positive, the state is updated according to the probability given by

$$P(\Delta E; T) = \exp\left(-\frac{\Delta E}{T}\right), \tag{3.10}$$

whereas the state is not updated according to the probability $1 - P(\Delta E; T)$. In the Metropolis method, we obtain finite-temperature equilibrium properties of systems after updating many times.

Simulated annealing is realized by introducing temperature into combinatorial optimization problems and by decreasing the temperature slowly enough. Simulated annealing consists of three parts. The first one is the melting process at high temperatures. Each element in the system we considered is randomly oriented and changed rapidly. The second one is the self-organizing process at moderate temperatures. In this part, the system feels the interactions from each surrounding element and seeks the (meta)stable states. The third one is the freezing process at low temperatures. The state of the system cannot change quickly because the probability $P(\Delta E; T)$ becomes small for positive ΔE. Thus, when the temperature decreases slow enough so as to reach the equilibrium state at each temperature, simulated annealing succeeds to obtain not metastable states but the stable state. This fact was proved by Geman and Geman mathematically as will be described in Section 3.4 (Geman and Geman, 1984). To realize the situation, we decrease temperature at slow speed near the temperature where the specific heat is large. This is because, the specific heat is defined by

$$C = \frac{d\langle E(T)\rangle}{dT}, \tag{3.11}$$

where $\langle E(T)\rangle$ represents the equilibrium value of the internal energy at T. Thus, when the specific heat is a large value, the internal energy changes rapidly as a function of temperature.

For simplicity, we considered ferromagnetic spin models although combinatorial optimization problems are not represented by ferromagnetic spin models but by random spin models with ferromagnetic and antiferromagnetic interactions. Review on random spin models was given in Sections 2.2–2.4 and mappings from combinatorial optimization problems to random spin systems will be explained in the next section. In random spin models, the density of states and dynamics become peculiar compared to ferromagnetic spin models because of the existence of frustration as shown in Section 2.2.1.

3.2 Representative Combinatorial Optimization Problems

As introduced in the previous section, combinatorial optimization problems involve finding the minimum/maximum solution of cost/objective functions. Combinatorial optimization problems are classified depending on the "difficulty" in computational complexity theory.

3.2.1 Computational complexity theory

Computational complexity theory classifies computational problems depending on the complexity of the problems. Most computational problems can be considered as decision problems of which the answer is *yes* or *no*. The computational complexity of the decision problem is measured by the time resource and memory resource needed to solve the problem. In the computational complexity theory, the lower bound of resources required by the best algorithm is clarified for a given computational problem. Furthermore, depending on the lower bound, the problem is classified into complexity classes P and NP. Classes P and NP are defined as follows.

P : The set of problems which can be solved by a *deterministic Turing machine* with polynomial time.

A deterministic Turing machine consists of an infinite memory tape and the head which can read and write the symbol of the memory tape. Furthermore, depending on the present internal state and the symbol at the head position, the machine performs the following three operations uniquely. (i) The head outputs the symbol to the memory tape, the head position. (ii) The head moves to the right or the left. (iii) The internal state is changed.

NP : The set of computational problems which can be solved by *non-deterministic Turing machine* with polynomial time.

In a non-deterministic Turing machine, depending on the present internal state and the symbol at the head position, the operations are not uniquely determined unlike the deterministic Turing machine.

The other definition of NP is as follows. When there is evidence of *yes* in the decision problem, the computational time is polynomial to judge whether the evidence is true or false. Besides, according to the definition of class NP, some computational problems are classified into some complexity classes such as NP-complete and NP-hard which are defined as follows.

NP-complete

NP-complete is the subset of computational problems which belong to the class NP. A problem in the subset can be translated into all problems belonging to the subset with polynomial time.

NP-hard

NP-hard is the set of computational problems which can be translated into any computational problem belonging to the class NP. Thus, NP-hard means the problem is at least as hard as any computational problems belonging to the class NP.

Let us discuss the relationship between the class P and the class NP. Whether P = NP or P $\neq$ NP is reasonable is an open problem. Here, it is clarified that the class P is a subset of the class NP. Then, by searching the computational problem which belongs to the class NP but does not belong to the classearching the computational problem which belongs to the class NP but does not belong to the class P, this problem will be clarified. Figure 3.2 shows the schematics of classification for P = NP and P $\neq$ NP. The computational complexity theory is reviewed in detail by Nielsen and Chuang (2000).

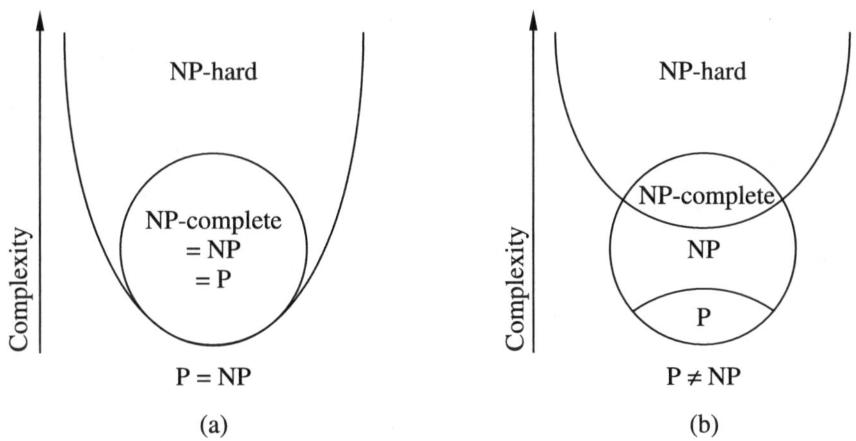

Fig. 3.2 Schematics of the computational complexity for (a) P = NP and (b) P $\neq$ NP.

3.2.2 Classification of combinatorial optimization problems

Most combinatorial optimization problems belong to the NP-complete or NP-hard classes in computational complexity theory. Representative combinatorial optimization problems are classified as follows.

NP-complete

- Number partitioning problem
- Satisfiability(SAT) problem
- Hamiltonian cycles and paths problem
- Vertex cover problem
- Minimum spanning tree
- Maximum cut problem
- Graph coloring problem
- Clique cover problem

NP-hard

- Traveling salesman problem
- Knapsack problem

In Section 3.1, we stated that combinatorial optimization problems can be mapped to the random Ising models explained in Section 2.2. That is, the cost function of combinatorial optimization problems can be written by the Hamiltonian of the random Ising models. In fact, these combinatorial optimization problems are mapped to the Ising model or the Potts model with random interactions as follows. Notice that the mapping from a given combinatorial optimization problem to the Ising/Potts model is not unique. Also, the problems which can be mapped into the Potts model can be expressed by the Ising model.

Ising model

- Number partitioning problem (NP-complete)
- Satisfiability(SAT) problem (NP-complete)
- Hamiltonian cycles and paths problem (NP-complete)
- Vertex cover problem (NP-complete)
- Minimum spanning tree (NP-complete)
- Maximum cut problem (NP-complete)
- Traveling salesman problem (NP-hard)
- Knapsack problem (NP-hard)

Potts model

- Graph coloring problem (NP-complete)
- Clique cover problem (NP-complete)

3.2.3 Examples of combinatorial optimization problems

Number partitioning problem (Ising model)

Problem

Here, we introduce the subset sum problem that is a general form of the number partitioning problem. The subset sum problem is given as follows. Let N be the number of given integers. Does at least one subset whose sum is M in the integer set $\{n_1, n_2, \cdots, n_N\}$ exist? If the number of integers N is small, we can easily obtain the answer, *yes* or *no*, by a brute force. However, as the number of integers increases, the subset sum problem becomes seriously difficult since the number of candidate solutions is $2^N - 1$.

Examples

1. Is there at least one subset whose sum is $M = 0$ in the set $\{-7, -3, -2, 4, 9\}$?

   The answer is *yes*, and the subset is $\{-7, -2, 9\}$.

2. Is there at least one subset whose sum is $M = 21$ in the set $\{2, 4, 6, 8, 12\}$?

   The answer is *no*.

Mapping to the Ising model

The cost function of the subset sum problem is given by

$$\mathcal{H} = \left(\sum_{i=1}^{N} n_i \varepsilon_i - M \right)^2, \quad (\varepsilon_i = 0 \text{ or } 1). \tag{3.12}$$

We will obtain the minimum value of $\mathcal{H}$ and the subset which minimizes $\mathcal{H}$. If at least one subset exists so that $\mathcal{H} = 0$, the answer is *yes*. On the other hand, if the cost function of all subsets is a positive value ($\mathcal{H} > 0$), the answer is *no*. This cost function of the subset sum problem can be mapped to the Hamiltonian of the Ising model with random interactions as follows. We introduce the Ising spin ($\sigma_i^z = \pm 1$) instead of ε_i as

$$\varepsilon_i = \frac{1}{2}(1 + \sigma_i^z). \tag{3.13}$$

Then, Eq. (3.12) is written as

$$\mathcal{H} = \left(\sum_{i=1}^{N} n_i \frac{1}{2}(1 + \sigma_i^z) - M \right)^2$$

$$= \left(\frac{1}{2} \sum_{i=1}^{N} n_i \sigma_i^z + \frac{1}{2} \sum_{i=1}^{N} n_i - M \right)^2$$

$$= \frac{1}{4} \sum_{1 \leq i,j \leq N} n_i n_j \sigma_i^z \sigma_j^z$$

$$+ \sum_{i=1}^{N} \left(\frac{1}{2} \sum_{j=1}^{N} n_j - M \right) n_i \sigma_i^z + \left(\frac{1}{2} \sum_{j=1}^{N} n_j - M \right)^2 . \tag{3.14}$$

Here, we introduce the random interaction J_{ij} and the magnetic field h_i as

$$J_{ij} := n_i n_j, \tag{3.15}$$

$$h_i := \left(\frac{1}{2} \sum_{j=1}^{N} n_j - M \right) n_i, \tag{3.16}$$

and obtain

$$\mathcal{H} = \frac{1}{4} \sum_{1 \leq i,j \leq N} J_{ij} \sigma_i^z \sigma_j^z + \sum_{i=1}^{N} h_i \sigma_i^z + \left(\frac{1}{2} \sum_{j=1}^{N} n_j - M \right)^2 , \tag{3.17}$$

where the third term of the right-hand side is constant. This is the Hamiltonian of the Ising model with random interactions and inhomogeneous magnetic fields.

Traveling salesman problem (Ising model)

Problem

The traveling salesman problem is to find the minimum path under the following four conditions.

1. There are N cities.
2. The salesman moves from the ith city to the jth city where the distance between them is $\ell_{i,j}$.
3. The salesman can pass through a city only once.
4. The salesman returns to the initial city after all the cities are passed through.

The length of a path, i.e., the cost function of the traveling salesman problem, is given by

$$\mathcal{H} = \sum_{\alpha=1}^{N} \ell_{c_\alpha, c_{\alpha+1}}, \tag{3.18}$$

where c_α denotes the city where we pass through at the αth step. The fourth condition can be written as

$$c_{N+1} = c_1. \tag{3.19}$$

The traveling salesman problem is to find the set $\{c_1, c_2, \cdots, c_N\}$ where the length of path $\mathscr{H}$ under the aforementioned four conditions becomes minimum value. If the number of cities N is small, we can easily obtain the best solution by a brute force. However, as the number of cities increases, the traveling salesman problem becomes seriously difficult since the number of candidate solutions is $(N-1)!/2$.

Example

Figure 3.3 (a) is the position of cities of a traveling salesman problem for $N = 6$. A non-best solution and the best solution where the length of the path $\mathscr{H}$ defined by Eq. (3.18) is the minimum value are shown in Figs. 3.3 (b) and (c), respectively.

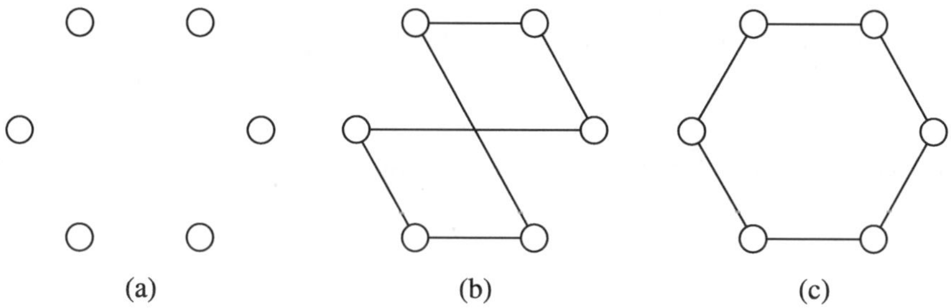

Fig. 3.3 (a) Position of cities of a traveling salesman problem for $N = 6$. (b) A non-best solution. The lines denote the selected path. (c) The best solution where the length of path is the minimum value.

Mapping to the Ising model

In order to map the problem to the Ising model, we introduce $n_{i,\alpha} = 0$ or 1 which represents the state at the ith city at the αth step (Hopfield and Tank, 1985). If the salesman passes through the ith city at the αth step, $n_{i,\alpha}$ is 1 whereas $n_{i,\alpha} = 0$ if the salesman does not pass through the ith city at the αth step. Thus, $n_{i,\alpha}$ is the microscopic state, and the number of elements $n_{i,\alpha}$ is $N \times N$. The third condition of the traveling salesman problem can be represented by

$$\sum_{\alpha=1}^{N} n_{i,\alpha} = 1, \quad \text{for } \forall i. \tag{3.20}$$

On top of that, the salesman cannot pass through more than one city at the same step, and thus, a following constraint should be satisfied:

$$\sum_{i=1}^{N} n_{i,\alpha} = 1, \quad \text{for } \forall \alpha. \tag{3.21}$$

Then, the length of the path $\mathscr{H}$ can be rewritten as

$$\mathscr{H} = \sum_{\alpha=1}^{N} \sum_{1 \leq i,j \leq N} \ell_{i,j} n_{i,\alpha} n_{j,\alpha+1} \tag{3.22}$$

$$= \frac{1}{4} \sum_{\alpha=1}^{N} \sum_{1 \leq i,j \leq N} \ell_{i,j} \sigma_{i,\alpha}^{z} \sigma_{j,\alpha+1}^{z} + \sum_{\alpha=1}^{N} \sum_{1 \leq i,j \leq N} \ell_{i,j} \sigma_{i,\alpha}^{z} + \text{const.}, \tag{3.23}$$

where the Ising spin variable $\sigma_{i,\alpha}^{z} = \pm 1$ is introduced as

$$\sigma_{i,\alpha}^{z} = 2n_{i,\alpha} - 1. \tag{3.24}$$

Here, the second term of Eq. (3.23) is constant from the two constraints obtained from Eq. (3.23). This is the Hamiltonian of the Ising model with random interactions and inhomogeneous magnetic fields.

There is another mapping to the Ising model (Chen et al., 2011). We introduce $U_{ij} = 0$ or 1 which is an element of an $(N-1) \times (N-1)$ upper triangle matrix. If the salesman goes through the path between the ith city and the jth city, U_{ij} is 1 whereas $U_{ij} = 0$ if the salesman does not through the path between the ith city and the jth city. Then, the length of the path $\mathscr{H}$ can be written as

$$\mathscr{H} = \sum_{\langle i,j \rangle} l_{i,j} U_{ij}, \tag{3.25}$$

where $\langle i, j \rangle$ signifies that each city is counted only once. By imposing some conditions of the traveling salesman problem, Eq. (3.25) can be represented by the Hamiltonian of the Ising model with random interactions and inhomogeneous magnetic fields.

As seen in the previous equation, there are some candidates for mapping to the Ising model from a given combinatorial optimization problem.

3-SAT problem (Ising model)

Problem

The 3-SAT problem consists of a set of N Boolean variables $\{x_1, x_2, \cdots, x_N\}$, where $x_i = 0$ "False" or 1 "True". Here, $\bar{x}_i$ returns a negated x_i, i.e., $\bar{x}_i = 0$ if $x_i = 1$, while $\bar{x}_i = 1$ if $x_i = 0$. The Boolean variables follow the logical OR ($\vee$) and logical AND ($\wedge$) which are defined by

$$x_i \vee x_j = \begin{cases} 0 & \text{for} \quad (x_i, x_j) = (0,0) \\ 1 & \text{for} \quad (x_i, x_j) = (0,1), (1,0), (1,1) \end{cases}, \tag{3.26}$$

$$x_i \wedge x_j = \begin{cases} 0 & \text{for} \quad (x_i, x_j) = (0,0), (0,1), (1,0) \\ 1 & \text{for} \quad (x_i, x_j) = (1,1) \end{cases}. \tag{3.27}$$

Furthermore, the 3-clause $\mathscr{C}$ is defined by the logical OR of three Boolean variables which are selected from $\{x_1, x_2, \cdots, x_N, \bar{x}_1, \bar{x}_2, \cdots, \bar{x}_N\}$ randomly, for example, $(x_i \vee x_j \vee x_k)$ or $(x_i \vee \bar{x}_j \vee \bar{x}_k)$. In the 3-SAT problem, we prepare the logical operation which consists of the M 3-clauses $\mathscr{C}_a$ $(a = 1, \cdots, M)$ and consider the logical AND of M different 3-clauses, that is, $\mathscr{C}_1 \wedge \mathscr{C}_2 \wedge \cdots \wedge \mathscr{C}_M$. By using the logical operations, the 3-SAT problem is given as follows. Is there a set of Boolean variables $(x_1, x_2, \cdots, x_N)$ such that the given logical operation $\mathscr{C}_1 \wedge \mathscr{C}_2 \wedge \cdots \wedge \mathscr{C}_M = 1$? The answer to this problem depends on the control parameter $\alpha := M/N$. The probability that the logical operation $\mathscr{C}_1 \wedge \mathscr{C}_2 \wedge \cdots \wedge \mathscr{C}_M$ outputs 1 (SAT formulas) becomes unity in the thermodynamic limit $(N \to \infty)$ for $\alpha < \alpha_c \simeq 4.267$ (Mézard et al., 2002; Mézard and Zecchina, 2002). On the other hand, when $\alpha > \alpha_c$, this probability becomes zero (UNSAT formulas). Moreover, in the region $\alpha_c > \alpha > \alpha_g \simeq 4.15$, many ground states which are separated from each other exist. Then, it is known that the replica symmetry breaking (see Section 2.3.3) occurs in this region (Mézard et al., 1986). Note that if we prepare the K-clauses and consider the logical AND of different K-clauses, the problem is referred to as a K-SAT problem. When the number of Boolean variables is N, the number of candidate solutions is 2^N in the K-SAT problem regardless of K.

Example

1. We consider the 3-SAT problem for $N = 3$ and $M = 2$ $(\alpha \simeq 0.667)$. Is there a set of Boolean variables (x_1, x_2, x_3) such that the logical operation defined by

$$(x_1 \vee x_2 \vee \bar{x}_3) \wedge (\bar{x}_1 \vee \bar{x}_2 \vee x_3) \tag{3.28}$$

outputs 1?

The answer is *yes*, and one of the sets is $(x_1, x_2, x_3) = (0, 1, 1)$.

2. We consider the 3-SAT problem for $N = 3$ and $M = 8$ $(\alpha \simeq 2.667)$. Is there a set of Boolean variables (x_1, x_2, x_3) such that the logical operation defined by

$$\begin{aligned}&(x_1 \vee x_2 \vee x_3) \wedge (\bar{x}_1 \vee x_2 \vee x_3) \wedge (x_1 \vee \bar{x}_2 \vee x_3) \wedge (x_1 \vee x_2 \vee \bar{x}_3)\\ &\wedge(\bar{x}_1 \vee \bar{x}_2 \vee x_3) \wedge (x_1 \vee \bar{x}_2 \vee \bar{x}_3) \wedge (\bar{x}_1 \vee x_2 \vee \bar{x}_3) \wedge (\bar{x}_1 \vee \bar{x}_2 \vee \bar{x}_3)\end{aligned} \tag{3.29}$$

outputs 1?

The answer is *no*. Of course, the result does not mean that any logical operators for $N = 3$ and $M = 8$ do not output 1, and there is another logical operator which outputs 1.

Mapping to the Ising model

When all 3-clauses output 1, the logical operation $\mathscr{C}_1 \wedge \mathscr{C}_2 \wedge \cdots \wedge \mathscr{C}_M$ outputs 1. Thus, for each $\mathscr{C}_a$, we introduce E_a defined by

$$E_a = \frac{(1 + J_{a,i}\sigma_i^z)(1 + J_{a,j}\sigma_j^z)(1 + J_{a,k}\sigma_k^z)}{8}, \quad (\sigma_i^z = (-1)^{x_i} = \pm 1), \tag{3.30}$$

where

$$J_{a,i} = \begin{cases} +1 & \text{if selected Boolean variable is } x_i \\ -1 & \text{if selected Boolean variable is } \bar{x}_i \end{cases}. \tag{3.31}$$

By the definition of E_a, when $\mathscr{C}_a$ outputs 1, $E_a = 0$, whereas when $\mathscr{C}_a$ outputs 0, $E_a = 1$. Then, the cost function of the 3-SAT problem can be written as

$$\mathscr{H} = \sum_{a=1}^{M} E_a. \tag{3.32}$$

This is the Hamiltonian of the Ising model with random interactions, inhomogeneous magnetic fields, and three-body interactions, that is, $\sigma_i^z \sigma_j^z \sigma_k^z$. When there is a solution such that $\mathscr{H} = 0$, the answer to the 3-SAT problem is *yes*. On the other hand, if the minimum value of $\mathscr{H}$ is positive, there is no variable set such that the given logical operation outputs 1.

Graph coloring problem (Potts model)

Problem

Here, we explain the vertex coloring problem which is a graph coloring problems. The vertex coloring problem is to find a way of coloring the vertices of a graph such that two adjacent vertices do not share the same color. The coloring problem for fixed k colors is called the k-coloring problem. When the number of vertices is N, the number of candidate solutions is k^N in the vertex k-coloring problem.

Example

Figure 3.4 (a) is the positions of vertices of a vertex coloring problem for $N = 6$. The answer of the three-coloring problem ($k = 3$) is shown in Fig. 3.4 (b). If $k = 2$, there is no set such that two adjacent vertices do not share the same color, and the "best" solution is shown in Fig. 3.4 (c). Here the "best" solution means wherein so that the cost function is minimized. Notice that the situation indicated in Fig. 3.4 (c) is an example of frustration the explained in Section 2.2.1.

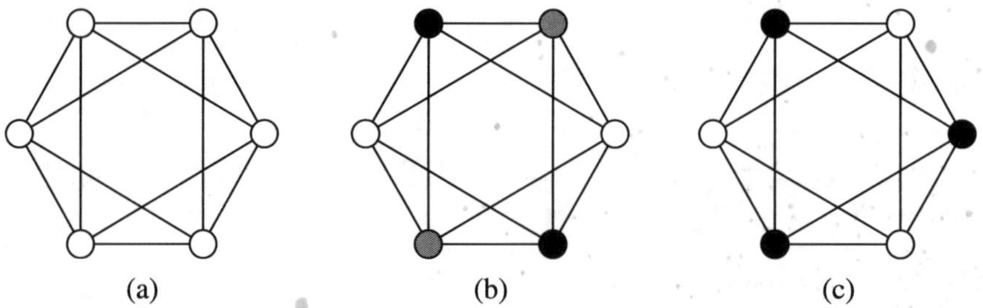

Fig. 3.4 (a) Position of vertices of a vertex coloring problem for $N = 6$. (b) The answer for the three-coloring problem ($k = 3$). (c) The "best" solution for the two-coloring problem ($k = 2$). This solution does not satisfy the constraint: two adjacent vertices do not share the same color.

Mapping to the Potts model

The cost function of the vertex k-coloring problem is given by

$$\mathcal{H} = -\sum_{1\le i,j\le N} J_{ij}\delta_{s_i,s_j}, \quad (s_i = 1,\cdots,k), \tag{3.33}$$

where

$$J_{ij} = \begin{cases} -1 & \text{if } i\text{th vertex and } j\text{th vertex are adjacent.} \\ 0 & \text{if } i\text{th vertex and } j\text{th vertex are not adjacent.} \end{cases} \tag{3.34}$$

This is the Hamiltonian of the k-state Potts model with random interactions (see Section 2.A.1). Each state of s_i corresponds to each color. When there is a set of $\{s_i\}$ so that $\mathcal{H} = 0$, the obtained set $\{s_i\}$ is the answer of the vertex k-coloring problem. On the other hand, if the minimum value of $\mathcal{H}$ is positive, there is no set such that two adjacent vertices do not share the same color in the vertex k-coloring problem.

3.3 Conventional Computation Methods Including Simulated Annealing

3.3.1 Monte Carlo methods

In this subsection, we explain the Monte Carlo methods which have been used to obtain the equilibrium state and dynamical properties in strongly correlated systems. The methods are based on stochastic dynamics called the master equation.

Master equation

The master equation represents the dynamics of probability distribution using the transition matrix. Let $P(\Sigma_i,t)$ be the probability where the macroscopic state takes Σ_i at time t ($\Sigma_i = \{\sigma_1 = +1, \sigma_2 = -1, \sigma_3 = -1, \sigma_4 = +1, \cdots\}$, for example). The normalization of the probability distribution is given by

$$\sum_{i=1}^{M} P(\Sigma_i,t) = 1, \tag{3.35}$$

where M is the number of macroscopic states. For example, $M = 2^N$ for the Ising model with N spins. The transition probability from the state Σ_i to the state Σ_j per unit time is represented by $w_{i\to j}$. Here we concentrate on the Markov process. The master equation is given by

$$P(\Sigma_i,t+\Delta t) = -\sum_{j\neq i} w_{i\to j}\Delta t P(\Sigma_i,t) + \sum_{j\neq i} w_{j\to i}\Delta t P(\Sigma_j,t) + P(\Sigma_i,t) \tag{3.36}$$

$$= \left(1 - \sum_{j \neq i} w_{i \to j} \Delta t\right) P(\Sigma_i, t) + \sum_{j \neq i} w_{j \to i} \Delta t P(\Sigma_j, t). \tag{3.37}$$

For convenience, we use the vector representation of the probability distribution:

$$\vec{P}(t) = \begin{pmatrix} P(\Sigma_1, t) \\ P(\Sigma_2, t) \\ \vdots \\ P(\Sigma_M, t) \end{pmatrix}. \tag{3.38}$$

By using the vector representation, the master equation can be expressed as

$$\vec{P}(t + \Delta t) = \mathscr{L} \vec{P}(t), \tag{3.39}$$

where $\mathscr{L}$ is the $M \times M$ matrix whose matrix elements are given by

$$\mathscr{L}_{ii} = 1 - \sum_{j \neq i} w_{i \to j} \Delta t \ (\geq 0), \tag{3.40}$$

$$\mathscr{L}_{ij} = w_{j \to i} \Delta t \ (\geq 0), \qquad (j \neq i). \tag{3.41}$$

Note that $\mathscr{L}$ is a non-negative matrix. Since we consider the Markov process, the probability distribution at $t + n\Delta t$ is represented as

$$\vec{P}(t + n\Delta t) = \mathscr{L} \vec{P}(t + (n-1)\Delta t) = \cdots = \mathscr{L}^n \vec{P}(t). \tag{3.42}$$

Next, in order to examine how to reach the stable state, the matrix $\mathscr{L}$ is analyzed. Let λ_m and $\vec{\phi}_m$ $(m = 1, \cdots, M)$ be the right-hand eigenvalues and corresponding eigenvectors, respectively. That is, the following relations are satisfied:

$$\mathscr{L} \vec{\phi}_m = \lambda_m \vec{\phi}_m. \tag{3.43}$$

Using the right-hand eigenvalues and eigenvectors, the probability distribution at time $n\Delta t$ is expressed as

$$\vec{P}(n\Delta t) = \sum_{m=1}^{M} C_m(n\Delta t) \vec{\phi}_m = \sum_{m=1}^{M} C_m(0) \lambda_m^n \vec{\phi}_m. \tag{3.44}$$

Because the normalization is given by Eq. (3.35), the following relation is satisfied:

$$\sum_{m=1}^{M} C_m(0) \lambda_m^n \sum_{i=1}^{M} \left(\vec{\phi}_m\right)_i = 1. \tag{3.45}$$

The left-hand side of Eq. (3.44) does not depend on n in the long time limit, $\lambda_{\max}(=1)$, where $\lambda_{\max}$ is the maximum value of λ_m, i.e., $\lambda_{\max} = \max\{\lambda_m | 1 \le m \le M\}$. Let $\vec{\phi}_{\mathrm{st}}$ be the corresponding eigenvector of the eigenvalue $\lambda_{\max}$. Thus, from Eq. (3.45), we obtain

$$\sum_{i=1}^{M} \left(\vec{\phi}_{\mathrm{st}}\right)_i = 1. \tag{3.46}$$

Here we used the fact that $0 \le \lambda_m < 1$ for $\lambda_m \neq \lambda_{\max}$, which can be proved by the Perron–Frobenius theorem. For any eigenvector except $\vec{\phi}_{\mathrm{st}}$,

$$\sum_{i=1}^{M} \left(\vec{\phi}_m\right)_i = 0. \tag{3.47}$$

Thus, the dynamics of the probability distribution is represented by

$$\vec{P}(n\Delta t) = \vec{\phi}_{\mathrm{st}} + \sum_m C_m(0) \mathrm{e}^{-n|\ln \lambda_m|} \vec{\phi}_m, \tag{3.48}$$

where the sum runs over $1 \le m \le M$ for $\lambda_m \neq \lambda_{\max}$. From Eq. (3.48), the dominant relaxation time is expressed as

$$\tau = \frac{1}{|\ln \lambda_{\mathrm{second}}|}, \tag{3.49}$$

where $\lambda_{\mathrm{second}}$ represents the second largest eigenvalue.

Actual algorithms

In Monte Carlo methods, we prepare the transition probability so that $\vec{\phi}_{\mathrm{st}} = \vec{P}_{\mathrm{eq}}$, where $\vec{P}_{\mathrm{eq}}$ is the probability distribution in the equilibrium state. $\vec{P}_{\mathrm{eq}}$ is given by

$$\vec{P}_{\mathrm{eq}} = \frac{1}{Z} \begin{pmatrix} \mathrm{e}^{-\beta \mathscr{H}(\Sigma_1)} \\ \mathrm{e}^{-\beta \mathscr{H}(\Sigma_2)} \\ \vdots \\ \mathrm{e}^{-\beta \mathscr{H}(\Sigma_M)} \end{pmatrix}, \tag{3.50}$$

where Z is the partition function which is the normalization factor of the probability distribution. Since $\vec{P}_{\mathrm{eq}}$ is the equilibrium probability distribution, the following relation is satisfied:

$$\mathscr{L} \vec{P}_{\mathrm{eq}} = \vec{P}_{\mathrm{eq}}. \tag{3.51}$$

Using Eqs. (3.40) and (3.41), we obtain

$$\sum_{j=1}^{M} \mathscr{L}_{ij} \left(\vec{P}_{\mathrm{eq}}\right)_j = \left(\vec{P}_{\mathrm{eq}}\right)_i \tag{3.52}$$

$$\Rightarrow \sum_{j=1}^{M} (w_{j\to i}\Delta t) \left(\vec{P}_{\mathrm{eq}}\right)_j = \left(\vec{P}_{\mathrm{eq}}\right)_i \tag{3.53}$$

$$\Rightarrow \sum_{j\neq i} w_{j\to i}\Delta t \left(\vec{P}_{\mathrm{eq}}\right)_j + \left(1 - \sum_{j\neq i} w_{i\to j}\Delta t\right) \left(\vec{P}_{\mathrm{eq}}\right)_i = \left(\vec{P}_{\mathrm{eq}}\right)_i. \tag{3.54}$$

From this equation, the following relation called the balance condition is derived:

$$\sum_{j\neq i} w_{j\to i} \left(\vec{P}_{\mathrm{eq}}\right)_j = \sum_{j\neq i} w_{i\to j} \left(\vec{P}_{\mathrm{eq}}\right)_i. \tag{3.55}$$

The following condition called the detailed balance condition (DBC) is a sufficient condition to satisfy Eq. (3.55):

$$w_{j\to i} \left(\vec{P}_{\mathrm{eq}}\right)_j = w_{i\to j} \left(\vec{P}_{\mathrm{eq}}\right)_i. \tag{3.56}$$

Since the equilibrium probability distribution is given by Eq. (3.50),

$$\frac{w_{i\to j}}{w_{j\to i}} = \mathrm{e}^{-\beta\left(\mathscr{H}(\Sigma_i)-\mathscr{H}(\Sigma_j)\right)} \tag{3.57}$$

is obtained.

There are two famous and well-used ways to determine $w_{i\to j}$: the heat-bath method and the Metropolis method. The heat-bath method is expressed by

$$w_{i\to j} = \frac{\mathrm{e}^{-\beta\mathscr{H}(\Sigma_j)}}{\mathrm{e}^{-\beta\mathscr{H}(\Sigma_i)} + \mathrm{e}^{-\beta\mathscr{H}(\Sigma_j)}}, \tag{3.58}$$

and the Metropolis method is expressed by

$$w_{i\to j} = \begin{cases} \mathrm{e}^{-\beta\left(\mathscr{H}(\Sigma_i)-\mathscr{H}(\Sigma_j)\right)} & (\mathscr{H}(\Sigma_j) > \mathscr{H}(\Sigma_i)) \\ 1 & (\mathscr{H}(\Sigma_j) \leq \mathscr{H}(\Sigma_i)) \end{cases}. \tag{3.59}$$

In actual simulations in the Monte Carlo method for N spin systems, the following procedures are performed:

Step 1 Prepare the initial spin state.

Step 2 Choose one spin from N spins randomly or sequentially.

step 3 Update the chosen spin according to the transition probability which satisfies Eq. (3.55). Well-known methods are given by Eq. (3.58) (heat-bath method) and Eq. (3.59) (Metropolis method).

Step 4 The procedures of Steps 2–3 are repeatedly performed N times, which is called one Monte Carlo step.

Step 5 If equilibrium physical quantities are calculated, we continue Steps 2–4 the until physical quantities converge. If dynamical properties are calculated, we continue Steps 2–4 while changing the external parameters, such as the temperature and the magnetic field.

Here, the DBC is imposed. The DBC means that every elementary transition balances with the corresponding inverse transition. When we impose the DBC, transition probability can be obtained easily. In fact, to reduce the autocorrelation time, some methods within the limit of the DBC were developed (Liu, 1996; Pollet et al., 2004). It should be noted that, however, the DBC is not a necessary condition but only a sufficient condition to realize the Monte Carlo methods. Recently, based on the fact, novel methods were proposed (Suwa and Todo, 2010; Ichiki and Ohzeki, 2013, 2015; Ohzeki and Ichiki, 2015). On top of that, the microscopic analysis to clarify the reason why these methods without DBC are efficient was reported (Sakai and Hukushima, 2013; Takahashi and Ohzeki, 2016).

3.3.2 Mean-field methods

In this subsection, we explain how to treat the Ising model with random interactions and inhomogeneous magnetic fields in terms of mean-field analysis. The Hamiltonian of the model we considered is

$$\mathscr{H} = -\sum_{\langle i,j\rangle} J_{ij}\sigma_i^z\sigma_j^z - \sum_{i=1}^{N} h_i\sigma_i^z, \qquad (\sigma_i^z = \pm 1), \tag{3.60}$$

where $\langle i, j\rangle$ represents the nearest-neighbor pairs on a given lattice. The density matrix of the system in the equilibrium state at a finite temperature T is expressed as

$$\rho = \frac{1}{Z}\mathrm{e}^{-\beta\mathscr{H}}, \qquad Z = \mathrm{Tr}\,\mathrm{e}^{-\beta\mathscr{H}}, \tag{3.61}$$

where β is the inverse temperature: $\beta = 1/T$ and Z is the partition function. Here we set the Boltzmann constant $k_\mathrm{B} = 1$ for simplicity. In the Bethe–Peierls–Weiss approximation, i.e., mean-field approximation, we assume that the density matrix of the total system can be expressed by the product of decoupled local density matrices:

$$\rho \sim \rho_{\mathrm{MF}} = \prod_{i=1}^{N} \rho_i, \tag{3.62}$$

where ρ_{MF} is the approximated density matrix and ρ_i is the local density matrix at site i. Here, local density matrices are normalized as

$$\mathrm{tr}_i \rho_i = 1, \tag{3.63}$$

where tr_i denotes the operator to trace out the degrees of freedom at site i. Using the local density matrix, the local magnetization is defined by

$$m_i := \mathrm{Tr}\left(\sigma_i^z \rho_{\mathrm{MF}}\right) = \mathrm{tr}_i\left(\sigma_i^z \rho_i\right). \tag{3.64}$$

A trial variational free energy is given by

$$\tilde{F} = \mathrm{Tr}\left(\mathscr{H} \rho_{\mathrm{MF}}\right) - TS_{\mathrm{MF}} \tag{3.65}$$

$$= -\sum_{\langle i,j \rangle} J_{ij} m_i m_j - \sum_{i=1}^{N} h_i m_i + T\,\mathrm{Tr}\,\rho_{\mathrm{MF}} \ln \rho_{\mathrm{MF}} \tag{3.66}$$

$$= -\sum_{\langle i,j \rangle} J_{ij} m_i m_j - \sum_{i=1}^{N} h_i m_i + T \sum_{i=1}^{N} \mathrm{tr}_i \rho_i \ln \rho_i. \tag{3.67}$$

Here we used the relation:

$$\mathrm{Tr}\left(\rho_{\mathrm{MF}} \sum_{i=1}^{N} \ln \rho_i\right) = \sum_{i=1}^{N} \mathrm{tr}_i \rho_i \ln \rho_i. \tag{3.68}$$

The Gibbs–Bogoliubov–Feynman inequality is given by $\tilde{F} \geq F$, where F is the free energy of the original Hamiltonian defined by Eq. (3.60). Thus, using the Lagrange multiplier λ and applying the variational method to

$$\tilde{F} - \lambda(\rho_{\mathrm{MF}} - 1), \tag{3.69}$$

the local density matrix can be obtained as

$$\rho_i = \frac{1}{2\cosh \beta \sum_{j \in \partial i} (J_{ij} m_j + h_i)} \times \begin{pmatrix} \exp\left[\beta\left(\sum_{j\in\partial i} J_{ij} m_j + h_i\right)\right] & 0 \\ 0 & \exp\left[-\beta\left(\sum_{j\in\partial i} J_{ij} m_j + h_i\right)\right] \end{pmatrix}, \tag{3.70}$$

where ∂i is the set of nearest-neighbor sites of site i. Using Eq. (3.64), the local magnetization is given by

$$m_i = \mathrm{tr}_i\left(\sigma_i^z \rho_i\right) = \tanh\left[\beta\left(\sum_{j\in\partial i} J_{ij}m_j + h_i\right)\right], \tag{3.71}$$

which is called the self-consistent equation. We can obtain the stable or metastable states by solving the self-consistent equation.

In actual simulations in the mean-field calculation, the following procedures are performed:

Step 1 Decide N independent spin variables.

Step 2 Prepare the initial values of local magnetization $\{m_i\}_{i=1,\cdots,N}$.

Step 3 Calculate $\{m_i\}_{i=1,\cdots,N}$ by solving the self-consistent equations defined by Eq. (3.71).

Step 4 By using the solution of $\{m_i\}_{i=1,\cdots,N}$, the physical quantities can be calculated, e.g., the internal energy is obtained by

$$E = -\sum_{\langle i,j\rangle} J_{ij}m_im_j - \sum_{i=1}^{N} h_im_i. \tag{3.72}$$

The obtained local magnetizations are continuous values $(-1 \leq m_i \leq 1)$. If we would like to obtain a typical snapshot, we can decide the spin configuration by the following equation:

$$\sigma_i^z = \mathrm{sign}[m_i]. \tag{3.73}$$

3.3.3 Preceding theoretical studies on simulated annealing

Huse–Fisher (1986)

To estimate the time required for equilibration of strongly correlated systems such as spin glasses is an important topic in statistical physics and condensed matter physics. Of course, to quantify the performance of simulated annealing, the estimation of the equilibration time is needed. The equilibration times in random Ising models with frustration exceed the time scales of the experiment at low temperatures. Thus, we cannot directly observe the equilibrium state at low temperatures and should get some information from the nonequilibrium behavior. Thus, it is significant to consider nonequilibrium behaviors under decreasing temperature. Here, we review the study performed by Huse and Fisher (Huse and Fisher, 1986).

They considered the internal energy in quenched random Ising spin models. We decrease the temperature from high temperature to zero temperature in the finite time τ.

The single-spin flip was adopted as the rule of dynamics. They focused on the residual energy per spin defined by

$$\varepsilon(\tau) := \langle E \rangle_\tau - E_0, \tag{3.74}$$

where $\langle E \rangle_\tau$ is the expectation value of the internal energy per spin at τ, i.e., the final state after cooling from high temperature to zero temperature. E_0 represents the exact ground state energy per spin. Huse and Fisher concluded that the residual energy per spin for large τ for frustrated systems obeys

$$\varepsilon(\tau) = (\ln \tau)^{-\zeta}, \tag{3.75}$$

where the exponent ζ depends on the detail of systems and is expected to be $\zeta \leq 2$ as will be described here. It was emphasized that the result is independent regardless of the complexity of problems, which is the opposite claim with the result reported by Grest et al. (Grest et al., 1986).

Before we review the study by Huse and Fisher, we show the study done by Grest et al. (Grest et al., 1986). Grest et al. considered the residual energy in random Ising models by using Monte Carlo simulations. They concluded that $\varepsilon(\tau) \sim (\ln \tau)^{-1}$ for systems in which the ground state cannot be obtained easily, i.e., NP-complete problems. Besides, they found that $\varepsilon(\tau)$ is an inverse power of τ for systems where the ground state can be obtained easily, i.e., P-class problems. As mentioned earlier, Huse and Fisher, however, pointed out that the result reported by Grest et al. is incorrect.

In their study, Huse and Fisher (1986) first considered two-level systems in which the energy difference between each state is given by Δ and an energy barrier B exists between the two states (see Fig. 3.5).

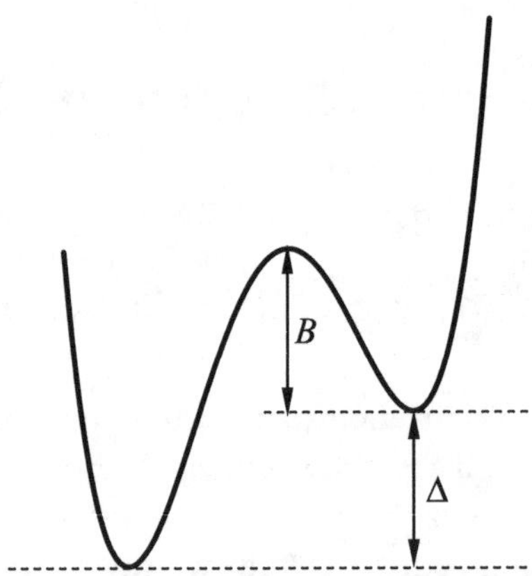

Fig. 3.5 Two-level systems considered by Huse and Fisher (1986).

Let $p(t)$ be the probability of excited state (the right valley in Fig. 3.5) at time t. The master equation is given by

$$\gamma^{-1}\frac{dp(t)}{dt} = [1-p(t)]\exp[-(\Delta+B)/T(t)] - p(t)\exp[-B/T(t)], \quad (0 \leq t \leq \tau), \qquad (3.76)$$

where γ is the attempt frequency and $T(t)$ is the temperature at time t. Here $T(0)$ is a high temperature compared with B and Δ, and $T(\tau) = 0$. Since $T(0)$ is large enough, the initial probability is $p(0) = 1/2$.

The system cannot keep the equilibrium state at temperature T^* such that $\gamma e^{-B/T^*} = 1/\tau$. When the temperature decreases below T^*, the dynamics is almost stopped, in other words, the state freezes. Thus, the probability of the excited state at τ is expressed as

$$p(\tau) \sim \exp\left(-\frac{\Delta}{T^*}\right) = (\gamma\tau)^{-\Delta/B}. \qquad (3.77)$$

This result can be obtained by the asymptotic analysis of the master equation in Eq. (3.76). Using $p(\tau)$, the residual energy can be expressed as

$$\varepsilon(\tau) = \Delta p(\tau) = \Delta(\gamma\tau)^{-\Delta/B}. \qquad (3.78)$$

Under the fixed B and τ, the energy gap Δ which maximizes the residual energy is given by

$$\Delta = \frac{B}{\ln(\gamma\tau)}. \qquad (3.79)$$

Here we used the condition $d\varepsilon(\tau)/d\Delta = 0$. This result means that the dynamics of two-level systems with a small Δ are dominant for large τ.

Quenched random Ising models with frustration include two-level systems, which consist of small clusters. When the distribution function of randomness is continuous, there are no exact degeneracies and excitations in the limit of $\Delta \to 0$. The density of states is ideally positive constant for small Δ. The aforementioned picture relates to the droplet picture in spin glasses, which was proposed by Fisher and Huse (Fisher and Huse, 1986).

In their study, Huse and Fisher considered the density of two-level systems with energy gap Δ and energy barrier B, $\Phi(\Delta, B)$. The residual energy density is given by

$$\varepsilon(\tau) = \int_0^\infty dB \int_0^\infty d\Delta \Phi(\Delta, B)\Delta p(\tau; \Delta, B), \qquad (3.80)$$

where $p(\tau; \Delta, B)$ is the probability of an excited state in the system with energy gap Δ and energy barrier B. $\Phi(\Delta, B)$ should be smoother than $\Delta p(\tau; \Delta, B)$, and $\Delta p(\tau; \Delta, B)$ has a local maximum at $\Delta = B/\ln(\gamma\tau)$. Thus, $\varepsilon(\tau)$ can be expressed by

$$\varepsilon(\tau) \sim \int_0^\infty dB\Phi\left[\frac{B}{\ln(\gamma\tau)}, B\right]\int_0^\infty d\Delta\Delta p(\tau;\Delta,B). \tag{3.81}$$

For large τ, the following approximation is satisfied:

$$\Phi\left[\frac{B}{\ln(\gamma\tau)}, B\right] \sim \Phi(0,B). \tag{3.82}$$

Thus, the residual energy is expressed as

$$\varepsilon(\tau) \sim \frac{c}{[\ln(\gamma\tau)]^2}\int_0^\infty dB\Phi(0,B), \tag{3.83}$$

where c depends on the schedule of decreasing the temperature and $c = \pi^2/12$ when the temperature decreases linearly. From this fact, $\zeta \leq 2$ in Eq. (3.75) is expected when the distribution function of randomness is continuous.

Huse and Fisher studied concrete examples of quenched random Ising models in d-dimension and found the exponent of the residual energy ζ. The obtained results (Huse and Fisher, 1986) are as follows:

- Spin glasses with $T_c > 0$: $\zeta = \dfrac{d-\theta}{d_s/2+\phi-\theta}$
- Spin glasses with $T_c = 0$: $\zeta = \min\left\{2, \dfrac{2(d-\theta)}{d_s/2+2\phi-\theta}\right\}$
- Random field Ising model: $\zeta = 1$
- Bond disordered ferromagnets: $\zeta = 2$ or an inverse power of ζ

Here, d_s is the fractal dimension of the domain wall and ϕ is the exponent of the energy barrier of the effective two-level systems which represents the droplet excitation, say, L^ϕ for large L. θ introduced by Fisher and Huse (1986) is the exponent of the density of states at zero energy for coherently flipped spins (droplet), that is, $L^{-\theta}$.

Shinomoto–Kabashima model (1991)

Shinomoto and Kabashima investigated a finite-time scaling of the internal energy in simulated annealing (Shinomoto and Kabashima, 1991). They found that the relation between the annealing time τ and the internal energy at τ, $E(\tau)$, is given by

$$E(\tau) = \varepsilon + c(\ln\tau)^{-1}, \tag{3.84}$$

where c is a constant. Note that ε is not necessary for the ground-state energy. They constructed a model to obtain analytically the finite-time scaling of the internal energy. The model they considered is a one-dimensional system with a large number of local minima shown in Fig. 3.6.

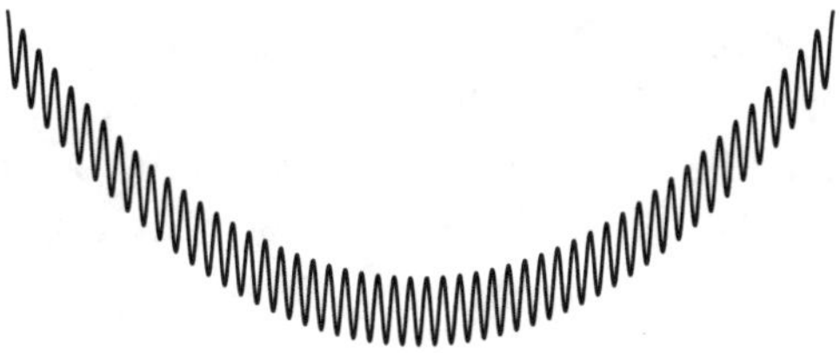

Fig. 3.6 Energy landscape of the model considered by Shinomoto and Kabashima (1991).

Shinomoto and Kabashima (1991) studied the random walk in their model. The system has two energy scales; the first one is the energy barriers of $\mathcal{O}(B)$ surrounding each local minimum, and the second one is a larger basin $E = x^2/2$, where x is the one-dimensional distance. They assumed that the local minima are arranged at periodic intervals of distance a and the position of the ith local minimum is $x_i = ai$. The master equation representing the random walk in the energy landscape at temperature T is given by

$$\frac{dP_i}{dt} = \mathrm{e}^{-B/T} P_{i+1} + \mathrm{e}^{-(B+\Delta_{i-1})/T} P_{i-1} + [e^{-B/T} + \mathrm{e}^{-(B+\Delta_i)/T}] P_i, \tag{3.85}$$

where

$$\Delta_i = \frac{a^2}{2}[(i+1)^2 - i^2], \tag{3.86}$$

and P_i is the probability of the existence of the particle, and Δ_i is the energy difference between adjacent local minima. Under the assumption that $a^2/T \ll 1$ which is valid at high temperatures, from Eq. (3.85), the Fokker–Planck equation is derived:

$$\frac{\partial P(x,t)}{\partial t} = \gamma(T)\frac{\partial x P(x,t)}{\partial x} + D(T)\frac{\partial^2 P(x,t)}{\partial x^2}, \tag{3.87}$$

$$\gamma(T) = \frac{a^2 \mathrm{e}^{-B/T}}{T}, \quad D(T) = T\gamma(T). \tag{3.88}$$

Using the Fokker–Planck equation, the time evolution of the average of the internal energy at t can be calculated as:

$$\langle E(t) \rangle = \int dx \frac{x^2}{2} P(x,t). \tag{3.89}$$

As a result, the internal energy scales as $\langle E(\tau) \rangle \sim B/\ln\tau$ in the asymptotic limit. Thus, the optimal annealing schedule is then $T(t) \sim B/\ln t$, which is consistent with the lower bound obtained by Geman and Geman (Geman and Geman, 1984) (see next section).

Shinomoto and Kabashima (1991) also studied the case that the larger basin is given by $E = x^{\alpha}/\alpha$ for $\alpha > 1$. When $\alpha \neq 2$, the solution of the differential equation of the average of the internal energy cannot be obtained in the closed form. According to the asymptotic analysis, the same asymptotic form as in the case of $\alpha = 2$, that is, $\langle E(\tau)\rangle \sim B/\ln\tau$ is obtained. The relation is satisfied even when $E = x^{\alpha}/\alpha$ without cusp or $\alpha > 1$.

3.4 Convergence Theorem for Simulated Annealing

The convergence theorem for simulated annealing proved by Geman and Geman (Geman and Geman, 1984) is given by

$$T(t) \geq \frac{\Delta}{\log(t+2)}, \tag{3.90}$$

where Δ is a positive constant which is proportional to the number of spins N. When we decrease the temperature $T(t)$ by the schedule satisfying this convergence theorem, the ground state at $T(t) = 0$ is definitely obtained in the limit of $t \to \infty$ in simulated annealing. Let us prove the Geman–Geman theorem in this section.

3.4.1 Inhomogeneous Markov chain

Suppose that the stochastic dynamics based on Monte Carlo methods realizes the time evolution. In simulated annealing, the Markov chain generates states and the transition probability changes at each time. This the temperature changes at each time, which is called the inhomogeneous Markov chain. Our purpose is to consider how to converge the probability distribution in simulated annealing. Thus, the convergence behavior in the inhomogeneous Markov chain is considered.

Let $\mathscr{S}$ be the set of all possible states. We first define the transition probability. The transition probability from a state x to another state y at time t, $G(y,x;t)$, is defined by

$$G(y,x;t) = \begin{cases} P(y,x)A(y,x;t) & (x \neq y) \\ 1 - \sum_{z\in\mathscr{S}} P(z,x)A(z,x;t) & (x = y) \end{cases}, \tag{3.91}$$

where $P(y,x)$ is the generation probability of the state y from the state x. Here, $A(y,x;t)$ is the acceptance probability where the transition is from the state x to the state y. The generation probability $P(y,x)$ has the following properties:

$$\forall x,y \in \mathscr{S}, \quad P(y,x) = P(x,y) \geq 0, \tag{3.92}$$

$$\forall x \in \mathscr{S}, \quad P(x,x) = 0, \tag{3.93}$$

$$\forall x \in \mathscr{S}, \quad \sum_{z\in\mathscr{S}} P(z,x) = 1. \tag{3.94}$$

Assume that an arbitrary state can be generated by the transition from another arbitrary state with finite steps. The acceptance probability is decided by an update rule such as the heat-bath method and the Metropolis method. Let $\mathscr{G}(t)$ be the transition matrix whose matrix elements are given by the transition probability $G(y,x;t)$.

Next, let $\mathscr{P}$ be the whole space of the probability distribution of each state. The probability of the state x at $t = t_0$ is given by $q(x;t_0) \in \mathscr{P}$. $\mathscr{Q}(t_0)$ is the vertical vector whose elements are given by $q(x;t_0)$. $q(x;t,t_0)$ is the probability of the state x generated by the inhomogeneous Markov chain at time t, where t_0 is the initial time. $\mathscr{Q}(t,t_0)$ is the vertical vector whose elements are given by $q(x;t,t_0)$, which represents the probability distribution. Assume that the time advances from t_0 to t_1 in the single step and $\Delta t := t_1 - t_0$. The probability distribution is given by

$$\mathscr{Q}(t_1,t_0) = \mathscr{G}(t_0)\mathscr{Q}(t_0). \tag{3.95}$$

Using $\mathscr{G}(t)$, the probability distribution $\mathscr{Q}(t,t_0)$ for arbitrary time $t = n\Delta t$, where n the arbitrary integer can be written by

$$\mathscr{Q}(t,t_0) = \mathscr{G}(t-\Delta t)\mathscr{G}(t-2\Delta t)\cdots\mathscr{G}(t_0)\mathscr{Q}(t_0) \tag{3.96}$$

$$=: \mathscr{G}^{t,t_0}\mathscr{Q}(t_0). \tag{3.97}$$

Here $\mathscr{G}^{t,t_0}$ is the transition matrix when the time advances from t_0 to t. In simulated annealing, the time evolution realizes by the inhomogeneous Markov chain defined earlier.

3.4.2 Ergodicity

When the Markov chain realizes the time evolution of the probability distribution, ergodicity guarantees that the probability distribution converges to the unique probability distribution in the limit of $t \to \infty$. Weak ergodicity and strong ergodicity will be explained here. In simulated annealing, the Markov chain satisfying strong ergodicity should be used.

Weak ergodicity

Weak ergodicity means that the probability distribution generated in the limit of $t \to \infty$ does not depend on the initial probability distribution, that is,

$$\lim_{t\to\infty} \sup\{\|\mathscr{Q}(t,t_0) - \mathscr{Q}'(t,t_0)\| \mid \mathscr{Q}(t_0), \mathscr{Q}'(t_0) \in \mathscr{P}\} = 0. \tag{3.98}$$

$\mathscr{Q}(t,t_0)$ and $\mathscr{Q}'(t,t_0)$ are the probability distributions generated from the initial probability distributions $\mathscr{Q}(t_0)$ and $\mathscr{Q}'(t_0)$, respectively. The norm of Eq. (3.98) is defined by

$$\|\mathscr{Q}(t,t_0) - \mathscr{Q}'(t,t_0)\| = \sum_{x\in\mathscr{S}} |q(x;t,t_0) - q'(x;t,t_0)|. \tag{3.99}$$

The necessary and sufficient condition that an inhomogeneous Markov chain is weak ergodicity is the following. A positive monotonically increasing sequence $\{t_k\}$ $(k = 0, 1, 2, \cdots)$ exists and the relation

$$\sum_{k=0}^{\infty} \left(1 - \alpha\left(\mathscr{G}^{t_{k+1},t_k}\right)\right) \to \infty \tag{3.100}$$

is satisfied. Here, $\alpha(\mathscr{G}^{t_{k+1},t_k})$ is called the ergodic coefficient defined by

$$\alpha\left(\mathscr{G}^{t,t'}\right) = 1 - \min_{x,y\in\mathscr{S}} \left\{ \sum_{z\in\mathscr{S}} \min\left\{\mathscr{G}^{t,t'}(z,x), \mathscr{G}^{t,t'}(z,y)\right\}\right\}, \tag{3.101}$$

where $\mathscr{G}^{t,t'}(z,x)$ is the (z,x)-element of the matrix $\mathscr{G}^{t,t'}$.

Strong ergodicity

Strong ergodicity means that the probability distribution generated by the Markov chain converges to the unique distribution in the limit of $t \to \infty$ for an arbitrary initial probability distribution. Strong ergodicity is given by

$$\lim_{t\to\infty} \sup\{\|\mathscr{Q}(t,t_0) - r\| \mid \mathscr{Q}(t_0) \in \mathscr{P}\} = 0, \tag{3.102}$$

where r is a unique probability distribution. The necessary and sufficient conditions that the inhomogeneous Markov chain is strong ergodicity are as follows:

1. The inhomogeneous Markov chain is weak ergodicity.
2. A steady distribution exists at each time t. That is, the probability distribution $\mathscr{Q}_t$ exists satisfying

$$\mathscr{Q}_t = \mathscr{G}(t)\mathscr{Q}_t. \tag{3.103}$$

3. The steady distribution $\mathscr{Q}_t$ is satisfied by the following relation:

$$\sum_{t=0}^{\infty} \|\mathscr{Q}_{t+1} - \mathscr{Q}_t\| < \infty. \tag{3.104}$$

Here the initial time is set to $t_0 = 0$.

3.4.3 Proof of the convergence theorem

Acceptance probability in simulated annealing

Herein we consider the Ising model. The Hamiltonian of the Ising model with random interactions and magnetic fields is defined as

$$\mathscr{H} = -\sum_{1\le i,j\le N} J_{ij}\sigma_i^z\sigma_j^z - \sum_{i=1}^{N} h_i\sigma_i^z. \tag{3.105}$$

The partition function of the Hamiltonian is expressed as

$$\mathscr{Z}(t) = \sum_{x \in \mathscr{S}} \exp\left[-\frac{E(x)}{T(t)}\right], \tag{3.106}$$

where x is a state of all spin variables $\{\sigma_i^z\}$ and the internal energy for the given state x is defined by

$$E(x) := -\sum_{1 \le i,j \le N} J_{ij}\sigma_i^z\sigma_j^z - \sum_{i=1}^{N} h_i\sigma_i^z. \tag{3.107}$$

Here, $T(t)$ is the time-dependent temperature at time t. In simulated annealing, $T(t)$ monotonically decreases with t and $T(t) = 0$ in the limit of $t \to \infty$. The acceptance probability $A(y,x;t)$ at time t is defined by

$$A(y,x;t) = f\left(\frac{g(y;t)}{g(x;t)}\right), \tag{3.108}$$

$$g(x;t) = \frac{1}{\mathscr{Z}(t)} \exp\left[-\frac{E(x)}{T(t)}\right], \tag{3.109}$$

where $g(x;t)$ is the equilibrium Boltzmann distribution at the instantaneous temperature $T(t)$. $f(u)$ in Eq. (3.109) is defined for $u \ge 0$ as follows:

$$f(u) = \frac{u}{1+u} \quad : \text{Heat-bath method}, \tag{3.110}$$

$$f(u) = \min\{1,u\} \quad : \text{Metropolis method}, \tag{3.111}$$

where $f(u)$ satisfies

$$f(1/u) = f(u)/u. \tag{3.112}$$

$f(u)$ is a monotonically increasing function and depends on the update rule, i.e., spin-flip rule. Moreover, the definition of $f(u)$ leads to the steady distribution $g(x;t)$ generated by the transition probability $G(y,x;t)$. Thus, for a fixed time t, the following relation is satisfied:

$$g(y;t) = \sum_{x \in \mathscr{S}} G(y,x;t)g(x;t). \tag{3.113}$$

Simulated annealing is performed by the inhomogeneous Markov chain by using the transition probability given by Eq. (3.91).

Confirmation for ergodicity

The sufficient condition for the convergence of simulated annealing is given by Eq. (3.90). Under this condition, since the inhomogeneous Markov chain has strong ergodicity, the probability distribution converges to the unique distribution in the limit of $t \to \infty$.

Definitions of states and quantities

To prove the convergence theorem, we define the set of states. Let $\mathscr{S}_x$ be the set of the states that can be generated by the single step from the state x. $\mathscr{S}_x$ is written as

$$\mathscr{S}_x = \{y | y \in \mathscr{S}, P(y,x) > 0\}, \tag{3.114}$$

where $\mathscr{S}$ is the set of the states. Let $\mathscr{S}_{\rm m}$ be the set of x such that $E(x) \geq E(y)$ for all states $y \in \mathscr{S}_x$. $\mathscr{S}_{\rm m}$ is written as

$$\mathscr{S}_{\rm m} = \{x | x \in \mathscr{S}, \forall y \in \mathscr{S}_x, E(x) \geq E(y)\}. \tag{3.115}$$

This is called the maximum state set of $E(x)$.

Next we define some quantities. Let $d(y,x)$ be the minimum steps which are needed to transition from the state x to the state y. Let $\mathscr{S} \backslash \mathscr{S}_{\rm m}$ be the set obtained by removing $\mathscr{S}_{\rm m}$ from $\mathscr{S}$. $D(x)$ is the maximum value of $d(y,x)$ for the states $x \in \mathscr{S} \backslash \mathscr{S}_{\rm m}$. $D(x)$ is the necessary steps that are required to transition from the state x, except the maximum state set of $E(x)$, to an arbitrary state. Let R be the minimum value of $D(x)$, that is,

$$R = \min_{x \in \mathscr{S} \backslash \mathscr{S}_{\rm m}} \left\{ \max_{y \in \mathscr{S}} [d(y,x)] \right\}. \tag{3.116}$$

The state x^* is defined by

$$x^* = \arg \min_{x \in \mathscr{S} \backslash \mathscr{S}_{\rm m}} \left\{ \max_{y \in \mathscr{S}} [d(y,x)] \right\}. \tag{3.117}$$

Arbitrary states can be generated from the states $x^* \in \mathscr{S} \backslash \mathscr{S}_{\rm m}$ by at most R steps. Here R increases with N linearly. In other words, the states $x^* \in \mathscr{S} \backslash \mathscr{S}_{\rm m}$ can be generated from an arbitrary state by at most R steps. R and x^* are used to estimate the ergodic coefficient.

In order to represent the change of $E(x)$ in the single step, L is introduced as follows:

$$L := \max\{|E(y) - E(x)| \mid x, y \in \mathscr{S}, P(y,x) > 0\}. \tag{3.118}$$

That is, the change of $E(x)$ is smaller than or equal to L. The non-zero minimum value of the generation probability $P(y,x)$ is defined by

$$w := \min\{P(y,x) \mid x, y \in \mathscr{S}, P(y,x) > 0\}. \tag{3.119}$$

Lower bound of the transition probability

Here we find the lower bound of the transition probability $G(y,x;t)$, which will be used in the proof of the convergence theorem. The lower bound of $G(y,x;t)$ is given by

- For $x, y \in \mathcal{S}$, where $P(y,x) > 0$,

$$G(y,x;t) \geq wf(1)\exp\left[-\frac{L}{T(t)}\right], \quad (t > 0). \tag{3.120}$$

- For $x \in \mathcal{S} \backslash \mathcal{S}_{\mathrm{m}}$,

$$G(x,x;t) \geq wf(1)\exp\left[-\frac{L}{T(t)}\right], \quad (t \geq t_1). \tag{3.121}$$

Here the initial time is set to $t = 0$, and t_1 will be introduced later.

Let us derive Eq. (3.120). When $x \neq y$, the transition probability is written as

$$G(y,x;t) = P(y,x)A(y,x;t) \tag{3.122}$$

$$= P(y,x)f\left(\exp\left[-\frac{E(y)-E(x)}{T(t)}\right]\right) \tag{3.123}$$

$$= P(y,x)f\left(\mathrm{e}^{-a}\right), \tag{3.124}$$

where a is defined by

$$a := \frac{E(y)-E(x)}{T(t)}. \tag{3.125}$$

For $a \geq 0$, $G(y,x;t)$ becomes

$$G(y,x;t) \geq wf\left(\mathrm{e}^{-a}\right) \tag{3.126}$$

$$= wf\left(\mathrm{e}^{a}\right)\mathrm{e}^{-a} \tag{3.127}$$

$$\geq wf(1)\mathrm{e}^{-a} \tag{3.128}$$

$$= wf(1)\exp\left[-\frac{E(y)-E(x)}{T(t)}\right] \tag{3.129}$$

$$\geq wf(1)\exp\left[-\frac{L}{T(t)}\right]. \tag{3.130}$$

On the other hand, for $a<0$, $G(y,x;t)$ becomes

$$G(y,x;t) \geq wf\left(\mathrm{e}^{-a}\right) \tag{3.131}$$

$$\geq wf(1) \tag{3.132}$$

$$\geq wf(1)\exp\left[-\frac{L}{T(t)}\right]. \tag{3.133}$$

Thus, Eq. (3.120) is satisfied for arbitrary a.

Next we derive Eq. (3.121). For the states $x \in \mathscr{S}\backslash\mathscr{S}_{\mathrm{m}}$, the states $y \in \mathscr{S}_x$ so that $E(y)-E(x)>0$ exist. If this is the case, for the states y, we obtain

$$\lim_{t\to\infty}\exp\left[-\frac{E(y)-E(x)}{T(t)}\right]=0. \tag{3.134}$$

Here, we used the fact that $T(t)\to 0$ in the limit of $t\to\infty$. Using the relation given by

$$f(u)=uf(1/u)\leq u, \tag{3.135}$$

we obtain

$$\lim_{t\to\infty}A(y,x;t)=\lim_{t\to\infty}f\left(\exp\left[-\frac{E(y)-E(x)}{T(t)}\right]\right) \tag{3.136}$$

$$\leq \lim_{t\to\infty}\exp\left[-\frac{E(y)-E(x)}{T(t)}\right] \tag{3.137}$$

$$=0. \tag{3.138}$$

To satisfy the relation, it is necessary that the acceptance probability $A(y,x;t)$ becomes an infinitely small positive value δ at a certain time $t(\geq t_1)$. That is, the time t_1 exists satisfying

$$A(y,x;t)\leq\delta, \quad (t\geq t_1). \tag{3.139}$$

Here we consider at $t\geq t_1$. According to the relation, for the states $x\in\mathscr{S}\backslash\mathscr{S}_{\mathrm{m}}$, $y\in\mathscr{S}_x$, the following relation is satisfied:

$$\sum_{z\in\mathscr{S}}P(z,x)A(z,x;t)=P(y,x)A(y,x;t)+\sum_{z\in\mathscr{S}\backslash y}P(z,x)A(z,x;t) \tag{3.140}$$

$$\leq P(y,x)\delta+\sum_{z\in\mathscr{S}\backslash y}P(z,x) \tag{3.141}$$

$$= -(1-\delta)P(y,x) + \sum_{z\in\mathscr{S}\backslash y} P(z,x) + P(y,x) \tag{3.142}$$

$$= -(1-\delta)P(y,x) + 1. \tag{3.143}$$

Here we used $A(z,x;t) \le 1$ and $\sum_{z\in\mathscr{S}} P(z,x) = 1$. Thus, when $t \ge t_1$, $G(x,x;t)$ is written as

$$G(x,x;t) = 1 - \sum_{z\in\mathscr{S}} P(z,x)A(z,x;t) \tag{3.144}$$

$$\ge 1 - \{-(1-\delta)P(y,x) + 1\} \tag{3.145}$$

$$= (1-\delta)P(y,x) > 0. \tag{3.146}$$

In addition, in the limit of $t \to \infty$, the following relation is satisfied:

$$wf(1)\exp\left[-\frac{L}{T(t)}\right] \to 0. \tag{3.147}$$

For large enough t_1, we obtain

$$G(x,x;t) \ge (1-\varepsilon)P(y,x) \tag{3.148}$$

$$\ge wf(1)\exp\left[-\frac{L}{T(t)}\right], \quad (t \ge t_1). \tag{3.149}$$

Thus, the lower bound of the transition probability was obtained.

Proof of weak ergodicity

In order to prove the weak ergodicity, we have to estimate the ergodic coefficient defined by Eq. (3.101). By the definition of x^*, the states $x^* \in \mathscr{S}\backslash\mathscr{S}_{\mathrm{m}}$ can be generated by at most R steps from arbitrary states $x \in \mathscr{S}$. Here, we consider the Markov chain that changes from the state x at $t-R$ to the state x^* at t. In this case, at least one integer $l\,(0 \le l \le R)$ exists and the transition process from x to x^* is satisfied:

$$x := x'_0 \ne x'_1 \ne x'_2 \ne \cdots \ne x'_l = x'_{l+1} = \cdots = x'_R =: x^*. \tag{3.150}$$

When t is large enough and $t \ge t_1$, for $i\,(0 \le i \le R-1)$, the following relations are satisfied.

- For $i < l$,

$$G(x'_{i+1}, x'_i; t-R+i) \ge wf(1)\exp\left[-\frac{L}{T(t-R+i)}\right]. \tag{3.151}$$

- For $i \geq l$,

$$G(x'_{i+1}, x'_i; t-R+i) = G(x^*, x^*; t-R+i) \tag{3.152}$$

$$\geq wf(1)\exp\left[-\frac{L}{T(t-R+i)}\right]. \tag{3.153}$$

Here we used Eqs. (3.120) and (3.121). Thus, when $t \geq t_1$, we obtain the (x^*, x)-element of the transition matrix $\mathcal{G}^{t,t-R}$ as

$$\mathcal{G}^{t,t-R}(x^*, x) = \sum_{x_1, x_2, \cdots, x_{R-1}} G(x^*, x_{R-1}; t-1) G(x_{R-1}, x_{R-2}; t-2)$$

$$\cdots G(x_2, x_1; t-R+1) G(x_1, x; t-R) \tag{3.154}$$

$$\geq G(x^*, x'_{R-1}; t-1) G(x'_{R-1}, x'_{R-2}; t-2)$$

$$\cdots G(x'_2, x'_1; t-R+1) G(x'_1, x; t-R) \tag{3.155}$$

$$\geq w^R f(1)^R \exp\left[-\sum_{i=0}^{R-1} \frac{L}{T(t-R+i)}\right] \tag{3.156}$$

$$\geq w^R f(1)^R \exp\left[-\frac{RL}{T(t-1)}\right]. \tag{3.157}$$

Here we used the fact that $T(t)$ is a monotonically decreasing function with t in the final inequality. By using the relation, we estimate the ergodic coefficient. The following is satisfied:

$$\sum_{z \in \mathcal{S}} \min\left\{\mathcal{G}^{t,t-R}(z, x), \mathcal{G}^{t,t-R}(z, y)\right\}$$

$$\geq \min\left\{\mathcal{G}^{t,t-R}(x^*, x), \mathcal{G}^{t,t-R}(x^*, y)\right\} \tag{3.158}$$

$$\geq w^R f(1)^R \exp\left[-\frac{RL}{T(t-1)}\right]. \tag{3.159}$$

From this inequality, the following is satisfied:

$$\min_{x, y \in \mathcal{S}} \left\{\sum_{z \in \mathcal{S}} \min\left[\mathcal{G}^{t,t-R}(z, x), \mathcal{G}^{t,t-R}(z, y)\right]\right\}$$

$$\geq w^R f(1)^R \exp\left[-\frac{RL}{T(t-1)}\right]. \tag{3.160}$$

Thus, the ergodic coefficient defined by Eq. (3.101) is given by

$$1-\alpha\left(\mathscr{G}^{t,t-R}\right) \geq w^R f(1)^R \exp\left[-\frac{RL}{T(t-1)}\right]. \tag{3.161}$$

Here we put: $t = kR\,(k = 0,1,2,\cdots)$. k_1 exists satisfying $t_1 = k_1 R$, and for $k_1 \leq k$, we obtain

$$1-\alpha\left(\mathscr{G}^{kR,kR-R}\right) \geq w^R f(1)^R \exp\left[-\frac{RL}{T(kR-1)}\right]. \tag{3.162}$$

To satisfy Eq. (3.100), it is enough to use the schedule given by the condition as

$$T(t) \geq \frac{RL}{\ln(t+2)}. \tag{3.163}$$

By using $t = kR-1$, the condition is rewritten as

$$\exp\left[-\frac{RL}{T(kR-1)}\right] \geq \frac{1}{kR+1}. \tag{3.164}$$

Thus, we obtain

$$\sum_{k=0}^{\infty}\left[1-\alpha\left(\mathscr{G}^{kR,kR-R}\right)\right]$$

$$\geq \sum_{k=k_1}^{\infty}\left[1-\alpha\left(\mathscr{G}^{kR,kR-R}\right)\right] \tag{3.165}$$

$$\geq w^R f(1)^R \sum_{k=k_1}^{\infty} \frac{1}{kR+1} \tag{3.166}$$

$$\to \infty. \tag{3.167}$$

As a result, it is proved that Eq. (3.100) is satisfied. In addition, we define Δ as

$$\Delta := RL \tag{3.168}$$

and obtain

$$T(t) \geq \frac{\Delta}{\ln(t+2)}. \tag{3.169}$$

Here, R is proportional to N, and thus Δ is a constant which is proportional to N.

Proof of strong ergodicity

In order to prove strong ergodicity, it is necessary to satisfy the three conditions mentioned earlier. The first condition is that the stochastic dynamics shows weak ergodicity when the annealing schedule represented by Eq. (3.90) is used, which has been already proved earlier. The second condition is satisfied since the Boltzmann distribution $g(x;t)$ is the steady distribution at t due to the definition of the acceptance probability $A(y,x;t)$. Then, the elements of the steady distribution $\mathscr{Q}_t$ are given by $g(x;t)$. Thus, in order to show strong ergodicity, all we have to do is to prove the third condition.

First, we focus on the monotonicity of $g(x;t)$. Let $E^{\min}$ be the minimum value of $E(x)$, and $\mathscr{S}^{\min}$ be the set of x such that $E(x) = E^{\min}$. In this case, $g(x;t)$ has the following properties:

- For $x \in \mathscr{S}^{\min}$,

$$g(x;t+1) \geq g(x;t), \quad (t > 0), \tag{3.170}$$

where the initial time is $t = 0$.

- For $x \in \mathscr{S} \backslash \mathscr{S}^{\min}$,

$$g(x;t+1) \leq g(x;t), \quad (t > t_2), \tag{3.171}$$

where t_2 will be introduced later.

In addition, we introduce

$$\delta E(x) = E(x) - E^{\min}. \tag{3.172}$$

Let us consider Eq. (3.170). For $x \in \mathscr{S}^{\min}$, the following is satisfied:

$$g(x;t) = \frac{\exp\left[-\dfrac{E^{\min}}{T(t)}\right]}{\sum\limits_{y \in \mathscr{S}^{\min}} \exp\left[-\dfrac{E^{\min}}{T(t)}\right] + \sum\limits_{y \in \mathscr{S} \backslash \mathscr{S}^{\min}} \exp\left[-\dfrac{E(y)}{T(t)}\right]} \tag{3.173}$$

$$= \frac{1}{\sum\limits_{y \in \mathscr{S}^{\min}} 1 + \sum\limits_{y \in \mathscr{S} \backslash \mathscr{S}^{\min}} \exp\left[-\dfrac{\delta E(y)}{T(t)}\right]}. \tag{3.174}$$

Here, since $T(t+1) < T(t)$, $\delta E(y) > 0$, and if $y \in \mathscr{S} \backslash \mathscr{S}^{\min}$, the following inequality is obtained:

$$\exp\left[-\frac{\delta E(y)}{T(t+1)}\right] < \exp\left[-\frac{\delta E(y)}{T(t)}\right]. \tag{3.175}$$

Thus, by using Eq. (3.174), we can obtain

$$g(x;t+1) \geq g(x;t). \tag{3.176}$$

Next let us consider Eq. (3.171). For $x \in \mathscr{S}\backslash\mathscr{S}^{\min}$, the following is satisfied:

$$g(x;t) = \frac{\exp\left[-\frac{E(x)}{T(t)}\right]}{\sum_{y\in\mathscr{S}^{\min}} \exp\left[-\frac{E^{\min}}{T(t)}\right] + \sum_{y\in\mathscr{S}\backslash\mathscr{S}^{\min}} \exp\left[-\frac{E(y)}{T(t)}\right]} \tag{3.177}$$

$$= \frac{\exp\left[-\frac{\delta E(x)}{T(t)}\right]}{\sum_{y\in\mathscr{S}^{\min}} 1 + \sum_{y\in\mathscr{S}\backslash\mathscr{S}^{\min}} \exp\left[-\frac{\delta E(y)}{T(t)}\right]}. \tag{3.178}$$

We differentiate both sides this equation with respect to t and obtain

$$\frac{d}{dt}g(x;t) = \frac{dT(t)}{dt}\frac{d}{dT(t)}g(x;t) \tag{3.179}$$

$$= \frac{dT(t)}{dt}\left[\delta E(x) - \sum_{y\in\mathscr{S}\backslash\mathscr{S}^{\min}} \delta E(y)g(y;t)\right]\frac{g(x;t)}{T(t)^2}. \tag{3.180}$$

Here for large enough t, the Boltzmann distribution $g(y;t)$ becomes very small value at the states $y \in \mathscr{S}\backslash\mathscr{S}^{\min}$. Thus, the time t_2 which is large enough exists satisfying

$$\delta E(x) - \sum_{y\in\mathscr{S}\backslash\mathscr{S}^{\min}} \delta E(y)g(y;t) > 0, \quad (t \geq t_2). \tag{3.181}$$

Also, since $dT(t)/dt < 0$, the following inequality is satisfied:

$$\frac{d}{dt}g(x;t) < 0, \tag{3.182}$$

and we obtain

$$g(x;t+1) \leq g(x;t). \tag{3.183}$$

Thus, for $t \geq t_2$, $\|\mathscr{Q}_{t+1} - \mathscr{Q}_t\|$ is written as

$$\|\mathscr{Q}_{t+1} - \mathscr{Q}_t\| = \sum_{x \in \mathscr{S}} |g(x;t+1) - g(x;t)| \tag{3.184}$$

$$= \sum_{x \in \mathscr{S}^{\min}} [g(x;t+1) - g(x;t)] - \sum_{x \in \mathscr{S} \setminus \mathscr{S}^{\min}} [g(x;t+1) - g(x;t)] \tag{3.185}$$

$$= 2 \sum_{x \in \mathscr{S}^{\min}} [g(x;t+1) - g(x;t)]. \tag{3.186}$$

Here we used the relation given as

$$\|\mathscr{Q}_t\| = \sum_{x \in \mathscr{S}} g(x;t) = \sum_{x \in \mathscr{S}^{\min}} g(x;t) + \sum_{x \in \mathscr{S} \setminus \mathscr{S}^{\min}} g(x;t) = 1. \tag{3.187}$$

Thus, we obtain

$$\sum_{t=t_2}^{\infty} \|\mathscr{Q}_{t+1} - \mathscr{Q}_t\| = 2 \sum_{x \in \mathscr{S}^{\min}} [g(x;\infty) - g(x;t_2)] \tag{3.188}$$

$$\leq 2. \tag{3.189}$$

In addition, since $\|\mathscr{Q}_t\| = 1$ for arbitrary t, the following inequality is satisfied:

$$\|\mathscr{Q}_{t+1} - \mathscr{Q}_t\| \leq 2. \tag{3.190}$$

Thus, we finally obtain

$$\sum_{t=0}^{\infty} \|\mathscr{Q}_{t+1} - \mathscr{Q}_t\| = \sum_{t=0}^{t_2-1} \|\mathscr{Q}_{t+1} - \mathscr{Q}_t\| + \sum_{t=t_2}^{\infty} \|\mathscr{Q}_{t+1} - \mathscr{Q}_t\| \tag{3.191}$$

$$\leq 2t_2 + 2 \tag{3.192}$$

$$< \infty, \tag{3.193}$$

where the initial time is $t = 0$. Thus, the third condition is shown. From these facts, it is proved that the inhomogeneous Markov chain represented by Eq. (3.90) guarantees strong ergodicity. This discussion is the proof of the Geman–Geman theorem.

References

Chen, H., X. Kong, B. Chong, G. Qin, X. Zhou, X. Peng, and J. Du. 2011. 'Experimental demonstration of a quantum annealing algorithm for the traveling salesman problem in a nuclear-magnetic-resonance quantum simulator'. *Phys. Rev. A* 83: 032314.

Dunham, B., D. Fridshal, R. Fridshal, and J. H. North. 1963. 'Design by natural selection'. *Synthese* 15: 254.

Fisher, D. S., and D. A. Huse. 1986. 'Ordered phase of short-range Ising spin-glasses'. *Phys. Rev. Lett.* 56: 1601.

Geman, S., and D. Geman. 1984. 'Stochastic relaxation, Gibbs distributions, and the Bayesian restoration of images'. *IEEE Trans. Patt. Anal. Machine Intel.* 6: 721.

Grest, G. S., C. M. Soukoulis, and K. Levin. 1986. 'Cooling-rate dependence for the spin-glass ground-state energy: Implications for optimization by simulated annealing'. *Phys. Rev. Lett.* 56: 1148.

Hopfield, J. J., and D. W. Tank. 1985. '"Neural" computation of decisions in optimization problems'. *Biol. Cybern.* 52: 141.

Huse, D. A., and D. S. Fisher. 1986. 'Residual energies after slow cooling of disordered systems'. *Phys. Rev. Lett.* 57: 2203.

Ichiki, A., and M. Ohzeki. 2013. 'Violation of detailed balance accelerates relaxation'. *Phys. Rev. E* 88: 020101(R).

Ichiki, A., and M. Ohzeki. 2015. 'Full-order fluctuation-dissipation relation for a class of nonequilibrium steady states'. *Phys. Rev. E* 91: 062105.

Karp, R. M. 1977. 'Probabilistic analysis of partitioning algorithms for the traveling-salesman problem in the plane'. *Math. Oper. Res.* 2: 209.

Kirkpatrick, S., C. D. Gelatt Jr., , and M. P. Vecchi. 1983. 'Optimization by simulated annealing'. *Science* 220: 671.

Lin, S. 1975. 'Heuristic programming as an aid to network design'. *Networks* 5: 33.

Liu, J. S. 1996. 'Metropolized independent sampling with comparisons to rejection sampling and importance sampling'. *Stat. Comput.* 6: 113.

Metropolis, N., A. W. Rosenbluth, M. N. Rosenbluth, A. H. Teller, and E. Teller. 1953. 'Equation of state calculations by fast computing machines'. *J. Chem. Phys.* 21: 1087.

Mézard, M., and R. Zecchina. 2002. 'Random K-satisfiability problem: From an analytic solution to an efficient algorithm'. *Phys. Rev. E* 66: 056126.

Mézard, M., G. Parisi, and M. Virasoro. 1986. *Spin Glass Theory and Beyond.* World Scientific Press.

Mézard, M., G. Parisi, and R. Zecchina. 2002. 'Analytic and algorithmic solution of random satisfiability problems. *Science* 297: 812.

Nielsen, M. A., and I. L. Chuang. 2000. *Quantum Computation and Quantum Information.* Cambridge University Press.

Ohzeki, M., and A. Ichiki. 2015. 'Langevin dynamics neglecting detailed balance condition'. *Phys. Rev. E* 92: 012105.

Pollet, L., S. M. A. Rombouts, K. V. Houcke, and K. Heyde. 2004. 'Optimal Monte Carlo updating'. *Phys. Rev. E* 70: 056705.

Sakai, Y., and K. Hukushima. 2013. 'Dynamics of one-dimensional Ising model without detailed balance condition'. *J. Phys. Soc. Jpn.* 82: 064003.

Shinomoto, S., and Y. Kabashima. 1991. 'Finite time scaling of energy in simulated annealing'. *J. Phys. A: Math. Gen.* 24: L141.

Suwa, H., and S. Todo. 2010. 'Markov chain Monte Carlo method without detailed balance'. *Phys. Rev. Lett.* 105: 120603.

Takahashi, K., and M. Ohzeki. 2016. 'Conflict between fastest relaxation of a Markov process and detailed balance condition'. *Phys. Rev. E* 93: 012129.

CHAPTER 4

Quantum Spin Glass

In the previous chapters, the classical Ising spin models and simulated annealing were explained. The background and basics of quantum annealing will be described in the following chapters. In this chapter, the quantum fluctuation effect which plays a significant role in quantum annealing is reviewed. In most cases, the transverse field is introduced into the Ising model in quantum annealing. The transverse field induces the quantum fluctuation effect, i.e., the quantum tunneling. Section 4.1 explains the quantum fluctuation effect in ferromagnetic Ising models in a transverse field. Herein, we refer to the Ising model in a transverse field as the quantum Ising model. Next, we show the quantum fluctuation effect in the Ising model with random interactions. In some quantum models with random interactions, the spin glass phase could appear. In Section 4.2, the basic concept of quantum spin glass will be described. Subsequently, the phase diagram of the Sherrington–Kirkpatrick model in a transverse field obtained by mean-field calculations and numerical simulations will be shown in Section 4.3. In the Sherrington–Kirkpatrick model in a transverse field, the spin glass phase appears at small temperatures and transverse fields. Next, in Section 4.4, we will introduce the quantum phase transition, i.e., the zero-temperature phase transition, between the paramagnetic phase and the spin glass phase in the Edwards–Anderson model in a transverse field. In Section 4.5, the existence of the replica symmetric spin glass phase in the quantum model with random interactions will be discussed.

4.1 Ferromagnetic Ising Model in a Transverse Field

In Chapter 2, order–disorder transitions caused by thermal fluctuation, i.e., the temperature, were shown. In general, thermal fluctuation breaks orders which appear due to interactions at low temperatures. The order–disorder transition is called the thermal phase transition or classical phase transition. As mentioned in Chapter 2, the Ising model is a prototypical model to study thermal phase transitions.

Another important fluctuation exists in nature – quantum fluctuation, which is observed mainly at low temperatures. The quantum fluctuation effect could cause exotic quantum orders such as quantum spin liquid and so on (Anderson, 1997; Auerbach, 1998; Shimizu et al., 2003; Schollwöck et al., 2004; Nakatsuji et al., 2005, 2006; Lee, 2007; Yamashita et al., 2008; Parkinson and Farnell, 2010). Thus, quantum fluctuation is a source of ordering, which is different role from thermal effect. In fact, this is one side of the quantum fluctuation effect. However, from the other side, the quantum fluctuation effect breaks orders as well, similar to the thermal fluctuation effect. As a result, order–disorder phase transition due to the quantum fluctuation effect could occur as well as the classical phase transition (see Fig. 4.1). The Ising model in a transverse field, in other words, the quantum Ising model, is a prototypical model to study order–disorder phase transitions due to the quantum fluctuation effect.

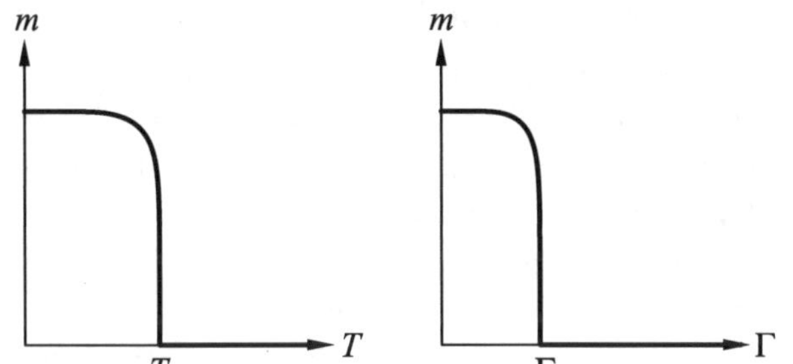

Fig. 4.1 (Left panel) A schematic of classical phase transition. The horizontal and vertical axes represent the temperature and the order parameter, respectively. (Right panel) A schematic of quantum phase transition. The horizontal and vertical axes represent the quantum field such as the transverse field and the order parameter, respectively.

The Hamiltonian of the quantum Ising model is given by

$$\mathscr{H} = -\sum_{\langle i,j \rangle} J_{ij}\hat{\sigma}_i^z\hat{\sigma}_j^z - \sum_{i=1}^{N} h_i\hat{\sigma}_i^z - \Gamma\sum_{i=1}^{N}\hat{\sigma}_i^x, \tag{4.1}$$

where N is the number of spins, and $\hat{\sigma}_i^\alpha$ ($\alpha = x, y, z$) denote the Pauli matrices defined by

$$\hat{\sigma}_i^x = \begin{pmatrix} 0 & 1 \\ 1 & 0 \end{pmatrix}, \quad \hat{\sigma}_i^y = \begin{pmatrix} 0 & -i \\ i & 0 \end{pmatrix}, \quad \hat{\sigma}_i^z = \begin{pmatrix} 1 & 0 \\ 0 & -1 \end{pmatrix}. \tag{4.2}$$

In this case, the bases are defined by

$$|\uparrow\rangle = |+1\rangle = \begin{pmatrix} 1 \\ 0 \end{pmatrix}, \tag{4.3}$$

$$|\downarrow\rangle = |-1\rangle = \begin{pmatrix} 0 \\ 1 \end{pmatrix}, \tag{4.4}$$

which are the eigenvectors of $\hat{\sigma}_i^z$. Furthermore, these states indicate the up-spin state and the down-spin state in the Ising model, respectively. Notice that the ground state of $\hat{\sigma}_i^x$ is represented by

$$|\rightarrow\rangle := \frac{1}{\sqrt{2}}(|\uparrow\rangle + |\downarrow\rangle), \tag{4.5}$$

The first and second terms in Eq. (4.1) respectively correspond to the classical Ising model and the third term is the transverse field. The quantum Ising model has been used to analyze phase transition caused by the quantum fluctuation effect for several decades. For example, in 1963, de Gennes studied the order–disorder transition in double-well ferroelectric systems such as KDP or KH_2PO_4 by the mean-field analysis of the quantum Ising model (Blinc, 1960; de Gennes, 1963). Another historical study was done by Katsura in 1962; the study examined the quantum Ising model in one dimension as a special case of the anisotropic Heisenberg model in a magnetic field (Katsura, 1962). After that, the quantum Ising model and its generalization have been widely used not only for the analysis of phase transition associated with the quantum fluctuation effect but also for exploring the topological nature and the consideration of quantum information theory (Suzuki, 1971; Kitaev, 2001; Raussendorf and Briegel, 2001; Raussendorf et al., 2003; Leijnse and Flensberg, 2012; Mourik et al., 2012; Franz, 2013; Fujii et al., 2013; Greiter et al., 2014). In addition, the model defined by Eq. (4.1) with $h_i = 0$ and EA-type random exchange interactions was first studied by Chakrabarti (Chakrabarti, 1981). Recently, the many-body localization in quantum random Ising chains has been studied (Kjäll et al., 2014).

In the following, the semiclassical analysis will be explained to show the order–disorder phase transition in the quantum Ising model. Next, a couple of results of the d-dimensional quantum Ising model will be shown. In addition, properties of the Husimi–Temperley–Curie–Weiss model (infinite-range Ising model) in a transverse field will be reviewed.

4.1.1 Semiclassical analysis and a simple mean-field approximation of the ferromagnetic Ising model in a transverse field

Here we consider the transverse ferromagnetic Ising model without an external longitudinal field. The Hamiltonian of the model is given by

$$\hat{\mathscr{H}} = -\sum_{\langle i,j\rangle} J_{ij}\hat{\sigma}_i^z\hat{\sigma}_j^z - \Gamma\sum_{i=1}^{N}\hat{\sigma}_i^x, \qquad (J_{ij} \geq 0). \tag{4.6}$$

When interactions are absent, i.e., $J_{ij} = 0$ for all i, j, the classical order parameter $\langle\hat{\sigma}^z\rangle$ should be zero. In contrast, in the presence of interactions, $\langle\hat{\sigma}^z\rangle$ could be a nonzero value.

On top of this, as the transverse field Γ increases, the classical order parameter $\langle\hat{\sigma}^z\rangle$ decreases. This is a simple explanation of the ferromagnetic Ising model in a transverse field.

Next, in order to consider the behavior of physical quantities at zero temperature, we perform a semiclassical approximation given by

$$\hat{\sigma}_i^z = \langle\hat{\sigma}^z\rangle = \cos\theta, \tag{4.7}$$

$$\hat{\sigma}_i^x = \langle\hat{\sigma}^x\rangle = \sin\theta. \tag{4.8}$$

Using the approximation, the internal energy E is given by

$$\frac{E}{N} = -\Gamma\sin\theta - \frac{1}{2}J(0)\cos^2\theta, \tag{4.9}$$

where N is the number of spins and $J(0)$ is defined by

$$J(0) = \sum_{j\in\partial i} J_{ij}, \tag{4.10}$$

where ∂i is the set of sites which are the nearest neighbors interacting with the spin at site i. The condition given by

$$\sin\theta = \Gamma/J(0) \tag{4.11}$$

minimizes the internal energy. From this fact, if $\Gamma \geq J(0)$, $\langle\hat{\sigma}^z\rangle = 0$ and $\langle\hat{\sigma}^x\rangle = 1$, that is, the ground state is polarized in the x-direction, which is often called the quantum paramagnetic state. On the contrary, if $\Gamma < J(0)$, both $\langle\hat{\sigma}^z\rangle$ and $\langle\hat{\sigma}^x\rangle$ take nonzero values. In addition, if $\Gamma = 0$, $\langle\hat{\sigma}^z\rangle = 1$ and $\langle\hat{\sigma}^x\rangle = 0$, i.e., a classically ordered state appears. Figure 4.2 shows $\langle\hat{\sigma}^z\rangle$ and $\langle\hat{\sigma}^x\rangle$ as a function of $\Gamma/J(0)$.

Next, we further consider the ferromagnetic Ising model in a transverse field at the mean-field level (Brout et al., 1966; Stinchcombe, 1973). In the mean-field approximation, we assume that the Hamiltonian is decoupled by local interactions. Thus, the approximated Hamiltonian is expressed by the summation of local interactions as follows:

$$\mathcal{H} = -\sum_{i=1}^{N} \vec{h}_i \cdot \vec{\sigma}_i, \tag{4.12}$$

where $\vec{h}_i$ is referred to as the molecular field at site i and is defined by

$$\vec{h}_i = \sum_{j\in\partial i} J_{ij}\langle\hat{\sigma}_j^z\rangle\vec{e}_z + \Gamma\vec{e}_x, \tag{4.13}$$

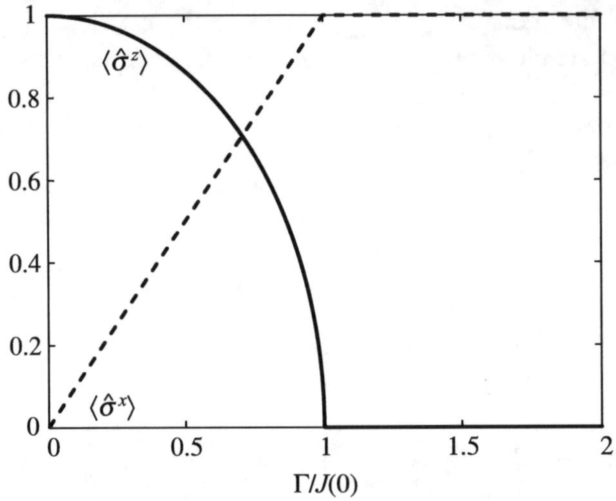

Fig. 4.2 $\Gamma/J(0)$-dependence of $\langle\hat{\sigma}^z\rangle$ (solid line) and $\langle\hat{\sigma}^x\rangle$ (dotted line). When $\Gamma \geq J(0)$, quantum paramagnetic state ($\langle\hat{\sigma}^x\rangle = 1$) appears.

where $\vec{e}_z$ and $\vec{e}_x$ are the unit vectors in the spin space along the z and x directions, respectively. By using $\vec{e}_z$ and $\vec{e}_x$, $\vec{\sigma}_i$ is written as

$$\vec{\sigma}_i = \langle\hat{\sigma}_i^z\rangle\vec{e}_z + \langle\hat{\sigma}_i^x\rangle\vec{e}_x. \tag{4.14}$$

Under the mean-field approximation, for all i, $\vec{h}_i = \vec{h}$. Here, $\vec{h}$ is defined by

$$\vec{h} = J(0)\langle\hat{\sigma}^z\rangle\vec{e}_z + \Gamma\vec{e}_x. \tag{4.15}$$

The mean-field equation for the spontaneous magnetization is expressed as

$$\langle\vec{\sigma}\rangle = \tanh\left(\beta|\vec{h}|\right)\frac{\vec{h}}{|\vec{h}|}, \tag{4.16}$$

$$|\vec{h}| = \sqrt{[J(0)\langle\hat{\sigma}^z\rangle]^2 + \Gamma^2}, \tag{4.17}$$

where $\beta = 1/T$. Here, the Boltzmann constant k_B is set to unity. From these equations, respective components are represented as

$$\langle\hat{\sigma}^z\rangle = \tanh\left(\beta|\vec{h}|\right)\left[\frac{J(0)\langle\hat{\sigma}^z\rangle}{|\vec{h}|}\right], \tag{4.18}$$

$$\langle\hat{\sigma}^x\rangle = \tanh\left(\beta|\vec{h}|\right)\left(\frac{\Gamma}{|\vec{h}|}\right). \tag{4.19}$$

By solving Eqs. (4.18)–(4.19), we obtain the phase diagram (see Fig. 4.3). In the ferromagnetic phase, the classical order parameter $\langle\hat{\sigma}^z\rangle$ is a non-zero value while $\langle\hat{\sigma}^z\rangle = 0$ in the paramagnetic phase. At zero temperature, if $\Gamma < J(0)$, the classically ordered phase appears whereas if $\Gamma \geq J(0)$, the quantum paramagnetic phase appears, which is consistent with the results shown in Fig. 4.2. At zero quantum field, i.e., $\Gamma = 0$, the classical phase transition occurs below the finite temperature $T_c/J(0) = 1$.

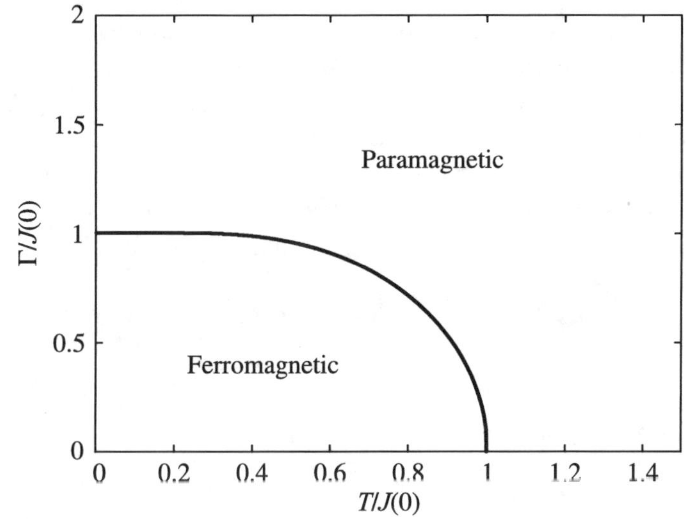

Fig. 4.3 Phase diagram of the ferromagnetic Ising model in a transverse field at the mean-field level.

4.1.2 Phase transition in the finite-dimensional ferromagnetic Ising models in a transverse field

In the previous subsection, to show the quantum fluctuation effect in a simple way, we reviewed a semiclassical analysis and a mean-field approximation of the ferromagnetic Ising model in a transverse field. In this subsection, we explain phase transitions in d-dimensional ($d = 1,2,3$) ferromagnetic Ising models in a transverse field. For simplicity, we focus here on a homogeneous ferromagnetic Ising model in a transverse field. The Hamiltonian of the model is given by

$$\hat{\mathscr{H}} = \hat{\mathscr{H}}_0 + \hat{\mathscr{H}}_q, \tag{4.20}$$

$$\hat{\mathscr{H}}_0 = -J \sum_{\langle i,j\rangle} \hat{\sigma}_i^z \hat{\sigma}_j^z, \tag{4.21}$$

$$\hat{\mathscr{H}}_q = -\Gamma \sum_{i=1}^{N} \hat{\sigma}_i^x, \tag{4.22}$$

where $\langle i, j \rangle$ denotes the nearest-neighbor spin pairs on the chain (for $d = 1$), the square lattice (for $d = 2$), and the cubic lattice (for $d = 3$). The ferromagnetic Ising model in a transverse field in one dimension ($d = 1$) has no finite-temperature phase transition and zero-temperature phase transition at $\Gamma = J$. Notice that the one-dimensional transverse Ising model can be solved using the Jordan–Wigner transformation and the Bogoliubov transformation. Through these transformations, the spin model is transformed to a free fermion model (Pfeuty, 1970; Suzuki et al., 2013). By using these transformations, some models related to the one-dimensional transverse Ising model can be solved analytically. Suzuki summarized the relation among exactly solvable models of critical phenomena (Suzuki, 1971). After that, related models were revived in the context of topological nature and quantum information science (Smacchia et al., 2011; Ohta et al., 2015; Ohta and Totsuka, 2016; Ohta et al., 2016). Some of them will be reviewed in Chapter 8. In contrast, the ferromagnetic Ising model in a transverse field in two or more higher dimensions has finite-temperature phase transition. Figure 4.4 summarizes this fact.

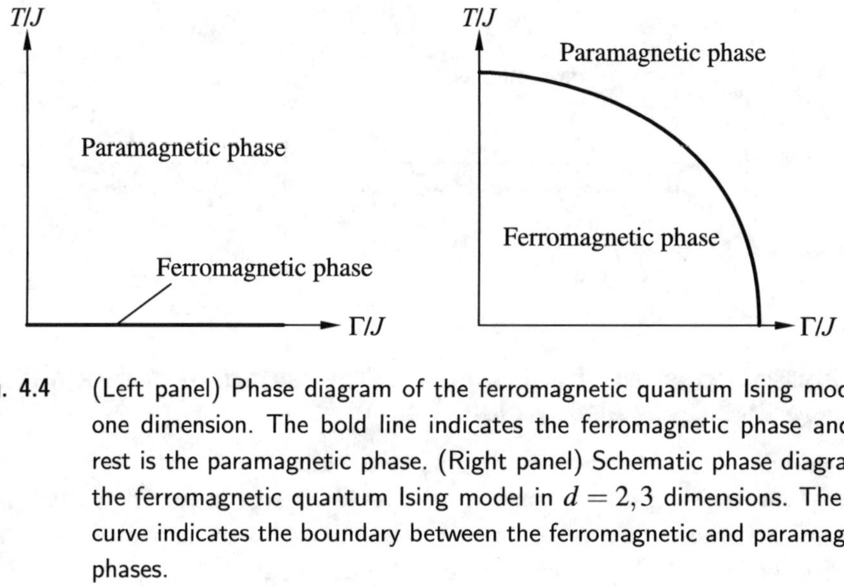

Fig. 4.4 (Left panel) Phase diagram of the ferromagnetic quantum Ising model in one dimension. The bold line indicates the ferromagnetic phase and the rest is the paramagnetic phase. (Right panel) Schematic phase diagram of the ferromagnetic quantum Ising model in $d = 2, 3$ dimensions. The bold curve indicates the boundary between the ferromagnetic and paramagnetic phases.

When we use the z-component of spins as the basis set, the Hamiltonian of the transverse Ising model is not a diagonal matrix. If this is the case, we cannot easily obtain finite-temperature properties because of the difficulty in calculating the free energy or the partition function. The difficulty appears in Monte Carlo methods which are widely used for analysis of finite-temperature properties (see Section 3.3.1). Thus, to overcome the difficulty, we use the path-integral formalism, that is, the Suzuki–Trotter decomposition (Trotter, 1959; Suzuki, 1976a,b, 1977). The Suzuki–Trotter decomposition replaces the quantum Hamiltonian on a d-dimensional lattice with the diagonal Hamiltonian on a

$d+1$-dimensional lattice. The Hamiltonian of the Ising model in a transverse field is defined by

$$\hat{\mathscr{H}} = \hat{\mathscr{H}}_0 + \hat{\mathscr{H}}_{\mathrm{q}}, \tag{4.23}$$

$$\hat{\mathscr{H}}_0 = -J\sum_{\langle i,j\rangle} \hat{\sigma}_i^z \hat{\sigma}_j^z - h\sum_{i=1}^{N} \hat{\sigma}_i^z, \tag{4.24}$$

$$\hat{\mathscr{H}}_{\mathrm{q}} = -\Gamma\sum_{i=1}^{N} \hat{\sigma}_i^x. \tag{4.25}$$

Here, we introduced the longitudinal magnetic field term in $\hat{\mathscr{H}}_0$. Then, the partition function is given by

$$Z = \mathrm{Tr}\,\mathrm{e}^{-\beta\hat{\mathscr{H}}} = \sum_{\Sigma} \left\langle \Sigma \middle| \mathrm{e}^{-\beta(\hat{\mathscr{H}}_0+\hat{\mathscr{H}}_{\mathrm{q}})} \middle| \Sigma \right\rangle, \tag{4.26}$$

where

$$|\Sigma\rangle := |\sigma_1^z\rangle \otimes |\sigma_2^z\rangle \otimes \cdots \otimes |\sigma_N^z\rangle. \tag{4.27}$$

$|\sigma_i^z\rangle$ denotes the spin state at site i and $|\sigma_i^z\rangle$ takes $|+1\rangle$ or $|-1\rangle$ values. To calculate the partition function, we should obtain the exponential of a non-diagonal matrix. The exponential of a matrix A is defined by

$$\mathrm{e}^A := \sum_{m=0}^{\infty} \frac{A^m}{m!}. \tag{4.28}$$

If we can diagonalize A, the calculation of A^m for an arbitrary m is easy, whereas it is difficult to obtain A^m in general. Thus, we should perform an ingenious technique. The Suzuki–Trotter decomposition is given as

$$\mathrm{e}^{-\beta\hat{\mathscr{H}}} = \left(\mathrm{e}^{-\beta\hat{\mathscr{H}}/P}\right)^P, \tag{4.29}$$

where P is referred to as Trotter number. Using the following relation

$$\begin{aligned}\exp\left(-\frac{\beta}{P}\hat{\mathscr{H}}\right) &= \exp\left[-\frac{\beta}{P}(\hat{\mathscr{H}}_0 + \hat{\mathscr{H}}_{\mathrm{q}})\right] \\ &= \exp\left(-\frac{\beta}{P}\hat{\mathscr{H}}_0\right)\exp\left(-\frac{\beta}{P}\hat{\mathscr{H}}_{\mathrm{q}}\right) + \mathscr{O}\left[\left(\frac{\beta}{P}\right)^2\right],\end{aligned} \tag{4.30}$$

we obtain

$$\begin{aligned} Z = \lim_{P\to\infty} \sum_{\{\Sigma_k\},\{\Sigma'_k\}} & \left\langle \Sigma_1 \left| e^{-\beta\hat{\mathscr{H}}_0/P} \right| \Sigma'_1 \right\rangle \left\langle \Sigma'_1 \left| e^{-\beta\hat{\mathscr{H}}_q/P} \right| \Sigma_2 \right\rangle \\ & \times \left\langle \Sigma_2 \left| e^{-\beta\hat{\mathscr{H}}_0/P} \right| \Sigma'_2 \right\rangle \left\langle \Sigma'_2 \left| e^{-\beta\hat{\mathscr{H}}_q/P} \right| \Sigma_3 \right\rangle \\ & \times \cdots \\ & \times \left\langle \Sigma_P \left| e^{-\beta\hat{\mathscr{H}}_0/P} \right| \Sigma'_P \right\rangle \left\langle \Sigma'_P \left| e^{-\beta\hat{\mathscr{H}}_q/P} \right| \Sigma_1 \right\rangle. \end{aligned} \tag{4.31}$$

Here, we define $|\Sigma_k\rangle$ as

$$|\Sigma_k\rangle := \left|\sigma^z_{1,k}\right\rangle \otimes \left|\sigma^z_{2,k}\right\rangle \otimes \cdots \otimes \left|\sigma^z_{N,k}\right\rangle, \tag{4.32}$$

where i and k in $\left|\sigma^z_{i,k}\right\rangle$ indicate the site index on the d-dimensional lattice (real space) and the site index in the Trotter axis (imaginary time), respectively (see Fig. 4.5).

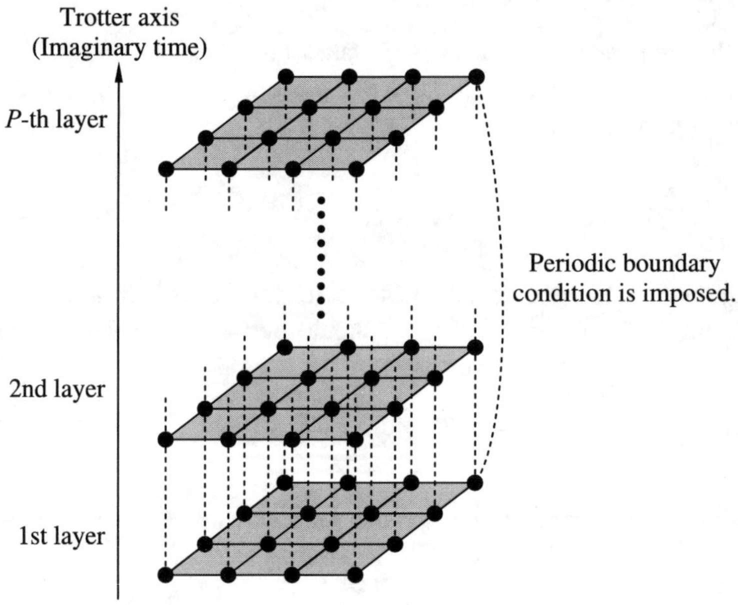

Fig. 4.5 Schematic representations of the Suzuki–Trotter decomposition.

Each term in Eq. (4.31) is calculated as follows:

$$\left\langle \Sigma_k \left| \mathrm{e}^{-\beta \hat{\mathscr{H}}_0/P} \right| \Sigma'_k \right\rangle$$

$$= \exp\left[\frac{\beta}{P}\left(J\sum_{\langle i,j\rangle}\sigma^z_{i,k}\sigma^z_{j,k} + h\sum_{i=1}^{N}\sigma^z_{i,k}\right)\right]\prod_{i=1}^{N}\delta(\sigma^z_{i,k},\sigma'^z_{i,k}), \tag{4.33}$$

$$\left\langle \Sigma'_k \left| \mathrm{e}^{-\beta \hat{\mathscr{H}}_\mathrm{q}/P} \right| \Sigma_{k+1} \right\rangle$$

$$= \left[\frac{1}{2}\sinh\left(\frac{2\beta\Gamma}{P}\right)\right]^{N/2}\exp\left[\frac{1}{2}\ln\coth\left(\frac{\beta\Gamma}{P}\right)\sum_{i=1}^{N}\sigma'^z_{i,k}\sigma^z_{i,k+1}\right], \tag{4.34}$$

where $\sigma^z_{i,k} = \pm 1$. By using the relation

$$\mathrm{e}^{-\beta\Gamma\hat{\sigma}^x/P} = \begin{pmatrix}\cosh(\beta\Gamma/P) & \sinh(\beta\Gamma/P)\\ \sinh(\beta\Gamma/P) & \cosh(\beta\Gamma/P)\end{pmatrix}, \tag{4.35}$$

we obtain

$$Z = \lim_{P\to\infty}\left[\frac{1}{2}\sinh\left(\frac{2\beta\Gamma}{P}\right)\right]^{NP/2}$$

$$\times \sum_{\{\sigma^z_{i,k}=\pm 1\}}\exp\left\{\sum_{k=1}^{P}\left[\frac{\beta J}{P}\sum_{\langle i,j\rangle}\sigma^z_{i,k}\sigma^z_{j,k}\right.\right.$$

$$\left.\left. + \frac{\beta h}{P}\sum_{i=1}^{N}\sigma^z_{i,k} + \frac{1}{2}\ln\coth\left(\frac{\beta\Gamma}{P}\right)\sum_{i=1}^{N}\sigma^z_{i,k}\sigma^z_{i,k+1}\right]\right\}. \tag{4.36}$$

Thus, an effective Hamiltonian corresponding to Eq. (4.20) is derived as

$$\mathscr{H}_\mathrm{eff} = -\frac{J}{P}\sum_{\langle i,j\rangle}\sum_{k=1}^{P}\sigma^z_{i,k}\sigma^z_{j,k} - \frac{h}{P}\sum_{i=1}^{N}\sum_{k=1}^{P}\sigma^z_{i,k}$$

$$-\frac{1}{2\beta}\ln\coth\left(\frac{\beta\Gamma}{P}\right)\sum_{i=1}^{N}\sum_{k=1}^{P}\sigma^z_{i,k}\sigma^z_{i,k+1}. \tag{4.37}$$

Since $\sigma^z_{i,k} = \pm 1$, this Hamiltonian can be regarded as a $d+1$-dimensional classical Ising model. Here, the interactions between spins in the Trotter axis are given by

$$\frac{1}{2\beta} \ln \coth(\beta\Gamma/P) > 0, \tag{4.38}$$

which is ferromagnetic. By using the representation, we can perform the Monte Carlo simulations explained in Section 3.3.1 of the system given by Eq. (4.23). The aforementioned procedure is called the quantum Monte Carlo method (Suzuki, 1987; Sandvik, 2010; Gubernatis et al., 2016).

Next we explain the results obtained by the quantum Monte Carlo simulations according to Nagai et al. (1986). Figure 4.6 shows the internal energy and the specific heat as a function of temperature for the transverse Ising chain ($d = 1$) at $\Gamma/J = 1/2$ (Nagai et al., 1986). Both of them were obtained by quantum Monte Carlo simulations in which the Trotter number was set to $P = 4, 8$, and 12. As mentioned before, the one-dimensional transverse Ising model has no finite-temperature phase transition. Thus, no singularities are observed in the specific heat.

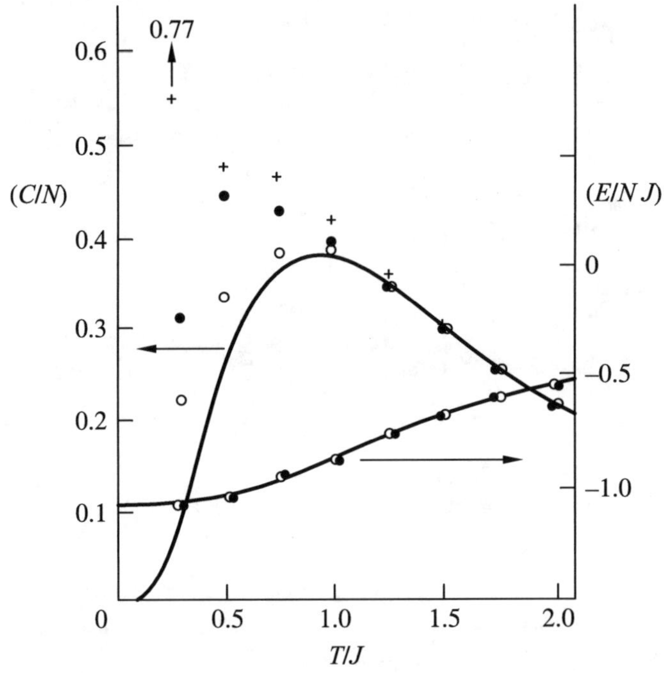

Fig. 4.6 Specific heat C and the internal energy E as a function of temperature for a transverse Ising chain at $\Gamma/J = 1/2$. The diamond, solid, and open circles represent the quantum Monte Carlo data for $P = 4, 8$, and 12, respectively. The solid curves are exact solutions (from Nagai et al., 1986).

Figure 4.7 shows the specific heat and longitudinal susceptibility as a function of temperature for a ferromagnetic transverse Ising model on the square lattice ($d = 2$) whose linear dimension is 24 (Nagai et al., 1986). Both of them were obtained by quantum Monte Carlo simulations in which the Trotter number was set to $P = 12$. The singularity in these physical quantities is observed, which indicates the existence of the phase transition.

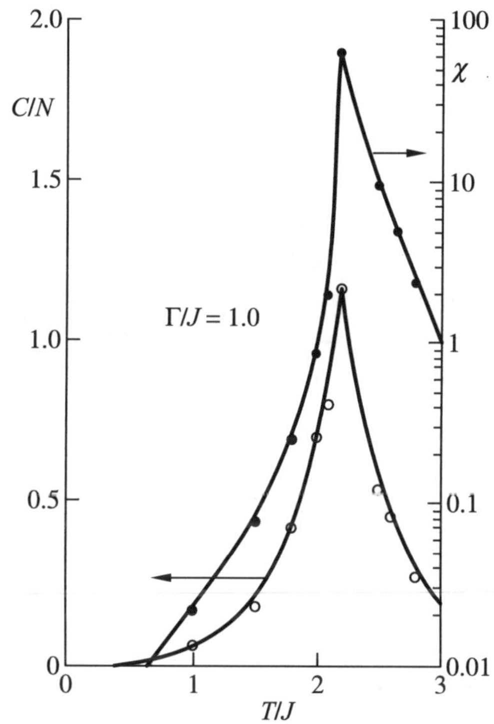

Fig. 4.7 Specific heat C and the longitudinal susceptibility χ for the square lattice whose lattice size is 24×24 at $\Gamma/J = 1$. The data was obtained by quantum Monte Carlo simulations for $P = 24$. The solid curves are a guide to the eye (from Nagai et al., 1986).

Figure 4.8 summarizes the phase boundaries for two-dimensional ($d = 2$) and three-dimensional ($d = 3$) transverse Ising models, which were obtained by quantum Monte Carlo simulations (Nagai et al., 1986).

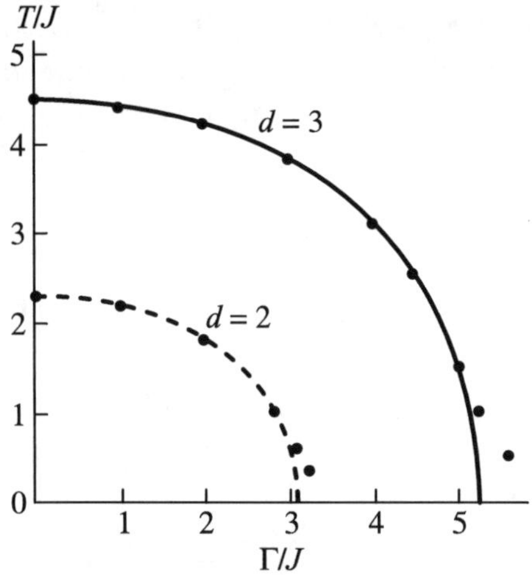

Fig. 4.8 Phase boundaries for the transverse Ising models on square ($d = 2$) and cubic ($d = 3$) lattices. The solid and broken curves represent results obtained by the series expansion (from Nagai et al., 1986).

4.1.3 Hushimi–Temperely–Curie–Weiss model in a transverse field

In this section, we will explain the nature of the infinite-range Ising model introduced in Section 2.1.2 in a transverse field–called the Husimi–Temperely–Curie–Weiss model in a transverse field (Stratt, 1986; Chayes et al., 2008; Krzakala et al., 2008). The Hamiltonian of the model is given by

$$\mathscr{H} = -\frac{zJ}{N}\sum_{1\le i<j\le N}\hat{\sigma}_i^z\hat{\sigma}_j^z - h\sum_{i=1}^{N}\hat{\sigma}_i^z - \Gamma\sum_{i=1}^{N}\hat{\sigma}_i^x \tag{4.39}$$

$$\simeq -\frac{zJ}{2N}\left(\sum_{i=1}^{N}\hat{\sigma}_i^z\right)^2 - h\sum_{i=1}^{N}\hat{\sigma}_i^z - \Gamma\sum_{i=1}^{N}\hat{\sigma}_i^x, \tag{4.40}$$

where the approximation is valid for $N \to \infty$. Here we introduced the longitudinal magnetic field term. By using the Suzuki–Trotter decomposition, the partition function is expressed as

$$Z = \lim_{P\to\infty} A^{PN}\sum_{\{\sigma_{i,k}^z=\pm1\}}\exp\left\{\sum_{k=1}^{P}\left[\frac{z\beta J}{2NP}\left(\sum_{i=1}^{N}\sigma_{i,k}^z\right)^2 + \frac{\beta h}{P}\sum_{i=1}^{N}\sigma_{i,k}^z + K_P\sum_{i=1}^{N}\sigma_{i,k}^z\sigma_{i,k+1}^z\right]\right\}, \tag{4.41}$$

where

$$A := \left[\frac{1}{2}\sinh\left(\frac{2\beta\Gamma}{P}\right)\right]^{\frac{1}{2}}, \tag{4.42}$$

$$K_{\rm P} := \frac{1}{2} \ln \coth \left(\frac{\beta \Gamma}{P} \right). \tag{4.43}$$

Applying the Hubbard–Stratonovitch transformation given by Eq. (2.76), we obtain

$$\exp\left[\frac{Nz\beta J}{2P} \left(\frac{1}{N} \sum_{i=1}^{N} \sigma_{i,k}^z \right)^2 \right]$$

$$= \left(\frac{Nz\beta J}{2\pi P} \right)^{\frac{1}{2}} \int_{-\infty}^{\infty} dm_k \exp\left(-\frac{Nz\beta J}{2P} m_k^2 + \frac{z\beta J}{P} m_k \sum_{i=1}^{N} \sigma_{i,k}^z \right). \tag{4.44}$$

Using this relation, the partition function is written as

$$Z = \lim_{P\to\infty} A^{PN} \left(\frac{Nz\beta J}{2\pi P} \right)^{\frac{P}{2}} \int_{-\infty}^{\infty} \prod_{k=1}^{P} dm_k \exp\left(-\frac{Nz\beta J}{2P} \sum_{k=1}^{P} m_k^2 \right)$$

$$\times \sum_{\{\sigma_{i,k}^z = \pm 1\}} \exp\left[\sum_{i-1}^{N} \sum_{k-1}^{P} \left(\frac{z\beta J}{P} m_k \sigma_{i,k}^z + \frac{\beta h}{P} \sigma_{i,k}^z + K_{\rm P} \sigma_{i,k}^z \sigma_{i,k+1}^z \right) \right] \tag{4.45}$$

$$= \lim_{P\to\infty} A^{PN} \left(\frac{Nz\beta J}{2\pi P} \right)^{\frac{P}{2}} \int_{-\infty}^{\infty} \prod_{k=1}^{P} dm_k \exp\left(-\frac{Nz\beta J}{2P} \sum_{k=1}^{P} m_k^2 \right)$$

$$\times \left\{ \sum_{\{\sigma_k^z = \pm 1\}} \exp\left[\sum_{k=1}^{P} \left(\frac{z\beta J}{P} m_k \sigma_k^z + \frac{\beta h}{P} \sigma_k^z + K_{\rm P} \sigma_k^z \sigma_{k+1}^z \right) \right] \right\}^N \tag{4.46}$$

$$= \lim_{P\to\infty} \left(\frac{Nz\beta J}{2\pi P} \right)^{\frac{P}{2}} \int_{-\infty}^{\infty} \prod_{k=1}^{P} dm_k \exp\Bigg\{ PN \ln A - \frac{Nz\beta J}{2P} \sum_{k=1}^{P} m_k^2$$

$$+ N \ln \sum_{\{\sigma_k^z = \pm 1\}} \exp\left[\sum_{k=1}^{P} \left(\frac{z\beta J}{P} m_k \sigma_k^z + \frac{\beta h}{P} \sigma_k^z + K_{\rm P} \sigma_k^z \sigma_{k+1}^z \right) \right] \Bigg\}. \tag{4.47}$$

Since the power of the exponential is proportional to N, we perform the integral with respect to m_k by the saddle-point method. We obtain

$$Z \simeq \lim_{P\to\infty} \exp\left\{ PN\ln A - \frac{Nz\beta J}{2P}\sum_{k=1}^{P} m_k^2 \right.$$
$$\left. +N\ln \sum_{\{\sigma_k^z=\pm1\}} \exp\left[\sum_{k=1}^{P}\left(\frac{z\beta J}{P} m_k\sigma_k^z + \frac{\beta h}{P}\sigma_k^z + K_{\rm P}\sigma_k^z\sigma_{k+1}^z\right)\right]\right\}, \quad (4.48)$$

where m_k is decided by the saddle-point condition given by

$$0 = \frac{\partial}{\partial m_k}\left\{ P\ln A - \frac{z\beta J}{2P}\sum_{k=1}^{P} m_k^2 \right.$$
$$\left. +\ln \sum_{\{\sigma_k^z=\pm1\}} \exp\left[\sum_{k=1}^{P}\left(\frac{z\beta J}{P} m_k\sigma_k^z + \frac{\beta h}{P}\sigma_k^z + K_{\rm P}\sigma_k^z\sigma_{k+1}^z\right)\right]\right\}. \quad (4.49)$$

As a result, m_k is obtained as

$$m_k = \frac{\sum_{\{\sigma_k^z=\pm1\}} \sigma_k^z \exp\left[\sum_{k=1}^{P}\left(\frac{z\beta J}{P} m_k\sigma_k^z + \frac{\beta h}{P}\sigma_k^z + K_{\rm P}\sigma_k^z\sigma_{k+1}^z\right)\right]}{\sum_{\{\sigma_k^z=\pm1\}} \exp\left[\sum_{k=1}^{P}\left(\frac{z\beta J}{P} m_k\sigma_k^z + \frac{\beta h}{P}\sigma_k^z + K_{\rm P}\sigma_k^z\sigma_{k+1}^z\right)\right]}. \quad (4.50)$$

m_k indicates the magnetization for each Trotter index k. In this case, m_k becomes independent of k because the present effective classical model is translationally invariant along the Trotter direction. Then, we put $m_k = m$ for all k, which is called the static approximation. Under the static approximation, we obtain

$$Z = \lim_{P\to\infty} \exp\left\{ PN\ln A - \frac{Nz\beta J}{2} m^2 \right.$$
$$\left. +N\ln \sum_{\{\sigma_k^z=\pm1\}} \exp\left[\left(\frac{\beta(zJm+h)}{P}\sum_{k=1}^{P}\sigma_k^z + K_{\rm P}\sum_{k=1}^{P}\sigma_k^z\sigma_{k+1}^z\right)\right]\right\}, \quad (4.51)$$

where

$$m = \frac{\sum_{\{\sigma_k^z=\pm1\}} \sigma_k^z \exp\left[\left(\frac{\beta(zJm+h)}{P}\sum_{k=1}^{P}\sigma_k^z + K_{\rm P}\sum_{k=1}^{P}\sigma_k^z\sigma_{k+1}^z\right)\right]}{\sum_{\{\sigma_k^z=\pm1\}} \exp\left[\left(\frac{\beta(zJm+h)}{P}\sum_{k=1}^{P}\sigma_k^z + K_{\rm P}\sum_{k=1}^{P}\sigma_k^z\sigma_{k+1}^z\right)\right]}. \quad (4.52)$$

Furthermore, the free energy per spin can be written as

$$f = -\frac{1}{\beta N}\ln Z \tag{4.53}$$

$$= -\frac{1}{\beta}\lim_{P\to\infty}\Bigg\{P\ln A - \frac{z\beta J}{2}m^2$$

$$+\ln\sum_{\{\sigma_k^z=\pm1\}}\exp\left[\left(\frac{\beta(zJm+h)}{P}\sum_{k=1}^{P}\sigma_k^z + K_{\mathrm{P}}\sum_{k=1}^{P}\sigma_k^z\sigma_{k+1}^z\right)\right]\Bigg\}. \tag{4.54}$$

Here, we take the limit $N\to\infty$ before we take the limit $P\to\infty$ so that $P/N\to 0$. In order to take the trace of $\{\sigma_k^z\}$, we use the transfer matrix defined by

$$V = \begin{pmatrix} \exp\left[\frac{\beta(zJm+h)}{P}+K_{\mathrm{P}}\right] & \exp\left[-K_{\mathrm{P}}\right] \\ \exp\left[-K_{\mathrm{P}}\right] & \exp\left[\frac{\beta(zJm+h)}{P}-K_{\mathrm{P}}\right] \end{pmatrix}, \tag{4.55}$$

and

$$\sum_{\{\sigma_k^z=\pm1\}}\exp\left[\left(\frac{\beta(zJm+h)}{P}\sum_{k=1}^{P}\sigma_k^z + K_{\mathrm{P}}\sum_{k=1}^{P}\sigma_k^z\sigma_{k+1}^z\right)\right] = \sum_{\{\sigma_k^z=\pm1\}}V^P. \tag{4.56}$$

Let λ_+ and λ_- be the eigenvalues of V so that $\lambda_+ > \lambda_-$:

$$\lambda_\pm = \mathrm{e}^{K_{\mathrm{P}}}\cosh\left(\frac{\beta(zJm+h)}{P}\right)$$

$$\pm\sqrt{\mathrm{e}^{2K_{\mathrm{P}}}\sinh^2\left(\frac{\beta(zJm+h)}{P}\right)+\mathrm{e}^{-2K_{\mathrm{P}}}}. \tag{4.57}$$

Here, the following relation is satisfied:

$$A\lambda_\pm = \cosh\left(\frac{\beta\Gamma}{P}\right)\cosh\left(\frac{\beta(zJm+h)}{P}\right)$$

$$\pm\sqrt{\cosh^2\left(\frac{\beta\Gamma}{P}\right)\sinh^2\left(\frac{\beta(zJm+h)}{P}\right)+1} \tag{4.58}$$

$$\underset{P\to\infty}{\to} \; 1 \pm \frac{\beta}{P}\sqrt{(zJm+h)^2+\Gamma^2} + \mathcal{O}\left(\frac{1}{P^2}\right). \tag{4.59}$$

In addition, we take the limit of $P \to \infty$ for

$$P\ln A + \ln \sum_{\{\sigma_k^z=\pm 1\}} V^P = (A\lambda_+)^P + (A\lambda_-)^P \tag{4.60}$$

$$\underset{P\to\infty}{\to} \; e^{\beta\sqrt{(zJm+h)^2+\Gamma^2}} + e^{-\beta\sqrt{(zJm+h)^2+\Gamma^2}} \tag{4.61}$$

$$= 2\cosh\left[\beta\sqrt{(zJm+h)^2+\Gamma^2}\right]. \tag{4.62}$$

Thus, the free energy per spin is given by

$$f = \frac{zJ}{2}m^2 - \frac{1}{\beta}\ln\left\{2\cosh\left[\beta\sqrt{(zJm+h)^2+\Gamma^2}\right]\right\}. \tag{4.63}$$

From the free energy, we can calculate the longitudinal magnetization m_z and the transverse magnetization m_x as

$$m_z = -\frac{\partial f}{\partial h} \tag{4.64}$$

$$= \frac{zJm+h}{\sqrt{(zJm+h)^2+\Gamma^2}}\tanh\left[\beta\sqrt{(zJm+h)^2+\Gamma^2}\right], \tag{4.65}$$

$$m_x = -\frac{\partial f}{\partial \Gamma} \tag{4.66}$$

$$= \frac{\Gamma}{\sqrt{(zJm+h)^2+\Gamma^2}}\tanh\left[\beta\sqrt{(zJm+h)^2+\Gamma^2}\right]. \tag{4.67}$$

Here, it can be easily understood that m_z corresponds to the m given by Eq. (4.52).

Next, we consider the phase diagram for $h \to +0$. In the limit of $h \to +0$, Eq. (4.65) for $m = m_z$ is written as

$$m_z = \frac{zJm_z}{\sqrt{(zJm_z)^2+\Gamma^2}}\tanh\left(\beta\sqrt{(zJm_z)^2+\Gamma^2}\right). \tag{4.68}$$

At the phase transition point, m_z changes from zero to a nonzero value. We then consider the limit of $m_z \to 0$ and obtain

$$\Gamma_c/zJ = \tanh(\beta_c\Gamma_c). \tag{4.69}$$

Therefore, the transition temperature T_c is obtained as

$$T_c = 1/\beta_c = \frac{\Gamma_c}{\tanh^{-1}(\Gamma_c/zJ)}. \tag{4.70}$$

In the region where $T < T_c$ and $\Gamma < \Gamma_c$, the longitudinal magnetization is a nonzero value, which means that the ferromagnetic phase appears. Figure 4.9 is the phase diagram of the Husimi–Temperely–Curie–Weiss model in a transverse field when $h \to +0$. Finally, we consider the quantum phase transition at $T = 0$. In this case, Eq. (4.68) can be written as

$$m_z = \sqrt{1 - (\Gamma/zJ)^2}. \tag{4.71}$$

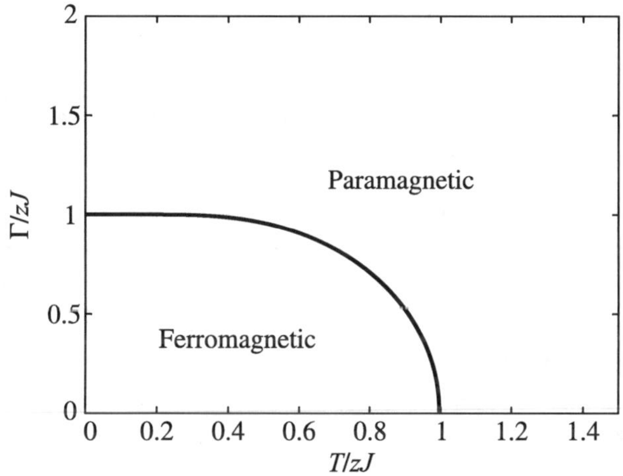

Fig. 4.9 Phase diagram of the Husimi–Temperely–Curie–Weiss model in a transverse field Γ for $h \to +0$.

The transverse field dependence of m_z is shown in Fig. 4.10. The quantum phase transition occurs at $\Gamma_c/zJ = 1$. Since m_z is not discontinuous at the phase transition point, the quantum phase transition is of the second-order.

4.2 Introduction of Quantum Spin Glass

Studies on quantum spin glass began at the same time as studies on classical spin glass, introduced in Chapter 2. Quantum spin glass can be categorized into two types. One is the random Ising model in a transverse field. The other is the random vector spin model in a quantum field studied by Bray and Moore (Bray and Moore, 1980). Analysis of the random vector spin model is more complicated than that of the random Ising model since controlling the quantum field is difficult. Thus, in order to understand the nature of the quantum spin glass, the random Ising model in a transverse field is mainly studied.

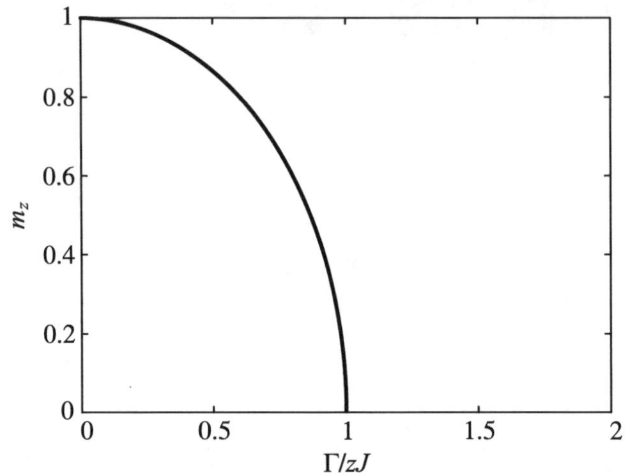

Fig. 4.10 Transverse field, Γ, dependence of the longitudinal magnetization m_z of the Husimi–Temperely–Curie–Weiss model in a transverse field Γ when $T = 0$. The quantum phase transition occurs at $\Gamma_c/zJ = 1$.

The most interesting aspect of quantum spin glass is that there are some possibilities of the existence of a replica symmetric spin glass phase. In the classical spin glass, there are barriers to the free energy landscape. The height of the barriers becomes infinite in the thermodynamic limit explained in Section 2.3.3. Therefore, it is impossible to escape from local minima in the free energy landscape by thermal fluctuations. As described in Section 2.3.3, this behavior means that there is non-ergodicity, i.e., replica symmetry breaking occurs in classical spin glass. On the other hand, quantum fluctuation could prevent trapping at local minima with the help of the quantum tunneling effect. Thus, an ergodic spin glass phase, that is, the replica symmetric spin glass phase, would appear in quantum spin glass. However, for the presence or absence of the replica symmetric spin glass phase, conflicting results have been reported (Ray et al., 1989; Thirumalai et al., 1989; Gold-schmidt and Lai, 1990; Lai and Goldschmidt, 1990), and this problem is still unresolved. With this background, the phase diagrams of the Sherrington–Kirkpatrick model in a transverse field (Pirc et al., 1985; Yamamoto and Ishii, 1987; Kopec, 1988a,b; Büttner and Usadel, 1991), the short-range Edwards–Anderson model in a transverse field (dos Santos et al., 1985; Walasek and Lukierska-Walasek, 1986; Guo et al., 1994; Rieger and Young, 1994; Guo et al., 1996; Rieger and Young, 1996), and some other models have been investigated.

A schematic phase diagram of the random Ising model in a transverse field is shown in Fig. 4.11 (Thill and Huse, 1995). In the small T and small Γ region, the spin glass phase appears. When the transverse field Γ is small, the thermal fluctuation is dominant. Thus, in the area indicated by $O_{\rm cl}$, the spin glass phase is characterized by the classical spin glass order. Furthermore, the solid line is the boundary of the phase transition induced by the thermal fluctuation between the paramagnetic phase and the spin glass phase. On the other hand, when the temperature T is small, quantum fluctuation is dominant. Thus, in the area indicated by O_q, there are possibilities of the existence of the replica symmetric spin glass phase. Furthermore, the dotted line denotes the boundary of the phase transition induced by the quantum fluctuation. Note that the boundaries between the paramagnetic phase and the spin glass phase are second-order phase transitions.

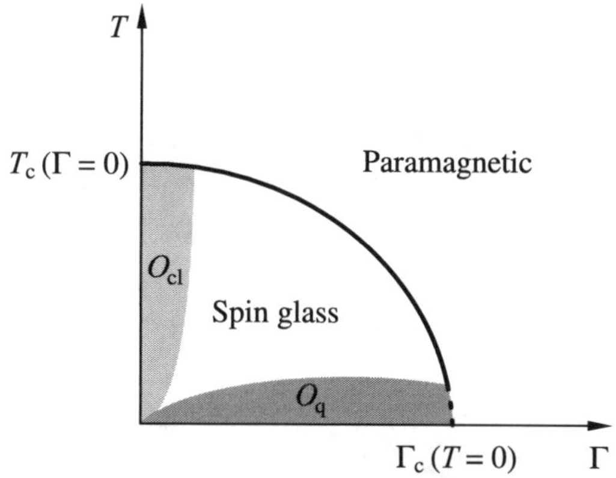

Fig. 4.11 Schematic phase diagram of the Ising model with random interactions in a transverse field Γ. O_cl, the spin glass phase, is characterized by the classical spin glass order. In the region indicated by O_q, there are possibilities of the existence of the replica symmetric spin glass phase due to the quantum fluctuation effect.

4.3 Sherrington–Kirkpatrick Model in a Transverse Field

The Hamiltonian of the Ising model in a transverse field Γ is written as:

$$\mathscr{H} = -\sum_{1\le i<j\le N} J_{ij}\hat{\sigma}_i^z\hat{\sigma}_j^z - \Gamma\sum_{i=1}^{N}\hat{\sigma}_i^x, \tag{4.72}$$

where

$$\hat{\sigma}_i^z = \begin{pmatrix}1 & 0\\ 0 & -1\end{pmatrix}, \tag{4.73}$$

$$\hat{\sigma}_i^x = \begin{pmatrix}0 & 1\\ 1 & 0\end{pmatrix}. \tag{4.74}$$

Here, if the interactions between all spin pairs exist, the probability distribution of the interactions is the Gaussian distribution, i.e., $\{J_{ij}\}$ is distributed as

$$P(J_{ij}) = \left(\frac{N}{2\pi J^2}\right)^{\frac{1}{2}} \exp\left[-\frac{N}{2J^2}\left(J_{ij}-\frac{J_0}{N}\right)^2\right]. \tag{4.75}$$

This model is called the Sherrington–Kirkpatrick (SK) model in a transverse field. The phase diagram of the model for temperature versus transverse field for $J_0 = 0$ has been studied by mean-field calculations and Monte Carlo simulations, which is explained in this section.

4.3.1 Mean-field calculations

Single-spin approximation

We study the phase diagram of the SK model in a transverse field by using the single-spin approximation. Under the approximation, we start from the replica symmetric solution for $\Gamma = 0$ explained in Section 2.3.2, that is,

$$m = \frac{1}{\sqrt{2\pi}} \int_{-\infty}^{\infty} dr e^{-\frac{r^2}{2}} \tanh\left[\beta h^z(r)\right], \tag{4.76}$$

$$q = \frac{1}{\sqrt{2\pi}} \int_{-\infty}^{\infty} dr e^{-\frac{r^2}{2}} \tanh^2\left[\beta h^z(r)\right], \tag{4.77}$$

where m and q are the order parameters representing the ferromagnetic phase and spin glass phase, respectively. Furthermore, $h^z(r)$ means the internal field at each spin when $\Gamma = 0$. Thus, if the transverse field is finite, the single-spin Hamiltonian of the SK model can be written as follows:

$$\hat{\mathscr{H}} = -h^z(r)\hat{\sigma}^z - \Gamma\hat{\sigma}^x. \tag{4.78}$$

Taking the semiclassical approximation described in Section 4.1.1, the effective field at each site is expressed as,

$$\vec{h}_0(r) = -h^z(r)\vec{e}_z - \Gamma\vec{e}_x, \tag{4.79}$$

$$h_0(r) := |\vec{h}_0(r)| = \sqrt{h^z(r)^2 + \Gamma^2}, \tag{4.80}$$

where $\vec{e}_z$ and $\vec{e}_x$ are the unit vectors in the z direction and x direction. By using the variables, the order parameters of the SK model in a transverse field under the semiclassical approximation are written as follows:

$$m = \frac{1}{\sqrt{2\pi}} \int_{-\infty}^{\infty} dr e^{-\frac{r^2}{2}} \tanh\left[\beta h^z(r)\right] \frac{|h^z(r)|}{h_0(r)}, \tag{4.81}$$

$$q = \frac{1}{\sqrt{2\pi}} \int_{-\infty}^{\infty} dr e^{-\frac{r^2}{2}} \tanh^2\left[\beta h^z(r)\right] \left[\frac{|h^z(r)|}{h_0(r)}\right]^2. \tag{4.82}$$

At the phase boundary of the spin glass phase, the spin glass order parameter changes from zero to a finite value. In the limit of $q \to 0$, i.e., $h_0(r) \simeq \Gamma$, Eq. (4.82) is approximated as follows:

$$q \simeq \frac{1}{\sqrt{2\pi}} \int_{-\infty}^{\infty} dr e^{-\frac{r^2}{2}} \tanh^2(\beta\Gamma) \frac{J^2 q r^2}{\Gamma^2}, \tag{4.83}$$

and we obtain

$$\frac{\Gamma}{J} = \tanh\left(\frac{\Gamma}{T}\right). \tag{4.84}$$

From Eq. (4.84), we obtain the phase diagram of the SK model in a transverse field under the single-spin approximation and the semi-classical approximation (see Fig. 4.12). In this analysis, when $T = 0$, the quantum phase transition between the paramagnetic phase and the spin glass phase occurs at $\Gamma_c/J = 1$.

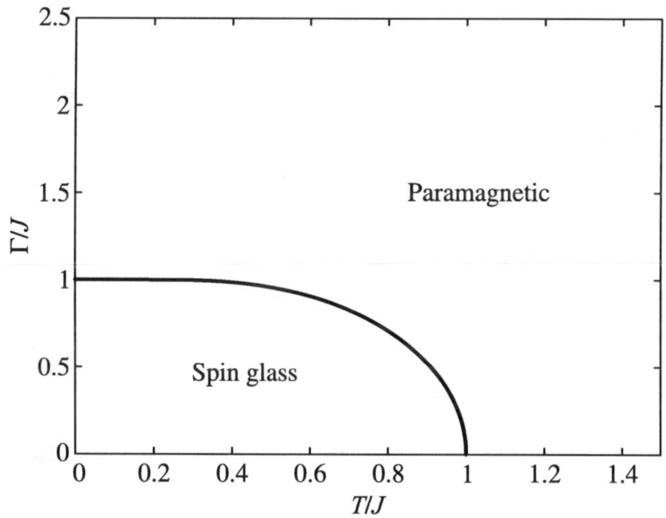

Fig. 4.12 Phase diagram of the SK model in a transverse field under the single-spin approximation.

Replica symmetric solution

We introduce the result by the replica symmetric approximation and static approximation. Here, we divide the Hamiltonian as follows:

$$\hat{\mathcal{H}} = \hat{\mathcal{H}}_0 + \hat{\mathcal{H}}_q, \tag{4.85}$$

$$\hat{\mathcal{H}}_0 = -\sum_{1 \leq i < j \leq N} J_{ij} \hat{\sigma}_i^z \hat{\sigma}_j^z, \tag{4.86}$$

$$\hat{\mathscr{H}}_{\mathrm{q}} = -\Gamma \sum_{i=1}^{N} \hat{\sigma}_i^x, \tag{4.87}$$

where N is the number of spins. In this case, the imaginary time representation of the partition function is written as follows:

$$Z = \mathrm{Tr}\exp\left(-\beta \hat{\mathscr{H}}_{\mathrm{q}}\right) \mathscr{T} \exp\left[\int_0^\beta d\tau \sum_{1 \le i < j \le N} J_{ij} \hat{\sigma}_i^z(\tau) \hat{\sigma}_j^z(\tau)\right], \tag{4.88}$$

where $\hat{\sigma}_i^z(\tau)$ indicates the imaginary time τ dependent spin variable, and $\mathscr{T}$ is the time-ordered product operator. $\mathscr{T}$ arranges the operators according to the imaginary time. For example, the time-ordered product operator between two imaginary time dependent operators $\hat{O}(\tau)$ is given by

$$\mathscr{T}\left[\hat{O}(\tau_1), \hat{O}(\tau_2)\right] = \begin{cases} \hat{O}(\tau_1)\hat{O}(\tau_2) & (\tau_1 > \tau_2) \\ \hat{O}(\tau_2)\hat{O}(\tau_1) & (\tau_2 > \tau_1) \end{cases}. \tag{4.89}$$

In this case, all operators in the partition function are commutative, and we can perform the replica method. The partition function obtained by the replica method is

$$\begin{aligned} Z^n &= \mathrm{Tr}\exp\left(-\beta \sum_{\alpha=1}^{n} \hat{\mathscr{H}}_{\mathrm{q}\alpha}\right) \\ &\quad \times \mathscr{T} \exp\left[\int_0^\beta d\tau \sum_{1 \le i < j \le N} J_{ij} \sum_{\alpha=1}^{n} \hat{\sigma}_{i\alpha}^z(\tau) \hat{\sigma}_{j\alpha}^z(\tau)\right], \end{aligned} \tag{4.90}$$

where α is the index of the replica. The partition function where the configurational average of $\{J_{ij}\}$ is performed based on Eq. (4.75) for $J_0 = 0$ is written as

$$\begin{aligned} \overline{Z^n} &= \int_{-\infty}^{\infty} \prod_{1 \le i < j \le N} dJ_{ij} P(J_{ij}) \mathrm{Tr}\exp\left(-\beta \sum_{\alpha=1}^{n} \hat{\mathscr{H}}_{\mathrm{q}\alpha}\right) \\ &\quad \times \mathscr{T} \exp\left[\int_0^\beta d\tau \sum_{1 \le i < j \le N} J_{ij} \sum_{\alpha=1}^{n} \hat{\sigma}_{i\alpha}^z(\tau) \hat{\sigma}_{j\alpha}^z(\tau)\right] \\ &= \mathrm{Tr}\exp\left(-\beta \sum_{\alpha=1}^{n} \hat{\mathscr{H}}_{\mathrm{q}\alpha}\right) \end{aligned} \tag{4.91}$$

$$\times \mathscr{T} \exp\left[\frac{J^2}{2N} \sum_{1\le i<j\le N} \int_0^\beta d\tau \int_0^\beta d\tau' \right.$$

$$\left. \times \sum_{1\le \alpha,\beta\le n} \hat{\sigma}^z_{i\alpha}(\tau)\hat{\sigma}^z_{j\alpha}(\tau)\hat{\sigma}^z_{i\beta}(\tau')\hat{\sigma}^z_{j\beta}(\tau')\right]. \tag{4.92}$$

Let us consider the case for a large enough N. The power of the exponential in Eq. (4.92) becomes

$$\frac{J^2}{2N} \sum_{1\le i<j\le N} \int_0^\beta d\tau \int_0^\beta d\tau' \sum_{1\le \alpha,\beta\le n} \hat{\sigma}^z_{i\alpha}(\tau)\hat{\sigma}^z_{j\alpha}(\tau)\hat{\sigma}^z_{i\beta}(\tau')\hat{\sigma}^z_{j\beta}(\tau') \tag{4.93}$$

$$\simeq \frac{J^2}{2N}\frac{1}{2} \sum_{1\le i,j\le N} \int_0^\beta d\tau \int_0^\beta d\tau'$$

$$\times \left[2 \sum_{1\le \alpha<\beta\le n} \hat{\sigma}^z_{i\alpha}(\tau)\hat{\sigma}^z_{j\alpha}(\tau)\hat{\sigma}^z_{i\beta}(\tau')\hat{\sigma}^z_{j\beta}(\tau')\right.$$

$$\left. + \sum_{\alpha=1}^{n} \hat{\sigma}^z_{i\alpha}(\tau)\hat{\sigma}^z_{j\alpha}(\tau)\hat{\sigma}^z_{i\alpha}(\tau')\hat{\sigma}^z_{j\alpha}(\tau')\right] \tag{4.94}$$

$$= \frac{J^2}{2N} \int_0^\beta d\tau \int_0^\beta d\tau' \sum_{1\le \alpha<\beta\le n} \left[\sum_{i=1}^{N} \hat{\sigma}^z_{i\alpha}(\tau)\hat{\sigma}^z_{i\beta}(\tau')\right]^2$$

$$+ \frac{J^2}{4N} \int_0^\beta d\tau \int_0^\beta d\tau' \sum_{\alpha=1}^{n} \left[\sum_{i=1}^{N} \hat{\sigma}^z_{i\alpha}(\tau)\hat{\sigma}^z_{i\alpha}(\tau')\right]^2. \tag{4.95}$$

Besides, for $\hat{\mathscr{H}}_{\mathrm{q}\alpha}$, we also use the imaginary time representation given by

$$\beta \hat{\mathscr{H}}_{\mathrm{q}\alpha} = -\Gamma \int_0^\beta d\tau \sum_{i=1}^{N} \hat{\sigma}^x_{i\alpha}(\tau). \tag{4.96}$$

Thus, the partition function is written as follows:

$$\overline{Z^n} = \mathrm{Tr} \exp\left[\Gamma \int_0^\beta d\tau \sum_{\alpha=1}^{n} \sum_{i=1}^{N} \hat{\sigma}^x_{i\alpha}(\tau)\right]$$

$$\times \mathscr{T} \exp \left\{ \frac{J^2}{2N} \int_0^\beta d\tau \int_0^\beta d\tau' \sum_{1\leq\alpha<\beta\leq n} \left[\sum_{i=1}^N \hat{\sigma}_{i\alpha}^z(\tau)\hat{\sigma}_{i\beta}^z(\tau') \right]^2 \right.$$

$$\left. + \frac{J^2}{4N} \int_0^\beta d\tau \int_0^\beta d\tau' \sum_{\alpha=1}^n \left[\sum_{i=1}^N \hat{\sigma}_{i\alpha}^z(\tau)\hat{\sigma}_{i\alpha}^z(\tau') \right]^2 \right\}. \tag{4.97}$$

Next, by the Hubbard–Stratonovich transformation, we obtain

$$\exp\left[\frac{J^2}{2N} \int_0^\beta d\tau \int_0^\beta d\tau' \left(\sum_{i=1}^N \hat{\sigma}_{i\alpha}^z(\tau)\hat{\sigma}_{i\beta}^z(\tau') \right)^2 \right]$$

$$= \sqrt{\frac{N\beta J^2}{2\pi}} \int_{-\infty}^{\infty} dQ^{\alpha\beta}(\tau,\tau') \exp\left[-\frac{N}{2} J^2 \int_0^\beta d\tau \int_0^\beta d\tau' Q^{\alpha\beta}(\tau,\tau')^2 \right.$$

$$\left. + J^2 \int_0^\beta d\tau \int_0^\beta d\tau' Q^{\alpha\beta}(\tau,\tau') \left(\sum_{i=1}^N \hat{\sigma}_{i\alpha}^z(\tau)\hat{\sigma}_{i\beta}^z(\tau') \right) \right], \tag{4.98}$$

$$\exp\left[\frac{J^2}{4N} \int_0^\beta d\tau \int_0^\beta d\tau' \left(\sum_{i=1}^N \hat{\sigma}_{i\alpha}^z(\tau)\hat{\sigma}_{i\alpha}^z(\tau') \right)^2 \right]$$

$$= \sqrt{\frac{N\beta J^2}{4\pi}} \int_{-\infty}^{\infty} dR^{\alpha\alpha}(\tau,\tau') \exp\left[-\frac{N}{4} J^2 \int_0^\beta d\tau \int_0^\beta d\tau' R^{\alpha\alpha}(\tau,\tau')^2 \right.$$

$$\left. + \frac{1}{2} J^2 \int_0^\beta d\tau \int_0^\beta d\tau' R^{\alpha\alpha}(\tau,\tau') \left(\sum_{i=1}^N \hat{\sigma}_{i\alpha}^z(\tau)\hat{\sigma}_{i\alpha}^z(\tau') \right) \right]. \tag{4.99}$$

By using the aforementioned two relations, the partition function is written as follows:

$$\overline{Z^n} = \int_{-\infty}^{\infty} \prod_{1\leq\alpha<\beta\leq n} dQ^{\alpha\beta}(\tau,\tau') \int_{-\infty}^{\infty} \prod_{\alpha=1}^n dR^{\alpha\alpha}(\tau,\tau')$$

$$\times \mathrm{Tr} \exp\left[\Gamma \int_0^\beta d\tau \sum_{\alpha=1}^n \sum_{i=1}^N \hat{\sigma}_{i\alpha}^x(\tau) \right]$$

$$\times \mathscr{T} \exp\left[-\frac{N}{2}J^2 \int_0^\beta d\tau \int_0^\beta d\tau' \sum_{1\le\alpha<\beta\le n} Q^{\alpha\beta}(\tau,\tau')^2\right.$$

$$-\frac{N}{4}J^2 \int_0^\beta d\tau \int_0^\beta d\tau' \sum_{\alpha=1}^{n} R^{\alpha\alpha}(\tau,\tau')^2$$

$$+J^2 \int_0^\beta d\tau \int_0^\beta d\tau' \sum_{1\le\alpha<\beta\le n} Q^{\alpha\beta}(\tau,\tau')\left(\sum_{i=1}^{N}\hat{\sigma}_{i\alpha}^z(\tau)\hat{\sigma}_{i\beta}^z(\tau')\right)$$

$$\left.+\frac{1}{2}J^2 \int_0^\beta d\tau \int_0^\beta d\tau' \sum_{\alpha=1}^{n} R^{\alpha\alpha}(\tau,\tau')\left(\sum_{i=1}^{N}\hat{\sigma}_{i\alpha}^z(\tau)\hat{\sigma}_{i\alpha}^z(\tau')\right)\right] \tag{4.100}$$

$$= \int_{-\infty}^{\infty} \prod_{1\le\alpha<\beta\le n} dQ^{\alpha\beta}(\tau,\tau') \int_{-\infty}^{\infty} \prod_{\alpha=1}^{n} dR^{\alpha\alpha}(\tau,\tau')$$

$$\times \exp\left[-\frac{N}{2}J^2 \int_0^\beta d\tau \int_0^\beta d\tau' \sum_{1\le\alpha<\beta\le n} Q^{\alpha\beta}(\tau,\tau')^2\right.$$

$$\left.-\frac{N}{4}J^2 \int_0^\beta d\tau \int_0^\beta d\tau' \sum_{\alpha=1}^{n} R^{\alpha\alpha}(\tau,\tau')^2\right]$$

$$\times\left\{\mathrm{Tr}\exp\left[\Gamma\int_0^\beta d\tau \sum_{\alpha=1}^{n}\hat{\sigma}_\alpha^x(\tau)\right]\right.$$

$$\times \mathscr{T} \exp\left[J^2 \int_0^\beta d\tau \int_0^\beta d\tau' \sum_{1\le\alpha<\beta\le n} Q^{\alpha\beta}(\tau,\tau')\hat{\sigma}_\alpha^z(\tau)\hat{\sigma}_\beta^z(\tau')\right.$$

$$\left.\left.+\frac{1}{2}J^2 \int_0^\beta d\tau \int_0^\beta d\tau' \sum_{\alpha=1}^{n} R^{\alpha\alpha}(\tau,\tau')\hat{\sigma}_\alpha^z(\tau)\hat{\sigma}_\alpha^z(\tau')\right]\right\}^N. \tag{4.101}$$

Here, we define ϕ by

$$\phi := \frac{1}{2}J^2 \int_0^\beta d\tau \int_0^\beta d\tau' \sum_{1\le\alpha<\beta\le n} Q^{\alpha\beta}(\tau,\tau')^2$$

$$+\frac{1}{4}J^2\int_0^\beta d\tau\int_0^\beta d\tau'\sum_{\alpha=1}^n R^{\alpha\alpha}(\tau,\tau')^2$$

$$-\ln\left\{\mathrm{Tr}\mathscr{T}\exp\left[\Gamma\int_0^\beta d\tau\sum_{\alpha=1}^n \hat{\sigma}_\alpha^x(\tau)\right]\right.$$

$$\times\exp\left[J^2\int_0^\beta d\tau\int_0^\beta d\tau'\sum_{1\le\alpha<\beta\le n} Q^{\alpha\beta}(\tau,\tau')\hat{\sigma}_\alpha^z(\tau)\hat{\sigma}_\beta^z(\tau')\right.$$

$$\left.\left.+\frac{1}{2}J^2\int_0^\beta d\tau\int_0^\beta d\tau'\sum_{\alpha=1}^n R^{\alpha\alpha}(\tau,\tau')\hat{\sigma}_\alpha^z(\tau)\hat{\sigma}_\alpha^z(\tau')\right]\right\}. \tag{4.102}$$

Using ϕ, we obtain

$$\overline{Z^n}=\int_{-\infty}^{\infty}\prod_{1\le\alpha<\beta\le n} dQ^{\alpha\beta}(\tau,\tau')\int_{-\infty}^{\infty}\prod_{\alpha=1}^n dR^{\alpha\alpha}(\tau,\tau')\mathrm{e}^{-N\phi}. \tag{4.103}$$

Since the power of the exponential in Eq. (4.103) is proportional to N, we can calculate these integrals by the saddle-point method. Then, we obtain

$$\overline{Z^n}\simeq \mathrm{e}^{-N\phi}\simeq 1+Nn\left(\frac{1}{n}\phi\right). \tag{4.104}$$

By using $\overline{Z^n}$, the free energy per spin is expressed as

$$-\beta f=\frac{1}{N}\lim_{n\to 0}\frac{\overline{Z^n}-1}{n}=\lim_{n\to 0}\left(\frac{1}{n}\phi\right), \tag{4.105}$$

where $Q^{\alpha\beta}$ and $R^{\alpha\alpha}$ are decided by the saddle-point conditions. The saddle-point conditions are given by

$$R^{\alpha\alpha}(\tau,\tau')=\frac{\mathrm{Tr}\,\mathscr{T}\hat{\sigma}_\alpha^z(\tau)\hat{\sigma}_\alpha^z(\tau')\mathrm{e}^L}{\mathrm{Tr}\mathscr{T}\mathrm{e}^L}, \tag{4.106}$$

$$Q^{\alpha\beta}(\tau,\tau')=\frac{\mathrm{Tr}\,\mathscr{T}\hat{\sigma}_\alpha^z(\tau)\hat{\sigma}_\beta^z(\tau')\mathrm{e}^L}{\mathrm{Tr}\mathscr{T}\mathrm{e}^L}, \tag{4.107}$$

where L is defined as

$$\mathrm{e}^L := \exp\left[\Gamma\int_0^\beta d\tau\sum_{\alpha=1}^n \hat{\sigma}_\alpha^x(\tau)\right]$$

$$\times \exp\left[J^2 \int_0^\beta d\tau \int_0^\beta d\tau' \sum_{1\le\alpha<\beta\le n} Q^{\alpha\beta}(\tau,\tau')\hat{\sigma}_\alpha^z(\tau)\hat{\sigma}_\beta^z(\tau')\right.$$

$$\left.+\frac{1}{2}J^2 \int_0^\beta d\tau \int_0^\beta d\tau' \sum_{\alpha=1}^{n} R^{\alpha\alpha}(\tau,\tau')\hat{\sigma}_\alpha^z(\tau)\hat{\sigma}_\alpha^z(\tau')\right]. \qquad (4.108)$$

Next, we take the replica symmetric approximation and static approximation given by

$$R^{\alpha\alpha}(\tau,\tau') = R, \qquad (4.109)$$

$$Q^{\alpha\beta}(\tau,\tau') = Q, \qquad (4.110)$$

that is, the dependences of the replica and imaginary time τ are eliminated. Under the approximations, ϕ is written as

$$\begin{aligned}
\phi &= \frac{1}{4}\beta^2 J^2 n(n-1)Q^2 + \frac{1}{4}\beta^2 J^2 nR^2 \\
&\quad - \ln\left\{ \mathrm{Tr}\mathscr{T} \exp\left[\Gamma \int_0^\beta d\tau \sum_{\alpha=1}^{n} \hat{\sigma}_\alpha^x(\tau)\right]\right. \\
&\quad \times \exp\left[\frac{1}{2}J^2 Q \int_0^\beta d\tau \int_0^\beta d\tau' \sum_{1\le\alpha\ne\beta\le n} \hat{\sigma}_\alpha^z(\tau)\hat{\sigma}_\beta^z(\tau')\right. \\
&\quad \left.\left. + \frac{1}{2}J^2 R \int_0^\beta d\tau \int_0^\beta d\tau' \sum_{\alpha=1}^{n} \hat{\sigma}_\alpha^z(\tau)\hat{\sigma}_\alpha^z(\tau')\right]\right\} \qquad (4.111) \\
&= \frac{1}{4}\beta^2 J^2 n(n-1)Q^2 + \frac{1}{4}\beta^2 J^2 nR^2 \\
&\quad - \ln\left\{ \mathrm{Tr}\mathscr{T} \exp\left[\Gamma \int_0^\beta d\tau \sum_{\alpha=1}^{n} \hat{\sigma}_\alpha^x(\tau)\right]\right. \\
&\quad \times \exp\left[\frac{1}{2}\left(J\sqrt{Q}\sum_{\alpha=1}^{n}\int_0^\beta d\tau \hat{\sigma}_\alpha^z(\tau)\right)^2\right.
\end{aligned}$$

$$+\frac{1}{2}\sum_{\alpha=1}^{n}\left(J\sqrt{R-Q}\int_0^\beta d\tau\hat{\sigma}_\alpha^z(\tau)\right)^2\Bigg]\Bigg\}. \tag{4.112}$$

By using the Hubbard–Stratonovich transformation, we obtain

$$\exp\left[\frac{1}{2}\left(J\sqrt{Q}\sum_{\alpha=1}^{n}\int_0^\beta d\tau\hat{\sigma}_\alpha^z(\tau)\right)^2\right]$$

$$=\frac{1}{\sqrt{2\pi}}\int_{-\infty}^{\infty}dxe^{-\frac{x^2}{2}}\exp\left[J\sqrt{Q}\sum_{\alpha=}^{n}\int_0^\beta d\tau\hat{\sigma}_\alpha^z(\tau)x\right], \tag{4.113}$$

$$\exp\left[\frac{1}{2}\left(J\sqrt{R-Q}\int_0^\beta d\tau\hat{\sigma}_\alpha^z(\tau)\right)^2\right]$$

$$=\frac{1}{\sqrt{2\pi}}\int_{-\infty}^{\infty}dye^{-\frac{y^2}{2}}\exp\left[J\sqrt{R-Q}\int_0^\beta d\tau\hat{\sigma}_\alpha^z(\tau)y\right]. \tag{4.114}$$

From these relations, ϕ can be expressed as

$$\phi = \frac{1}{4}\beta^2J^2n(n-1)Q^2+\frac{1}{4}\beta^2J^2nR^2$$

$$-\ln\left[\frac{1}{\sqrt{2\pi}}\int_{-\infty}^{\infty}dxe^{-\frac{x^2}{2}}\left(\frac{1}{\sqrt{2\pi}}\int_{-\infty}^{\infty}dye^{-\frac{y^2}{2}}\right.\right.$$

$$\left.\left.\times\mathrm{Tr}\exp\left\{\int_0^\beta d\tau\left[\Gamma\hat{\sigma}^x(\tau)+J\sqrt{Q}\hat{\sigma}^z(\tau)x+J\sqrt{R-Q}\hat{\sigma}^z(\tau)y\right]\right\}\right)^n\right] \tag{4.115}$$

$$=\frac{1}{4}\beta^2J^2n(n-1)Q^2+\frac{1}{4}\beta^2J^2nR^2$$

$$-\ln\left(\frac{1}{\sqrt{2\pi}}\int_{-\infty}^{\infty}dxe^{-\frac{x^2}{2}}\left\{\frac{1}{\sqrt{2\pi}}\int_{-\infty}^{\infty}dye^{-\frac{y^2}{2}}\right.\right.$$

$$\left.\left.\times2\cosh\left[\beta\sqrt{\Gamma^2+\left(J\sqrt{Q}x+J\sqrt{R-Q}y\right)^2}\right]\right\}^n\right). \tag{4.116}$$

We define $h(x,y)$ as

$$h(x,y) := J\sqrt{Q}x + J\sqrt{R-Q}y, \tag{4.117}$$

and the logarithmic part in ϕ (Eq. (4.116)) can be calculated as follows:

$$\ln\left(\frac{1}{\sqrt{2\pi}}\int_{-\infty}^{\infty} dx e^{-\frac{x^2}{2}} \times\left\{\frac{1}{\sqrt{2\pi}}\int_{-\infty}^{\infty} dy e^{-\frac{y^2}{2}} 2\cosh\left[\beta\sqrt{\Gamma^2+h(x,y)^2}\right]\right\}^n\right) \tag{4.118}$$

$$= \ln\left[\frac{1}{\sqrt{2\pi}}\int_{-\infty}^{\infty} dx e^{-\frac{x^2}{2}} \times \exp\left(n\ln\left\{\frac{1}{\sqrt{2\pi}}\int_{-\infty}^{\infty} dy e^{-\frac{y^2}{2}} 2\cosh\left[\beta\sqrt{\Gamma^2+h(x,y)^2}\right]\right\}\right)\right]$$

$$\simeq \ln\left[\frac{1}{\sqrt{2\pi}}\int_{-\infty}^{\infty} dx e^{-\frac{x^2}{2}} \times \left(1+n\ln\left\{\frac{1}{\sqrt{2\pi}}\int_{-\infty}^{\infty} dy e^{-\frac{y^2}{2}} 2\cosh\left[\beta\sqrt{\Gamma^2+h(x,y)^2}\right]\right\}\right)\right] \tag{4.119}$$

$$\simeq n\frac{1}{\sqrt{2\pi}}\int_{-\infty}^{\infty} dx e^{-\frac{x^2}{2}} \times \ln\left\{\frac{1}{\sqrt{2\pi}}\int_{-\infty}^{\infty} dy e^{-\frac{y^2}{2}} 2\cosh\left[\beta\sqrt{\Gamma^2+h(x,y)^2}\right]\right\}, \tag{4.120}$$

where the approximation is valid for very small n. On top of that, the free energy per spin can be written as

$$-\beta f = \frac{1}{4}\beta^2 J^2 (R^2 - Q^2) - \frac{1}{\sqrt{2\pi}}\int_{-\infty}^{\infty} dx e^{-\frac{x^2}{2}} \times \ln\left\{\frac{1}{\sqrt{2\pi}}\int_{-\infty}^{\infty} dy e^{-\frac{y^2}{2}} 2\cosh\left[\beta\sqrt{\Gamma^2+h(x,y)^2}\right]\right\}. \tag{4.121}$$

The saddle-point conditions of R and Q are given as follows:

$$R = \frac{1}{\sqrt{2\pi}}\int_{-\infty}^{\infty} dx e^{-\frac{x^2}{2}} Y^{-1}$$

$$\times \frac{1}{\sqrt{2\pi}}\int_{-\infty}^{\infty} dy e^{-\frac{y^2}{2}} \left\{ \frac{h(x,y)^2}{h'(x,y)^2}\cosh\left[\beta h'(x,y)\right]\right.$$

$$\left. + \frac{\Gamma^2}{\beta h'(x,y)^3}\sinh\left[\beta h'(x,y)\right]\right\}, \tag{4.122}$$

$$Q = \frac{1}{\sqrt{2\pi}}\int_{-\infty}^{\infty} dx e^{-\frac{x^2}{2}} Y^{-2}$$

$$\times \left\{ \frac{1}{\sqrt{2\pi}}\int_{-\infty}^{\infty} dy e^{-\frac{y^2}{2}} \frac{h(x,y)}{h'(x,y)}\sinh\left[\beta h'(x,y)\right]\right\}^2, \tag{4.123}$$

where

$$Y := \frac{1}{\sqrt{2\pi}}\int_{-\infty}^{\infty} dy e^{-\frac{y^2}{2}} \cosh\left[\beta h'(x,y)\right], \tag{4.124}$$

$$h'(x,y) := \sqrt{h(x,y)^2 + \Gamma^2}. \tag{4.125}$$

Figure 4.13 is the phase diagram of the SK model in a transverse field under the replica symmetric approximation and static approximation. The phase boundaries of the phase diagram are obtained by Eqs. (4.122) and (4.123). In this case, when $T = 0$, the quantum phase transition occurs at $\Gamma_c/J = 2$ which is larger than the value obtained by the single-spin approximation described earlier. In addition, the low-temperature behavior obtained here is quite different from that obtained under the single-spin approximation.

In the following discussions, this phase diagram obtained from the aforementioned equation followed by Ishii and Yamamoto (1985) will be used. On the other hand, Fedorov and Shender (1986) and Usadel (1986) derived an other type of equation of state for the quantum SK model:

$$\int_{-\infty}^{\infty} dx e^{-\frac{x^2}{2}} (x^2 - 2)\cosh\left(\sqrt{\beta J x^2 + \beta^2\Gamma^2}\right) = 0. \tag{4.126}$$

Of course, the phase diagram obtained by this equation is the same as Fig. 4.13.

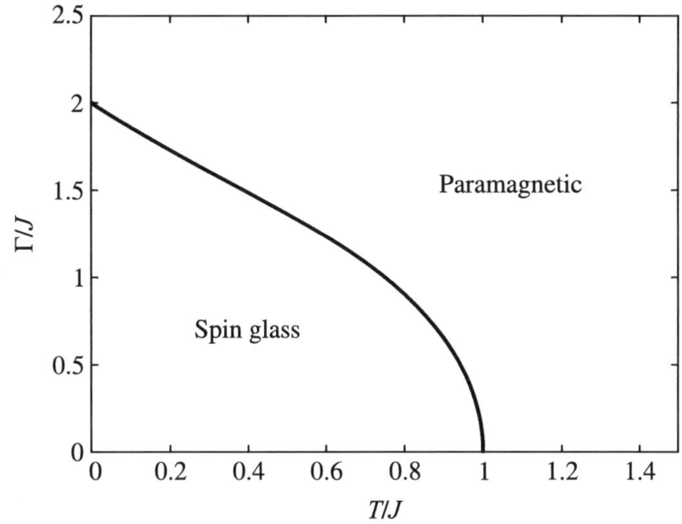

Fig. 4.13 Phase diagram of the SK model in a transverse field under the replica symmetric approximation and static approximation.

Analysis beyond the static approximation
In the static approximation, the imaginary time dependence of the order parameter is eliminated. This imaginary time dependence is strongly related to the quantum effect. Since the quantum fluctuation is dominant at low- temperatures, we should take care of the imaginary time dependence of the order parameter. Analyses beyond the static approximation for the SK model in a transverse field were done (Walasek and Lukierska-Walasek, 1988; Goldschmidt and Lai, 1990; Usadel et al., 1991; Takahashi, 2007). Figure 4.14 shows the phase diagrams obtained by the analysis beyond the static approximation with or without the replica symmetry breaking. At low temperatures, the behavior of these results is different from that obtained by the static approximation shown in Fig. 4.13. Furthermore, the critical transverse field at $T/J = 0$ was estimated as $\Gamma_c \simeq 1.6$ in these phase diagrams. This value is consistent with that obtained by Monte Carlo simulations and exact diagonalization, which will be explained in the next subsection.

4.3.2 Simulation results

We now review the phase diagram of the SK model in a transverse field obtained by Monte Carlo simulations and exact diagonalization methods. Monte Carlo simulations can study the finite-temperature properties of large size systems. On the other hand, zero-temperature properties of small size systems can be exactly calculated by exact diagonalization methods.

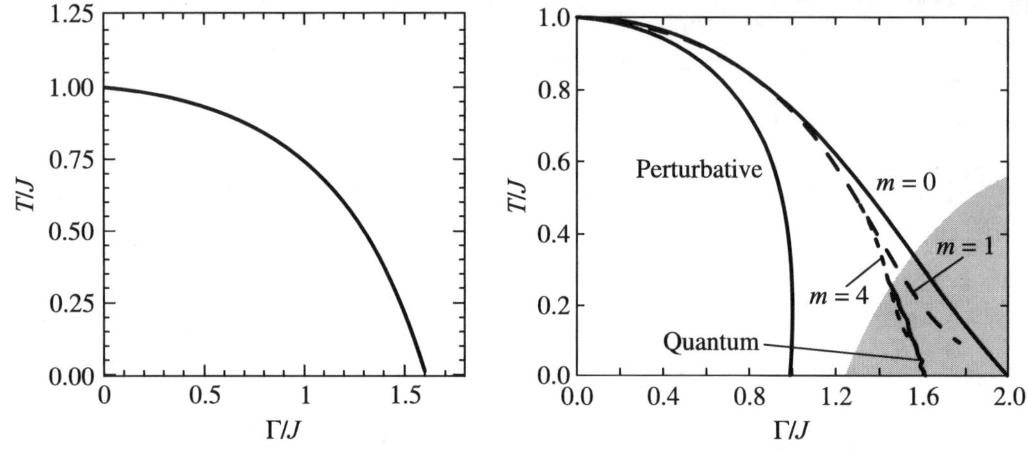

Fig. 4.14 (Left panel) Phase diagram of the SK model in a transverse field obtained by the replica symmetry breaking solution without using static approximation (from Goldschmidt and Lai, 1990). (Right panel) Phase diagram of the SK model in a transverse field obtained by the replica symmetric solution without using static approximation. "Perturbative" denotes the naive Landau expansion result which is the same as in Fig. 4.12. m stands for the number of modes expressing the dynamical nature; $m=0$ is the result of static approximation. "Quantum" denotes the result of the improved Landau expansion in a quantum regime where the quantum fluctuation is dominant. The quantum regime is shown by the shaded portion (from Takahashi, 2007).

Monte Carlo simulation results

We explain the Monte Carlo simulation results of the SK model in a transverse field. The Hamiltonian of the SK model in a transverse field is a non-diagonal matrix, when we use the eigenvectors of $\hat{\sigma}_i^z$ as the basis. Thus, we can use the path-integral formalism explained in Section 4.1.2. The partition function of the SK model in a transverse field is obtained as

$$Z = \lim_{P\to\infty} \left[\frac{1}{2}\sinh\left(\frac{2\beta\Gamma}{P}\right)\right]^{NP/2} \times \sum_{\{\sigma_{i,k}^z=\pm 1\}} \exp\left\{\sum_{k=1}^{P}\left[\sum_{1\le i,j\le N} \frac{\beta J_{ij}}{P}\sigma_{i,k}^z\sigma_{j,k}^z + \frac{1}{2}\ln\coth\left(\frac{\beta\Gamma}{P}\right)\sum_{i=1}^{N}\sigma_{i,k}^z\sigma_{i,k+1}^z\right]\right\}. \quad (4.127)$$

Thus, an effective Hamiltonian which is the $d+1$ -dimensional classical Ising model is given as

$$\mathscr{H}_{\text{eff}} = -\sum_{1\le i,j\le N}\sum_{k=1}^{P}\frac{J_{ij}}{P}\sigma_{i,k}^z\sigma_{j,k}^z - \frac{1}{2\beta}\ln\coth\left(\frac{\beta\Gamma}{P}\right)\sum_{i=1}^{N}\sum_{k=1}^{P}\sigma_{i,k}^z\sigma_{i,k+1}^z. \quad (4.128)$$

Using this representation, Monte Carlo simulations can calculate finite-temperature properties of the system given by Eq. (4.72) as explained in Section 3.3.1.

To quantify the spin glass order, we define the spin glass order parameter as

$$q := \frac{1}{NP}\sum_{i=1}^{N}\sum_{k=1}^{P}(\sigma_{i,k}^{z})_{\alpha}(\sigma_{i,k}^{z})_{\beta}, \tag{4.129}$$

where $(\sigma_{i,k}^{z})_{\alpha}$ is the spin variable at replica α. Furthermore, the Binder ratio for the spin glass order parameter is given as

$$g = \frac{1}{2}\left[3 - \overline{\left(\frac{\langle q^4\rangle}{\langle q^2\rangle^2}\right)}\right], \tag{4.130}$$

where $\langle\cdot\rangle$ and the bar indicate the thermal average and the configurational average, respectively. As described in Section 2.1.3, the second-order phase transition point and a critical exponent are obtained from the Binder ratio for several system sizes. The Binder ratio is scaled as

$$g = g(L/\xi, P/L^{z}), \tag{4.131}$$

near the critical point (Guo et al., 1994). Here, L is the linear dimension of the system, ξ is the correlation length defined by Eq. (2.11), and z is the dynamical exponent. Furthermore, the correlation length is scaled as

$$\xi \sim (T - T_{\mathrm{c}})^{-\nu_T} \sim (\Gamma - \Gamma_{\mathrm{c}})^{-\nu_\Gamma}, \tag{4.132}$$

where ν_T and ν_Γ are the critical exponents for correlation length. However, the linear dimension is not well-defined since the interactions exist between all spins in the SK model in a transverse field. Then, we introduce the effective dimension of the system d_{c} such that $N = L^{d_{\mathrm{c}}}$. By using the effective dimension, the Binder ratio is scaled as

$$g \sim g\left[(T - T_{\mathrm{c}})N^{x_T}, P/N^{z/d_{\mathrm{c}}}\right] \tag{4.133}$$

$$\sim g\left[(\Gamma - \Gamma_{\mathrm{c}})N^{x_\Gamma}, P/N^{z/d_{\mathrm{c}}}\right], \tag{4.134}$$

where N is the number of spins, $x_T := 1/(\nu_T d_{\mathrm{c}})$, and $x_\Gamma := 1/(\nu_\Gamma d_{\mathrm{c}})$. At $T = T_{\mathrm{c}}$ or $\Gamma = \Gamma_{\mathrm{c}}$, the Binder ratio does not depend on the lattice size, which means that the Binder ratios cross at the critical point for different system size N.

There have been some studies which investigated finite-temperature properties of the SK model in a transverse field by Monte Carlo simulations (Ishii and Yamamoto, 1986; Ray et al., 1989; Lai and Goldschmidt, 1990; Mukherjee et al., 2015). We explain the recent results by Mukherjee et al. (2015) for $J_0 = 0$ in Eq. (4.75). Figure 4.15 shows the Binder ratio at high temperatures where the quantum effect is small. The appropriate effective dimension and dynamical exponent are used in the classical SK model ($d_{\mathrm{c}} = 6$

and $z = 4$) (Billoire and Campbell, 2011), since the quantum effect is small at high temperatures. The crossing points in Figs. 4.15 (a), (c), and (e) correspond to the critical point (T_c, Γ_c). Furthermore, in all cases, the Binder ratios for different system sizes cross at the same value, $g_c \simeq 0.22$, which is called the critical Binder ratio. In addition, Figs. 4.15 (b), (d), and (f) show that the Binder ratio is well scaled by the scaling function (Eqs. (4.133) and (4.134)) using $x_T = x_\Gamma \simeq 0.31$. At high temperatures, the critical exponent of the correlation length is $\nu_T = \nu_\Gamma \simeq 0.5$. On the other hand, Fig. 4.15 shows the Binder ratio at low temperatures where the quantum effect is dominant. Thus, the appropriate effective dimension and dynamical exponent are used in the SK model in a transverse field ($d_c = 8$ and $z = 2$) (Read et al., 1995; Lancaster and Ritort, 1997). The crossing points in Figs. 4.15 (a) and (c) correspond to the critical point (T_c, Γ_c). Furthermore, in all cases, the Binder ratios for different system sizes cross at the same value $g_c \simeq 0$. In addition, Figs. 4.15 (b) and (d) show that the Binder ratio is well scaled by the scaling function (Eq. (4.134)) using $x_T = x_\Gamma \simeq 0.50$. Thus, at low temperatures, the critical exponent of the correlation length is $\nu_T = \nu_\Gamma \simeq 0.25$. These results mean that the critical behavior changes between the high-temperature region ($\nu_T = \nu_\Gamma \simeq 0.5$) and the low-temperature region ($\nu_T = \nu_\Gamma \simeq 0.25$). Mukherjee et al. (2015) concluded that the boundary is $T_c/J \simeq 0.45$ (see Fig. 4.18).

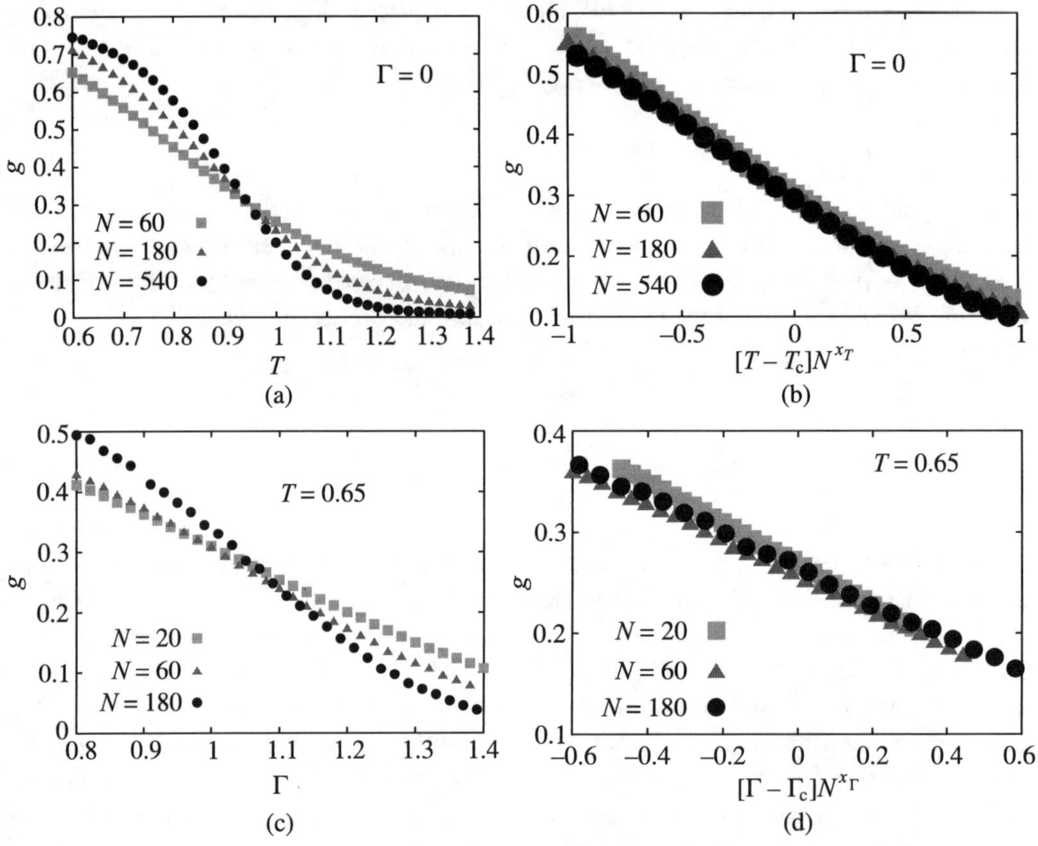

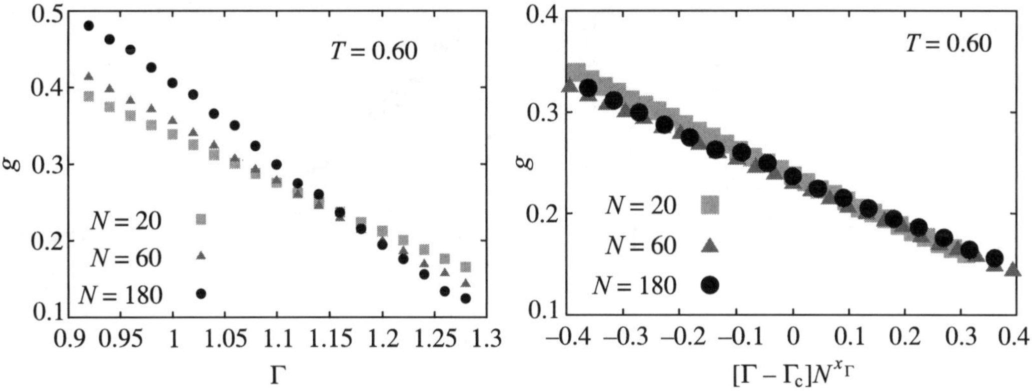

(e) (f)

Fig. 4.15 Monte Carlo simulation results of the SK model in a transverse field in a high-temperature region for $J = 1$. In these cases, the classical fluctuation is dominant. (a) Temperature, T, dependence of the Binder ratio g for $\Gamma = 0$. (b) Scaling plot of the Binder ratio g when $x_T = 0.31$ for $\Gamma = 0$. (c) Transverse field, Γ, dependence of g for $T = 0.65$. (d) Scaling plot of g when $x_\Gamma = 0.31$ and $T = 0.65$. (e) Γ-dependence of g for $T = 0.6$. (f) Scaling plot of g when $x_\Gamma = 0.31$ and $T = 0.6$ (from Mukherjee et al., 2015).

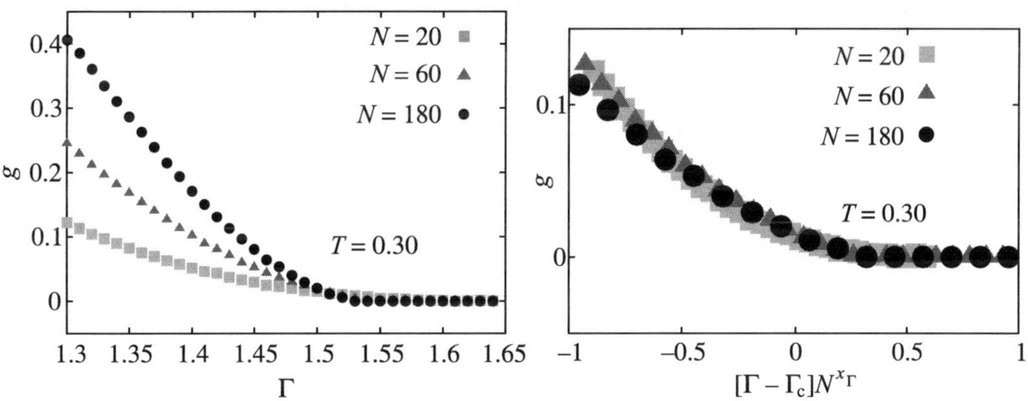

(a) (b)

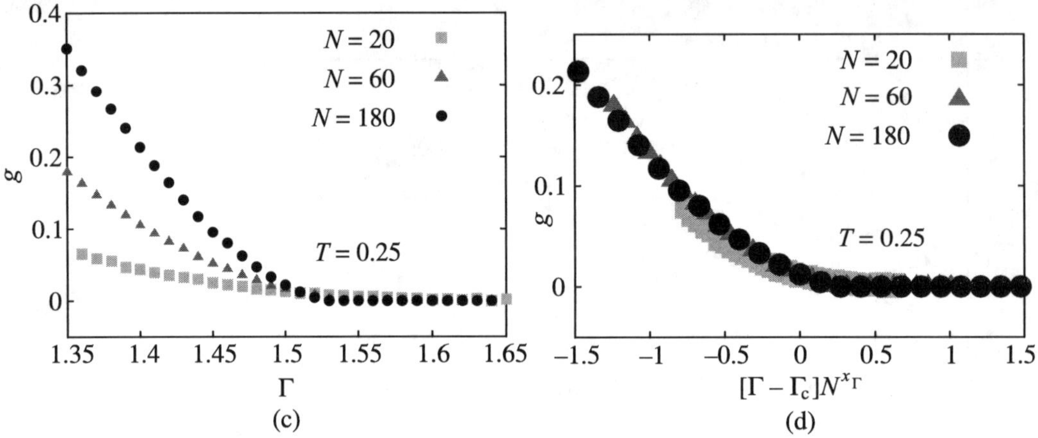

Fig. 4.16 Monte Carlo simulation results of the SK model in a transverse field in a low-temperature region for $J = 1$. In these cases, the quantum fluctuation is dominant. (a) Transverse field, Γ, dependence of the Binder ratio g for $T = 0.30$. (b) Scaling plot of the Binder ratio g when $x_\Gamma = 0.50$ for $T = 0.30$. (c) Γ-dependence of g for $T = 0.25$. (d) Scaling plot of g when $x_\Gamma = 0.50$ for $T = 0.25$ (from Mukherjee et al., 2015).

Exact diagonalization results

In order to clarify properties of the quantum phase transition at $T = 0$ in the SK model in a transverse field, some studies by exact diagonalization methods have been done (Sen and Chakrabarti, 1992; Mukherjee et al., 2015). By performing exact diagonalization, all eigenvalues $(E_0, E_1, \cdots, E_m)$ and corresponding eigenvectors $(\psi_0, \psi_1, \cdots, \psi_m)$ of the finite-size system at $T = 0$ can be obtained. Here, $m = 2^N - 1$, where N is the number of spins. In order to capture the properties of phase transition, we use the following physical quantities given by

$$\Delta := \overline{E_1 - E_0}, \tag{4.135}$$

$$E_{\mathrm{g}} := \overline{\left\langle \psi_0 \left| \sum_{1 \le i,j \le N} J_{ij} \sigma_i^z \sigma_j^z \right| \psi_0 \right\rangle}, \tag{4.136}$$

$$C := \frac{dE_{\mathrm{g}}}{d\Gamma}, \tag{4.137}$$

$$q := \frac{1}{N} \sum_{i=1}^{N} \overline{\langle \psi_0 | \sigma_i^z | \psi_0 \rangle^2}, \tag{4.138}$$

where Δ is the mass gap between the ground-state energy E_g and the first excited state energy, C is a quantity which is a quantum version of the "specific heat", and q is the spin glass order parameter. Furthermore, the bar indicates the configurational average. Sen et al. investigated zero-temperature properties of the SK model in a transverse field when the maximum size is $N = 8$ (Sen and Chakrabarti, 1992). The authors found that as the transverse field decreases, Δ disappears and q appears near the transverse field located at the peak position of C. These results suggest that the quantum phase transition occurs at the peak position of C.

In addition, Mukherjee et al. (2015) studied the Binder ratio of the SK model in a transverse field calculated by the exact diagonalization method for $N \leq 22$. They introduced the k-spin correlation function given by

$$q_k = \frac{1}{N^k} \sum_{i_1=1}^{N} \sum_{i_2=1}^{N} \cdots \sum_{i_k=1}^{N} \langle \psi_0 | \sigma_{i_1}^z \sigma_{i_2}^z \cdots \sigma_{i_k}^z | \psi_0 \rangle^2. \tag{4.139}$$

Using the k-spin correlation function, the Binder ratio is defined as:

$$g = \frac{1}{2} \left[3 - \overline{\left(\frac{q_4}{q_2^2} \right)} \right]. \tag{4.140}$$

Furthermore, the Binder ratio is scaled as:

$$g \sim g\left[(\Gamma - \Gamma_c) N^{x_\Gamma} \right]. \tag{4.141}$$

Figure 4.17 (a) shows the transverse field dependence of the Binder ratio for different system sizes. In order to consider the finite-size effect, the extrapolation of the critical transverse field Γ_c and the critical Binder ratio g_c for the limit of $N \to \infty$ was investigated. The authors obtained $x_\Gamma = 0.51$ from the extrapolation. Furthermore, according to the scaling, $\Gamma_c/J = 1.62(3)$ in the limit of $N \to \infty$ was obtained. Using the obtained values of Γ_c and x_Γ, the Binder ratio is well scaled by Eq. (4.141) which is shown in Fig. 4.17 (b). When we use the effective dimension of the quantum SK model $d_c = 8$, we obtain $\nu_\Gamma = 0.25$ which is the same value at low temperatures obtained by Monte Carlo simulations. In addition, from the extrapolation of the critical Binder ratio for the limit of $N \to \infty$, the authors found $g_c = 0$, which is also the same value at low temperatures obtained by Monte Carlo simulations. Summarizing the aforementioned results, the SK model in a transverse field at zero temperature exhibits the quantum phase transition at $\Gamma_c/J = 1.62$, and the critical phenomena are characterized by $\nu_\Gamma = 0.25$ and $g_c = 0$.

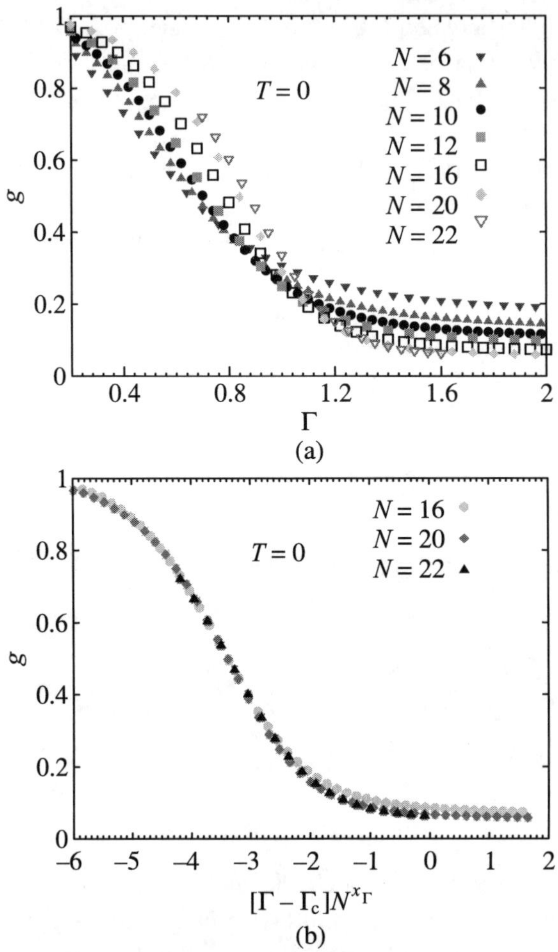

Fig. 4.17 Exact diagonalization results of the SK model in a transverse field at zero temperature for $J = 1$. (a) Transverse field, Γ, dependence of the Binder ratio g. (b) Scaling plot of the Binder ratio g when $x_\Gamma = 0.51$ and $\Gamma_c = 1.62$ (from Mukherjee et al., 2015).

Phase diagram by numerical simulations

The phase diagram of the SK model in a transverse field can also be obtained from the numerical simulation results. Figure 4.18 is the schematic phase diagram. At $T/J \simeq 0.45$, the crossover from the critical behavior characterized by the quantum fluctuation and that by the classical fluctuation occurs. For a confirmation of this crossover point using finite temperature exact diagonalization analysis, see, Yao et al. (2016). This suggests that the quantum fluctuation dominated spin glass phase exists at low temperatures. In this case, when $T = 0$, the quantum phase transition occurs at $\Gamma_c/J = 1.62$, which is the same result obtained by the analysis beyond the static approximation in the mean-field calculation (see Section 4.3.1).

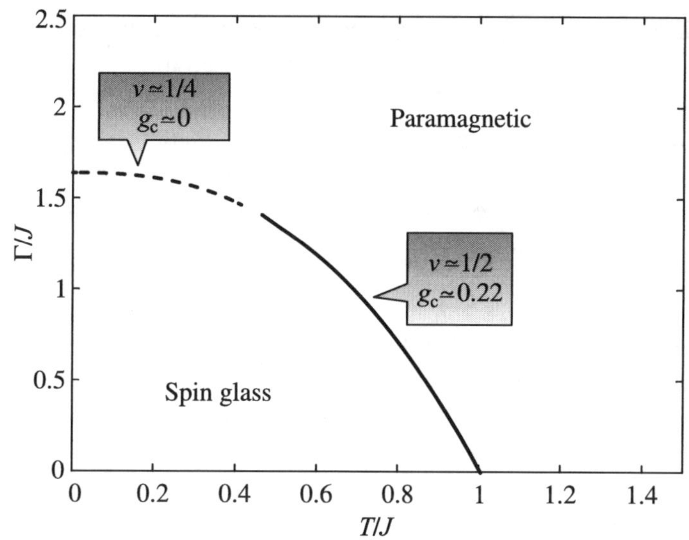

Fig. 4.18 Schematic phase diagram of the SK model in a transverse field obtained by simulation results. The solid and dotted curves indicate the phase boundary dominated by the classical fluctuation and that by the quantum fluctuation, respectively. At $T/J \simeq 0.45$, crossover occurs.

4.4 Edwards–Anderson Model in a Transverse Field

In the Edwards–Anderson(EA) model, interactions only exist between nearest-neighbor spin pairs. Note that the SK model is a particular version of the EA model, that is, the EA model having infinite-ranged interactions is the SK model. The Hamiltonian of the EA model in a transverse field is given by

$$\hat{\mathscr{H}} = -\sum_{\langle i,j\rangle} J_{ij}\hat{\sigma}_i^z\hat{\sigma}_j^z - \Gamma\sum_{i=1}^{N}\hat{\sigma}_i^x, \tag{4.142}$$

where $\langle i,j\rangle$ indicates the nearest-neighbor spin pairs on the given lattice. Furthermore, the interactions J_{ij} are distributed as Eq. (4.75). Chakrabarti (1981) first investigated the physical properties including the quantum phase transition of the model. In this subsection, we explain the quantum phase transition nature of the EA model in a transverse field for the $d = 1$, 2, and 3 when $J_0 = 0$ in Eq. (4.75). Note that the $d = \infty$ case of the EA model in a transverse field corresponds to the SK model in a transverse field described in Section 4.3.

One-dimensional case

The Hamiltonian of the EA model in a transverse field on the one-dimensional lattice, that is, on the chain is given by

$$\hat{\mathscr{H}} = -\sum_{i=1}^{N} J_i\hat{\sigma}_i^z\hat{\sigma}_{i+1}^z - \Gamma\sum_{i=1}^{N}\hat{\sigma}_i^x. \tag{4.143}$$

Applying the Suzuki–Trotter decomposition, the model can be exactly mapped to the McCoy–Wu model, i.e., a randomly layered Ising model on the square lattice (McCoy and Wu, 1968, 1969; McCoy, 1969, 1970). Many analytical results of the McCoy–Wu model have been obtained. The critical phenomenon of the quantum phase transition at zero temperature is equivalent to that of the two-dimensional classical Ising model. Furthermore, slow relaxation appears in this model around the critical point Γ_c. This slow relaxation behavior is called the Griffith singularity (Fisher, 1992).

Two-dimensional case

The quantum phase transition at zero temperature of the EA model on the two-dimensional lattice has been investigated by quantum Monte Carlo simulations. By the Suzuki–Trotter decomposition, the EA model in a transverse field in d-dimension can be mapped to the following $d+1$-dimensional classical Ising model:

$$\beta \mathcal{H}_{\text{eff}} = -\sum_{\langle i,j \rangle} \sum_{k=1}^{P} K_{ij} \sigma_{i,k}^z \sigma_{j,k}^z - \sum_{i=1}^{N} \sum_{k=1}^{P} K \sigma_{i,k}^z \sigma_{i,k+1}^z, \tag{4.144}$$

where

$$K_{ij} = \frac{\beta}{P} J_{ij}, \tag{4.145}$$

$$K = \frac{1}{2} \ln \coth \left(\frac{\beta \Gamma}{P} \right). \tag{4.146}$$

Here, in the limit of $P \to \infty$, the classical model on the $d+1$-dimensional lattices is equivalent to the quantum model on the d-dimensional lattice. However, this limit is inconvenient to calculate the physical properties of Eq. (4.144) by numerical simulations, because $K_{ij} \to 0$ and $K \to \infty$ in the limit of $P \to \infty$. Thus, we assume that the universal properties do not heavily depend on β/P, and β/P is fixed to 1 in general.

In order to clarify the quantum phase transition at zero temperature, the Binder ratio g for spin glass order parameter q defined by Eqs. (4.129) and (4.130) is used. The Binder ratio g is scaled by Eq. (4.131) near the critical point. Here, we consider the K-dependence of physical quantities instead of the Γ-dependence. By using K, the correlation length is scaled as

$$\xi \sim (K - K_c)^{-\nu}, \tag{4.147}$$

where $K_c = 0.5 \ln \coth(\Gamma_c)$ is the critical point. The Binder ratio is scaled as

$$g \sim g \left[(K - K_c) L^{1/\nu}, L_\tau / L^z \right], \tag{4.148}$$

where L_τ is the linear dimension along the Trotter axis, which corresponds to the Trotter number P. In the disordered phase in the thermodynamic limit ($L \to \infty$), the value of g becomes zero. Thus, in the limit of $L_\tau/L^z \to 0$, $g = 0$. On the other hand, in the limit of $L_\tau/L^z \to \infty$, the system becomes independent classical Ising chains along the Trotter axis. In this case also, g becomes zero, since the ordered phase does not exist in the classical Ising chain. Therefore, when we fix the value of L, the single peak of g as a function of L_τ is observed. Furthermore, the maximum value at the peak does not depend on L at the critical point K_c. From this fact, the value of K_c can be determined. Figure 4.19 shows the L_τ-dependence of g for various K for the effective classical model given by Eq. (4.144) corresponding to the EA model in a transverse field on the square lattice. Around $K^{-1} = 3.30$, the maximum values at the peak become about the same value for various L. Thus, the phase transition point is determined as $K_c^{-1} \simeq 3.275$. In addition, the value of the dynamical exponent z can be determined by the scaling plot of the Binder ratio as a function of L_τ/L^z, which is given by Eq. (4.148). Figure 4.20 (a) is the scaling plot of the Binder ratio g, which means that $z \simeq 1.5$. The value of the critical exponent ν can be determined by the first derivative of the Binder ratio with respect to K at K_c; $g'(K = K_c) \sim L^{1/\nu}$. Figure 4.20 (b) shows the L-dependence of $g'(K = K_c)$. The slope of the fitting line in Fig. 4.20 (b) is $1/\nu$, and $1/\nu = 1.0$.

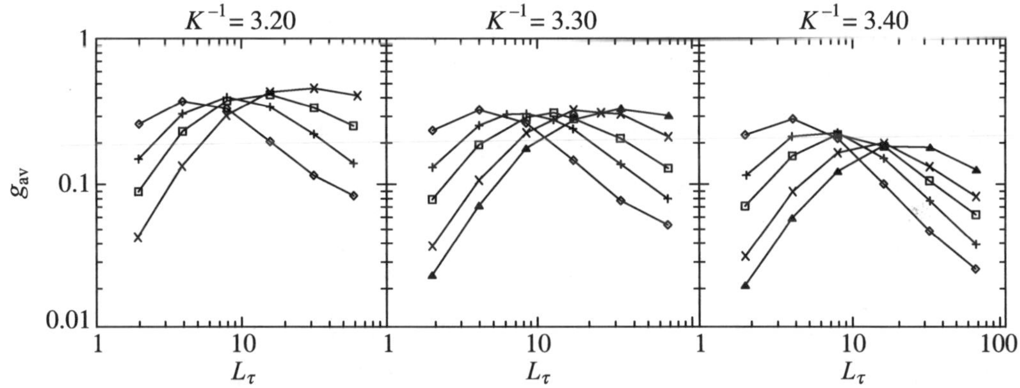

Fig. 4.19 L_τ-dependence of g for various K for the effective classical model corresponding to the EA model in a transverse field on the square lattice. The diamond, plus, square, cross, and triangular points indicate results for $L = 4$, 6, 8, 12, and 16, respectively. At $K^{-1} = 3.30$ (center figure), the maximum values at the peak become about the same value for various L. This result was obtained by Monte Carlo simulations (from Rieger and Young, 1994).

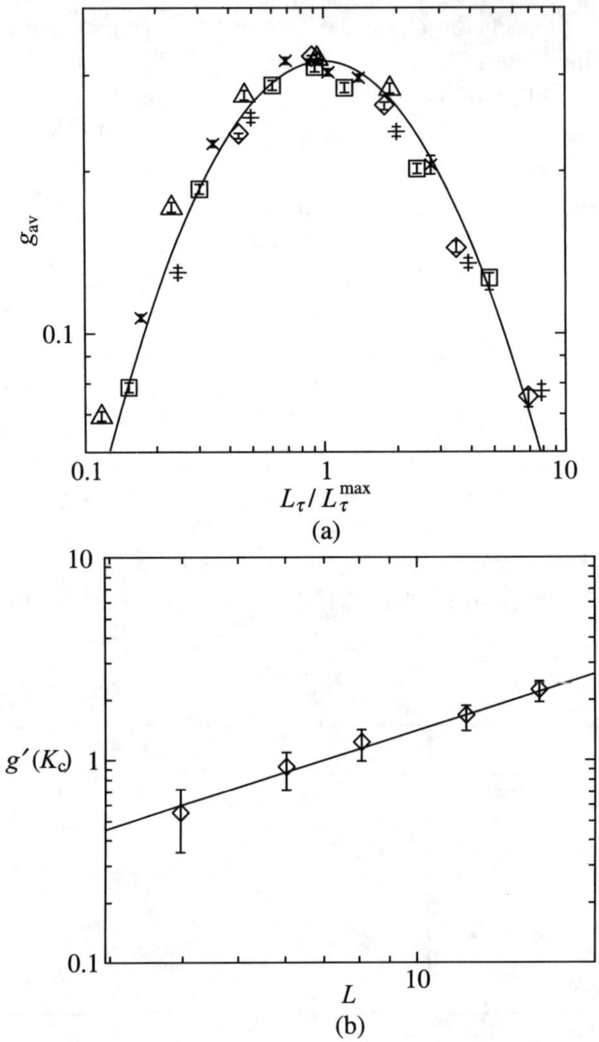

Fig. 4.20 (a) Scaling plot of the Binder ratio g when the value of z is 1.5 for the effective classical model corresponding to the EA model in a transverse field on the square lattice. (b) The first derivative of Binder ratio with respect to K at K_c : $g'(K = K_c)$ depending on L. The slope of the fitting line is $1/\nu = 1.0$. This result was obtained by Monte Carlo simulations (from Rieger and Young, 1994).

Summarizing these results, the two-dimensional EA model in a transverse field exhibits the quantum phase transition at zero temperature and the quantum critical point is given by

$$K_c^{-1} = \left(\frac{1}{2}\ln\coth\Gamma_c\right)^{-1} \simeq 3.275. \tag{4.149}$$

Furthermore, the critical exponents of the quantum phase transition are

$$1/\nu \simeq 1.0, \qquad z \simeq 1.5, \tag{4.150}$$

which were obtained by the Monte Carlo simulations (Rieger and Young, 1994).

Three-dimensional case

The quantum phase transition at zero temperature of the three-dimensional EA model in a transverse field was investigated by Monte Carlo simulations (Guo et al., 1994) as well as the two-dimensional case. By the scaling plot of the Binder ratio g shown in Fig. 4.21 for the effective classical model given by Eq. (4.144) corresponding to the EA model in a transverse field on the cubic lattice, the critical point K_c^{-1} and the dynamical exponent z can be determined as $K_c^{-1} \simeq 4.3$ and $z \simeq 1.3$. In addition, by using Eq. (4.148), the critical exponent ν can be determined by the scaling plot of the Binder ratio as a function of K^{-1} as $1/\nu \simeq 1.3$ (see Fig. 4.22).

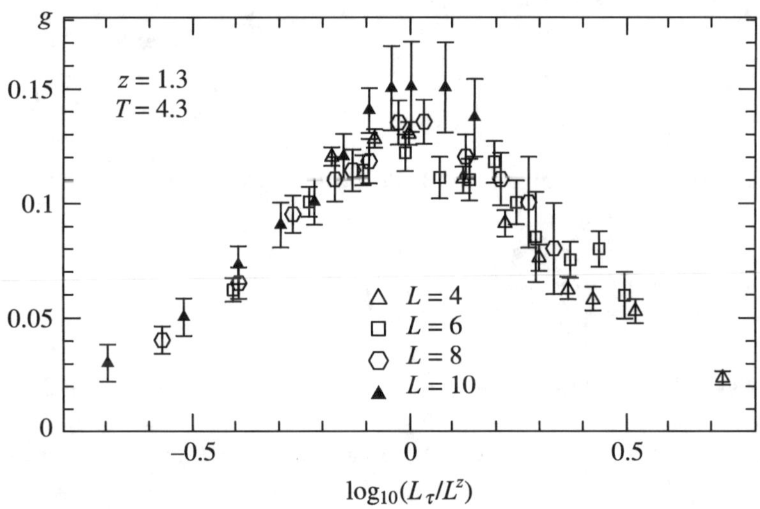

Fig. 4.21 Scaling plot of the Binder ratio g when $K^{-1} = T = 4.3$ and $z = 1.3$ for the effective classical model corresponding to the EA model in a transverse field on the cubic lattice. This result was obtained by Monte Carlo simulations (from Guo et al., 1994).

Summarizing these results, the three-dimensional EA model in a transverse field exhibits the quantum phase transition at zero temperature and the quantum critical point is given by

$$K_c^{-1} = \left(\frac{1}{2}\ln\coth\Gamma_c\right)^{-1} \simeq 4.3. \tag{4.151}$$

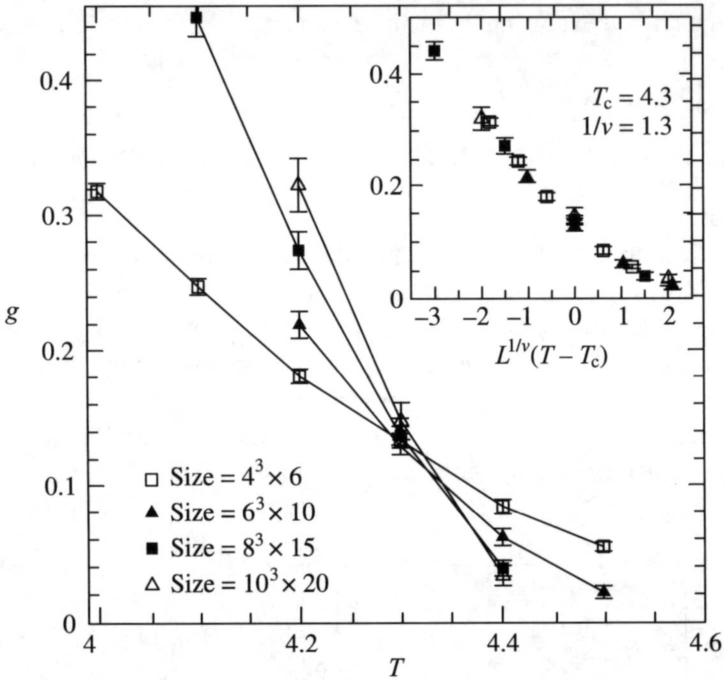

Fig. 4.22 K^{-1}-dependence of the Binder ratio g for the effective classical model corresponding to the EA model in a transverse field on the cubic lattice. Inset is the scaling plot of the Binder ratio g when $K^{-1} = T = 4.3$ and $1/\nu = 1.3$. This result was obtained by Monte Carlo simulations (from Guo et al., 1994).

This result means that the quantum critical point Γ_c of the cubic lattice is higher than that of the square lattice. This fact is similar to the critical point of the classical Ising model (see Section 2.1.3). Furthermore, the critical exponents of the quantum phase transition are

$$1/\nu \simeq 1.3, \qquad z \simeq 1.3, \tag{4.152}$$

which were obtained by the Monte Carlo simulations (Guo et al., 1994).

4.5 Existence of Replica Symmetric Spin Glass Phase in the Random Quantum Ising Model

We review some discussions for the replica symmetric spin glass phase in the random quantum Ising model. As mentioned in Section 4.2, it is expected that the replica symmetric spin glass phase appears because of the quantum tunneling effect. To study the nature of the spin glass phase in the region where the quantum fluctuation is dominant, an overlap distribution function $P(q)$ is used. The overlap distribution function $P_N(q)$ of the finite-size system where the number of spins is N is defined as

$$P_N(q) = \overline{\langle \delta(q - q^{\alpha\beta}) \rangle}, \tag{4.153}$$

$$q^{\alpha\beta} = \frac{1}{N} \sum_{i=1}^{N} \sigma_{i\alpha}^z \sigma_{i\beta}^z, \tag{4.154}$$

where $\langle \cdot \rangle$ indicates the thermal average, the bar indicates the configurational average, and $\delta(\cdot)$ is the Dirac delta function. Here, $\sigma_{i\alpha}^z$ is the Ising spin in the classical effective Hamiltonian obtained by the Suzuki–Trotter decomposition. As shown in Chapter 2, $q^{\alpha\beta}$ is the order parameter which indicates the correlation between replica α and β (Edwards and Anderson, 1975; Parisi, 1983). The overlap distribution function $P(q)$ is defined by the thermodynamic limit of $P_N(q)$:

$$P(q) = \lim_{N \to \infty} P_N(q). \tag{4.155}$$

In the replica symmetric spin glass phase, $q^{\alpha\beta}$ does not depend on the α and β, and $q^{\alpha\beta} = q'$, where the value of q' characterizes the replica symmetric spin glass phase. Thus, in this case, we expect that $P_N(q)$ becomes the delta function around $q = q'$ in the thermodynamic limit. On the other hand, in the paramagnetic phase, we expect that $P_N(q)$ becomes the delta function around $q = 0$ in the thermodynamic limit.

Ray et al. studied the overlap distribution function in the SK model in a transverse field by Monte Carlo simulations (Ray et al., 1989). The effective Hamiltonian of the SK model in a transverse field is given by Eq. (4.128). In this study, $q^{\alpha\beta}$ is defined by

$$q^{\alpha\beta} = \frac{1}{NP} \sum_{i=1}^{N} \sum_{k=1}^{P} \sigma_{i,k\alpha}^z \sigma_{i,k\beta}^z, \tag{4.156}$$

where P is the Trotter number. Figure 4.23 shows the lattice size, N, dependence of $P_N(q)$ at $\Gamma/J = 0.15$ and $T/J = 0.4$ which is located in the spin glass phase. The oscillation behavior of $P_N(q)$ is observed as a function of q. When the lattice size N increases, the amplitude of oscillation at low q becomes small. Furthermore, the dotted curves which indicate the upper profiles of $P_N(q)$ seem to approach the delta function as the lattice size increases. Thus, Ray et al. concluded that the replica symmetric spin glass phase exists in the SK model in a transverse field by the quantum tunneling effect.

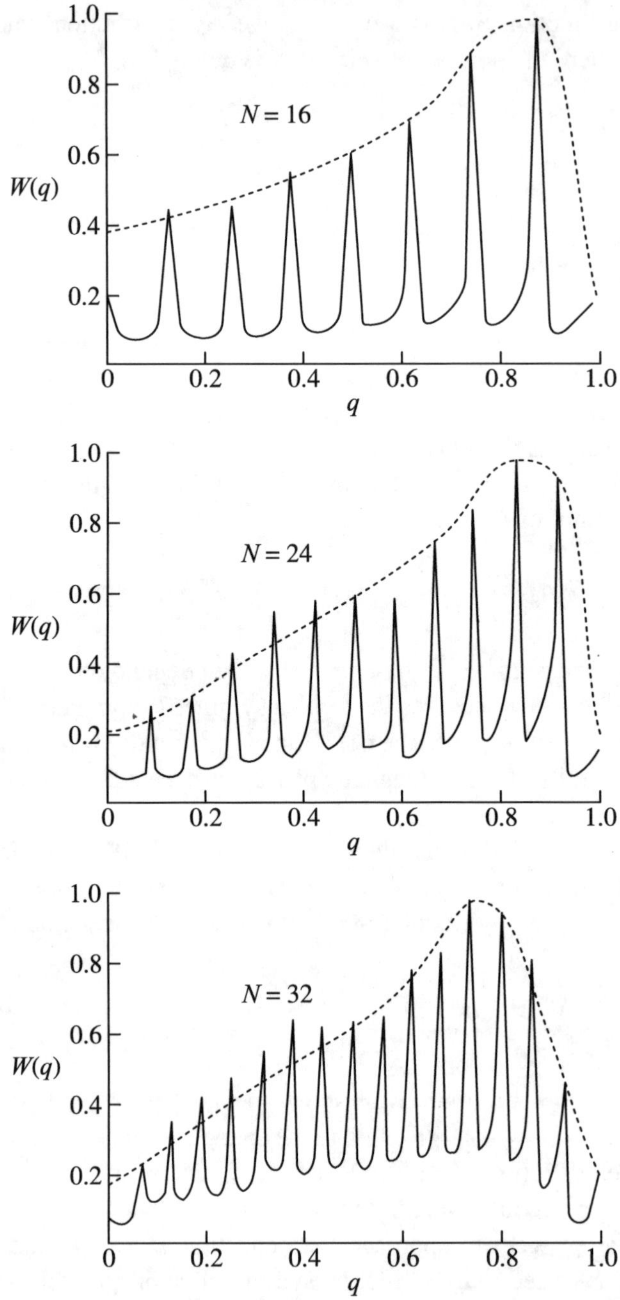

Fig. 4.23 Overlap distribution function $P_N(q)$ which is indicated by $W(q)$ in this figure as a function of q for various lattice sizes N at $\Gamma/J = 0.15$ and $T/J = 0.4$, which is located in the spin glass phase of the SK model in a transverse field. The dotted curve indicates the upper profiles of $P_N(q)$ (from Ray et al., 1989).

On the other hand, Lai and Goldschmidt studied the overlap distribution function using a different definition from the definition used by Ray et al. in the SK model in a transverse field (Lai and Goldschmidt, 1990). In this study, $q^{\alpha\beta}$ is defined as

$$q_{kk'}^{\alpha\beta} = \frac{1}{N}\sum_{i=1}^{N} \sigma_{i,k\alpha}^{z}\sigma_{i,k'\beta}^{z}, \tag{4.157}$$

where $k \neq k'$ are arbitrary. That is, $q_{kk'}^{\alpha\beta}$ is defined by the correlation between different sites along the Trotter axis. Figure 4.24 shows that the lattice size, N, dependence of $P_N(q) = \langle\delta(q - q_{kk'}^{\alpha\beta})\rangle$ at $\Gamma/J = 0.4$ and $T/J = 0.4$, which is located in the spin glass phase, and the oscillation behavior is not observed. Furthermore, at $q = 0$, $P_N(q)$ has a tail even if the lattice size N increases. Thus, Lai and Goldschmidt considered that $P_N(q)$ does not become the delta function and concluded that the replica symmetric spin glass phase does not exist in the SK model in a transverse field by the quantum tunneling effect.

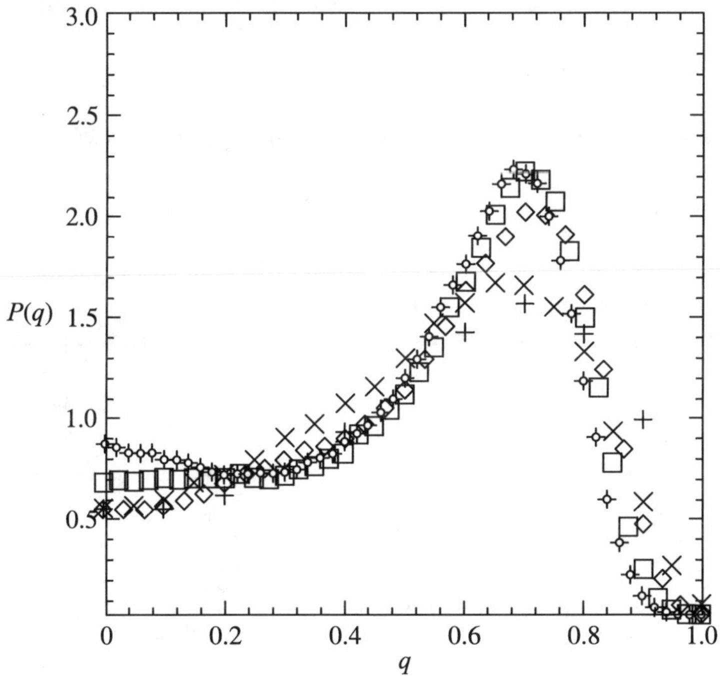

Fig. 4.24 Overlap distribution function $P_N(q)$ which is indicated by $P(q)$ in this figure as a function of q for various lattice sizes N at $\Gamma/J = 0.4$ and $T/J = 0.4$, which is located in the spin glass phase of the SK model in a transverse field. The plus, cross, diamond, square, and circle points indicate results for $N = 20$, 40, 60, 80, and 100, respectively (from Lai and Goldschmidt, 1990).

Many other contradictory results for the replica symmetric spin glass phase have been found by simulations and analytical calculations (Ray et al., 1989; Thirumalai et al., 1989; Lai and Goldschmidt, 1990; Büttner and Usadel, 1990a,b; Yokota, 1991; Sen and Chakrabarti, 1992; Guo et al., 1994; Rieger and Young, 1994; Read et al., 1995). Thus, even now, it is an open question whether replica symmetric spin glass phase exists in the random quantum Ising model.

Recently, Katzgraber and his coauthors pointed out that the distribution function of the spin glass order parameter $P(q)$ characterizes the hardness of given problems (Katzgraber et al., 2015).

References

Anderson, P. W. 1997. *The Theory of Superconductivity in the High-T_c Cuprate Superconductors*. Princeton University Press.

Auerbach, A. 1998. *Interacting Electrons and Quantum Magnetism*. Springer.

Billoire, A., and I. A. Campbell. 2011. 'Dynamics in the Sherrington-Kirkpatrick Ising spin glass at and above T_g. *Phys. Rev. B* 84: 054442.

Blinc, R. 1960. 'On the isotopic effects in the ferroelectric behaviour of crystals with short hydrogen bonds'. *J. Phys. Chem. Solids* 13: 204.

Bray, A. J., and M. A. Moore. 1980. 'Replica theory of quantum spin glasses'. *J. Phys. C: Solid State Phys.* 13: L655.

Brout, R., K. A. Müller, and H. Thomas. 1966. 'Tunnelling and collective excitations in a microscopic model of ferroelectricity'. *Solid State Commun.* 4: 507.

Büttner, G., and K. D. Usadel. 1990a. 'Replica-symmetry breaking for the Ising spin glass in a transverse field'. *Phys. Rev. B* 42: 6385.

Büttner, G., and K. D. Usadel. 1990b. 'Stability analysis of an Ising spin glass with transverse field'. *Phys. Rev. B* 41: 428.

Büttner, G., and K. D. Usadel.1991. 'The exact phase diagram of the quantum XY spin glass model in a transverse field'. *Z. Phys. B, Condens. Matter* 83: 131.

Chakrabarti, B. K. 1981. 'Critical behavior of the Ising spin-glass models in a transverse field'. *Phys. Rev. B* 24: 4062.

Chayes, L., N. Crawford, D. Ioffe, and A. Levit. 2008. 'The phase diagram of the quantum Curie Weiss model'. *J. Stat. Phys.* 133: 131.

de Gennes, P. G. 1963. 'Collective motions of hydrogen bonds'. *Solid State Commun.* 1: 132.

dos Santos, R. R., R. Z. dos Santos, and M. Kischinhevsky. 1985. 'Transverse Ising spin-glass model'. *Phys. Rev. B* 31: 4694.

Edwards, S. F., and P. W. Anderson. 1975. 'Theory of spin glasses'. *J. Phys. F: Metal Phys.* 5: 965.

Fedorov, Y. V., and E. F. Shender. 1986. 'Quantum spin glasses in the Ising model with transverse field'. *Pis'ma Zh. Eksp. Teor. Fiz.* 43: 526.

Fisher, D. S. 1992. 'Random transverse field Ising spin chains'. *Phys. Rev. Lett.* 69: 534.

Franz, M. 2013. 'Majorana's wires'. *Nature Nanotech.* 8: 149.

Fujii, K., Y. Nakata, M. Ohzeki, and M. Murao. 2013. 'Measurement-based quantum computation on symmetry breaking thermal states'. *Phys. Rev. Lett.* 110: 120502.

Goldschmidt, Y. Y., and P. Y. Lai. 1990. 'Ising spin glass in a transverse field: Replica-symmetry-breaking solution'. *Phys. Rev. Lett.* 64: 2467.

Greiter, M., V. Schnells, and R. Thomale. 2014. 'The 1D Ising model and the topological phase of the Kitaev chain'. *Ann. Phys.* 351: 1026.

Gubernatis, J., N. Kawashima, and P. Werner. 2016. *Quantum Monte Carlo methods: Algorithms for Lattice Models*. Cambridge University Press.

Guo, M., R. N. Bhatt, and D. A. Huse. 1994. 'Quantum critical behavior of a three-dimensional Ising spin glass in a transverse magnetic field'. *Phys. Rev. Lett.* 72: 4137.

Guo, M., R. N. Bhatt, and D. A. Huse. 1996. 'Quantum Griffiths singularities in the transverse-field Ising spin glass'. *Phys. Rev. B* 54: 3336.

Ishii, H., and T. Yamamoto. 1985. 'Effect of a transverse field on the spin glass freezing in the Sherrington-Kirkpatrick model'. *J. Phys. C: Solid State Phys.* 18: 6225.

Ishii, H., and T. K. Yamamoto. 1986. *Quantum Monte Carlo Methods*. Springer.

Katsura, S. 1962. 'Statistical mechanics of the anisotropic linear Heisenberg model'. *Phys. Rev.* 127: 1508.

Katzgraber, H. G., F. Hamze, Z. Zhu, A. J. Ochoa, and H. Munoz-Bauza. 2015. 'Seeking quantum speedup through spin glasses: The good, the bad, and the ugly'. *Phys. Rev. X* 5: 031026.

Kitaev, A. Y. 2001. 'Unpaired Majorana fermions in quantum wires'. *Phys.-Usp.* 44: 131.

Kjäll, J. A., J. H. Bardarson, and F. Pollmann. 2014. 'Many-body localization in a disordered quantum Ising chain'. *Phys. Rev. Lett.* 113: 107204.

Kopec, T. K. 1988a. 'A dynamic theory of transverse freezing in the Sherrington-Kirkpatrick Ising model'. *J. Phys. C: Solid State Phys.* 21: 6053.

Kopec, T. K. 1988b. 'Transverse freezing in the quantum Ising spin glass: A thermofield dynamic approach'. *J. Phys. C: Solid State Phys.* 21: 297.

Krzakala, F., A. Rosso, G. Semerjian, and F. Zamponi. 2008. 'Path-integral representation for quantum spin models: Application to the quantum cavity method and Monte Carlo simulations'. *Phys. Rev. B* 78: 134428.

Lai, P. Y., and Y. Y. Goldschmidt. 1990. 'Monte Carlo studies of the Ising spin-glass in a transverse field'. *Europhys. Lett.* 13: 289.

Lancaster, D., and F. Ritort. 1997. 'Solving the Schrödinger equation for the Sherrington-Kirkpatrick model in a transverse field'. *J. Phys. A: Math. Gen.* 30: L41.

Lee, P. A. 2007. 'From high temperature superconductivity to quantum spin liquid: Progress in strong correlation physics'. *Rep. Prog. Phys.* 71: 012501.

Leijnse, M., and K. Flensberg. 2012. 'Introduction to topological superconductivity and Majorana fermions'. *Semicond. Sci. Technol.* 27: 124003.

McCoy, B. M. 1969. 'Theory of a two-dimensional Ising model with random impurities. iii. Boundary effects'. *Phys. Rev.* 188: 1014.

McCoy, B. M. 1970. 'Theory of a two-dimensional Ising model with random impurities. iv. Generalizations'. *Phys. Rev. B* 2: 2795.

McCoy, B. M., and T. T. Wu. 1968. 'Theory of a two-dimensional Ising model with random impurities. i. Thermodynamics'. *Phys. Rev.* 176: 631.

McCoy, B. M., and T. T. Wu. 1969. 'Theory of a two-dimensional Ising model with random impurities. ii. Spin correlation functions'. *Phys. Rev.* 188: 982.

Mourik, V., K. Zuo, S. M. Frolov, S. R. Plissard, E. P. A. M. Bakkers, and L. P. Kouwenhoven. 2012. 'Signatures of Majorana fermions in hybrid superconductor-semiconductor nanowire devices'. *Science*, 336: 1003.

Mukherjee, S., A. Rajak, and B. K. Chakrabarti. 2015. 'Classical-to-quantum crossover in the critical behavior of the transverse-field Sherrington-Kirkpatrick spin glass model'. *Phys. Rev. E* 92: 042107.

Nagai, O., Y. Yamada, and Y. Miyatake. 1986. â^ŸMonte Carlo simulations of three-dimensional Heisenberg and transverse-Ising magnetsâ^ in *Quantum Monte Carlo Methods in Equilibrium and Nonequilibrium systems*, p. 95. Springer.

Nakatsuji, S., Y. Nambu, H. Tonomura., O. Sakai, S. Jonas, C. Broholm, H. Tsunetsugu, Y. Qiu, and Y. Maeno. 2005. 'Spin disorder on a triangular lattice'. *Science* 309: 1697.

Nakatsuji, S., Y. Machida, Y. Maeno, T. Tayama, T. Sakakibara, J. van Duijn, L. Balicas, J. N. Millican, R. T. Macaluso, and J. Y. Chan., 2006. 'Metallic spin-liquid behavior of the geometrically frustrated Kondo lattice $Pr_2Ir_2O_7$'. *Phys. Rev. Lett.* 96: 087204.

Ohta, T., and K. Totsuka, 2016. 'Topological quantum phase transitions in a Majorana chain with spatial modulation'. *J. Phys. Soc. Jpn.* 85: 074003.

Ohta, T., S. Tanaka, I. Danshita, and K. Totsuka. 2015. 'Phase diagram and sweep dynamics of a one-dimensional generalized cluster model'. *J. Phys. Soc. Jpn.* 84: 063001.

Ohta, T., S. Tanaka, I. Danshita, and T. Totsuka. 2016. 'Topological and dynamical properties of a generalized cluster model in one dimension'. *Phys. Rev. B* 93: 165423.

Parisi, G. 1983. 'Order parameter for spin-glasses'. *Phys. Rev. Lett.* 50: 1946.

Parkinson, J. B., and D. J. J. Farnell. 2010. *An Introduction to Quantum Spin Systems*. Springer.

Pfeuty, P. 1970. 'The one-dimensional Ising model with a transverse field'. *Ann. Phys.* 57: 79.

Pirc, R., B. Tadić, and R. Blinc. 1985. 'Tunneling model of proton glasses'. *Z. Phys. B, Condens. Matter* 61: 69.

Raussendorf, R., and H. J. Briegel. 2001. 'A one-way quantum computer'. *Phys. Rev. Lett.* 86: 5188.

Raussendorf, R., D. E. Browne, and H. J. Briegel. 2003. 'Measurement-based quantum computation on cluster states'. *Phys. Rev. A* 68: 022312.

Ray, P., B. K. Chakrabarti, and A. Chakrabarti. 1989. 'Sherrington-Kirkpatrick model in a transverse field: Absence of replica symmetry breaking due to quantum fluctuations'. *Phys. Rev. B* 39: 11828.

Read, N., S. Sachdev, and J. Ye. 1995. 'Landau theory of quantum spin glasses of rotors and Ising spins'. *Phys. Rev. B* 52: 384.

Rieger, H., and A. P. Young. 1994. 'Zero-temperature quantum phase transition of a two-dimensional Ising spin glass'. *Phys. Rev. Lett.* 72: 4141.

Rieger, H., and A. P. Young. 1996. 'Griffiths singularities in the disordered phase of a quantum Ising spin glass'. *Phys. Rev. B* 54: 3328.

Sandvik, A. W. 2010. 'Computational studies of quantum spin systems'. *AIP Conf. Proc.* 1297: 135.

Schollwöck, U., J. Richter, D. J. J. Farnell, and R. F Bishop, eds. 2004. *Quantum Magnetism*. Springer.

Sen, P., and B. K. Chakrabarti. 1992. 'Frustrated transverse Ising models: A class of frustrated quantum system'. *Int. J. Mod. Phys. B* 6: 2439.

Shimizu, Y., K. Miyagawa, K. Kanoda, M. Maesato, and G. Saito. 2003. 'Spin liquid state in an organic Mott insulator with a triangular lattice'. *Phys. Rev. Lett.* 91: 107001.

Smacchia, P., L. Amico, P. Facchi, R. Fazio, G. Florio, S. Pascazio, and V. Vedral. 2011. 'Statistical mechanics of the cluster Ising model'. *Phys. Rev. A* 84: 022304.

Stinchcombe, R. B. 1973. 'Ising model in a transverse field. I. Basic theory'. *J. Phys. C: Solid State Phys.* 6: 2459.

Stratt, R. M. 1986. 'Path-integral methods for treating quantal behavior in solids: Mean-field theory and the effects of fluctuations'. *Phys. Rev. B* 33: 1921.

Suzuki, M. 1971. 'Relationship among exactly soluble models of critical phenomena. I 2D Ising model, dimer problem and the generalized XY-model'. *Prog. Theory. Phys.* 46: 1337.

Suzuki, M. 1976a. 'Generalized Trotter's formula and systematic approximants of exponential operators and inner derivations with applications to many-body problems'. *Commun. Math. Phys.* 51: 183.

Suzuki, M. 1976b. 'Relationship between *d*-dimensional quantal spin systems and $(d+1)$-dimensional Ising systems'. *Prog. Theor. Phys.* 56: 1454.

Suzuki, M. 1977. 'On the convergence of exponential operators—the Zassenhaus formula, BCH formula and systematic approximants'. *Comm. Math. Phys.* 57: 193.

Suzuki, M., ed. 1987. *Quantum Monte Carlo Methods in Equilibrium and Nonequilibrium Systems Ilibrium Systems*. Springer.

Suzuki, S., J. Inoue, and B. K. Chakrabarti. 2013. *Quantum Ising Phases and Transitions in Transverse Ising Models*. Springer.

Takahashi, K. 2007. 'Quantum fluctuations in the transverse Ising spin glass model: A field theory of random quantum spin systems'. *Phys. Rev. B* 76: 184422.

Thill, M. J., and D. A. Huse. 1995. 'Equilibrium behaviour of quantum Ising spin glass'. *Phys. A, Stat. Mech. Appl.* 214: 321.

Thirumalai, D., Q. Li, and T. R. Kirkpatrick. 1989. 'Infinite-range Ising spin glass in a transverse field'. *J. Phys. A: Math. Gen.* 22: 3339.

Trotter, H. F. 1959. 'On the product of semi-groups of operators'. *Proc. Am. Math. Soc.* 10: 545.

Usadel, K. D. 1986. Spin glass transition in an Ising spin system with transverse field. *Solid State Commun.*, 58, 629.

Usadel, K. D., G. Büttner, and T. K. Kopeć. 1991. 'Phase diagram of the quantum Ising spin glass in a transverse field'. *Phys. Rev. B* 44: 12583.

Walasek, K., and K. Lukierska-Walasek. 1986. 'Quantum transverse Ising spin-glass model in the mean-field approximation'. *Phys. Rev. B* 34: 4962.

Walasek, K., and K. Lukierska-Walasek. 1988. 'Cluster-expansion method for the infinite-range quantum transverse Ising spin-glass model'. *Phys. Rev. B* 38: 725.

Yamamoto, T., and H. Ishii. 1987. 'A perturbation expansion for the Sherrington-Kirkpatrick model with a transverse field'. *J. Phys. C: Solid State Phys.* 20: 6053.

Yamashita, S., Y. Nakazawa, M. Oguni, Y. Oshima, H. Nojiri, Y. Shimizu, and K. Miyagawa. 2008. 'Thermodynamic properties of a spin-1/2 spin-liquid state in a κ-type organic salt'. *Nature Phys.* 4: 459.

Yao, N. Y., Grusdt, F., Swingle, B., Lukin, M. D., Stamper-Kurn, D. M., Moore, J. L., and Demler, E. A., 2016. *Interferometric approach to probing fast scrambling*, arXiv:1607.01801

Yokota, T. 1991. 'Numerical study of the SK spin glass in a transverse field by the pair approximation'. *J. Phys. Condens. Matter* 3: 7039.

CHAPTER

5

Quantum Dynamics

In quantum annealing, we obtain the solution of a combinatorial optimization problem by decreasing the quantum fluctuation such as a transverse field to zero to decreasing the quantum fluctuation such as a transverse field to zero. Before we explain quantum annealing, we will describe the dynamics of quantum systems where the control parameter such as a transverse field changes with the time. As mentioned in Chapter 4, at zero temperature, quantum phase transition occurs in quantum systems such as the Ising model in a transverse field. When the quantum phase transition appears, properties of the ground states change at the quantum phase transition point. The quantum phase transition is characterized by an energy gap between the ground state and the excited states. At the quantum phase transition point, the energy gap vanishes in the thermodynamic limit. The quantum state cannot follow adiabatically around the phase transition point in the thermodynamic limit when the sweep speed of the control parameter is finite. Furthermore, even away from the phase transition point, transition from the ground state to an excited state often occurs if the energy gap is small. These facts are discrepancies to the theory of quantum annealing which will be described in more detail in Chapter 6. Thus, it is important to understand the dynamics of quantum systems when there is a change in the control parameters. In Section 5.1 we will review the Landau–Zener formula which expresses the transition probability from the ground state to the excited state in a two-level quantum system. According to the Landau–Zener formula, by decreasing the sweep speed of the control parameter or by increasing the energy gap, the transition probability from the ground state to the excited state decreases. In addition, in Section 5.2, the crossover between the adiabatic and nonadiabatic time evolutions will be shown. This occurs under the dynamic process that passes through the second-order phase transition point. The crossover phenomenon is called the Kibble–Zurek mechanism. From the Kibble–Zurek mechanism, the creation of a domain wall, i.e., topological defect, can be understood.

5.1 Landau–Zener Transition

In this section, we consider the nonadiabatic transition due to the finite speed sweep of a model parameter. The simplest example is the Landau–Zener transition that is the transition in a two-level system (Landau, 1932; Zener, 1932; Stueckelberg, 1932; Majorana, 1932). The transition probability from the ground state to the excited state of the two-level system can be analytically obtained. The Landau–Zener transition has been used for analysis of experimental results in single molecular magnets or other quantum systems (Rubbmark et al., 1981;Wernsdorfer et al., 2000, 2002; Chiorescu et al., 2003; Wernsdorfer et al., 2005; Koizumi et al., 2007).

Here we consider a two-level system whose time-dependent Hamiltonian is given by

$$\mathscr{H}_0(t) = \varepsilon_1(t)\,|1\rangle\,\langle 1| + \varepsilon_2(t)\,|2\rangle\,\langle 2|, \tag{5.1}$$

where $\varepsilon_i(t)$ and $|i\rangle$ $(i = 1,2)$ are instantaneous eigenvalues and eigenvectors at the time t, respectively. In this model, no coupling between energy levels exists. Here, we assume that the difference between eigenvalues is changed linearly as a function of time, i.e.,

$$\varepsilon_1(t) - \varepsilon_2(t) = \alpha t, \qquad (\alpha > 0). \tag{5.2}$$

Next, a time-independent Hamiltonian that represents the interaction between energy levels is given by

$$\mathscr{H}_1 = -\Delta\left(|1\rangle\,\langle 2| + |2\rangle\,\langle 1|\right), \qquad (\Delta > 0). \tag{5.3}$$

Note that $d\Delta/dt = 0$ for all t since the Hamiltonian shown in Eq. (5.3) does not depend on time. The total Hamiltonian we consider is

$$\mathscr{H}(t) = \mathscr{H}_0(t) + \mathscr{H}_1 = \begin{pmatrix} \varepsilon_1(t) & -\Delta \\ -\Delta & \varepsilon_2(t) \end{pmatrix}. \tag{5.4}$$

Notice that the Hamiltonian is equivalent to the Hamiltonian given by

$$\mathscr{H}(t) = -\frac{\alpha}{2} t\hat{\sigma}^z - \Delta\hat{\sigma}^x. \tag{5.5}$$

The instantaneous energy difference is expressed as

$$\sqrt{[\varepsilon_1(t) - \varepsilon_2(t)]^2 + 4\Delta^2}, \tag{5.6}$$

and takes the minimum value 2Δ at $t = 0$.

Let $|\psi(t)\rangle$ be the wavefunction at t. Using Eq. (5.4), the time-dependent Schrödinger equation is given by

$$i\frac{d}{dt}|\psi(t)\rangle = \begin{pmatrix} \varepsilon_1(t) & -\Delta \\ -\Delta & \varepsilon_2(t) \end{pmatrix} |\psi(t)\rangle, \tag{5.7}$$

where the Planck constant $\hbar$ is set to unity. Using $|1\rangle$ and $|2\rangle$, we expand the wavefuntion as follows:

$$|\psi(t)\rangle = \tilde{c}_1(t)\,|1\rangle + \tilde{c}_2(t)\,|2\rangle, \tag{5.8}$$

$$|\tilde{c}_1(t)|^2 + |\tilde{c}_2(t)|^2 = 1, \qquad (\text{for } \forall t). \tag{5.9}$$

Substituting Eq. (5.8) into Eq. (5.7), we obtain

$$\begin{aligned} & i\frac{d\tilde{c}_1(t)}{dt}|1\rangle + i\frac{d\tilde{c}_2(t)}{dt}|2\rangle \\ & = [\varepsilon_1(t)\tilde{c}_1(t) - \Delta\tilde{c}_2(t)]\,|1\rangle + [-\Delta\tilde{c}_1(t) + \varepsilon_2(t)\tilde{c}_2(t)]\,|2\rangle. \end{aligned} \tag{5.10}$$

From this equation, the following simultaneous equations are derived:

$$i\frac{d\tilde{c}_1(t)}{dt} = \varepsilon_1(t)\tilde{c}_1(t) - \Delta\tilde{c}_2(t), \tag{5.11}$$

$$i\frac{d\tilde{c}_2(t)}{dt} = \varepsilon_2(t)\tilde{c}_2(t) - \Delta\tilde{c}_1(t). \tag{5.12}$$

Here we take the following transformation from $\tilde{c}_i(t)$ to $c_i(t)$ $(i = 1, 2)$ given by

$$\tilde{c}_1(t) = c_1(t)\exp\left[-i\int_{-\infty}^{t} dt'\varepsilon_1(t')\right], \tag{5.13}$$

$$\tilde{c}_2(t) = c_2(t)\exp\left[-i\int_{-\infty}^{t} dt'\varepsilon_2(t')\right]. \tag{5.14}$$

Using the transformations given by Eqs. (5.13) and (5.14), Eqs. (5.11) and (5.12) are represented as

$$i\frac{dc_1(t)}{dt} = -\Delta c_2(t)\exp\left\{i\int_{-\infty}^{t} dt'\,[\varepsilon_1(t') - \varepsilon_2(t')]\right\} \tag{5.15}$$

$$= -\Delta c_2(t) \exp\left(i \int_{-\infty}^{t} dt' \alpha t' \right), \tag{5.16}$$

$$i\frac{dc_2(t)}{dt} = -\Delta c_1(t) \exp\left\{ -i \int_{-\infty}^{t} dt' \left[\varepsilon_1(t') - \varepsilon_2(t') \right] \right\} \tag{5.17}$$

$$= -\Delta c_1(t) \exp\left(-i \int_{-\infty}^{t} dt' \alpha t' \right). \tag{5.18}$$

Note that here we used the relation:

$$\frac{d}{dt}\left\{ \exp\left[-i \int_{-\infty}^{t} dt' \varepsilon_i(t') \right] \right\} = -i\varepsilon_i(t) \exp\left[-i \int_{-\infty}^{t} dt' \varepsilon_i(t') \right]. \tag{5.19}$$

From Eqs. (5.15) and (5.17), $c_1(t)$ is eliminated and the derivative equation of $c_2(t)$ is obtained:

$$\frac{d^2 c_2(t)}{dt^2} + i\alpha t \frac{dc_2}{dt} + \Delta^2 c_2(t) = 0. \tag{5.20}$$

Here we used the relation $d\Lambda/dt = 0$ for all t. Next we take the transformation from $c_2(t)$ to $U(t)$ defined by

$$c_2(t) = \exp\left\{ -\frac{i}{2} \int_{-\infty}^{t} dt' \left[\varepsilon_1(t') - \varepsilon_2(t') \right] \right\} U(t) \tag{5.21}$$

$$= \exp\left(-\frac{i}{2} \int_{-\infty}^{t} dt' \alpha t' \right) U(t), \tag{5.22}$$

and using $U(t)$, $dc_2(t)/dt$ and $d^2 c_2(t)/dt^2$ are expressed as

$$\frac{dc_2(t)}{dt} = \exp\left(-\frac{i}{2} \int_{-\infty}^{t} dt' \alpha t' \right) \left[-\frac{i}{2} \alpha t U(t) + \frac{dU(t)}{dt} \right], \tag{5.23}$$

$$\frac{d^2 c_2(t)}{dt^2} = \exp\left(-\frac{i}{2} \int_{-\infty}^{t} dt' \alpha t' \right)$$

$$\times \left[-\frac{\alpha^2 t^2}{4} U(t) - \frac{i}{2} \alpha U(t) - i\alpha t \frac{dU(t)}{dt} + \frac{d^2 U(t)}{dt^2} \right]. \tag{5.24}$$

Substituting Eqs. (5.23) and (5.24) into Eq. (5.20), the following differential equation is derived:

$$\frac{d^2U(t)}{dt^2}+\left(\Delta^2+\frac{\alpha^2t^2}{4}-\frac{i}{2}\alpha\right)U(t)=0. \tag{5.25}$$

Here we introduce a new variable z defined by

$$z=\mathrm{e}^{-i\frac{\pi}{4}}\alpha^{\frac{1}{2}}t. \tag{5.26}$$

Equation (5.25) is rewritten using z:

$$\frac{d^2\tilde{U}(z)}{dz^2}+\left(n-\frac{z^2}{4}+\frac{1}{2}\right)\tilde{U}(z)=0, \tag{5.27}$$

where $\tilde{U}(z):=U(\mathrm{e}^{i\pi/4}\alpha^{-1/2}z)$ and $n:=i\Delta^2/\alpha$. Here, we used the relations given by

$$\frac{dU(t)}{dt}=\mathrm{e}^{-i\frac{\pi}{4}}\alpha^{\frac{1}{2}}\frac{dU(t)}{dz}, \tag{5.28}$$

$$\frac{d^2U(t)}{dt^2}=\mathrm{e}^{-i\frac{\pi}{2}}\alpha\frac{d^2U(t)}{dz^2}. \tag{5.29}$$

Equation (5.27) is the Weber differential equation.

Before we consider the Landau–Zener transition, we derive the solution of the Weber differential equation. The Weber differential equation given by Eq. (5.27) has twofold solutions, $D_n(F-z)$ and $D_{-n-1}(-iz)$. $D_\lambda(z)$ are called the parabolic cylinder functions (Whittaker and Watson, 1990) and are given by

$$D_\lambda(z)=2^{\lambda/2}\sqrt{\pi}e^{-z^2/4}\left[\frac{F(-\frac{\lambda}{2},\frac{1}{2};\frac{z^2}{2})}{\Gamma((1-\lambda)/2)}-\frac{\sqrt{2}zF(\frac{1-\lambda}{2},\frac{3}{2};\frac{z^2}{2})}{\Gamma(-\frac{\lambda}{2})}\right], \quad (|\arg z|\geq 3\pi/4), \tag{5.30}$$

$$D_\lambda(z)=\frac{\sqrt{\pi}2^{\lambda/2+1/4}M_{\lambda/2+1/4,-1/4}(\frac{z^2}{2})}{\Gamma((1-\lambda)/2)z^{1/2}}-\frac{\sqrt{\pi}2^{\lambda/2+5/4}M_{\lambda/2+1/4,1/4}(\frac{z^2}{2})}{\Gamma(-\lambda/2)z^{1/2}}, \quad (|\arg z|<3\pi/4), \tag{5.31}$$

where the hypergeometric function $F(\alpha,\gamma;z)$ is defined by

$$F(\alpha,\gamma;z)=\sum_{n=0}^{\infty}\frac{\alpha(\alpha+1)\cdots(\alpha+n-1)z^n}{\gamma(\gamma+1)\cdots(\gamma+n-1)n!}, \tag{5.32}$$

and $M_{\kappa,\mu}(z)$ is the Whittaker function (Whittaker and Watson, 1990; Gradshteyn et al., 2000a,b) defined by

$$M_{\kappa,\mu}(z) = z^{\mu+1/2}e^{-z/2}F\left(\mu - \kappa + \frac{1}{2}, 2\mu + 1; z\right). \tag{5.33}$$

Here we assume that λ is a non-negative integer and $|z|$ is sufficiently large. In addition, $\arg z$ takes values between $\pm\pi/4$ and $\pm 5\pi/4$. Under the assumptions, $D_\lambda(z)$ is approximated as

$$D_\lambda(z) \sim \mathrm{e}^{-z^2/4} z^\lambda \sum_{n=0}^{\infty} (-1)^n \frac{\lambda(\lambda-1)\cdots(\lambda-2n+1)}{n!2^n z^{2n}}$$
$$- \frac{\sqrt{2\pi}}{\Gamma(-\lambda)} \mathrm{e}^{\pm\lambda\pi i + z^2/4} z^{-\lambda-1} \sum_{n=0}^{\infty} \frac{(\lambda+1)(\lambda+2)\cdots(\lambda+2n)}{n!2^n z^{2n}},$$
$$(|\arg z| \geq 3\pi/4), \tag{5.34}$$

$$D_\lambda(z) \sim \mathrm{e}^{-z^2/4} z^\lambda \sum_{n=0}^{\infty} \frac{\lambda(\lambda-1)(\lambda-2n+1)}{n!2^n z^{2n}}, \qquad (|\arg z| < 3\pi/4). \tag{5.35}$$

Thus, the asymptotics of $D_\lambda(z)$ is obtained in the limit of $R \to -\infty$ as follows:

$$D_n(\pm z) = D_n(\pm R\mathrm{e}^{-i\pi/4}) = D_n(\mp|R|\mathrm{e}^{-i\pi/4}) \to \text{const.}, \tag{5.36}$$

$$D_{-n-1}(-iz) = D_{-n-1}(R\mathrm{e}^{-i3\pi/4}) = D_{-n-1}(|R|\mathrm{e}^{i\pi/4}) \to 0, \tag{5.37}$$

$$D_{-n-1}(iz) = D_{-n-1}(R\mathrm{e}^{i\pi/4}) = D_{-n-1}(|R|\mathrm{e}^{i3\pi/4}) \to \text{const.} \tag{5.38}$$

By using these asymptotics, we derive the Landau–Zener formula. Here we assume that the initial condition is the ground state at $t \to -\infty$, that is,

$$|c_1(-\infty)| = 1, \qquad c_2(-\infty) = 0. \tag{5.39}$$

Thus, the corresponding boundary condition of Eq. (5.27) is given by

$$\lim_{R\to\infty} \tilde{U}\left(z = R\mathrm{e}^{i3\pi/4}\right) = 0. \tag{5.40}$$

A particular solution of the Weber differential equation, $D_{-n-1}(-iz)$, becomes zero on the axes given by $z = R\mathrm{e}^{i\pi/4}$ and $z = R\mathrm{e}^{i3\pi/4}$ in the limit of $R \to \infty$. This is consistent with the boundary condition we considered. Thus, the solution of Eq. (5.27) is represented as

$$\tilde{U}(z) = AD_{-n-1}(-iz), \tag{5.41}$$

where A is a normalized constant, which can be determined from one of the initial conditions, $|c_1(-\infty)| = 1$. The t-axis on the z plane is parallel to $z = \mathrm{e}^{i3\pi/4}$. Thus, we take the asymptotic expansion to $D_{-n-1}(iz)$ (Gradshtyen and Ryzhik, 2007). As a result, we obtain

$$D_{-n-1}(iR\mathrm{e}^{i3\pi/4}) \approx \mathrm{e}^{-i(n+1)\pi/4}\mathrm{e}^{-iR^2/4}R^{-n-1}, \tag{5.42}$$

$$\frac{d}{dR}D_{-n-1}(-iR\mathrm{e}^{i3\pi/4}) \approx \mathrm{e}^{-i(n+1)\pi/4}\left(-\frac{iR}{2}\right)\mathrm{e}^{-iR^2/4}R^{-n-1}. \tag{5.43}$$

Here we ignored $1/R^2$ or higher-order terms. By using Eqs. (5.17), (5.42), and (5.43), we obtain the asymptotic form of $|c_1(t)|$ as follows:

$$|C_1(t)| = \frac{1}{\Delta}\left|\frac{d}{dt}C_2(t)\right| \tag{5.44}$$

$$= \frac{1}{\Delta}|A|\left|\left[-\frac{1}{2}\alpha^{1/2}RD_{-n-1}(-iR\mathrm{e}^{i3\pi/4}) - i\alpha^{1/2}\frac{d}{dR}D_{-n-1}(-iR\mathrm{e}^{i3\pi/4})\right]\right| \tag{5.45}$$

$$\approx \frac{1}{\Delta}|A|\sqrt{\alpha}\left|e^{-i(n+1)\pi/4}\mathrm{e}^{iR^2/4}R^{-n}\right|. \tag{5.46}$$

Here we used the relation given by

$$\sqrt{\alpha}t = -\sqrt{\alpha}|t| = -R. \tag{5.47}$$

From Eq.(5.47), we obtain

$$1 = |C_1(-\infty)| = \frac{\sqrt{\alpha}}{\Delta}|A|\mathrm{e}^{\pi\Delta^2/4\alpha}. \tag{5.48}$$

Thus, the normalization factor is obtained as

$$|A| = \sqrt{\frac{\Delta^2}{\alpha}}\mathrm{e}^{-\pi\Delta^2/4\alpha}. \tag{5.49}$$

Next we find the solution at $t \to \infty$, which corresponds to $z = \lim_{R\to\infty} R\mathrm{e}^{-i\pi/4}$ as follows:

$$U(t) = \tilde{U}(\mathrm{e}^{-i\pi/4}\sqrt{\alpha}t) = \sqrt{\frac{\Delta^2}{\alpha}}\mathrm{e}^{-\pi\Delta^2/4\alpha}D_{-n-1}(R\mathrm{e}^{-i3\pi/4}). \tag{5.50}$$

Here we take the asymptotic expansion to $D_{-n-1}(z)$ along the axis represented by $z = e^{-i3\pi/4}$. As a result, we obtain

$$D_{-n-1}(Re^{Re^{-i3\pi/4}}) \approx e^{-iR^2/4}e^{3\pi(n+1)i/4}R^{-n-1}$$

$$+\frac{\sqrt{2\pi}}{\Gamma(n+1)}e^{iR^2/4}e^{\pi ni/4}R^n. \tag{5.51}$$

Since n is a pure imaginary number, the first term of the right-hand side of Eq.(5.51) can be ignored in the limit of $R \to \infty$. Thus, we obtain

$$|C_2(\infty)|^2 = \frac{\Delta^2}{\alpha}e^{-\pi\Delta^2/2\alpha}\frac{2\pi}{|\Gamma(i\gamma+1)|^2}e^{-\frac{\pi\gamma}{2}} \tag{5.52}$$

$$= 1 - e^{-2\pi\Delta^2/\alpha}. \tag{5.53}$$

Here we used the relation given by

$$|\Gamma(ix)|^{-2} = \frac{x}{2\pi}\left(e^{\pi x} - e^{-\pi x}\right). \tag{5.54}$$

The probability of the adiabatic time evolution is given by $|C_2(\infty)|^2$. Thus, the probability of the *non*adiabatic time evolution is represented as

$$p = 1 - |C_2(\infty)|^2 = \exp(-2\pi\Delta^2/\alpha) = \exp\left\{-2\pi\frac{\Delta^2}{\frac{d}{dt}[\varepsilon_1(t) - \varepsilon_2(t)]}\right\}. \tag{5.55}$$

This equation is called the Landau–Zener formula. As the sweep speed decreases or the energy gap increases, the transition probability decreases (see Fig. 5.1). Many further theoretical studies based on the Landau–Zener formula has been done (Wu and Niu, 2000; Liu et al., 2002; Saito et al., 2007b; Shevchenko et al., 2010; Nakamura, 2012).

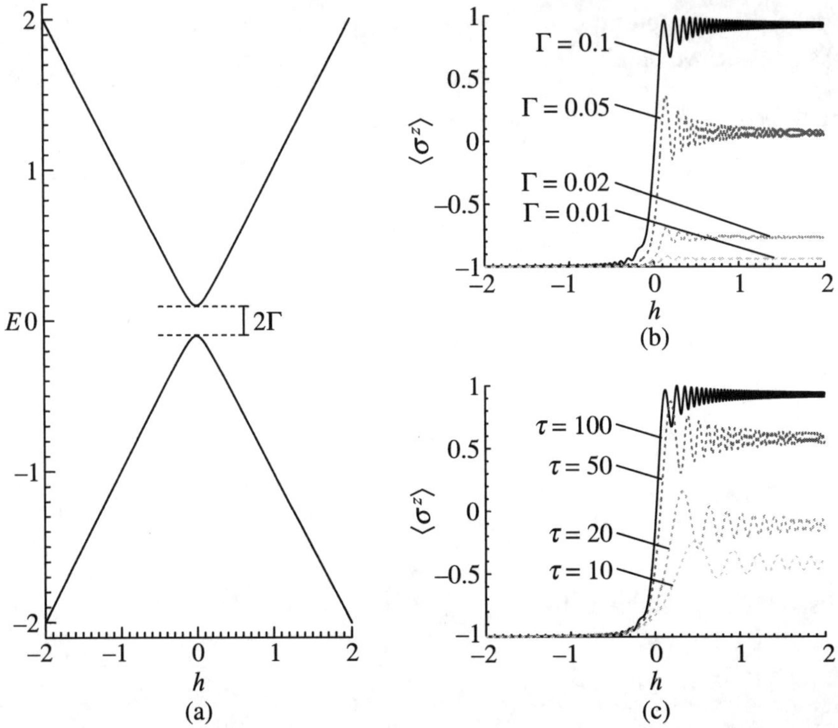

Fig. 5.1 Landau–Zener transition of a system whose Hamiltonian is given by $\mathscr{H} = -h(t)\sigma^z - \Gamma\sigma^x$, where $h(t) = t/\tau$. (a) The instantaneous eigenenergies of the system. (b) Γ-dependence of the expectation value of σ^z as a function of $h(t)$ with $\tau = 100$. Here, the initial state is set to the ground state for $h = -5$. (c) τ-dependence of the expectation value of σ^z as a function of $h(t)$ with $\Gamma = 0.1$. Here, the initial state is set to the ground state for $h = -5$.

5.2 Kibble–Zurek Mechanism

The Kibble–Zurek mechanism treats the dynamics which passes through the second-order phase transition point (Kibble, 1976, 1980; Zurek, 1985; Kibble, 1985). With the Kibble–Zurek mechanism, we can understand the crossover between adiabatic and nonadiabatic time evolutions in systems which exhibit the second-order phase transition. In Section 5.2.1, we will describe the basic of the Kibble–Zurek mechanism by focusing on the classical system applying simulated annealing. In contrast, in Section 5.2.2, we will show results of the Kibble–Zurek mechanism in the random ferromagnetic quantum Ising chain under change of temperature and the transverse field. This is the simplest example of simulated annealing and quantum annealing.

5.2.1 Basic concept of the Kibble–Zurek mechanism

Let us discuss the Kibble–Zurek mechanism which is a dynamical behavior observed in systems which exhibit second-order classical phase transitions (Dziarmaga, 2010; Polkovnikov et al., 2011). Here, we consider the dynamical properties under simulated annealing explained in Section 3.1. The second-order phase transition point is passed through when the temperature decreases. When the annealing speed is almost zero, i.e., slow enough, it is guaranteed that the time evolution follows the equilibrium state at instantaneous temperatures. On the other hand, the situation is changed when the annealing speed is finite. As explained in Section 2.1, the correlation length of spin in the equilibrium state diverges at the second-order phase transition point. Thus, the relaxation time which is the characteristic time to reach the equilibrium state becomes infinite. We cannot obtain the equilibrium state of systems where the second-order phase transition occurs, using simulated annealing with a finite annealing speed. Even near the second-order phase transition point, it is difficult to obtain the equilibrium state, since the relaxation time is extremely long—called the critical slowing down. This is because the growth of the correlation length stops below the temperature such that the equilibrium state cannot be reached. For slow annealing speeds, the system can reach the equilibrium state close to the second-order phase transition point, and the stopped correlation length is long. The stopped correlation length is scaled by the annealing speed as follows.

We introduce the dimensionless temperature g defined by

$$g := \frac{T - T_{\rm c}}{T_{\rm c}}, \tag{5.56}$$

where $T_{\rm c}$ is the second-order phase transition temperature. Near the second-order phase transition temperature, the correlation length $\xi(g)$ is scaled by

$$\xi(g) \propto |g|^{-\nu}, \tag{5.57}$$

where ν is the critical exponent, which was explained in Section 2.1. As mentioned before, the relaxation time becomes extremely long near the second-order phase transition point. This divergent behavior of relaxation time $\tau_{\rm relax}$ is expressed by

$$\tau_{\rm relax} \propto \xi^z, \tag{5.58}$$

where z is called the dynamical exponent. Then, the following relation is satisfied:

$$\tau_{\rm relax}(g) \propto [\xi(g)]^z \propto |g|^{-z\nu}. \tag{5.59}$$

Such critical behavior in dynamics is called the dynamical critical phenomena.

Here, we investigate dynamical properties under simulated annealing in which the schedule of the time-dependent temperature is given by

$$T(t) = T_{\rm c}\left(1 - \frac{t}{\tau_{\rm Q}}\right), \quad (-\infty < t \leq \tau_{\rm Q}), \tag{5.60}$$

where $\tau_{\rm Q}^{-1}$ is the annealing speed. As the value of $\tau_{\rm Q}$ increases, the temperature decreases slowly. Then, at the initial time $t = -\infty$, the temperature is infinitely large: $T(t = -\infty) = +\infty$, and at the final time $t = \tau_{\rm Q}$, the temperature is zero: $T(t = \tau_{\rm Q}) = 0$. Furthermore, at $t = 0$, the temperature is $T(t = 0) = T_{\rm c}$, which means that the second-order phase transition point is passed through at $t = 0$. The time-dependent dimensionless temperature is expressed as

$$g(t) = \frac{T(t) - T_{\rm c}}{T_{\rm c}} = -\frac{t}{\tau_{\rm Q}}. \tag{5.61}$$

Let $|t|$ be the remaining time to the second-order phase transition point. In the Kibble–Zurek mechanism, we assume that the system can reach the equilibrium state if $|t|$ is longer than the relaxation time $\tau_{\rm relax}(g(t))$. On the other hand, the system cannot reach the equilibrium state if $|t|$ is shorter than $\tau_{\rm relax}(g(t))$. In other words, the following relations are assumed:

$$\begin{cases} \tau_{\rm relax}(g(t)) < |t| & \text{(If the system can reach the equilibrium state.)} \\ \tau_{\rm relax}(g(t)) > |t| & \text{(If the system cannot reach the equilibrium state.)} \end{cases}$$

According to these relations, the time $t_{\rm s}$ at which the system cannot reach the equilibrium state is obtained as

$$\tau_{\rm relax}(g(t_{\rm s})) = |t_{\rm s}|, \tag{5.62}$$

which means that the growth of the correlation length is stopped at $t = t_{\rm s}$. Thus, below the temperature $T(t_{\rm s})$, the correlation length does not change at all and the value is fixed at $\xi(g(t_{\rm s}))$. Schematics of the temperature dependence of the correlation length for different annealing speeds $\tau_{\rm Q}^{-1}$ are shown in Fig. 5.2 (a). Furthermore, the dimensionless temperature at $t = t_{\rm s}$ is represented as

$$g(t_s) = \frac{|t_{\rm s}|}{\tau_{\rm Q}} = \frac{\tau_{\rm relax}(g(t_{\rm s}))}{\tau_{\rm Q}} \propto \frac{|g(t_{\rm s})|^{-z\nu}}{\tau_{\rm Q}}. \tag{5.63}$$

From Eq.(5.63), $g(t_{\rm s})$ is scaled by the annealing speed:

$$g(t_{\rm s}) \propto \tau_{\rm Q}^{-\frac{1}{1+z\nu}}. \tag{5.64}$$

Using the relation, we obtain the scaling law of the correlation length at $t = t_{\rm s}$:

$$\xi(g(t_{\rm s})) \propto \tau_{\rm Q}^{\frac{\nu}{1+z\nu}}. \tag{5.65}$$

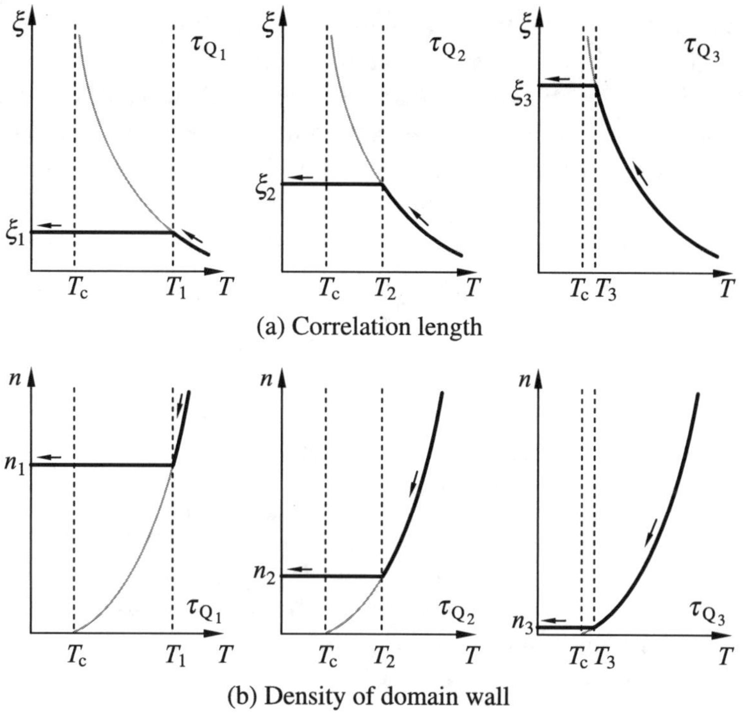

Fig. 5.2 (a) Schematics of the temperature dependence of correlation length $\xi(g(t))$ in simulated annealing where the annealing speed τ_{Q}^{-1} ($\tau_{\mathrm{Q}1} < \tau_{\mathrm{Q}2} < \tau_{\mathrm{Q}3}$). $T_i := T(t_{\mathrm{s}})$ for $i = 1,2,3$ denotes the temperature at which the system cannot reach the equilibrium state under simulated annealing whose annealing speed is τ_{Q_i}. The stopped correlation length is written as $\xi_i := \xi(g(t_{\mathrm{s}}))$ for $i = 1,2,3$. Here, $\xi_1 < \xi_2 < \xi_3$. (b) Schematics of the temperature dependence of the density of the domain wall $n(t)$ in simulated annealing where the annealing speed τ_{Q}^{-1} is different. The stopped density of the domain wall is written as $n_i := n(t_{\mathrm{s}})$ for $i = 1,2,3$. Here, $n_1 > n_2 > n_3$.

Moreover, the density of the domain wall $n(t)$ can be estimated by the correlation length $\xi(g(t))$ as:

$$n(t) \propto \xi(g(t))^{-d}, \tag{5.66}$$

where d is the spatial dimension. From the behavior of the correlation length shown in Fig. 5.2 (a), the temperature dependence of $n(t)$ can be obtained as Fig. 5.2 (b). Below $T(t_{\mathrm{s}})$, the density of the domain wall does not change at all and the value is fixed at $n(t_{\mathrm{s}})$. In addition, at $t = t_{\mathrm{s}}$, the density of the domain wall is scaled by

$$n(t_{\mathrm{s}}) \propto \tau_{\mathrm{Q}}^{-\frac{d\nu}{1+z\nu}}. \tag{5.67}$$

The correlation length and density of the domain wall, i.e., topological defect, below the second-order phase transition temperature are scaled by the annealing speed, which is the Kibble–Zurek mechanism. According to the Kibble–Zurek mechanism, the topological defect, i.e., domain wall, can be created when the temperature decreases rapidly in systems with second-order phase transition. The phenomenon relates to the creation of cosmic strings caused by the spontaneous symmetry breaking in the Big Bang theory (Kibble, 1976, 1980; Zurek, 1985; Kibble, 1985). The creation of topological defects in magnetic models (Damski, 2005; Zurek et al., 2005; Polkovnikov, 2005; Dziarmaga, 2005; Biroli et al., 2010), superfluid helium systems (Ruutu et al., 1996; Eltsov et al., 2000), and Bose–Einstein condensations (Saito et al., 2007a; Weiler et al., 2008) can also be described by the Kibble–Zurek mechanism.

5.2.2 Kibble–Zurek mechanism in the Ising chain with random ferromagnetic interactions

As an example of the Kibble–Zurek mechanism, we describe the Kibble–Zurek mechanism in the Ising chain with random ferromagnetic interactions. Here, the temperature and transverse field are control parameters. We consider the Kibble–Zurek mechanism when either the temperature or the transverse field decreases. The density of the domain wall will bc cstimated for each case.

Kibble–Zurek mechanism when the temperature decreases

Suzuki(2009) considered the density of the domain wall in a one-dimensional random ferromagnetic Ising chain. The Hamiltonian of the system is defined by

$$\mathcal{H} = -\sum_{i=1}^{N} J_i \sigma_i^z \sigma_{i+1}^z, \quad (\sigma_i^z = \pm 1), \tag{5.68}$$

where J_i is the magnetic interaction between the ith site and the $i+1$th site. Here, the value of J_i is distributed uniformly between $0 \leq J_i \leq 1$. Thus, the distribution function of the magnetic interaction $P(J_i)$ is given as

$$P(J_i) = \begin{cases} 1 & (0 \leq J_i \leq 1) \\ 0 & (J_i < 0 \text{ and } 1 < J_i) \end{cases}. \tag{5.69}$$

In this case, the ferromagnetic configuration is the ground state because the value of J_i is always positive. In the Ising chain, as discussed in Section 2.1.3, the finite-temperature phase transition does not occur. But phase transition from the paramagnetic phase to the ferromagnetic phase occurs at zero temperature.

The correlation function between two spins whose is distance r, is expressed as

$$\overline{\langle \sigma_i^z \sigma_{i+r}^z \rangle} = \left(\frac{1}{\beta} \ln \cosh \beta \right)^r, \tag{5.70}$$

where $\langle\cdot\rangle$ and the bar denote the thermal average and the configurational average, respectively. Using the correlation function, the correlation length ξ is represented by

$$\overline{\langle\sigma_i^z\sigma_{i+r}^z\rangle} = \mathrm{e}^{-r/\xi}. \tag{5.71}$$

Thus, the correlation length can be written as

$$\xi = -\frac{1}{\ln(\beta^{-1}\ln\cosh\beta)} \simeq \frac{\beta}{\ln 2}, \tag{5.72}$$

where the final approximation is valid in the low temperature limit. Since $T_\mathrm{c} = 0$ in the model, the behavior of ξ near zero temperature is important. By using the Glauber dynamics (Glauber, 1963), we obtain the relaxation time τ_relax:

$$\tau_\mathrm{relax} = \frac{1}{1-\tanh 2\beta} \simeq \frac{1}{2}\mathrm{e}^{4\beta} = \frac{1}{2}\mathrm{e}^{4\xi\ln 2}, \tag{5.73}$$

where the approximation is also valid at low temperatures.

To perform simulated annealing, we use the following schedule of the time-dependent temperature $T(t)$:

$$T(t) = -\frac{t}{\tau_\mathrm{Q}}, \quad (-\infty < t \leq 0). \tag{5.74}$$

In this case, the initial state is $T(-\infty) = +\infty$ at $t = -\infty$, while when $t = 0$, $T(0) = 0 (= T_\mathrm{c})$.

According to the Kibble–Zurek mechanism, we define t_s where the system cannot reach the equilibrium state as follows:

$$\tau_\mathrm{relax}(T(t_\mathrm{s})) = |t_\mathrm{s}|, \tag{5.75}$$

and we obtain

$$T(t_\mathrm{s}) = \frac{|t_\mathrm{s}|}{\tau_\mathrm{Q}} = \frac{\tau_\mathrm{relax}(T(t_\mathrm{s}))}{\tau_\mathrm{Q}}. \tag{5.76}$$

By using Eqs. (5.72), (5.73), and (5.76), the following relation is satisfied:

$$\frac{1}{\xi(T(t_\mathrm{s}))\ln 2} \simeq \frac{1}{2\tau_\mathrm{Q}}\mathrm{e}^{4\xi(T(t_\mathrm{s}))\ln 2}. \tag{5.77}$$

Here we consider the case that $\tau_\mathrm{Q} \gg 1$ where the annealing speed is very slow, and we obtain

$$\xi(T(t_\mathrm{s})) = \frac{\ln\tau_\mathrm{Q} + \ln 2 - \ln[\xi(T(t_\mathrm{s}))\ln 2]}{4\ln 2} \propto \frac{\ln\tau_\mathrm{Q}}{4\ln 2}. \tag{5.78}$$

Thus, the relation between the density of the domain wall $n_{\mathrm{SA}}(t_{\mathrm{s}})$ at $t = t_{\mathrm{s}}$ and the annealing speed τ_{Q}^{-1} is obtained as

$$n_{\mathrm{SA}}(t_{\mathrm{s}}) \propto \frac{4\ln 2}{\ln \tau_{\mathrm{Q}}}. \tag{5.79}$$

Incidentally, the density of the domain wall in the pure ferromagnetic Ising model (i.e., $J_i = J$ for all i) is given by

$$n_{\mathrm{SA}}(t_{\mathrm{s}}) \propto \frac{1}{\sqrt{\tau_{\mathrm{Q}}}}. \tag{5.80}$$

Eq. (5.80) is derived by the same scheme as the case of the random ferromagnetic Ising model (Suzuki, 2009).

Kibble–Zurek mechanism when the transverse field decreases

The Kibble–Zurek mechanism in the random ferromagnetic Ising chain in a transverse field $\Gamma(t)$ has been studied (Dziarmaga, 2006; Caneva et al., 2007; Suzuki, 2011). The time-dependent Hamiltonian of the model is given as

$$\mathscr{H}(t) = -\sum_{i=1}^{N} J_i \hat{\sigma}_i^z \hat{\sigma}_{i+1}^z - \Gamma(t) \sum_{i=1}^{N} \hat{\sigma}_i^x, \tag{5.81}$$

where the value of J_i is distributed uniformly between $0 \le J_i \le 1$ as well as the case of simulated annealing. We introduce the time-dependent dimensionless transverse field $g(t)$ as

$$g(t) := \frac{\Gamma(t) - \Gamma_{\mathrm{c}}}{\Gamma_{\mathrm{c}}}. \tag{5.82}$$

When $|g(t)| \ll 1$, the correlation length $\xi(g(t))$ is obtained by the renormalization group analysis (Shankar and Murthy, 1987) as

$$\xi(g(t)) \propto |g(t)|^{-\nu}, \quad (\nu = 2). \tag{5.83}$$

Furthermore, the coherence time τ_{coh} which corresponds to the relaxation time in simulated annealing is scaled by

$$\tau_{\mathrm{coh}}(g(t)) \propto [\xi(g(t))]^z \propto |g(t)|^{-\nu z}, \quad (\nu = 2), \tag{5.84}$$

where z is the dynamical exponent which is obtained by

$$z \sim \frac{1}{2|g(t)|}, \tag{5.85}$$

in systems with random interactions. This formula indicates that the dynamical exponent diverges at the transition point. Note that in the pure system, the dynamical exponent is constant and given by $z = 1$. The divergent behavior of the dynamical exponent does not appear in the pure system.

We use the following schedule for the transverse field $\Gamma(t)$:

$$\Gamma(t) = \Gamma_c\left(1 - \frac{t}{\tau_Q}\right), \quad (-\infty < t \leq \tau_Q). \tag{5.86}$$

Here, in the model, it is known that the quantum phase transition from the paramagnetic phase to the ferromagnetic phase occurs at $\Gamma(t) = \Gamma_c = \exp(\overline{\ln J_i})$ (Shankar and Murthy, 1987). That is, $\Gamma(t)$ gradually decreases from $\Gamma(-\infty) = +\infty$ at $t = -\infty$ to $\Gamma(t_s) = 0$ at $t = t_s$. The schedule is often used in quantum annealing (see Chapter 6).

According to the Kibble–Zurek mechanism, we define t_s where the system cannot reach the equilibrium state as follows:

$$\tau_{\mathrm{coh}}(g(t_s)) = |t_s|. \tag{5.87}$$

By using the coherence time, $g(t_s)$ is represented as

$$g(t_s) = \frac{|t_s|}{\tau_Q} = \frac{\tau_{\mathrm{coh}}(g(t_s))}{\tau_Q}. \tag{5.88}$$

By using Eqs. (5.83), (5.84), (5.85), and (5.88), the following relation is satisfied:

$$\frac{1}{\sqrt{\xi(g(t_s))}} \propto \frac{1}{\tau_Q}|\xi(g(t_s))|^z \simeq \frac{1}{\tau_Q}|\xi(g(t_s))|^{\frac{1}{2}\sqrt{\xi(g(t_s))}}, \tag{5.89}$$

and we obtain

$$\frac{1}{2}\left[\sqrt{\xi(g(t_s))} + 1\right]\ln\xi(g(t_s)) \propto \ln\tau_Q. \tag{5.90}$$

Furthermore, we consider the case that $\tau_Q \gg 1$, where the transverse field decreases slowly and the system can reach the equilibrium state even near Γ_c. In this limit, since $\xi(g(t_s))$ is large, the following approximation is valid:

$$\sqrt{\xi(g(t_s))} + 1 \simeq \sqrt{\xi(g(t_s))}. \tag{5.91}$$

In addition, since the change of $\ln\xi(g(t_s))$ is gradual in comparison with that of $\xi(g(t_s))$, we obtain

$$\xi(g(t_s)) \propto \left[\frac{\ln\tau_Q}{\ln\xi(g(t_s))}\right]^2 \propto (\ln\tau_Q)^2. \tag{5.92}$$

Thus, the relation between the density of the domain wall $n_{\mathrm{QA}}(t_s)$ at $t = t_s$ and the annealing speed τ_{Q} is obtained as

$$n_{\mathrm{QA}}(t_s) \propto (\ln \tau_{\mathrm{Q}})^{-2}. \tag{5.93}$$

By the way, the density of the domain wall in the pure ferromagnetic Ising chain (i.e., $J_i = J$ for all i) is scaled by

$$n_{\mathrm{QA}}(t_s) \propto \frac{1}{\sqrt{\tau_{\mathrm{Q}}}}. \tag{5.94}$$

This was shown by Zurek et al. (2005) and Dziarmaga (2005).

Comparison between the Kibble–Zurek mechanisms when the temperature and transverse field decreases

According to the obtained results, we can compare the efficiencies of simulated annealing and quantum annealing to find the ground state of the Ising chain with random ferromagnetic interactions. In the model, there is no domain wall in the ground state. Thus, the density of the domain wall $n(t)$ expresses the degree of excitation depending on the time t. In other words, $n(t_s)$ means the difference between states obtained by simulated annealing and quantum annealing. Furthermore, in general, when we solve optimization problems, we would like to obtain a better solution as fast as possible, that is, as small τ_{Q} as possible.

We have obtained the relations between the density of the domain wall and the annealing speed for the Ising chain with random ferromagnetic interactions as

$$n_{\mathrm{SA}}(t_s) \propto (\ln \tau_{\mathrm{Q}})^{-1}, \quad \text{(Simulated annealing)}, \tag{5.95}$$

$$n_{\mathrm{QA}}(t_s) \propto (\ln \tau_{\mathrm{Q}})^{-2}, \quad \text{(Quantum annealing)}. \tag{5.96}$$

This means that the decrease of $n_{\mathrm{QA}}(t_s)$ is faster than that of $n_{\mathrm{SA}}(t_s)$ with the value of τ_{Q}. Thus, according to the Kibble–Zurek mechanism, we conclude that quantum annealing is more efficient than simulated annealing to obtain the ground state of the Ising chain with random ferromagnetic interactions.

We can compare the performance of simulated annealing and quantum annealing from a viewpoint of the residual energy. The residual energy is given by

$$E_{\mathrm{res}}(\tau) = E_\tau - E_0, \tag{5.97}$$

where E_τ is the final energy after annealing and E_0 is the ground state energy. According to the Huse and Fisher's theory (Huse and Fisher, 1986) (see Section 3.3.3), the residual energy under simulated annealing is scaled as

$$E_{\text{res}}^{\text{SA}}(\tau) \sim \frac{1}{(\ln \tau_{\text{Q}})^2}. \tag{5.98}$$

On the other hand, two contradictory results were reported in quantum annealing. Caneva et al.(2007) concluded that

$$E_{\text{res}}^{\text{QA}}(\tau) \sim \frac{1}{(\ln \tau_{\text{Q}})^{\zeta}}, \qquad \zeta \sim 3.4. \tag{5.99}$$

However, Suzuki (2011) insisted that this result is affected by the finite-size effect and concluded that

$$E_{\text{res}}^{\text{QA}}(\tau) \sim \frac{1}{(\ln \tau_{\text{Q}})^4}. \tag{5.100}$$

Of course, since this is just an example, we cannot conclude that quantum annealing is appropriate for arbitrary optimization problems in comparison with simulated annealing in general. This is because the efficiency of annealing methods depends on the type of the optimization problem (see Chapter 6).

References

Biroli, G., L. F. Cugliandolo, and A. Sicilia. 2010. 'Kibble-Zurek mechanism and infinitely slow annealing through critical points'. *Phys. Rev. E* 81: 050101(R).

Caneva, T., R. Fazio, and G. E. Santoro. 2007. 'Adiabatic quantum dynamics of a random Ising chain across its quantum critical point'. *Phys. Rev. B* 76: 144427.

Chiorescu, I., W. Wernsdorfer, A. Müller, S. Miyashita, and B. Barbara. 2003. 'Adiabatic Landau-Zener-Stückelberg transition with or without dissipation in the low-spin molecular system V_{15}'. *Phys. Rev. B* 67: 020402(R).

Damski, B. 2005. 'The simplest quantum model supporting the Kibble-Zurek mechanism of topological defect production: Landau-Zener transitions from a new perspective'. *Phys. Rev. Lett.* 95: 035701.

Dziarmaga, J. 2005. 'Dynamics of a quantum phase transition: Exact solution of the quantum Ising model'. *Phys. Rev. Lett.* 95: 245701.

Dziarmaga, J. 2006. 'Dynamics of a quantum phase transition in the random Ising model: Logarithmic dependence of the defect density on the transition rate'. *Phys. Rev. B* 74: 064416.

Dziarmaga, J. 2010. 'Dynamics of a quantum phase transition and relaxation to a steady state'. *Adv. Phys.* 59: 1063.

Eltsov, V. B., T. W. B. Kibble, M. Krusius, V. M. H. Ruutu, and G. E. Volovik. 2000. 'Composite defect extends analogy between cosmology and 3He'. *Phys. Rev. Lett.* 85: 4739.

Glauber, R. J. 1963. 'Time-dependent statistics of the Ising model'. *J. Math. Phys.* 4: 294.

Gradshteyn, I. S., I. M. Ryzhik, A. Jeffrey, and D. Zwillinger. 2000a. 'Parabolic Cylinder Functions' in *Table of Integrals, Series, and Products*. Sixth edition . p. 835. Academic Press.

Gradshteyn, I. S., I. M. Ryzhik, A. Jeffrey, and D. Zwillinger. 2000b. 'Parabolic Cylinder Functions $D_p(z)$' in *Table of Integrals, Series, and Products.* Sixth edition. p. 1018. Academic Press.

Gradshtyen, I. S., and I. M. Ryzhik. 2007. *Table of Integrals, Series, and Products.* Seventh edition. Academic Press.

Huse, D. A., and D. S. Fisher. 1986. 'Residual energies after slow cooling of disordered systems'. *Phys. Rev. Lett.* 57: 2203.

Kibble, T. W. B. 1976. 'Topology of cosmic domains and strings'. *J. Phys. A: Math. Gen.* 9: 1387.

Kibble, T. W. B. 1980. 'Some implications of a cosmological phase transition'. *Phys. Rep.* 67: 183.

Kibble, T. W. B. 1985. 'Phase transitions: Cosmology in the laboratory'. *Nature* 317: 472.

Koizumi, S., M. Nihei, T. Shiga, M. Nakano, H. Nojiri, R. Bircher, O. Waldmann, S. T. Ochsenbein, H. U. Güdel, F. Fernandez-Alonso, and H. Oshio. 2007. 'A wheel-shaped single-molecule magnet of [MnII 3MnIII 4]: Quantum tunneling of magnetization under static and pulse magnetic fields'. *Chem. Eur. J* 13: 8445.

Landau, L. D. 1932. 'Zur theorie der energieübertragung. II'. *Physikalische Zeitschrift der Sowjetunion* 2: 46.

Liu, J., L. Fu, B.-Y. Ou, S.-G. Chen, D.-I. Choi, B. Wu, and Q. Niu. 2002. 'Theory of nonlinear Landau-Zener tunneling'. *Phys. Rev. A* 66: 023404.

Majorana, E. 1932. 'Atomi orientati in campo magnetico variabile'. *Il Nuovo Cimento* 9: 43.

Nakamura, H. 2012. *Nonadiabatic Transition: Concepts, Basic Theories and Applications.* World Scientific Press.

Polkovnikov, A. 2005. 'Universal adiabatic dynamics in the vicinity of a quantum critical point'. *Phys. Rev. B* 72: 161201(R).

Polkovnikov, A., K. Sengupta, A. Silva, and M. Vengalattore. 2011. 'Nonequilibrium dynamics of closed interacting quantum systems'. *Rev. Mod. Phys.* 83: 863.

Rubbmark, J. R., M. M. Kash, M. G. Littman, and D. Kleppner. 1981. 'Dynamical effects at avoided level crossings: A study of the Landau-Zener effect using Rydberg atoms'. *Phys. Rev. A* 23: 3107.

Ruutu, V. M. H., V. B. Eltsov, A. J. Gill, T. W. B. Kibble, M. Krusius, Y. G. Makhlin, B. Placais, G. E. Volovik, and W. Xu. 1996. 'Vortex formation in neutron-irradiated superfluid 3He as an analogue of cosmological defect formation'. *Nature* 382: 334.

Saito, H., Y. Kawaguchi, and M. Ueda. 2007a. 'Kibble-Zurek mechanism in a quenched ferromagnetic Bose-Einstein condensate'. *Phys. Rev. A* 76: 043613.

Saito, K., M. Wubs, S. Kohler, Y. Kayanuma, and P. Hänggi. 2007b. 'Dissipative Landau-Zener transitions of a qubit: Bath-specific and universal behavior'. *Phys. Rev. B* 75: 214308.

Shankar, R., and G. Murthy. 1987. 'Nearest-neighbor frustrated random-bond model in $d = 2$: Some exact results'. *Phys. Rev. B* 36: 536.

Shevchenko, S. N., S. Ashhab, and F. Nori. 2010. 'Landau–Zener–Stückelberg interferometry'. *Physics Reports* 492: 1.

Stueckelberg, E. C. G. 1932. 'Theorie der unelastischen Stösse zwischen Atomen'. *Helvetica Physica Acta* 5: 369.

Suzuki, S. 2009. 'Cooling dynamics of pure and random Ising chains'. *J. Stat. Mech.* P03032.

Suzuki, S. 2011. 'Kibble-Zurek mechanism in simulated annealing and quantum annealing'. *J. Phys.: Conf. Ser.* 302: 012046.

Weiler, C. N., T. W. Neely, D. R. Scherer, A. S. Bradley, M. J. Davis, and B. P. Anderson. 2008. 'Spontaneous vortices in the formation of Bose–Einstein condensates'. *Nature* 455: 948.

Wernsdorfer, W., R. Sessoli, A. Caneschi, D. Gatteschi, and A. Cornia. 2000. 'Nonadiabatic Landau-Zener tunneling in Fe_8 molecular nanomagnets'. *Europhys. Lett.* 50: 552.

Wernsdorfer, W., S. Bhaduri, C. Boskovic, G. Christou, and D. N. Hendrickson. 2002. 'spin-parity dependent tunneling of magnetization in single-molecule magnets'. *Phys. Lett. B* 65: 180403(R).

Wernsdorfer, W., N. E. Chakov, and G. Christou. 2005. 'Quantum phase interference and spin-parity in Mn_{12} single-molecule magnets'. *Phys. Rev. Lett.* 95: 037203.

Whittaker, E. T., and G. N. Watson. 1990. 'The parabolic cylinder function' in *A Course in Modern Analysis.* Fourth edition. p 347. Cambridge University Press.

Wu, B., and Q. Niu. 2000. 'Nonlinear Landau-Zener tunneling'. *Phys. Rev. A* 61: 023402.

Zener, C. 1932. 'Non-adiabatic crossing of energy levels'. *Proc. R. Soc. London, Ser. A* 137: 696.

Zurek, W. H. 1985. 'Cosmological experiments in superfluid helium?' *Nature* 317: 505.

Zurek, W. H., U. Dorner, and P. Zoller. 2005. 'Dynamics of a quantum phase transition'. *Phys. Rev. Lett.* 95: 105701.

CHAPTER

6

Quantum Annealing

In this chapter, we explain the quantum annealing method in detail. Quantum annealing is a quantum-information technology the aim of which is to solve the combinatorial optimization problems explained in Chapter 3. A pioneering paper on quantum annealing by Kadowaki and Nishimori in 1998 (Kadowaki and Nishimori, 1998) gave the final shape. As mentioned earlier (specifically in Chapters 1 and 4), the idea of the study is based on many preceding pieces of research on the thermal/quantum fluctuation effect in random Ising models (see Chapter 2 and Chapter 4) and quantum dynamics (see Chapter 5). First, in Section 6.1, we present the general concept of quantum annealing. Next, some methods which can perform quantum annealing are explained with examples the Schrödinger equation (Section 6.2), path-integral Monte Carlo method (Section 6.3), Green's function Monte Carlo method (Section 6.4), density matrix renormalization group method (Section 6.5), and mean-field approximation (Section 6.6). Furthermore, in Section 6.7, toward the realization of quantum annealing by experiments, experimental results for quantum field response to magnetic materials are presented. In Section 6.8, quantum adiabatic evolution is explained, which is used to prove the convergence theorem of quantum annealing by the Schrödinger equation. In Section 6.9, the convergence theorem of quantum annealing by the Schrödinger equation, path-integral Monte Carlo method and Green's function Monte Carlo method are reviewed. Owing to the convergence theorem, quantum annealing is a mathematically guaranteed algorithm to obtain the ground state of random Ising models.

6.1 General Concept of Quantum Annealing

Quantum annealing is an efficient method to find the best solution of optimization problems by using the quantum effect. It is a cousin of the simulated annealing proposed by Kirkpatrick et al. (Kirkpatrick et al., 1983). In order to solve optimization problems, simulated annealing uses the thermal fluctuation as explained in Section 3.1. Thermal fluctuation induces transitions between states, which avoids stabilization in local minima and efficiently finds the global minimum. In principle, simulated annealing is successful in obtaining the best solution definitely when the temperature decreases, keeping the equilibrium state at instantaneous temperatures as described in Section 3.4.

Kadowaki and Nishimori proposed an alternative method to solve optimization problems by using the quantum tunneling process – quantum annealing (Kadowaki and Nishimori, 1998). Here, we explain the concept of quantum annealing by using a transverse field as a quantum effect in the Ising model. In this case, quantum annealing is performed by the following time-dependent Hamiltonian:

$$\hat{\mathcal{H}}(t) = \hat{\mathcal{H}}_0 + \hat{\mathcal{H}}_{\mathrm{q}}(t), \tag{6.1}$$

$$\hat{\mathcal{H}}_0 = -\sum_{1\le i,j\le N} J_{ij}\hat{\sigma}_i^z\hat{\sigma}_j^z - \sum_{i=1}^{N} h_i\hat{\sigma}_i^z, \tag{6.2}$$

$$\hat{\mathcal{H}}_{\mathrm{q}}(t) = -\Gamma(t)\sum_{i=1}^{N}\hat{\sigma}_i^x, \tag{6.3}$$

where i and j indicate the indices of sites, N is the number of spins, and t denotes the time. Here, $\hat{\sigma}_i^z$ and $\hat{\sigma}_i^x$ are respectively the Pauli matrices defined by

$$\hat{\sigma}_i^z = \begin{pmatrix} 1 & 0 \\ 0 & -1 \end{pmatrix}, \tag{6.4}$$

$$\hat{\sigma}_i^x = \begin{pmatrix} 0 & 1 \\ 1 & 0 \end{pmatrix}. \tag{6.5}$$

In this case, the bases are defined by

$$|\uparrow\rangle = \begin{pmatrix} 1 \\ 0 \end{pmatrix}, \tag{6.6}$$

$$|\downarrow\rangle = \begin{pmatrix} 0 \\ 1 \end{pmatrix}, \tag{6.7}$$

and these states indicate the up-spin state and down-spin state of the classical Ising model, respectively. $\hat{\mathcal{H}}_0$ is a time-independent Hamiltonian of the classical Ising model and describes optimization problems as mentioned in Secs. 3.1 and 3.2. Thus, the ground state of $\hat{\mathcal{H}}_0$ corresponds to the best solution of the corresponding optimization problem. $\hat{\mathcal{H}}_{\mathrm{q}}(t)$ is a time-dependent Hamiltonian which causes the quantum tunneling effect. When $\Gamma(t)$ is large, the quantum tunneling effect is strong. In other words, $\Gamma(t)$ represents the magnitude of the quantum fluctuation.

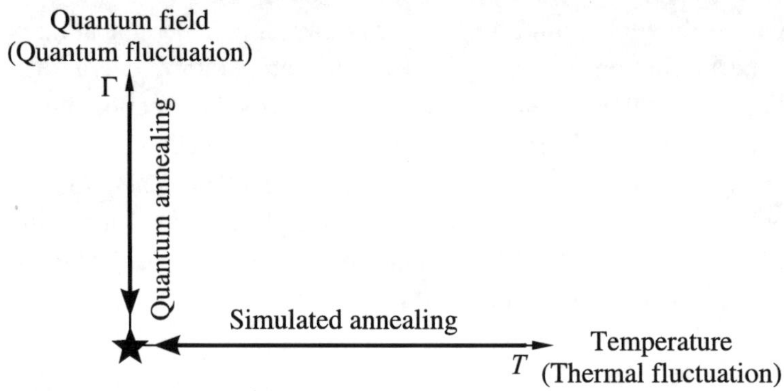

Fig. 6.1 Schematic of simulated annealing and quantum annealing. The purpose of these methods is to obtain the best solution at the position indicated by the star ($T = 0$, $\Gamma = 0$).

Figure 6.1 shows a schematic of simulated annealing and quantum annealing. Our purpose is to obtain the stable state at zero temperature and zero quantum field. In simulated annealing, we start with high temperature. Since the probability of the realization of states with energy E is proportional to $\exp(-\beta E)$, all states appear with almost the same probability at high temperatures (i.e., $\beta \to 0$). Next, the temperature is decreased gradually and we obtain the ground state of $\hat{\mathscr{H}}_0$ at $T = 0$ finally. The protocol of quantum annealing is similar to that of simulated annealing. In quantum annealing, we start with large $\Gamma(0)$. At large $\Gamma(t)$, the ground state is approximately represented by a superposition of all states with almost the same weight. In the limit of $\Gamma(t) \to \infty$, each spin state is represented by

$$|\rightarrow\rangle := \frac{1}{\sqrt{2}}(|\uparrow\rangle + |\downarrow\rangle), \tag{6.8}$$

which is the ground state of $\hat{\sigma}^x$. Next, we decrease $\Gamma(t)$ gradually and obtain the ground state of $\hat{\mathscr{H}}_0$ at time t such that $\Gamma(t) = 0$. In other words, quantum annealing is a method to obtain the nontrivial ground state at $\Gamma(t) = 0$ by starting a trivial ground state at large $\Gamma(0)$ and decreasing $\Gamma(t)$. Schematic of the concept of quantum annealing is shown in Fig. 6.2.

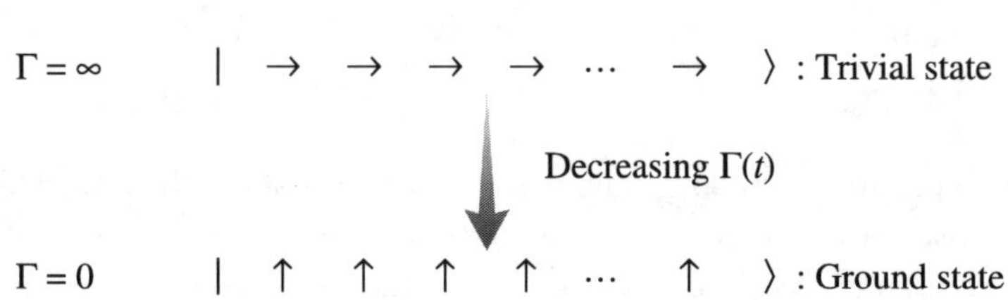

Fig. 6.2 Schematic of the concept of quantum annealing to search for the ferromagnetic ground state.

Quantum annealing for systems with degenerated ground states has interesting properties. Matsuda et al. studied the difference between simulated annealing and quantum annealing for geometrically frustrated systems and random systems (Matsuda et al., 2009). Models they considered have macroscopically degenerated ground states. Suppose we consider the case when the annealing speed is slow. When we perform simulated annealing to the models, the probability of each ground state is the same. This fact is consistent with the principle of equal weight. On the other hand, when we perform quantum annealing to the models by using the transverse field, the probability of a ground state is not always the same as that of another ground state. The difference was used as evidence of the realization of quantum annealing in quantum annealing machines.

The bottleneck of quantum annealing is the energy gap between the ground state and the excited states near the ground state. The Landau–Zener transition shown in Chapter 5 is the most typical transition between states. In addition, if a system exhibits a phase transition at a waypoint of annealing, the energy gap decreases at the phase transition point. The minimum gap which is the energy gap between the ground state and the first excited state decreases exponentially when the first-order phase transition occurs at the phase transition point. As a result, quantum annealing does not work well. Thus, it is an important issue to construct a way to escape the first-order phase transition. Seki and Nishimori considered the ferromagnetic p-spin model. The model with the transverse field exhibits a first-order phase transition at a finite transverse field. They found that when p is large, the first-order phase transition can be erased and the second-order phase transition appears by applying the antiferromagnetic transverse interaction (Seki and Nishimori, 2012). Seki and Nishimori then studied the effect of antiferromagnetic transverse interaction on the Hopfield model (Seki and Nishimori, 2015). Recently, anomalous behavior of the energy gap at the first-order phase transition point was reported by Tsuda et al. (2013) and Okuyama et al. (2015).

Before the theory of quantum annealing was conceptualized, Nishimori and Nonomura studied quantum effects in neural networks (Nishimori and Nonomura, 1996). They found that the ground-state phase diagram of the Hopfield model with the transverse field is similar to the finite-temperature phase diagram of the Hopfield model. Thus, the transverse field plays a similar role as temperature. After that, there has been extensive studies on the effect of quantum fluctuation in informatics. Tanaka and Horiguchi proposed the quantum statistical–mechanical iterative method to restore images (Tanaka and Horiguchi, 2000). In addition, quantum mean-field decoding algorithm for error-correcting codes (Inoue et al., 2009), the quantum fluctuation effect in error-correcting codes (Otsubo et al., 2012), and the code-division multiple-access multiuser demodulator by using quantum fluctuations (Otsubo et al., 2014) were studied from the viewpoint of statistical mechanical informatics.

Before the D-Wave, a quantum annealing machine, was evolved, quantum annealing for machine learning was investigated. Kurihara et al. studied quantum annealing for clustering (Kurihara et al., 2009). They found that the hybrid type annealing in which simulated annealing and quantum annealing were performed simultaneously is better than when quantum annealing and simulated annealing are performed separately. In addition,

Sato et al. considered quantum annealing for Bayes inference (Sato et al., 2009). They also found that the hybrid type annealing in which simulated annealing and quantum annealing were performed simultaneously is better than when quantum annealing and simulated annealing are performed separately. Sato et al. also considered network clustering by using quantum annealing (Sato et al., 2013). Wittek built the bridge between quantum information science and machine learning (Wittek, 2014).

Before finishing the section, let us mention some pedagogical review papers on quantum annealing (Suzuki and Okada, 2005; Santoro and Tosatti, 2006; Das and Chakrabarti, 2008; Ohzeki and Nishimori, 2011; Tanaka and Tamura, 2012; Suzuki et al., 2013; Cohen and Tamir, 2014), which may be consulted for details.

6.2 Quantum Annealing by the Schrödinger Equation

In this section, we will explain quantum annealing by using the Schrödinger equation. The time-dependent Schrödinger equation is described by

$$i\frac{d|\psi(t)\rangle}{dt} = \mathscr{H}(t)|\psi(t)\rangle. \tag{6.9}$$

Herein, we use $\hbar = 1$. $|\psi(t)\rangle$ is the wave function at time t.

We consider the solution of the Schrödinger equation. Let $\hat{U}(t_2, t_1)$ be the time-evolution operator from $t = t_1$ to $t_2(\geq t_1)$. The wave function at time $t(\geq 0)$ is given by

$$|\psi(t)\rangle = \hat{U}(t,0)|\psi(0)\rangle, \tag{6.10}$$

where

$$\hat{U}(t,t) = 1. \tag{6.11}$$

Here 1 means the identity operator. Using the Schrödinger equation, the time-evolution operator $\hat{U}(t,0)$ is

$$i\frac{d}{dt}\hat{U}(t,0) = \hat{\mathscr{H}}(t)\hat{U}(t,0). \tag{6.12}$$

By integrating both sides of Eq. (6.12), we obtain

$$\hat{U}(t,0) = 1 - i\int_0^t dt' \hat{\mathscr{H}}(t')\hat{U}(t',0). \tag{6.13}$$

Here we used the relation $\hat{U}(0,0) = 1$. The solution of the integral equation is given by the sequential equation:

$$\hat{U}(t,0) = 1 - i\int_0^t dt_1 \hat{\mathcal{H}}(t_1)$$

$$+(-i)^2 \int_0^t dt_1 \int_0^t dt_2 \hat{\mathcal{H}}(t_1)\hat{\mathcal{H}}(t_2)$$

$$+(-i)^3 \int_0^t dt_1 \int_0^t dt_2 \int_0^t dt_3 \hat{\mathcal{H}}(t_1)\hat{\mathcal{H}}(t_2)\hat{\mathcal{H}}(t_3)$$

$$+\cdots. \tag{6.14}$$

It should be noted that the Hamiltonian depends on time and the Hamiltonians at different times cannot commute, i.e., $[\hat{\mathcal{H}}(t_1), \hat{\mathcal{H}}(t_2)] \neq 0$. Thus, in order to calculate Eq. (6.14), we should arrange the Hamiltonians in a sequence of time. Let $\mathcal{T}$ be the time-ordered product operator. For example, the time-ordered product operator between two time-dependent Hamiltonians is given by

$$\mathcal{T}\left[\hat{\mathcal{H}}(t_1)\hat{\mathcal{H}}(t_2)\right] = \begin{cases} \hat{\mathcal{H}}(t_1)\hat{\mathcal{H}}(t_2) & (t_1 \geq t_2) \\ \hat{\mathcal{H}}(t_2)\hat{\mathcal{H}}(t_1) & (t_2 > t_1) \end{cases}. \tag{6.15}$$

By using the time-ordered product operator $\mathcal{T}$, the time-evolution operator $U(t,0)$ is represented by

$$U(t,0) = \sum_{n=0}^{\infty} \frac{1}{n!}(-i)^n \int_0^t dt_1 \int_0^t dt_2 \cdots \int_0^t dt_n$$

$$\times \mathcal{T}\left[\hat{\mathcal{H}}(t_1)\hat{\mathcal{H}}(t_2)\cdots\hat{\mathcal{H}}(t_n)\right] \tag{6.16}$$

$$= \mathcal{T}\exp\left[-i\int_0^t dt' \hat{\mathcal{H}}(t')\right]. \tag{6.17}$$

This is a formal solution of the Schrödinger equation. In quantum annealing, we start with a large $\Gamma(t)$ at $t = 0$, and the initial state is approximately represented as

$$|\psi(0)\rangle = |\rightarrow\rightarrow\rightarrow\cdots\rightarrow\rangle. \tag{6.18}$$

Furthermore, we decide the schedule of $\Gamma(t)$ which is a decreasing function with t, and calculate $U(t,0)$ defined by Eq. (6.17) for arbitrary t. We obtain the state $|\psi(t)\rangle$ at t such that $\Gamma(t) \simeq 0$ is the solution of the optimization problem. The convergence theorem of quantum annealing using the Schrödinger equation was proved from a viewpoint of the adiabatic theorem, which will be explained in Section 6.9.1.

Demonstration : Ferromagnetic ising model

We show the performance of quantum annealing in the ferromagnetic Ising model in contrast to simulated annealing. At first, we consider simulated annealing which is performed at $\Gamma(t) = 0$, i.e., $\hat{\mathscr{H}}(t) = \hat{\mathscr{H}}_0$. Thus, the Hamiltonian does not depend on time. Instead, the temperature depends on time. For comparison with the Schrödinger equation, the master equation is used for simulated annealing.

$$\frac{dP_i(t)}{dt} = \sum_{j=1}^{2^N} \mathscr{L}_{ij}(t) P_j(t), \tag{6.19}$$

where $P_i(t)$ is the probability of the ith state at time t and $\mathscr{L}_{ij}(t)$ is the transition probability from the jth state to the ith state. N is the number of spins, and the number of states is 2^N since we consider $S = 1/2$ spin systems. By using the time-dependent temperature $T(t)$, the time-dependent transition probability $\mathscr{L}_{ij}(t)$ is given by

$$\mathscr{L}_{ij}(t) = \begin{cases} \{1 + \exp[(E_i - E_j)/T(t)]\}^{-1} & \text{(single-spin difference)} \\ 1 - \sum_{k \neq j} \mathscr{L}_{kj}(t) & (i = j) \\ 0 & \text{(otherwise)} \end{cases} . \tag{6.20}$$

Here "single-spin difference" means the case when the state i is generated by flipping a single spin in the state j. E_i is the energy of state i, which is $(\hat{\mathscr{H}}_0)_{ii}$.

For simplicity, we assume that the ground state is not degenerate. In order to compare quantum annealing and simulated annealing, we use the time-dependent probabilities $P_{\mathrm{QA}}(t) = |\langle g|\psi(t)\rangle|^2$ for quantum annealing and $P_{\mathrm{SA}}(t) = P_g(t)$ for simulated annealing. Here $|g\rangle$ is the wave function of the ground state of $\hat{\mathscr{H}}_0$ and $P_g(t)$ is the probability of the ground state. It should be noted that $P_{\mathrm{QA}}(t)$ and $P_{\mathrm{SA}}(t)$ have been used to understand the performance only when the ground state can be determined. If we cannot find the ground state by any method, we should compare the performance of annealing methods by the difference between internal energies obtained by both methods at the final step. In addition, the probabilities $P_{\mathrm{SA}}^{\mathrm{st}}(T) = Z(T)^{-1}\exp(-E_g/T)$ and $P_{\mathrm{QA}}^{\mathrm{st}}(\Gamma) = |\langle g|\psi_\Gamma\rangle|^2$ are useful in order to understand the performance of annealing methods. E_g is the ground state energy of $\hat{\mathscr{H}}_0$ and $|\psi_\Gamma\rangle$ is the wave function of the ground state of the system given by $\hat{\mathscr{H}} = \hat{\mathscr{H}}_0 - \Gamma\sum_{i=1}^N \hat{\sigma}_i^x$. $Z(T) = \mathrm{Tr}\exp(-\hat{\mathscr{H}}_0/T)$ is the partition function of the system given by $\hat{\mathscr{H}}_0$ at the temperature T. Since the probabilities satisfy the following relations $P_{\mathrm{QA}}(t) = P_{\mathrm{QA}}^{\mathrm{st}}(\Gamma(t))$ and $P_{\mathrm{SA}}(t) = P_{\mathrm{SA}}^{\mathrm{st}}(T(t))$ in the adiabatic limit, the differences $P_{\mathrm{QA}}(t) - P_{\mathrm{QA}}^{\mathrm{st}}(\Gamma(t))$ and $P_{\mathrm{SA}}(t) - P_{\mathrm{SA}}^{\mathrm{st}}(T(t))$ quantify the distance between the results obtained by dynamical processes and the adiabatic limit.

Kadowaki and Nishimori (1998) studied the performance of quantum annealing for the Ising model with homogeneous ferromagnetic interactions. The number of spins in the model is $N = 8$. In the model, ferromagnetic interactions exist for all spin pairs, i.e., $J_{ij} = J$

for all pairs of i and j, and the magnetic field is applied for all spins, i.e., $h_i = 0.1J$ for all i to avoid trivial degeneracy when the transverse field is absent. Figure 6.3 shows the ground state configuration.

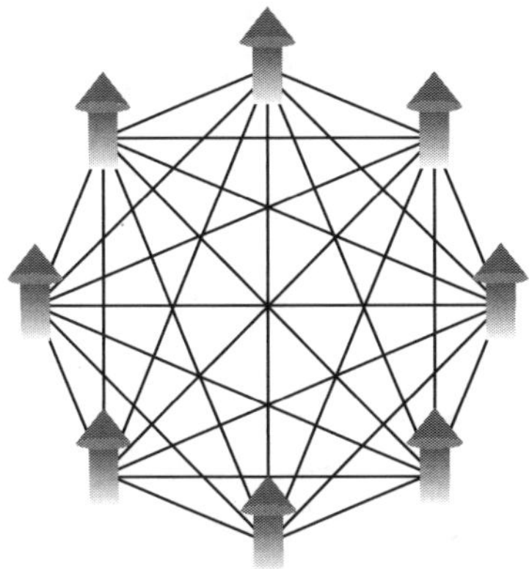

Fig. 6.3 Ground state spin configuration of the Ising model with homogeneous ferromagnetic interactions for $N = 8$. The magnetic field is applied for all spins, i.e., $h_i = 0.1J$ for all i to avoid the trivial degeneracy when the transverse field is absent.

Here, the transverse field $\Gamma(t)$ and the temperature $T(t)$ decrease with the following schedules:

$$\text{(a)} \; : \; \Gamma(t)/J = T(t)/J = \frac{3}{\ln(t+1)}, \tag{6.21}$$

$$\text{(b)} \; : \; \Gamma(t)/J = T(t)/J = \frac{3}{\sqrt{t}}, \tag{6.22}$$

$$\text{(c)} \; : \; \Gamma(t)/J = T(t)/J = \frac{3}{t}, \tag{6.23}$$

and the schematic of these schedules is shown in Fig. 6.4. Figure 6.5 shows the results obtained by quantum annealing based on the Schrödinger equation and the results obtained by simulated annealing based on the master equation when the schedules (a) and (b) were used. At first, we focus on schedule (a) which is the slowest schedule among the schedules mentioned here. The convergence of simulated annealing is guaranteed, as was explained in Section 3.4. Thus, $P_{\mathrm{SA}}(t) \simeq P_{\mathrm{SA}}^{\mathrm{st}}(T(t))$ is satisfied. Besides simulated annealing, $P_{\mathrm{QA}}(t) \simeq P_{\mathrm{QA}}^{\mathrm{st}}(\Gamma(t))$ is also satisfied in quantum annealing according to the convergence theorem of quantum annealing which will be shown in Section 6.9.1. Thus, the convergence to the ground state at large t can be confirmed both in simulated annealing and quantum annealing. In schedule (a), the relation $P_{\mathrm{QA}}(t) < P_{\mathrm{SA}}(t)$ is always satisfied, and quantum annealing is not better than simulated annealing.

Next, let us see the case of schedule (b). For simulated annealing, since the schedule is rapid compared with schedule (a), where the convergence to the ground state is guaranteed, there is a difference between $P_{SA}(t)$ and $P_{SA}^{st}(T(t))$. This means that the system is trapped in the local minima in simulated annealing.

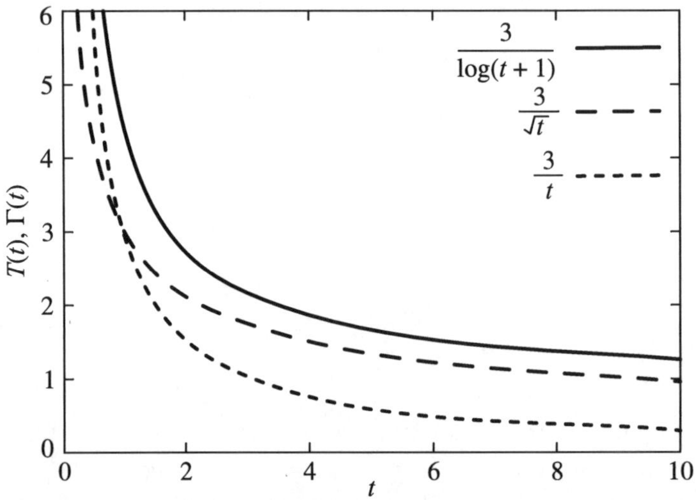

Fig. 6.4 Schematic of the schedules of annealing methods defined by Eqs. (6.21), (6.22), and (6.23).

On the other hand, $P_{QA}(t) \simeq P_{QA}^{st}(\Gamma(t))$ is confirmed and the convergence to the ground state is realized just as in schedule (a). In addition, $P_{QA}(t) > P_{SA}(t)$ for large t. From this viewpoint, quantum annealing is better than simulated annealing. Furthermore, the authors confirmed that $P_{QA}(t) > P_{SA}(t)$ realizes for a wide range of t when the fastest schedule (c) was used (Kadowaki and Nishimori, 1998). As explained here, quantum annealing is better than simulated annealing when the schedule is rapid for a homogeneous ferromagnetic Ising model.

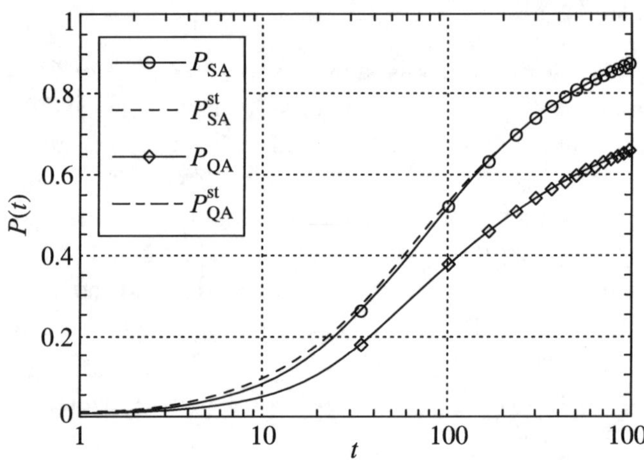

(a)

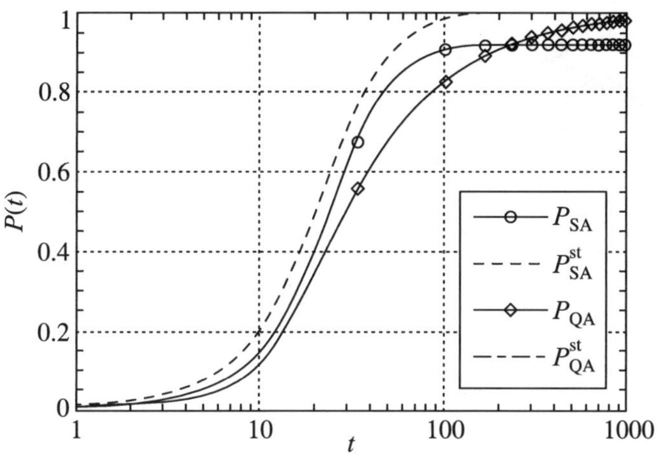

(b)

Fig. 6.5 Time dependence of $P_{\mathrm{QA}}(t)$, $P^{\mathrm{st}}_{\mathrm{QA}}(\Gamma(t))$, $P_{\mathrm{SA}}(t)$, and $P^{\mathrm{st}}_{\mathrm{SA}}(T(t))$ for the Ising model with homogeneous ferromagnetic interactions. (a) The schedule of annealing is $\Gamma(t)/J = T(t)/J = 3/\ln(t+1)$. (b) The schedule of annealing is $\Gamma(t)/J = T(t)/J = 3/\sqrt{t}$ (from Kadowaki and Nishimori, 1998).

Demonstration : Ising model with random interactions

We consider the performance of quantum annealing for an Ising model with random interactions according to the pioneering work by Kadowaki and Nishimori (1998). As a fundamental model with random interactions, the infinite-range interaction version of $\pm J$ Ising model is used. The Hamiltonian is given by Eq. (6.1). Interactions between all spin pairs exist; take $J_{ij} = +J$ or $J_{ij} = -J$ randomly. Just like in the previous case, let us take $h_i = 0.1J$ for all i to avoid trivial degeneracy when the transverse field is absent. Figure 6.6 displays the simulation results for the $N = 8$ case. Here, $P_{\mathrm{QA}}(t)$, $P^{\mathrm{st}}_{\mathrm{QA}}(\Gamma(t))$, $P_{\mathrm{SA}}(t)$, and $P^{\mathrm{st}}_{\mathrm{SA}}(T(t))$ are calculated. The schedule was selected as $\Gamma(t)/J = T(t)/J = 3/\sqrt{t}$. In this schedule, it was confirmed that quantum annealing is better than simulated annealing for the homogeneous ferromagnetic Ising model as described earlier. The schedule is faster than schedule (a) where the convergence is guaranteed for simulated annealing. Thus, the difference between $P_{\mathrm{SA}}(t)$ and $P^{\mathrm{st}}_{\mathrm{SA}}(T(t))$ becomes large, and the probability of converging to the ground state becomes small. On the other hand, $P_{\mathrm{QA}}(t) \simeq P^{\mathrm{st}}_{\mathrm{QA}}(\Gamma(t))$ is satisfied and $P_{\mathrm{QA}}(t) > P_{\mathrm{SA}}(t)$ for large t even for the Ising model with homogeneous ferromagnetic interactions. Thus, it is concluded that quantum annealing is better than simulated annealing for the Ising model with random interactions.

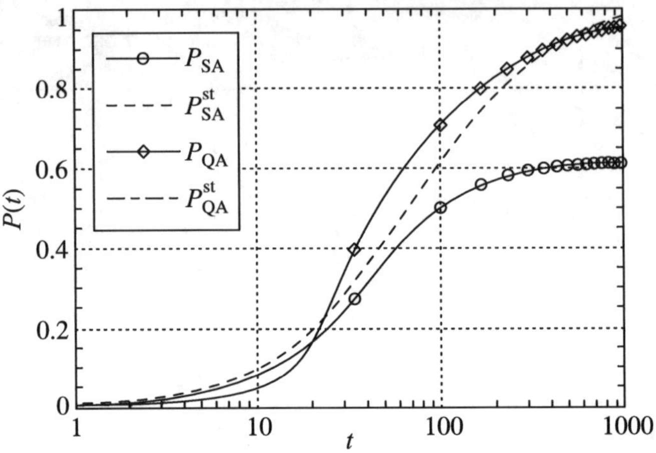

Fig. 6.6 Time dependence of $P_{\mathrm{QA}}(t)$, $P_{\mathrm{QA}}^{\mathrm{st}}(\Gamma(t))$, $P_{\mathrm{SA}}(t)$, and $P_{\mathrm{SA}}^{\mathrm{st}}(T(t))$ of an Ising model with random interactions. The schedule of annealing is $\Gamma(t)/J = T(t)/J = 3/\sqrt{t}$ (from Kadowaki and Nishimori, 1998).

6.3 Quantum Annealing by the Path-Integral Monte Carlo Method

Quantum annealing can be performed by the path integral Monte Carlo method explained in Secs. 4.1.2 and 4.3.2. By the path-integral formalism, namely, the Suzuki–Trotter decomposition (Trotter, 1959; Suzuki, 1976a,b, 1977), the partition function of the Ising model in a transverse field defined by Eqs. (6.1), (6.2), and (6.3) is obtained as

$$\begin{aligned} Z = & \lim_{P\to\infty}\left[\frac{1}{2}\sinh\left(\frac{2\beta\Gamma}{P}\right)\right]^{NP/2} \\ & \times \sum_{\{\sigma_{i,k}^z=\pm1\}} \exp\left\{\sum_{k=1}^{P}\left[\sum_{1\le i,j\le N}\frac{\beta J_{ij}}{P}\sigma_{i,k}^z\sigma_{j,k}^z - \sum_{i=1}^{N}\frac{\beta h_i}{P}\sigma_{i,k}^z\right.\right. \\ & \left.\left. +\frac{1}{2}\ln\coth\left(\frac{\beta\Gamma}{P}\right)\sum_{i=1}^{N}\sigma_{i,k}^z\sigma_{i,k+1}^z\right]\right\}. \end{aligned} \tag{6.24}$$

Thus, an effective Hamiltonian which is the $d+1$-dimensional classical Ising model is given as

$$\begin{aligned} \mathscr{H}_{\mathrm{eff}} = & -\sum_{1\le i,j\le N}\sum_{k=1}^{P}\frac{J_{ij}}{P}\sigma_{i,k}^z\sigma_{j,k}^z - \sum_{i=1}^{N}\sum_{k=1}^{P}\frac{h_i}{P}\sigma_{i,k}^z \\ & -\frac{1}{2\beta}\ln\coth\left[\frac{\beta\Gamma(t)}{P}\right]\sum_{i=1}^{N}\sum_{k=1}^{P}\sigma_{i,k}^z\sigma_{i,k+1}^z, \end{aligned} \tag{6.25}$$

where $\sigma_{i,k}^z = \pm 1$. By using this representation, the time evolution of the system given by Eqs. (6.1), (6.2), and (6.3) can be calculated by the Monte Carlo simulation shown in Section 3.3.1. In the time evolution, the value of $\Gamma(t)$ decreases with the Monte Carlo step. The convergence theorem of quantum annealing by using the path-integral quantum Monte Carlo method was proven (Morita and Nishimori, 2008) and will be shown in Section 6.9.3. The temperature T/J is set to be finite in this method. Das and Chakrabarti proposed a zero-temperature quantum Monte Carlo method which is based on the transfer matrix technique. This method is efficient for searching not only the ground state of classical random Ising models but also the ground state of quantum random Ising models. The method is also very efficient at low transverse fields (Das and Chakrabarti, 2008b).

Demonstration : Ising model with random interactions

The performance of quantum annealing using the Monte Carlo simulation is shown according to Santoro et al. (2002) and Martoňák et al. (2002). They used the Ising model with random interactions on the two-dimensional square lattice, whose Hamiltonian is defined by Eqs. (6.1), (6.2), and (6.3). The number of spins is 80×80. In the model, only the nearest neighbor interactions exist and $\{J_{ij}\}$ takes a random value in $-2 \leq J_{ij} \leq 2$, where the probability distribution of $\{J_{ij}\}$ is a uniform distribution. The magnetic field is $h_i = 0$ for all sites. Notice that the model has trivial degenerated ground states.

To investigate the performance of annealing methods, the authors used the residual energy defined by

$$\varepsilon_{\text{res}}(\tau) = E_{\text{final}}(\tau) - E_{\text{GS}}, \tag{6.26}$$

where $E_{\text{final}}(\tau)$ and E_{GS} represent the energy of the state obtained by annealing at τ which is the Monte Carlo step and the ground state energy, respectively. Notice that in the two-dimensional Ising model without magnetic field, the ground state can be found in polynomial time (Bieche et al., 1980; Barahona, 1982; Grest et al., 1986). $\varepsilon_{\text{res}}(\tau) = 0$ indicates that the annealing methods obtain the ground state. In simulated annealing (classical annealing: CA), the initial value of the temperature was set to $T(t = 0) = 3.0$ and the temperature decreases linearly toward $T(t = \tau) = 0$. In quantum annealing, the initial value of the transverse field was set to $\Gamma(t = 0) = 2.5$ and the transverse field decreases linearly toward $\Gamma(t = \tau) = 10^{-8}$. Thus, the annealing speed in both cases is characterized by τ^{-1}.

Figure 6.7 shows the Monte Carlo step, τ, dependence of the residual energy obtained by simulated annealing and quantum annealing where the product between the temperature T and the Trotter number P is fixed as $PT = 1$. This figure indicates that the residual energy of quantum annealing is smaller than that of simulated annealing, which is qualitatively similar to the results shown in Fig. 6.6. The obtained results by the path-integral Monte Carlo method suggest that quantum annealing is better than simulated annealing for the Ising model with random interactions.

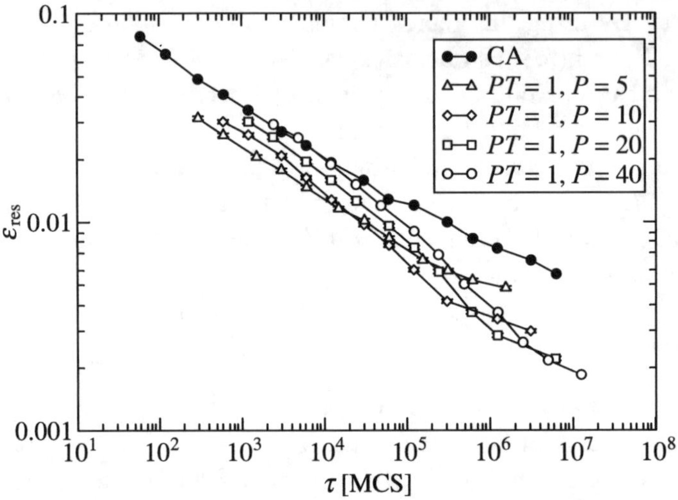

Fig. 6.7 Monte Carlo step, τ, dependence of residual energy $\varepsilon_{res}(\tau)$ obtained by simulated annealing (classical annealing: CA) and quantum annealing for various temperature T and the Trotter number P when $PT = 1$ (from Martoňák et al., 2002).

Santoro et al. (2002) discussed a scaling behavior of $\varepsilon_{res}(\tau)$ as a function of τ. Using the Landau–Zener transition and a semiclassical expression, Santoro et al. found the scaling of $\varepsilon_{res}(\tau)$ for the Ising model with random interactions as

$$\varepsilon_{res}(\tau) \sim \frac{1}{(\ln \tau)^6}. \tag{6.27}$$

Thus, the accuracy in quantum annealing logarithmically decreases with increasing τ which is not fundamentally different from the case of simulated annealing where $\varepsilon_{res}(\tau) \sim (\log \tau)^{-2}$. Furthermore, the logarithmic scaling in quantum annealing was confirmed by the path-integral quantum Monte Carlo method in the Ising model with a random magnetic field (Sarjala et al., 2006).

Demonstration : Traveling salesman problem

Martoňák et al. (2004) studied the performance of quantum annealing for the traveling salesman problem introduced in Section 3.2. Like most cases, the authors represented the traveling salesman problem by the random Ising model and used the transverse field as a quantum fluctuation. In this study, the traveling salesman problem with $N = 1002$ cities prepared from pr1002 of TSPLIB (http://elib.zib.de/pub/mp-testdata/tsp/tsplib/tsp/pr1002.tsp) was adopted. The optimal length of the path was exactly obtained as $L_{opt} = 259045$ in the problem. The mapping of the transverse field to the Ising model was introduced in Section 3.2.3.

Martoňák et al. compared the performances of quantum annealing by the path-integral quantum Monte Carlo method explained in Section 6.3 with the performances of simulated annealing by the Monte Carlo method reviewed in Section 3.3.1. In simulated annealing, the temperature decreases from $T = 100$ to zero linearly in the Monte Carlo step τ. On the other hand, in quantum annealing, the transverse field decreases from $\Gamma = 300$ to zero linearly in τ, the temperature is fixed as $T = 10/3$, and the Trotter number is set to $P = 30$. Figure 6.8 shows the Monte Carlo step, τ, dependence of the excess length after both the annealing methods. The excess length is defined by

$$\varepsilon_{\rm exc} = \frac{L_{\rm best}(\tau) - L_{\rm opt}}{L_{\rm opt}}, \tag{6.28}$$

where $L_{\rm best}(\tau)$ is the obtained length of the path in the annealing methods with τ Monte Carlo steps. Notice that the number of spins in quantum annealing is P times as large as that in simulated annealing. Then, in order to compare the results by the simulation time, the result of QA multiplied by P is also shown in Fig. 6.8. At large τ, better solutions than the solution obtained by the Lin–Kernigham algorithm, the standard local-search algorithms, is obtained (Johnson and McGeoch, 1997). The decay of the excess length as a function of τ in quantum annealing is faster than that in simulated annealing.

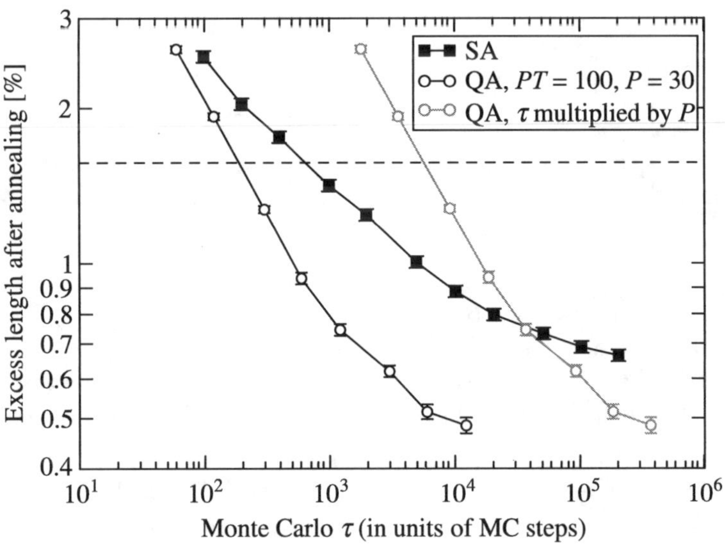

Fig. 6.8 Monte Carlo steps, τ, dependence of the excess length after annealing $\varepsilon_{\rm exc}$ obtained by simulated annealing (SA) and quantum annealing (QA). In order to compare the results by the simulation time, the result of QA multiplied by P is also shown. The dotted line indicates the value of $\varepsilon_{\rm exc} \simeq 1.57$ obtained by the Lin–Kernigham algorithm which is a standard local-search algorithm (from Martoňák et al., 2004).

Moreover, at large τ, the excess length obtained by quantum annealing is shorter than that in simulated annealing. These facts suggest that quantum annealing is better than simulated annealing for the traveling salesman problem.

Demonstration : 3-SAT problem

Battaglia et al. (2005) studied the performance of quantum annealing for the 3-SAT problem introduced in Section 3.2. Like most preceding studies, the authors introduced the time-dependent transverse field as a quantum fluctuation. In this study, the authors used a 3-SAT problem with $N = 10^4$ and $\alpha = 4.24$, where the replica symmetry breaking occurs in the ground state. A lower bound of the energy of the metastable states in a 3-SAT problem is provided by the Gardner energy (Montanari et al., 2004).

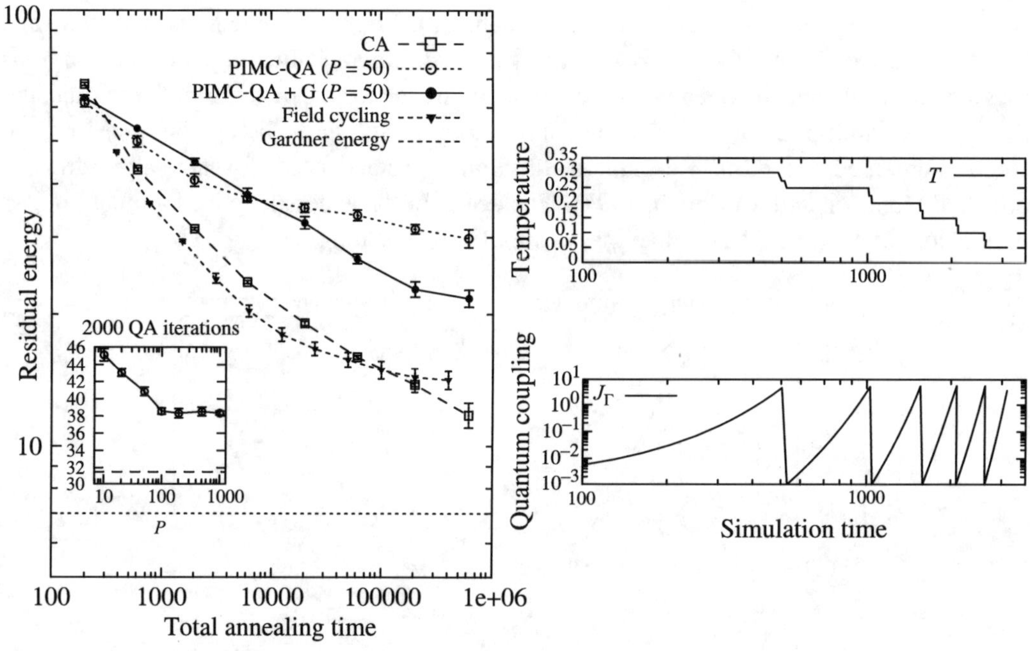

Fig. 6.9 (Left panel) Total annealing time (Monte Carlo step) dependence of the residual energy after simulated annealing (classical annealing, CA) and quantum annealing (QA). The dotted line indicates the Gardner energy for the 3-SAT problem, which is a lower bound of the energy of the metastable states. Inset is the Trotter number, P, dependence of the residual energy for the 3-SAT problem (Battaglia et al., 2004). (Right panel) Schedule of the temperature and the transverse field $J_\Gamma = \ln\coth(\beta\Gamma(t)/P)/2\beta$ in field cycling. (from Battaglia et al., 2005).

The time evolution of quantum annealing is realized by the path-integral quantum Monte Carlo method with the Trotter number $P = 50$. In quantum annealing, the initial transverse field is $\Gamma = 0.7$ which decreases to $\Gamma = 0.001$ with time t; the temperature is $T = 0.006$.

On the other hand, in simulated annealing (classical annealing: CA), the initial temperature is $T = 0.3$, after which it linearly decreases. Left panel of Fig. 6.9 is the total annealing time (Monte Carlo step) dependence of the residual energy.

The residual energy is defined by the difference between the energy defined by Eq. (3.30) by the annealing method and the known ground-state energy. In the left panel of Fig. 6.9, "PIMC-QA" and "PIMC-QA+G" indicate the results by the path-integral quantum Monte Carlo method without and with the global update, respectively. Furthermore, "field cycling" in the left panel of Fig. 6.9 means the results by quantum annealing when the transverse field is not a monotonically decreasing function with t and repeating the increase and decrease of the transverse field was performed (see the right panel of Fig. 6.9). The decays of the residual energy in quantum annealing is slower than that in simulated annealing. In addition, the obtained residual energy at large total annealing times ($\gtrsim 1000$ steps) by simulated annealing is smaller than that by quantum annealing. The result means that quantum annealing is worse than simulated annealing for the 3-SAT problem. Battaglia et al. pointed out the peculiar dynamical relaxation process behind the path-integral quantum Monte Carlo simulation for the 3-SAT problem as the reason why quantum annealing is worse than simulated annealing. Thus, the study (Battaglia et al., 2005) insisted that quantum annealing is not necessarily better than simulated annealing.

6.4 Quantum Annealing by the Green's Function Monte Carlo Method

We explain quantum annealing by the Green's function Monte Carlo method, which is a stochastic method to solve the imaginary-time Schrödinger equation. The imaginary-time Schrödinger equation is defined by

$$\frac{d|\psi(t)\rangle}{dt} = -\hat{\mathscr{H}}(t)|\psi(t)\rangle, \tag{6.29}$$

where t is the imaginary time. Let $\hat{U}(t_2, t_1)$ be the time-evolution operator of the dynamics from t_1 to t_2 ($t_1 \leq t_2$) given by Eq. (6.29). The dynamics is given by

$$|\psi(t)\rangle = \hat{U}(t,0)|\psi(0)\rangle \tag{6.30}$$

$$= \mathscr{T} \exp\left[-\int_0^t \hat{H}(t')dt'\right]|\psi(0)\rangle, \tag{6.31}$$

where $\mathscr{T}$ is the time-ordered product operator defined by Eq. (6.15). It should be noted that the wave function is not normalized since the matrix $\hat{U}(t,0)$ is not unitary. Here, the integral of the right-hand side of Eq. (6.31) is decomposed by the product among the time-evolution operator $\hat{G}(t)$ with time:

$$|\psi(t)\rangle = \lim_{n\to\infty} \hat{G}(t_{n-1})\hat{G}(t_{n-2})\cdots\hat{G}(t_1)\hat{G}(t_0)|\psi(0)\rangle, \tag{6.32}$$

where

$$t_k = k\Delta t \quad (k = 0, 1, \cdots, n-1), \tag{6.33}$$

$$\Delta t = t/n. \tag{6.34}$$

In addition, $\hat{G}(t)$ is defined by

$$|\psi(t_{k+1})\rangle = \hat{G}(t_k)|\psi(t_k)\rangle \tag{6.35}$$

$$= \hat{U}(t_{k+1}, t_k)|\psi(t_k)\rangle \tag{6.36}$$

$$= \mathscr{T}\exp\left[-\int_{t_k}^{t_{k+1}} \hat{H}(t')dt'\right]|\psi(t_k)\rangle. \tag{6.37}$$

For $\Delta t \ll 1$, the integral of this equation is approximated as

$$\hat{G}(t_k) \simeq \exp\left[-\Delta t \hat{H}(t_k)\right] \tag{6.38}$$

$$\simeq 1 - \Delta t \hat{H}(t_k). \tag{6.39}$$

In the Green's function Monte Carlo method, the time evolution is performed by replacing $\hat{G}(t_k)$ with $\hat{G}_1(t_k)$:

$$\hat{G}_1(t_k) = 1 - \Delta t\left[\hat{H}(t_k) - E_{\mathrm{T}}\right], \tag{6.40}$$

where E_{T} is called the reference energy. E_{T} is decided such that all the matrix elements of $\hat{G}_1(t_k)$ becomes positive. In the Green's function Monte Carlo method, we consider the dynamics given by

$$|\psi(t_{k+1})\rangle = \hat{G}_1(t_k)|\psi(t_k)\rangle \tag{6.41}$$

$$= \left\{1 - \Delta t\left[\hat{H}(t_k) - E_{\mathrm{T}}\right]\right\}|\psi(t_k)\rangle. \tag{6.42}$$

Next, the wave function is expanded by using the basis $|x\rangle$ as

$$|\psi(t_k)\rangle = \sum_x \psi_k(x)|x\rangle, \tag{6.43}$$

where

$$\psi_k(x) = \langle x|\psi(t_k)\rangle. \tag{6.44}$$

By using the basis $|x\rangle$, each matrix element of $\hat{G}_1(t_k)$ is represented as

$$G_1(y,x;t_k) = \langle y \left| 1 - \Delta t \left[\hat{H}(t_k) - E_{\rm T} \right] \right| x \rangle . \tag{6.45}$$

Thus, the time evolution of Eq. (6.42) is treated by

$$\psi_{k+1}(y) = \langle y | \psi(t_{k+1}) \rangle \tag{6.46}$$

$$= \left\langle y \left| \left\{ 1 - \Delta t \left[\hat{H}(t_k) - E_{\rm T} \right] \right\} \sum_x \psi_k(x) \right| x \right\rangle \tag{6.47}$$

$$= \sum_x G_1(y,x;t_k) \psi_k(x). \tag{6.48}$$

We can consider the time evolution by the transition matrix defined here as well as the Markov chain process. However, since the matrix $\hat{G}_1(t_k)$ is not normalized, the matrix itself is not a transition matrix. Thus, each element $G_1(y,x;t_k)$ of $\hat{G}_1(t_k)$ is written by using the matrix elements $G'_1(y,x;t_k)$ of the normalized matrix $\hat{G}'_1(t_k)$ and the normalization factor $w(x;t_k)$:

$$G_1(y,x;t_k) = G'_1(y,x;t_k) w(x;t_k), \tag{6.49}$$

where $w(x;t_k)$ and $G'_1(y,x;t_k)$ are defined as

$$w(x;t_k) := \sum_y G_1(y,x;t_k), \tag{6.50}$$

$$G'_1(y,x;t_k) := \frac{G_1(y,x;t_k)}{w(x;t_k)}. \tag{6.51}$$

From these equations, since $0 \leq G'_1(y,x;t_k) \leq 1$ and $\sum_y G'_1(y,x;t_k) = 1$, the matrix $\hat{G}'_1(t_k)$ can be regarded as the transition matrix. By using the matrix elements $G'_1(y,x;t_k)$ and the normalization factor $w(x;t_k)$, the wave function $\psi_k(y)$ at $t = t_k$ is represented as

$$\psi_k(y) = \sum_{x_{k-1}} G'_1(y,x_{k-1};t_{k-1}) w(x_{k-1};t_{k-1}) \psi_{k-1}(x_{k-1}) \tag{6.52}$$

$$\begin{aligned} &= \sum_{x_{k-1}} G'_1(y,x_{k-1};t_{k-1}) w(x_{k-1};t_{k-1}) \\ &\quad \times \sum_{x_{k-2}} G'_1(x_{k-1},x_{k-2};t_{k-2}) w(x_{k-2};t_{k-2}) \\ &\quad \times \cdots \times \sum_{x_0} G'_1(x_1,x_0;t_0) w(x_0;t_0) \psi_0(x_0) \end{aligned} \tag{6.53}$$

$$= \sum_{\{x_i\}_{i=0,\cdots,k}} \delta_{y,x_k} \prod_{i=1}^{k} G'_1(x_i, x_{i-1}; t_{i-1}) w(x_{i-1}; t_{i-1}) \psi_0(x_0). \tag{6.54}$$

In actual simulations in the Green's function Monte Carlo method, we can obtain $\psi_k(y)$ at $t = t_k$ by the following procedure:

Step 1 Prepare the initial wave function $\psi_0(x_0) \geq 0$.

Step 2 Decide the initial configuration x_0 by using the probability distribution which is proportional to $|\psi_0(x_0)|^2$.

Step 3 Decide x_1 according to the transition probabilities $G'_1(x_1, x_0; t_0)$. The weight is updated as $W_1 = w(x_0; t_0) W_0$, where the initial weight is $W_0 = 1$.

Step 4 The following scheme is continued for $i = 2, \cdots k$. Decide x_i according to the transition probability $G'_1(x_i, x_{i-1}; t_{i-1})$. The weight is changed to

$$W_i = w(x_{i-1}; t_{i-1}) W_{i-1}. \tag{6.55}$$

Step 5 The procedures of Steps 1–4 are independently performed M times, and $\psi_k(y)$ is obtained as

$$\psi_k(y) = \lim_{M\to\infty} \frac{1}{M} \sum_{j=1}^{M} W_k^j \delta_{y,x_k^j}. \tag{6.56}$$

When quantum annealing is performed, we use the Hamiltonian given by Eqs. (6.1), (6.2), and (6.3). In this case, the matrix defined by Eq. (6.45) is as follows:

$$G_1(y, x; t) = \begin{cases} 1 - \Delta t \, [E_0(x) - E_\mathrm{T}] & (x = y) \\ \Delta t \Gamma(t) & (\text{single-spin difference}) \\ 0 & (\text{otherwise}) \end{cases}, \tag{6.57}$$

where

$$E_0(x) := \langle x | \hat{\mathscr{H}}_0 | x \rangle. \tag{6.58}$$

The weight $w(x; t)$ is given by

$$w(x; t) = \sum_y G_1(y, x; t) \tag{6.59}$$

$$= 1 - \Delta t \, [E_0(x) - E_\mathrm{T}] + N \Delta t \Gamma(t). \tag{6.60}$$

The transition probability in the Green's function Monte Carlo method is given by

$$G'_1(y,x;t)$$

$$= \frac{G_1(y,x;t)}{w(x;t)} \tag{6.61}$$

$$= \begin{cases} \dfrac{1-\Delta t\,[E_0(x)-E_{\mathrm{T}}]}{1-\Delta t\,[E_0(x)-E_{\mathrm{T}}]+N\Delta t\Gamma(t)} & (x=y) \\ \dfrac{\Delta t\Gamma(t)}{1-\Delta t\,[E_0(x)-E_{\mathrm{T}}]+N\Delta t\Gamma(t)} & \text{(single-spin difference)} \\ 0 & \text{(otherwise)} \end{cases} . \tag{6.62}$$

Quantum annealing is performed when the transverse field $\Gamma(t)$ decreases with time under the time-evolution based on the aforementioned transition probability. The convergence of the time evolution by using the Green's function Monte Carlo simulation is guaranteed and will be shown in Section 6.9.4.

Demonstration : Ising model with random interactions

The performance of quantum annealing by using the Green's function Monte Carlo method is explained according to the study by Stella and Santoro (2007). In this study, the Ising model with random interactions on the two-dimensional square lattice is used, whose Hamiltonian is defined by Eqs. (6.1), (6.2), and (6.3). The number of spins is $N = 80 \times 80$, and $\{J_{ij}\}$ takes a random value from -2 to 2 between nearest-neighbor spin pairs. Furthermore, the magnetic field is $h_i = 0$. These conditions are the same as for the demonstration of the path-integral Monte Carlo method explained in Section 6.3.

Figure 6.10 shows the Monte Carlo step, τ, dependence of the residual energy $\varepsilon_{\mathrm{res}}(\tau)$ defined by Eq. (6.26). In this case, the transverse field $\Gamma(t)$ is decreased stepwise. The residual energy obtained by the Green's function Monte Carlo method when the trial wave function is constant (GFMC $\psi_T = 1$) is much larger than that obtained by the path-integral Monte Carlo method (PIMC-QA) and simulated annealing (classical annealing: CA). The performance of quantum annealing by the Green's function Monte Carlo method strongly depends on the trial wave function. The authors used the Boltzmann type wave function as a natural choice of the trial wave function. The Boltzmann type wave function is defined by

$$\psi_T^{(b)}(x) \propto \mathrm{e}^{-bE_0(x)/2}, \tag{6.63}$$

where $E_0(x)$ is the classical energy calculated by Eq. (6.2) of the state x and b is the variational parameter. The value of b which minimizes the variational energy $\langle \psi_T^{(b)}(x)|\mathcal{H}(t)|\psi_T^{(b)}(x)\rangle$ depends on $\Gamma(t)$. When the Boltzmann type wave function is used

(GFMC-QA), the obtained residual energies indicated by the lower diamonds in Fig. 6.10 are not worse. However, the slope of $\varepsilon_{\text{res}}(\tau)$ as a function of τ obtained by the Green's function Monte Carlo method is worse than that obtained by the path-integral Monte Carlo method (PIMC-QA) and about the same as that obtained with simulated annealing (classical annealing: CA). The computational cost of the Green's function Monte Carlo method is quite large. Thus, Stella and Santoro concluded that the Green's function Monte Carlo method is not a true competitor of the path-integral Monte Carlo method at present.

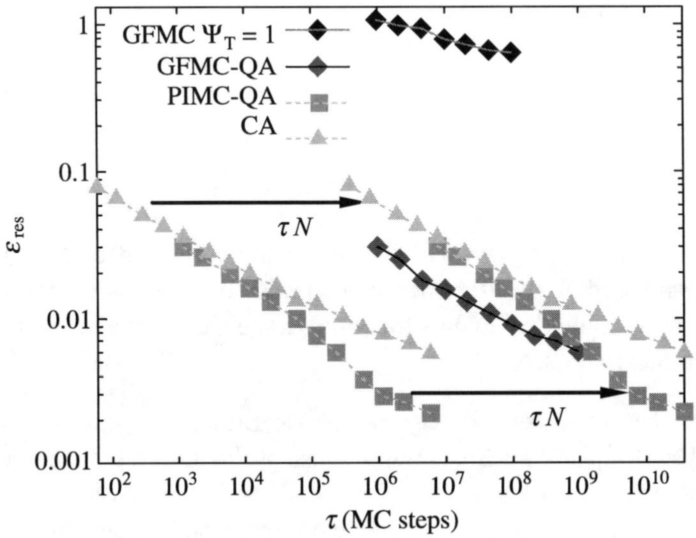

Fig. 6.10 Monte Carlo step, τ, dependence of the residual energy obtained by quantum annealing based on the Green's function Monte Carlo method (GFMC) and the path-integral Monte Carlo method (PIMC-QA), and simulated annealing (classical annealing: CA) for the Ising model with random interactions. The results of PIMC-QA and CA are from Fig. 6.7. Two types of simulations for GFMC were performed using different trial wave function. When the trial wave function is constant, results of upper diamonds are obtained (GFMC $\psi_T = 1$). On the other hand, when the trial wave function is the Boltzmann type, results of lower diamonds are obtained (GFMC-QA). The time unit in GFMC is a step that a single-spin flip event is tried, while in PIMC-QA and CA, time unit is a step that the update of all spins is tried. Thus, in order to compare these results when the time unit is a single-spin flip, the results of PIMC-QA and CA multiplied by N are also shown (from Stella and Santoro, 2007).

6.5 Quantum Annealing by the Density Matrix Renormalization Group

We explain quantum annealing by using the density matrix renormalization group (DMRG), which was developed by White (1992). By using DMRG, a time evolution of the Schrödinger equation of large quantum systems in one dimension or on ladders can be calculated (Nishino, 1995; Nishino and Okunishi, 1996; Nishino et al., 1999; Shibata and Tsunetsugu, 1999; Jeckelmann, 2002; Yoshikawa et al., 2004; Schollwöck, 2005; Hallberga, 2006; Nishino et al., 2007; Nishino and Okunishi, 2007). The time evolution by the Schrödinger equation is represented by

$$|\psi(t)\rangle = \hat{U}(t,0)|\psi(0)\rangle, \tag{6.64}$$

where $\hat{U}(t,0)$ is the time-evolution operator defined by Eq. (6.17) and t is the real time. When we approximate that the Hamiltonian does not depend on time from t to $t+\Delta t$, where $\Delta t \ll 1$, the time-evolution operator is described by

$$\hat{U}(t+\Delta t,t) \simeq \mathrm{e}^{-i\hat{\mathscr{H}}(t)\Delta t}. \tag{6.65}$$

From this fact, the wave function $|\psi(t)\rangle$ at $t = N_t\Delta t$ is given by

$$|\psi(t)\rangle = \hat{U}(N_t\Delta t,(N_t-1)\Delta t)\cdots\hat{U}(2\Delta t,\Delta t)\hat{U}(\Delta t,0)|\psi(0)\rangle \tag{6.66}$$

$$\simeq \mathrm{e}^{-i\hat{\mathscr{H}}((N_t-1)\Delta t)\Delta t}\cdots\mathrm{e}^{-i\hat{\mathscr{H}}(\Delta t)\Delta t}\mathrm{e}^{-i\hat{\mathscr{H}}(0)\Delta t}|\psi(0)\rangle. \tag{6.67}$$

This equation indicates that the wave function at t can be obtained by taking the infinitesimal time-evolution operators defined by Eq. (6.65) repeatedly to $|\psi(0)\rangle$.

Let us consider the one-dimensional Ising model in a transverse field with the open boundary condition. The Hamiltonian is represented by

$$\hat{\mathscr{H}}(t) = -\sum_{i=1}^{N-1} J_i\hat{\sigma}_i^z\hat{\sigma}_{i+1}^z - \sum_{i=1}^{N} h_i\hat{\sigma}_i^z - \Gamma(t)\sum_{i=1}^{N}\hat{\sigma}_i^x \tag{6.68}$$

$$=: \sum_{i=1}^{N}\hat{\mathscr{H}}_i(t), \tag{6.69}$$

where N is the number of spins. Here, the local Hamiltonians $\hat{\mathscr{H}}_i(t)$ are given by

$$\hat{\mathscr{H}}_i(t) = \begin{cases} -J_i\hat{\sigma}_i^z\hat{\sigma}_{i+1}^z - h_i\hat{\sigma}_i^z - \Gamma(t)\hat{\sigma}_i^x & (i \neq N) \\ -h_i\hat{\sigma}_i^z - \Gamma(t)\hat{\sigma}_i^x & (i = N) \end{cases}, \tag{6.70}$$

where the local Hamiltonians $\hat{\mathcal{H}}_i(t)$ include only the interaction between the ith and the $i+1$th sites and the external field at the ith site. In this case, the infinitesimal time-evolution operator at t is approximately given by

$$\mathrm{e}^{-i\mathcal{H}(t)\Delta t} \simeq \mathrm{e}^{-i\mathcal{H}_1(t)\Delta t/2}\mathrm{e}^{-i\mathcal{H}_2(t)\Delta t/2}\cdots\mathrm{e}^{-i\mathcal{H}_N(t)\Delta t/2}$$
$$\times\mathrm{e}^{-i\mathcal{H}_N(t)\Delta t/2}\cdots\mathrm{e}^{-i\mathcal{H}_2(t)\Delta t/2}\mathrm{e}^{-i\mathcal{H}_1(t)\Delta t/2}. \tag{6.71}$$

The decomposition of exponential operators are presented by Suzuki (1990, 1991), and have been widely used for solving differential equations in nature. In this way, thus, the time evolution from t to $t+\Delta t$ of the Schrödinger equation can be represented by operating the local infinitesimal time-evolution operators $\mathrm{e}^{-i\hat{\mathcal{H}}_i(t)\Delta t/2}$ $(i=1,\cdots,N)$ to $|\psi(t)\rangle$ repeatedly. The procedure can be done by using DMRG, which will be explained as follows.

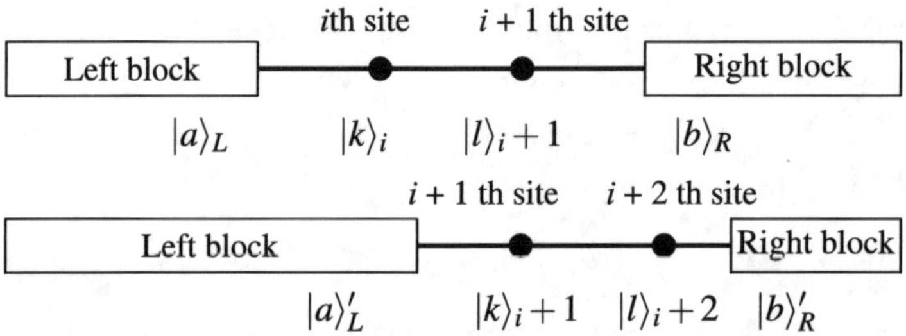

Fig. 6.11 Schematic of division of blocks in DMRG.

Let us consider that the local infinitesimal time-evolution operator $\mathrm{e}^{-i\hat{\mathcal{H}}_i(t)\Delta t/2}$ operates to the wave function $|\psi_i(t)\rangle$. In the procedure of DMRG, we divide one-dimensional systems into four parts; the left-handed block (from the first site to the $i-1$th site), the ith site, the $i+1$th site, and the right-handed block (from the $i+2$th site to the Nth site) as shown in the upper panel of Fig. 6.11. In addition, the bases in the respective parts are represented by $|a\rangle_L$, $|k\rangle_i$, $|l\rangle_{i+1}$, and $|b\rangle_R$. The wave function $|\psi_i(t)\rangle$ is represented by the superposition of these bases in respective parts:

$$|\psi_i(t)\rangle = \sum_{a,b,k,l} \phi_{abkl}|a\rangle_L|k\rangle_i|l\rangle_{i+1}|b\rangle_R. \tag{6.72}$$

Let the bases of $|k\rangle_i$, $|l\rangle_{i+1}$ be the eigenstates of $\hat{\sigma}^z$. In this case, the operation $\mathrm{e}^{-i\hat{\mathcal{H}}_i(t)\Delta t/2}|\psi_i(t)\rangle$ is represented by using the matrix elements of local infinitesimal time-evolution operators $[\mathrm{e}^{-i\hat{\mathcal{H}}_i(t)\Delta t/2}]_{kl,k'l'}$:

$$\mathrm{e}^{-i\hat{\mathcal{H}}_i(t)\Delta t/2}|\psi\rangle = \sum_{a,b,k,l} \phi'_{abkl}|a\rangle_L|k\rangle_i|l\rangle_{i+1}|b\rangle_R, \tag{6.73}$$

$$\phi'_{abkl} = \sum_{k',l'} \left[\mathrm{e}^{-i\hat{\mathcal{H}}_i(t)\Delta t/2}\right]_{kl,k'l'} \phi_{abk'l'}. \tag{6.74}$$

Next, in order to apply the local infinitesimal time-evolution operator $e^{-i\mathcal{H}_{i+1}(t)\Delta t/2}$ to $|\psi_{i+1}(t)\rangle := e^{-i\mathcal{H}_i(t)\Delta t/2}|\psi_i(t)\rangle$, we use other bases. We calculate the reduced density matrix which is obtained by tracing out the degree of freedom in the $i+1$th site and the right-handed block. The elements of the reduced density matrix are represented by

$$\rho_{ak,a'k'} = \sum_{b,l} \phi'_{abkl} \phi'^{*}_{a'bk'l}. \tag{6.75}$$

Here $\sum_{b,l}$ corresponds to the partial trace which means the trace out of the degree of freedom of the $i+1$th site and the right block. The reduced density matrix is defined by taking the partial trace of the total density matrix as we did. The reduced density matrix can be diagonalized by using a unitary matrix. Let the elements of the unitary matrix be $u_{c,ak}$ and the eigenvalues of the reduced density matrix be d_c. The reduced density matrix is given by

$$\rho_{ak,a'k'} = \sum_{c} u_{c,ak} d_c u^{*}_{c,a'k'}, \tag{6.76}$$

where $\sum_c$ denotes the summation over the dimension of the reduced density matrix. By using the unitary matrix, the basis of the updated left-handed block from the first site to the ith site is given by

$$|a'\rangle_{L'} = \sum_{a',k} u_{a',ak} |a\rangle_L |k\rangle_i, \tag{6.77}$$

which is the eigenstate of the reduced density matrix corresponding to the eigenvalue d_a. In addition, by the definition of the unitary matrix,

$$|a\rangle_L |k\rangle_i = \sum_{a'} u^{*}_{a',ak} |a'\rangle_{L'} \tag{6.78}$$

is satisfied. Subsequently, we prepare the basis of the right-handed block. We obtained the unitary matrix to make the basis of the updated right-handed block $|b'\rangle_R$. Let the elements of the unitary matrix be $v_{b,l'b'}$. $|b\rangle_R$ can be represented by using the updated basis at the $i+2$th site, $|l'\rangle_{i+2}$, and the basis of the updated right-handed block from the $i+3$th site to the Nth site, $|b'\rangle_{R'}$:

$$|b\rangle_R = \sum_{l',b'} v_{b,l'b'} |l'\rangle_{i+2} |b'\rangle_{R'}. \tag{6.79}$$

By using Eqs. (6.78) and (6.79), Eq. (6.73) is rewritten as

$$\begin{aligned} |\psi_{i+1}(t)\rangle &= e^{-i\mathcal{H}_i(t)\Delta t/2}|\psi(t)\rangle \\ &= \sum_{a',b',k',l'} \phi''_{a'b'k'l'} |a'\rangle_{L'} |k'\rangle_{i+1} |l'\rangle_{i+2} |b'\rangle_{R'}, \end{aligned} \tag{6.80}$$

$$\phi''_{a'b'k'l'} = \sum_{a'',b'',k''} \phi'_{a''b''k''k'} u^*_{a',a''k''} v_{b'',l'b'}. \tag{6.81}$$

In this step, we update the division of blocks as shown in the lower panel of Fig. 6.11. Since the local infinitesimal time-evolution operator $e^{-i\hat{\mathscr{H}}_{i+1}(t)\Delta t/2}$ acts only on $|k'\rangle_{i+1}|l'\rangle_{i+2}$ in $|\psi_{i+1}(t)\rangle$, it can be operated by changing the basis as explained earlier. We repeatedly change the bases according to the eigenstates of the reduced density matrix. By operating the local infinitesimal time-evolution operators repeatedly, we can calculate the time evolution by the infinitesimal time-evolution operator $e^{-i\hat{\mathscr{H}}(t)\Delta t}$, that is, the time evolution from t to $t+\Delta t$.

In principle, the number of bases is the same as the dimension of the density matrix. However, when we use DMRG, we focus on only the bases whose eigenvalues are not negligible and ignore the other bases. By performing this procedure, the computation cost decreases, and we can treat large-scale systems. In other words, if the number of significant bases whose eigenvalues are large increases, the precision of calculation becomes worse when we use the same strategy. Fortunately, however, the number of significant bases is small for one-dimensional quantum spin systems, which is explained in the language of entanglement (Amico et al., 2008; Eisert et al., 2010). Thus, DMRG is regarded as a robust algorithm to calculate the time evolution of the Schrödinger equation of large-scale systems on one-dimensional lattices. Like in Section 6.2, where we calculate a time evolution of the Schrödinger equation with decreasing $\Gamma(t)$ with the time t, by DMRG, quantum annealing is performed.

Demonstration : Random field Ising model

The performance of quantum annealing by using the density matrix renormalization group is explained according to Suzuki and Okada (2007). In their study, the Ising model with random fields and open boundary condition on a chain is used. Its Hamiltonian is defined by

$$\hat{\mathscr{H}}_0 = -J\sum_{i=1}^{N-1} \hat{\sigma}_i^z \hat{\sigma}_{i+1}^z - \sum_{i=1}^{N} h_i \hat{\sigma}_i^z, \tag{6.82}$$

where the number of spins is $N = 360$. For the random field h_i, eighty instances were prepared, and the residual energy is averaged out for eighty instances. Furthermore, as the quantum term, the transverse field is adopted, and the quantum Hamiltonian $\hat{\mathscr{H}}_\mathrm{q}$ is defined by

$$\hat{\mathscr{H}}_\mathrm{q} = -\sum_{i=1}^{N} \hat{\sigma}_i^x. \tag{6.83}$$

Here, the time-dependent Hamiltonian $\hat{\mathscr{H}}(t)$ in quantum annealing is expressed as

$$\hat{\mathscr{H}}(t) = \left(1 - \frac{t}{\tau}\right)\hat{\mathscr{H}}_\mathrm{q} + \frac{t}{\tau}\hat{\mathscr{H}}_0, \quad (1 \le t \le \tau). \tag{6.84}$$

This is the same as in the adiabatic quantum computation which will be explained in Section 6.8.1. In this case, when $t = \tau$, $\hat{\mathscr{H}}(t) = \hat{\mathscr{H}}_0$, and thus, the residual energy E_{res} is defined by

$$E_{\mathrm{res}} = \langle \psi(\tau)|\hat{\mathscr{H}}(\tau)|\psi(\tau)\rangle - E_{\mathrm{GS}}, \tag{6.85}$$

where $\psi(\tau)$ is the wave function obtained at $t = \tau$, and E_{GS} is the ground state energy.

Figure 6.12 shows the residual energy, E_{res}, dependence of τ when $J = 1.6$. This figure indicates that the scaling of E_{res} with τ by the density matrix renormalization group is given by

$$E_{\mathrm{res}} \sim \frac{1}{(\ln\tau)^4}, \tag{6.86}$$

for the random field Ising model in one dimension. The logarithmic scaling in quantum annealing is consistent with the results obtained by the path-integral Monte Carlo method shown in Section 6.3 and the Green's function Monte Carlo method shown in Section 6.4.

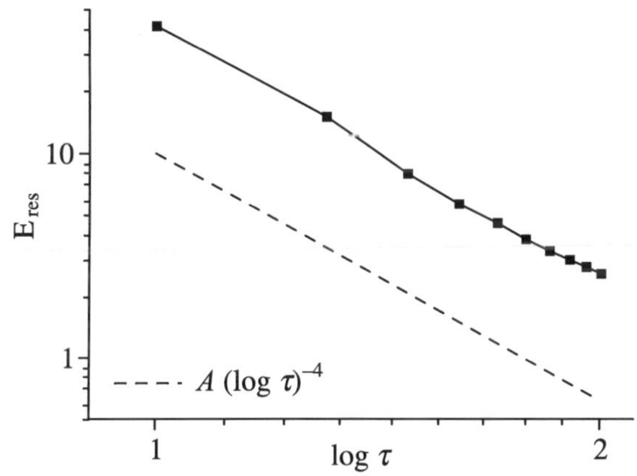

Fig. 6.12 Residual energy, E_{res}, dependence of τ when $J = 1.6$ obtained by the density matrix renormalization group for the random field Ising model. The dotted line is the guide to the eyes for logarithmic scaling (from Suzuki and Okada, 2007).

6.6 Quantum Annealing by the Mean-field Approximation

Quantum annealing using mean-field approximation is explained in this section. At first, we consider the mean-field approximation for the Ising model in a transverse field Γ. The Hamiltonian of this model is given by

$$\hat{\mathscr{H}} = -\sum_{1\le i,j\le N} J_{ij}\hat{\sigma}_i^z\hat{\sigma}_j^z - \sum_{i=1}^{N} h_i\hat{\sigma}_i^z - \Gamma\sum_{i=1}^{N}\hat{\sigma}_i^x, \tag{6.87}$$

where the number of spin is N. We begin with the free energy using the density matrix. The density matrix of the thermal equilibrium state $\hat{\rho}$, in other words, the canonical distribution, is defined by

$$\hat{\rho} := \frac{\exp\left(-\beta\hat{\mathscr{H}}\right)}{\mathrm{Tr}\exp\left(-\beta\hat{\mathscr{H}}\right)}, \tag{6.88}$$

which is normalized by

$$\mathrm{Tr}\hat{\rho} = 1. \tag{6.89}$$

Let λ_n and $|\psi_n\rangle$ $(n = 1, 2, \cdots, 2^N)$ be the eigenvalues and corresponding eigenstates of $\hat{\mathscr{H}}$, respectively. By using the notation, the density matrix can be written as

$$\hat{\rho} = \frac{\sum_{n=1}^{2^N} \exp(-\beta\lambda_n)\,|\psi_n\rangle\,\langle\psi_n|}{\sum_{n=1}^{2^N} \exp(-\beta\lambda_n)}. \tag{6.90}$$

Since the density matrix $\hat{\rho}$ is a $2^N \times 2^N$ matrix, it is difficult to obtain the eigenvalues and eigenstates when N is large. By using the density matrix, the thermal expectation value of the physical quantity representing the operator $\hat{A}$ is given by

$$\langle\hat{A}\rangle = \mathrm{Tr}\left(\hat{A}\hat{\rho}\right). \tag{6.91}$$

For example, the thermal expectation values of the x and z components of local magnetization, $m_i^x(t)$ and $m_i^z(t)$, are given by

$$m_i^x = \mathrm{Tr}\left(\hat{\sigma}_i^x\hat{\rho}\right), \tag{6.92}$$

$$m_i^z = \mathrm{Tr}\left(\hat{\sigma}_i^z\hat{\rho}\right). \tag{6.93}$$

Furthermore, the internal energy and entropy can be represented by using the density matrix as follows:

$$E = \mathrm{Tr}\left(\hat{\mathscr{H}}\hat{\rho}\right), \tag{6.94}$$

$$S = -\mathrm{Tr}\left(\hat{\rho}\ln\hat{\rho}\right). \tag{6.95}$$

By using E and S, the free energy is represented by

$$\tilde{F} = \mathrm{Tr}\left(\hat{\mathscr{H}}\hat{\rho} + \beta^{-1}\hat{\rho}\ln\hat{\rho}\right). \tag{6.96}$$

The Gibbs–Bogoliubov–Feynman inequality is given by $F \leq \tilde{F}$, where F denotes the free energy of the original Hamiltonian given by Eq. (6.87). Thus, applying the variational method to

$$\tilde{F} - \lambda \left(\mathrm{Tr}\hat{\rho} - 1 \right), \tag{6.97}$$

we obtain $\hat{\rho}$ and F, where λ is the Lagrange multiplier. However, it is difficult to directly solve this equation as well as obtain all the eigenvalues and eigenstates of the density matrix. Thus, we use a mean-field approximation. In the mean-field approximation, suppose the density matrix can be decomposed by the product of the local density matrix at the ith site, $\hat{\rho}_i$. That is,

$$\hat{\rho} \simeq \prod_{i=1}^{N} \hat{\rho}_i, \tag{6.98}$$

where $\hat{\rho}_i$ is given by

$$\hat{\rho}_i = \mathrm{Tr}'\hat{\rho}. \tag{6.99}$$

Here Tr' means the operation of tracing out the degrees of freedom except at the ith site. In addition, by using the local density matrix, the following relations are satisfied:

$$\mathrm{tr}_i \left(\hat{\rho}_i \right) = 1, \tag{6.100}$$

$$\mathrm{tr}_i \left(\hat{\sigma}_i^x \hat{\rho}_i \right) = m_i^x, \tag{6.101}$$

$$\mathrm{tr}_i \left(\hat{\sigma}_i^z \hat{\rho}_i \right) = m_i^z. \tag{6.102}$$

Here, tr_i denotes the operator to trace out the degrees of freedom at the ith site. By using these relations, the free energy given by Eq. (6.96) can be rewritten by the mean-field approximation:

$$\tilde{F}_{\mathrm{MF}} = \mathrm{Tr} \left(\hat{\mathscr{H}} \prod_{i=1}^{N} \hat{\rho}_i + \beta^{-1} \prod_{i=1}^{N} \hat{\rho}_i \ln \prod_{j=1}^{N} \hat{\rho}_j \right) \tag{6.103}$$

$$= - \sum_{1 \leq i,j \leq N} J_{ij} m_i^z m_j^z - \sum_{i=1}^{N} h_i m_i^z - \Gamma \sum_{i=1}^{N} m_i^x + \beta^{-1} \sum_{i=1}^{N} \mathrm{tr}_i \left(\hat{\rho}_i \ln \hat{\rho}_i \right). \tag{6.104}$$

Applying the variational method to

$$\tilde{F}_{\mathrm{MF}} - \lambda \left(\mathrm{tr}_i \hat{\rho}_i - 1 \right), \tag{6.105}$$

we obtain $\hat{\rho}_i$ which is given by

$$\hat{\rho}_i = \frac{\exp\left(-\beta\hat{\mathscr{H}}_i\right)}{\mathrm{Tr}\exp\left(-\beta\hat{\mathscr{H}}_i\right)}, \tag{6.106}$$

$$\hat{\mathscr{H}}_i = \begin{pmatrix} -\sum_{j\in\partial i} J_{ij}m_j^z - h_i & -\Gamma \\ -\Gamma & +\sum_{j\in\partial i} J_{ij}m_j^z + h_i \end{pmatrix}, \tag{6.107}$$

where $\sum_{j\in\partial i}$ is the summation over the sites which are interacted with the ith site. Let $\lambda_n^{(i)}$ and $|\psi_n^{(i)}\rangle$ $(n=1,2)$ be the eigenvalues and the corresponding eigenvectors of $\hat{\mathscr{H}}_i$. The local density matrix can be written by

$$\hat{\rho}_i = \frac{\sum_{n=1}^{2}\exp\left(-\beta\lambda_n^{(i)}\right)\left|\psi_n^{(i)}\right\rangle\left\langle\psi_n^{(i)}\right|}{\sum_{n=1}^{2}\exp\left(-\beta\lambda_n^{(i)}\right)}. \tag{6.108}$$

Now, since the local Hamiltonian $\hat{\mathscr{H}}_i$ is a 2×2 matrix, the eigenvalues and the corresponding eigenstates are obtained easily and $\hat{\rho}_i$ can be easily calculated. By using the local density matrix $\hat{\rho}_i$, the thermal expectation values of local magnetizations at the ith site is calculated by using Eq. (6.102).

Next, we consider the time evolution of the Hamiltonian defined by Eqs. (6.1), (6.2), and (6.3) by the mean-field approximation. In order to consider the time evolution, we use the following equation

$$m_i^z(t+1) = \mathrm{Tr}\left[\hat{\sigma}_i^z\hat{\rho}_i(t)\right]. \tag{6.109}$$

Here, $\hat{\rho}_i(t)$ is defined by

$$\hat{\rho}_i(t) = \frac{\exp\left[-\beta\hat{\mathscr{H}}_i(t)\right]}{\mathrm{Tr}\exp\left[-\beta\hat{\mathscr{H}}_i(t)\right]}, \tag{6.110}$$

$$\hat{\mathscr{H}}_i(t) = \begin{pmatrix} -\sum_{j\in\partial i} J_{ij}m_j^z(t) - h_i & -\Gamma(t) \\ -\Gamma(t) & +\sum_{j\in\partial i} J_{ij}m_j^z(t) + h_i \end{pmatrix}, \tag{6.111}$$

where $\sum_{j\in\partial i}$ is the summation over the sites which are interacted by the ith site. Thus, we obtain

$$\hat{\rho}_i(t) = \frac{\sum_{n=1}^{2} \exp\left[-\beta\lambda_n^{(i)}(t)\right] \left|\psi_n^{(i)}(t)\right\rangle \left\langle \psi_n^{(i)}(t)\right|}{\sum_{n=1}^{2} \exp\left[-\beta\lambda_n^{(i)}(t)\right]}, \tag{6.112}$$

where $\lambda_n^{(i)}(t)$ and $|\psi_n^{(i)}(t)\rangle$ $(n=1,2)$ are the eigenvalues and the corresponding eigenvectors of $\mathscr{H}_i(t)$. To judge the convergence of this equation, we define $\varepsilon(t)$ expressed as

$$\varepsilon(t) = \frac{1}{N}\sum_{i=1}^{N} |m_i^z(t+1) - m_i^z(t)|. \tag{6.113}$$

The iteration procedure is finalized when $\varepsilon(t)$ is smaller than a given certain value. By the aforementioned time-evolution procedure with decreasing $\Gamma(t)$, quantum annealing can be performed.

When we perform quantum annealing by the mean-field approximation, we should take care of the following points. By using this scheme, we obtain the local magnetizations $\{m_i^z(\tau)|i-1,2,\cdots N\}$, where τ is the annealing time. In this case, the final energy obtained by $\{m_i^z(\tau)\}$ is calculated by

$$E_{\text{final}}(\tau) = -\sum_{1\leq i,j\leq N} J_{ij} m_i^z(\tau) m_j^z(\tau) - \sum_{i=1}^{N} h_i m_i^z(\tau). \tag{6.114}$$

Although the spin configurations should be represented by $\sigma_i^z = \pm 1$ in the Ising model when the transverse field is zero, the local magnetizations $m_i^z(\tau)$ obtained are continuous values $(-1 \leq m_i^z(\tau) \leq 1)$. Thus, we decide the final spin configuration by the following equation:

$$\sigma_i^z = \text{sign}\left[m_i^z(\tau)\right]. \tag{6.115}$$

Furthermore, when we calculate the case under zero temperature ($T = 0$) in quantum annealing, the local density matrix is written by

$$\hat{\rho}_i(t) = \left|\psi_1^{(i)}(t)\right\rangle \left\langle \psi_1^{(i)}(t)\right|, \quad \left(\lambda_2^{(i)}(t) > \lambda_1^{(i)}(t)\right). \tag{6.116}$$

Demonstration : Random field Ising model

The performance of quantum annealing by using the mean-field approximation is shown by Suzuki et al. (Suzuki and Nishimori, 2007; Suzuki et al., 2007). In these studies, the Ising model with random fields on two-dimensional square lattice is used. The Hamiltonian is defined by Eq. (6.2) when $J_{ij} = J$ for all the nearest-neighbor spin pairs.

For each random field h_i, $+1$ or -1 is randomly chosen with the same probability. Furthermore, as the quantum term, two types of Hamiltonian were considered. One is the transverse field which is defined by Eq. (6.83). The other one is the transverse field and the ferromagnetic interactions of $\hat{\sigma}_i^x$, that is,

$$\hat{\mathscr{H}}_{\mathrm{q}} = -\sum_{i=1}^{N} \hat{\sigma}_i^x - \sum_{\langle i,j \rangle} \hat{\sigma}_i^x \hat{\sigma}_j^x. \tag{6.117}$$

Suzuki and Nishimori (2007) performed the mean-field calculations based on the Bragg–Williams approximation. In the approximation, when we construct a time-dependent local Hamiltonian $\hat{\mathscr{H}}_i(t)$, all spin variables except for the ith spin are replaced with their averages. Then, the following $\hat{\mathscr{H}}_i(t)$ is used instead of Eq. (6.111) when $\hat{\mathscr{H}}_{\mathrm{q}}$ is Eq. (6.117):

$$\hat{\mathscr{H}}_i(t) = \left(1 - \frac{t}{\tau}\right) \left[-1 - \sum_{j \in \partial i} m_j^x(t)\right] \hat{\sigma}_i^x + \frac{t}{\tau} \left[-\sum_{j \in \partial i} J_{ij} m_j^z(t) - h_i\right] \hat{\sigma}_i^z. \tag{6.118}$$

In quantum annealing, t changes by one from $t = 0$ to τ. On the other hand, in simulated annealing, we suppose $t = \tau$ in Eq. (6.118) and the temperature is decreased from an appropriate value to zero using τ steps. The authors found that the residual energy of quantum annealing by the transverse field defined by Eq. (6.83) is larger than that of simulated annealing at large τ, which is the opposite result of quantum annealing based on the Monte Carlo methods. On the other hand, the residual energy of quantum annealing by Eq. (6.117) is smaller than that of quantum annealing by Eq. (6.83) and simulated annealing.

Suzuki et al. (2007) performed quantum annealing by mean-field calculations based on the Bethe approximation. In the approximation, when we construct $\hat{\mathscr{H}}_i$, all spin variables except for the ith spin and its immediate neighbors are replaced with their averages. Figure 6.13 shows the residual energy depending on τ when (a) $J = 2.0$ and (b) $J = 0.6$. For $J = 2.0$, the residual energy of quantum annealing by Eq. (6.117) is smaller than that of quantum annealing by Eq. (6.83), which is consistent with the results obtained by the Bragg–Williams approximation. Notice that the difference between these residual energies diminishes as J decreases and the residual energy obtained by simulated annealing is smallest for $J = 0.6$. From these results, Suzuki et al. concluded that it is important to choose an appropriate quantum fluctuation in order to obtain high efficiency of quantum annealing.

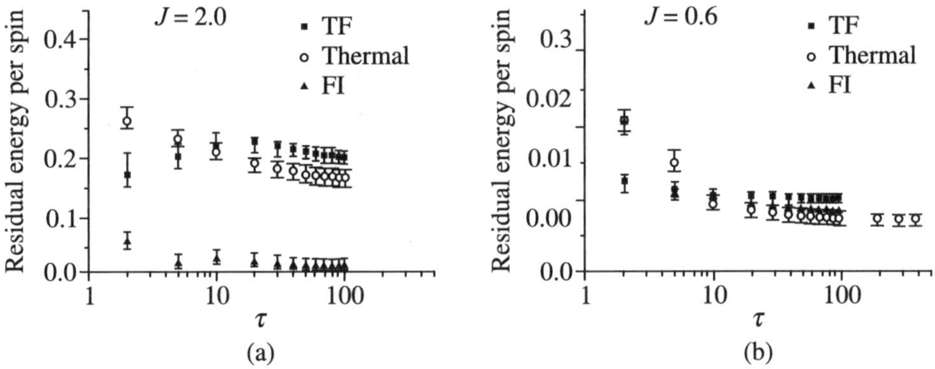

Fig. 6.13 Residual energy depending on the annealing time τ when (a) $J = 2.0$ and (b) $J = 0.6$ for the random field Ising model on the square lattice by mean-field calculation based on the Bethe approximation. The obtained results of quantum annealing by transverse field defined by Eq. (6.83) (TF) and by transverse field and ferromagnetic interaction defined by Eq. (6.117) (FI) and simulated annealing (thermal) are compared. These results are averaged out over eighty configurations of random fields (from Suzuki et al., 2007).

6.7 Quantum Field Response to Magnetic Materials

We review the quantum field response to random Ising spin systems, $LiHo_xY_{1-x}F_4$, which was done by Brooke et al. (1999). They realized the quantum annealing process by applying the transverse field to $LiHo_xY_{1-x}F_4$. Furthermore, they compared the state obtained by thermal annealing (simulated annealing: SA) and that obtained by quantum annealing through the alternating current (AC) susceptibility. Before we explain the details of the experimental demonstration of quantum annealing, physical properties of the material is reviewed. The ground state of a Ho^{3+} ion in the crystal field is a doublet that behaves as spin-1/2 eigenstates. As a result, Ho^{3+} ions in $LiHo_xY_{1-x}F_4$ are regarded as Ising spins. In addition, since Y^{3+} ions in $LiHo_xY_{1-x}F_4$ are nonmagnetic ions, the substitution of Ho^{3+} with Y^{3+} realizes the random dilution. The random dilution by the substitution of nonmagnetic ions causes random interactions between magnetic ions Ho^{3+}. That is, $LiHo_xY_{1-x}F_4$ can well be described by the Ising model with random interactions. In addition, the doubly degenerated energy levels are continuously split by applying the laboratory magnetic field H_t, which is perpendicular to the Ising axis of Ho^{3+}. In other words, H_t is regarded as the transverse field Γ in the Ising model. Thus, the quantum annealing process can be realized by applying the laboratory magnetic field H_t to $LiHo_xY_{1-x}F_4$. Figure 6.14 is the phase diagram of the materials. The horizontal axis is the concentration x of Ho^{3+} ions whereas the vertical axis is the temperature (Reich et al., 1990). In the phase diagram, the spin glass phase exists at low temperatures. Furthermore, Fig. 6.15 shows the schematic phase diagram of the temperature T versus the magnetic field H_t corresponding to Γ of $LiHo_xY_{1-x}F_4$ for $x = 0.56$. In the phase diagram, the spin glass phase appears at finite magnetic fields, and the lowest temperature phase under finite but not too large magnetic fields is the spin glass phase.

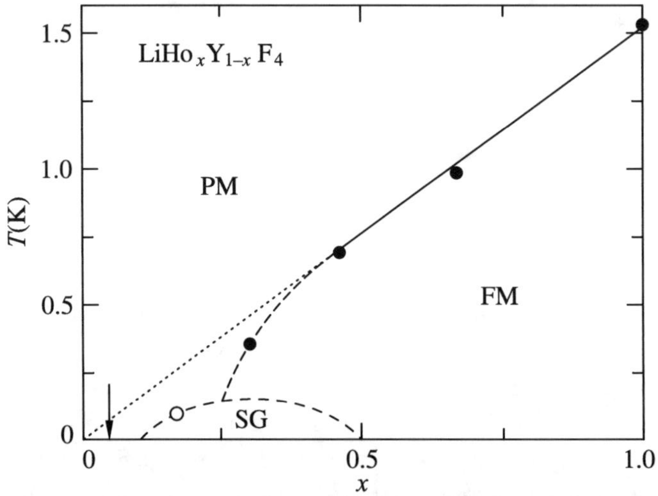

Fig. 6.14 Phase diagram of the magnetic materials $LiHo_xY_{1-x}F_4$. The horizontal axis corresponds to the concentration x of Ho^{3+} ions, and the vertical axis is the temperature. PM, FM, and SG denote the paramagnetic phase, ferromagnetic phase, and spin glass phase, respectively. The arrow shows the location of the decoupled cluster glass in which essentially isolated clusters with zero net moments appears at zero temperature. The coupling between these clusters increases with temperature; decoupled cluster glass does not appear at a finite temperature (from Reich et al., 1990).

Brooke et al. realized the thermal annealing process and the quantum annealing process in the spin glass phase ($H_t = 7.2$ kOe, $T = 0.03$ K) of the magnetic material. The processes of thermal annealing and quantum annealing are respectively indicated by arrows in Fig. 6.15, schematically. In thermal annealing, at first, the temperature decreases linearly from $T =$ 0.75 K to 0.03 K under $H_t = 0$ kOe. Next, the magnetic field H_t is linearly increased from 0 to 7.2 kOe under $T = 0.03$ K. On the other hand, in quantum annealing, the magnetic field H_t increases linearly from 0 kOe to 22 kOe under $T = 0.75$ K. Next, the temperature is linearly decreased from 0.75 K to 0.03 K under $H_t = 22$ kOe. Finally, H_t is linearly decreased from 22 kOe to 7.2 kOe under $T = 0.03$ K. Here the alternating current (AC) susceptibility was measured at $H_t = 7.2$ kOe under $T = 0.03$ K. The AC susceptibility is defined by

$$\chi(f) = \chi'(f) + i\chi''(f), \tag{6.119}$$

where f is the frequency. $\chi'(f)$ and $\chi''(f)$ are the real and imaginary parts of the AC susceptibility, respectively. The authors found that the values of $\chi'(f)$ at $H_t = 7.2$ kOe under $T = 0.03$ K obtained by quantum annealing are larger than that obtained by thermal annealing. The result means that the state obtained by quantum annealing is more sensitive than the state obtained by thermal annealing. In other words, the state obtained by quantum annealing has more rapid fluctuations than the state obtained by thermal annealing.

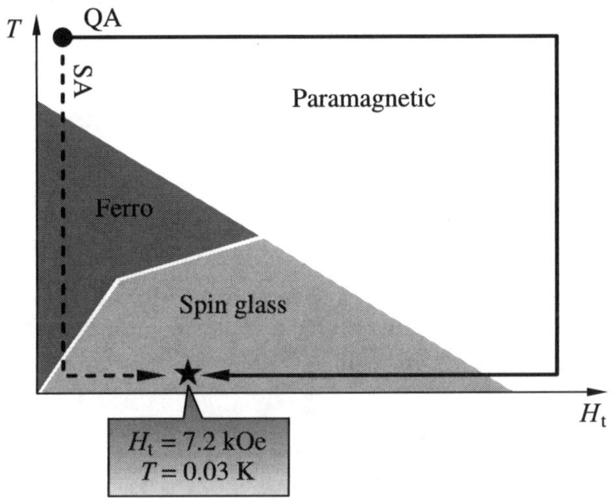

Fig. 6.15 Schematic phase diagram of the temperature T versus the magnetic field H_t corresponding to Γ of $LiHo_xY_{1-x}F_4$ for $x = 0.56$. The arrows indicate the quantum annealing (QA) process and the thermal annealing (simulated annealing: SA) process toward the point at $H_t = 7.2$ kOe under $T = 0.03$ K in the spin glass phase, respectively.

Next, the imaginary part of the AC susceptibility $\chi''(f)$ was considered for several H_t in the spin glass phase. The peak positions of $\chi''(f)$, f^*, represent the typical relaxation time $\tau = 1/(2\pi f^*)$ of the obtained state. The authors found that f^* for quantum annealing is larger than that for thermal annealing in the spin glass phase. This means that the relaxation time of the state obtained by quantum annealing is shorter than that obtained by thermal annealing. The shortening of the relaxation time is caused by the quantum tunneling effect. The result is consistent with the previous results for $\chi'(t)$. These experimental results suggest that quantum annealing can reach the ground state of the system more rapidly than thermal annealing.

6.8 Quantum Adiabatic Evolution

In this section, we first explain the concept of adiabatic quantum computation. Here we describe the adiabatic theorem and the adiabatic condition which are important for adiabatic quantum computation. The adiabatic theorem states that time evolution can be described by the eigenstates of the instantaneous Hamiltonian when the time-dependent Hamiltonian changes infinitely slowly. In addition, the adiabatic condition is a condition where the adiabatic theorem is asymptotically satisfied when the time-dependent Hamiltonian changes slow enough with the time.

6.8.1 Adiabatic quantum computation

In order to obtain the best solution for combinatorial optimization problems, Farhi et al. (2001) proposed a method called the adiabatic quantum computation which is based on the quantum adiabatic theorem (Messiah, 1976). In the adiabatic quantum computation, the Hamiltonian is evolved as

$$\hat{\mathscr{H}}(t) = \frac{t}{\tau}\hat{\mathscr{H}}_0 + \left(1 - \frac{t}{\tau}\right)\hat{\mathscr{H}}_{\mathrm{q}}, \quad (0 \leq t \leq \tau), \tag{6.120}$$

where $\hat{\mathscr{H}}_0$ and $\hat{\mathscr{H}}_{\mathrm{q}}$ are the Hamiltonian representing combinatorial optimization problems and the quantum tunneling effect, respectively. That is, the coefficient of the first term of Eq. (6.120) changes from zero to unity whereas the coefficient of the second term of Eq. (6.120) varies from unity to zero in the adiabatic quantum computation shown in Fig. 6.16.

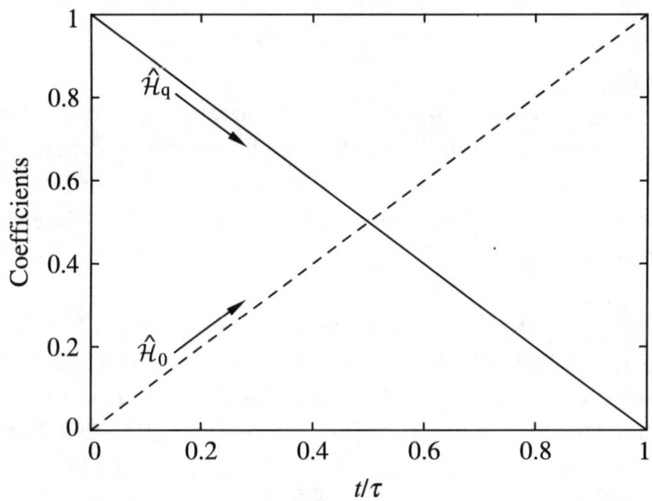

Fig. 6.16 Time evolution of the coefficients of $\hat{\mathscr{H}}_0$ and $\hat{\mathscr{H}}_{\mathrm{q}}$ in the adiabatic quantum computation.

The concepts of quantum annealing proposed by Kadowaki and Nishimori (1998) which was explained in Section 6.1 and the adiabatic quantum computation are the same but the methods to evolve the Hamiltonian were originally different. Notice that the adiabatic quantum computation is based on the adiabatic theorem or the adiabatic condition which will be explained later. However, quantum annealing is not necessarily restricted by the adiabatic theorem or the adiabatic condition. For example, quantum annealing with the Jarzynski equality (Jarzynski, 1997a,b) was proposed by Ohzeki (2010). The method, called the quantum Jarzynski annealing, is based on the quantum-classical correspondence developed by Somma et al. (2007) which will be explained in Section 6.9.1. In addition, through nonequilibrium work on random Ising spin systems in longitudinal and transverse fields, the Jarzynski equality for quantum annealing was derived (Ohzeki et al., 2011).

Suppose the system evolves according to Eq. (6.120) in the limit of $\tau \to \infty$. If the initial state is prepared as the ground state of $\hat{\mathscr{H}}_{\mathrm{q}}$ at $t = 0$, the state for each time is the ground state of the instantaneous Hamiltonian and the ground state of $\hat{\mathscr{H}}_0$ can be definitely obtained at $t = \tau$. The fact is called the adiabatic theorem and will be proved in Section 6.8.2. In a practical sense, we should consider the time evolution for finite τ. When the annealing time τ is large enough, the adiabatic theorem is satisfied asymptotically, which is called the adiabatic approximation. The conditions of the adiabatic approximation will be explained in Section 6.8.3, which is used in the proof of the convergence theorem of quantum annealing (see Section 6.9).

6.8.2 Quantum adiabatic theorem

In this subsection, we prove the adiabatic theorem. Suppose that the Hamiltonian is continuously evolved from $t = 0$ to $t = \tau$. Here a new variable is introduced as follows:

$$s := \frac{t}{\tau}. \tag{6.121}$$

When the variable t changes from zero to τ, the variable s changes from zero to unity. Thc Hamiltonian is rewritten by $\hat{\mathscr{H}}(s)$ by using the new variable s. The Schrödinger equation is expressed as

$$i\frac{d}{ds}|\psi_\tau(s)\rangle = \tau\hat{\mathscr{H}}(s)|\psi_\tau(s)\rangle, \tag{6.122}$$

where $|\psi_\tau(s)\rangle$ denotes the eigenstate which depends on τ and s. In addition, the time-evolution operator $\hat{U}_\tau(s)$ can be written as

$$|\psi_\tau(s)\rangle = \hat{U}_\tau(s)|\psi_\tau(0)\rangle, \tag{6.123}$$

where

$$\hat{U}_\tau(0) = 1. \tag{6.124}$$

According to the Schrödinger equation, the time-evolution operator $\hat{U}_\tau(s)$ is given by

$$i\frac{d}{ds}\hat{U}_\tau(s) = \tau\hat{\mathscr{H}}(s)\hat{U}_\tau(s). \tag{6.125}$$

Next, let $\lambda_i(s)$ $(i = 0, \cdots, N_{\mathrm{s}} - 1)$ be the eigenvalues of $\hat{\mathscr{H}}(s)$, where N_{s} is the number of states. The projection operators which project arbitrary state vector $|\phi\rangle$ onto respective eigenstates are represented by $\hat{P}_i(s)$ $(i = 0, \cdots, N_s - 1)$. The eigenstate corresponding to the eigenvalue $\lambda_i(s)$ is denoted by $\hat{P}_i(s)|\phi\rangle$ and the following relation is satisfied:

$$\mathscr{H}(s)\hat{P}_i(s)|\phi\rangle = \lambda_i(s)\hat{P}_i(s)|\phi\rangle. \tag{6.126}$$

In addition, the projection operators are Hermitian and have the following properties:

$$\hat{P}_i(s)\hat{P}_j(s) = \delta_{ij}\hat{P}_i(s), \tag{6.127}$$

$$\sum_{i=0}^{N_s-1} \hat{P}_i(s) = 1. \tag{6.128}$$

By using the projection operators, the Hamiltonian $\hat{\mathscr{H}}(s)$ can be represented by

$$\mathscr{H}(s) = \sum_{i=0}^{N_s-1} \lambda_i(s)\hat{P}_i(s). \tag{6.129}$$

Here, the adiabatic theorem is expressed by

$$\lim_{\tau\to\infty} \hat{U}_\tau(s)\hat{P}_i(0) = \hat{P}_i(s) \lim_{\tau\to\infty} \hat{U}_\tau(s), \quad \forall s. \tag{6.130}$$

We consider the meaning of the adiabatic theorem. At the initial time $s = 0$, let $|\psi_\tau^i(0)\rangle$ $(i = 0, \cdots, N_s - 1)$ be the eigenstates of $\mathscr{H}(0)$. The following equation is satisfied:

$$\hat{P}_i(0)|\psi_\tau^i(0)\rangle = |\psi_\tau^i(0)\rangle. \tag{6.131}$$

Thus, due to the adiabatic theorem, the relation:

$$\lim_{\tau\to\infty} \hat{U}_\tau(s)|\psi_\tau^i(0)\rangle = \hat{P}_i(s) \lim_{\tau\to\infty} \hat{U}_\tau(s)|\psi_\tau^i(0)\rangle, \quad \forall s, \tag{6.132}$$

is satisfied. According to the adiabatic theorem, the state vector $\hat{U}_\tau(s)|\psi_\tau^i(0)\rangle$ approaches the state projected by $\hat{P}_i(s)$ (i.e., the state corresponding to the eigenvalue $\lambda_i(s)$) in the limit of $\tau \to \infty$. That is, if the initial state is an eigenstate of the Hamiltonian at $s = 0$, the adiabatic theorem means that the state at each time is the same as the eigenstate of the instantaneous Hamiltonian.

Representation of rotating axis

To prove the adiabatic theorem which was explained just now, we introduce the representation of the rotating axis. Here, the operator of the rotating axis $\hat{A}(s)$ is defined by

$$\hat{P}_i(s) = \hat{A}(s)\hat{P}_i(0)\hat{A}^\dagger(s). \tag{6.133}$$

By introducing the operator $\hat{A}(s)$, each eigenstate of $\mathscr{H}(0)$ can be transformed to an eigenstate of $\mathscr{H}(s)$. In addition, $\hat{A}(s)$ is the unitary operator, that is,

$$\hat{A}(s)\hat{A}^{\dagger}(s) = \hat{A}^{\dagger}(s)\hat{A}(s) = 1. \tag{6.134}$$

Furthermore, from the initial condition, the following relation is obtained:

$$\hat{A}(0) = \hat{A}^{\dagger}(0) = 1. \tag{6.135}$$

Next we derive the differential equation of $\hat{A}(s)$. By taking the derivative of both sides of Eq. (6.134) with respect to s, we obtain

$$\frac{d\hat{A}(s)}{ds}\hat{A}^{\dagger}(s) + \hat{A}(s)\frac{d\hat{A}^{\dagger}(s)}{ds} = 0, \tag{6.136}$$

and then the following relation is satisfied:

$$i\frac{d\hat{A}(s)}{ds} = -i\hat{A}(s)\frac{d\hat{A}^{\dagger}(s)}{ds}\hat{A}(s) =: \hat{K}(s)\hat{A}(s). \tag{6.137}$$

Here, $\hat{K}(s)$ is a Hermitian operator and is defined as

$$\hat{K}(s) := -i\hat{A}(s)\frac{d\hat{A}^{\dagger}(s)}{ds}. \tag{6.138}$$

The necessary and sufficient condition to satisfy Eq. (6.133) for $\hat{K}(s)$ is given by

$$i\frac{d\hat{P}_i(s)}{ds} = \left[\hat{K}(s), \hat{P}_i(s)\right], \tag{6.139}$$

which will be proved as follows.

- Derivation from Eq. (6.133) to Eq. (6.139)

  By taking the derivative of both sides of Eq. (6.133) with respect to s,

$$\frac{d\hat{P}_i(s)}{ds} = \frac{d\hat{A}(s)}{ds}\hat{P}_i(0)\hat{A}^{\dagger}(s) + \hat{A}(s)\hat{P}_i(0)\frac{d\hat{A}^{\dagger}(s)}{ds} \tag{6.140}$$

$$= \frac{d\hat{A}(s)}{ds}\hat{A}^{\dagger}(s)\hat{A}(s)\hat{P}_i(0)\hat{A}^{\dagger}(s) + \hat{A}(s)\hat{P}_i(0)\hat{A}^{\dagger}(s)\hat{A}(s)\frac{d\hat{A}^{\dagger}(s)}{ds} \tag{6.141}$$

$$= \frac{d\hat{A}(s)}{ds}\hat{A}^{\dagger}(s)\hat{P}_i(s) + \hat{P}_i(s)\hat{A}(s)\frac{d\hat{A}^{\dagger}(s)}{ds} \tag{6.142}$$

$$= -\hat{A}(s)\frac{d\hat{A}^\dagger(s)}{ds}\hat{P}_i(s) + \hat{P}_i(s)\hat{A}(s)\frac{d\hat{A}^\dagger(s)}{ds} \tag{6.143}$$

$$= -i\left[\hat{K}(s), \hat{P}_i(s)\right]. \tag{6.144}$$

As a result, we can derive Eq. (6.139) from Eq. (6.133).

- Derivation from Eq. (6.139) to Eq. (6.133)

By using Eq. (6.139), we obtain

$$\frac{d}{ds}\left(\hat{A}^\dagger(s)\hat{P}_i(s)\hat{A}(s)\right)$$

$$= \frac{d\hat{A}^\dagger(s)}{ds}\hat{P}_i(s)\hat{A}(s) + \hat{A}^\dagger(s)\frac{d\hat{P}_i(s)}{ds}\hat{A}(s) + \hat{A}^\dagger(s)\hat{P}_i(s)\frac{d\hat{A}(s)}{ds} \tag{6.145}$$

$$= \hat{A}^\dagger(s)\hat{A}(s)\frac{d\hat{A}^\dagger(s)}{ds}\hat{P}_i(s)\hat{A}(s)$$
$$+\hat{A}^\dagger(s)\left(-i\left[\hat{K}(s), \hat{P}_i(s)\right]\right)\hat{A}(s) + \hat{A}^\dagger(s)\hat{P}_i(s)\frac{d\hat{A}(s)}{ds}\hat{A}^\dagger(s)\hat{A}(s) \tag{6.146}$$

$$= \hat{A}^\dagger(s)\left\{\hat{A}(s)\frac{d\hat{A}^\dagger(s)}{ds}\hat{P}_i(s)\right.$$
$$\left. -i\left[\hat{K}(s), \hat{P}_i(s)\right] - \hat{P}_i(s)\hat{A}(s)\frac{d\hat{A}^\dagger(s)}{ds}\right\}\hat{A}(s) \tag{6.147}$$

$$= \hat{A}^\dagger(s)\left\{i\hat{K}(s)\hat{P}_i(s) - i\left[\hat{K}(s), \hat{P}_i(s)\right] - i\hat{P}_i(s)\hat{K}(s)\right\}\hat{A}(s) \tag{6.148}$$

$$= 0. \tag{6.149}$$

Here, $\hat{A}^\dagger(s)\hat{P}_i(s)\hat{A}(s)$ is not a constant. By integrating this equation and using Eq. (6.135), the following relation is obtained:

$$\hat{A}^\dagger(s)\hat{P}_i(s)\hat{A}(s) = \hat{A}^\dagger(0)\hat{P}_i(0)\hat{A}(0) = \hat{P}_i(0). \tag{6.150}$$

As a result, Eq. (6.133) can be derived from Eq. (6.139).

Next we consider how to determine $\hat{K}(s)$ by using Eq. (6.139). Let $\hat{B}_i(s)$ be arbitrary operators depending on s. In this case, the following relation is satisfied:

$$\left[\hat{K}(s) + \sum_{j=0}^{N_s-1}\hat{P}_j(s)\hat{B}_j(s)\hat{P}_j(s), \hat{P}_i(s)\right] \tag{6.151}$$

$$= \left[\hat{K}(s), \hat{P}_i(s)\right] + \sum_{j=0}^{N_s-1} \left(\hat{P}_j(s)\hat{B}_j(s)\hat{P}_j(s)\hat{P}_i(s) - \hat{P}_i(s)\hat{P}_j(s)\hat{B}_j(s)\hat{P}_j(s)\right) \tag{6.152}$$

$$= \left[\hat{K}(s), \hat{P}_i(s)\right] + \sum_{j=0}^{N_s-1} \left(\hat{P}_j(s)\hat{B}_j(s)\delta_{ij}\hat{P}_i(s) - \delta ij\hat{P}_j(s)\hat{B}_j(s)\hat{P}_j(s)\right) \tag{6.153}$$

$$= \left[\hat{K}(s), \hat{P}_i(s)\right], \tag{6.154}$$

and thus, Eq. (6.139) can be replaced as

$$\hat{K}(s) \to \hat{K}(s) + \sum_{j=0}^{N_s-1} \hat{P}_j(s)\hat{B}_j(s)\hat{P}_j(s). \tag{6.155}$$

In this way, $\hat{K}(s)$ cannot be uniquely determined. To remove the ambiguity and uniquely determine $\hat{K}(s)$, we impose the following condition:

$$\hat{P}_i(s)\hat{K}(s)\hat{P}_i(s) = 0, \quad (i = 0, \cdots, N_s - 1). \tag{6.156}$$

By using this condition, we can decide $K(s)$ uniquely as follows:

$$\hat{K}(s) = i \sum_{i=0}^{N_s-1} \frac{d\hat{P}_i(s)}{ds}\hat{P}_i(s). \tag{6.157}$$

Now we will see that Eqs. (6.139) and (6.156) are satisfied in $\hat{K}(s)$ defined by Eq. (6.157). First, $\hat{K}(s)$ is a Hermitian operator, that is,

$$\hat{K}^{\dagger}(s) = -i \sum_{i=0}^{N_s-1} \hat{P}_i(s)\frac{d\hat{P}_i(s)}{ds} \tag{6.158}$$

$$= -i \sum_{i=0}^{N_s-1} \left(\frac{d\hat{P}_i(s)}{ds} - \frac{d\hat{P}_i(s)}{ds}\hat{P}_i(s)\right) \tag{6.159}$$

$$= i \sum_{i=0}^{N_s-1} \frac{d\hat{P}_i(s)}{ds}\hat{P}_i(s) = \hat{K}(s). \tag{6.160}$$

Here, we use the following equations:

$$\frac{d\hat{P}_i(s)}{ds}\hat{P}_i(s)+\hat{P}_i(s)\frac{d\hat{P}_i(s)}{ds}=\frac{d\hat{P}_i(s)}{ds}, \tag{6.161}$$

$$\sum_{i=0}^{N_s-1}\frac{d\hat{P}_i(s)}{ds}=0, \tag{6.162}$$

which are obtained by performing the derivative of Eqs. (6.127) and (6.128) with respect to s. In addition, by using Eq. (6.161), we obtain

$$\left[\hat{K}(s),\hat{P}_i(s)\right]=i\sum_{j=0}^{N_s-1}\frac{d\hat{P}_j(s)}{ds}\hat{P}_j(s)\hat{P}_i(s)+i\sum_{j=0}^{N_s-1}\hat{P}_i(s)\hat{P}_j(s)\frac{d\hat{P}_j(s)}{ds} \tag{6.163}$$

$$=i\left(\frac{d\hat{P}_i(s)}{ds}\hat{P}_i(s)+\hat{P}_i(s)\frac{d\hat{P}_i(s)}{ds}\right) \tag{6.164}$$

$$=i\frac{d\hat{P}_i(s)}{ds}. \tag{6.165}$$

Thus, it can be shown that Eq. (6.139) is satisfied in $\hat{K}(s)$ defined by Eq. (6.157). Furthermore, Eq. (6.161) is rewritten by

$$\hat{P}_i(s)\hat{K}(s)\hat{P}_i(s)= i\hat{P}_i(s)\sum_{j=0}^{N_s-1}\frac{d\hat{P}_j(s)}{ds}\hat{P}_j(s)\hat{P}_i(s) \tag{6.166}$$

$$= i\hat{P}_i(s)\frac{d\hat{P}_i(s)}{ds}\hat{P}_i(s) \tag{6.167}$$

$$= i\hat{P}_i(s)\left(\frac{d\hat{P}_i(s)}{ds}-\hat{P}_i(s)\frac{d\hat{P}_i(s)}{ds}\right) \tag{6.168}$$

$$=0. \tag{6.169}$$

Thus, Eq. (6.156) is satisfied in $\hat{K}(s)$ defined by Eq. (6.157).

Next, we consider the equation of the time-evolution operator in the rotating axis representation. Let $\mathscr{H}^{A}(s)$ be the Hamiltonian in the rotating axis representation, and $\hat{\mathscr{H}}^{A}(s)$ be defined as

$$\hat{\mathscr{H}}^{A}(s):=\hat{A}^{\dagger}(s)\hat{\mathscr{H}}(s)\hat{A}(s) \tag{6.170}$$

$$= \sum_{i=0}^{N_s-1} \lambda_i(s)\hat{A}^\dagger(s)\hat{P}_i(s)\hat{A}(s) \tag{6.171}$$

$$= \sum_{i=0}^{N_s-1} \lambda_i(s)\hat{P}_i(0). \tag{6.172}$$

As shown here, the Hamiltonian at each time can be represented by the projection operator at $s = 0$ by using the rotating axis representation. The Hamiltonians $\hat{\mathscr{H}}^A(s)$ and each projection operator $\hat{P}_i(0)$ at $s = 0$ are commuted. Thus, the Hamiltonians $\hat{\mathscr{H}}^A(s)$ and $\hat{\mathscr{H}}^A(s')$ for arbitrary s and s' are commuted, which is the reason why the rotating axis representation was introduced. Here, the Hermitian operator $\hat{K}(s)$ is given by

$$\hat{K}^A(s) := \hat{A}^\dagger(s)\hat{K}(s)\hat{A}(s). \tag{6.173}$$

In this case, the time-evolution operator in the rotating axis representation is defined by

$$\hat{U}_\tau^A(s) := \hat{A}^\dagger(s)\hat{U}_\tau(s), \tag{6.174}$$

and the equation of the time-evolution operator is derived as follows:

$$i\frac{d}{ds}\hat{U}_\tau^A(s) = i\frac{d\hat{A}^\dagger(s)}{ds}\hat{U}_\tau(s) + i\hat{A}^\dagger(s)\frac{d\hat{U}_\tau(s)}{ds} \tag{6.175}$$

$$= -\hat{A}^\dagger(s)\hat{K}(s)\hat{U}_\tau(s) + \tau\hat{A}^\dagger(s)\hat{\mathscr{H}}(s)\hat{U}_\tau(s) \tag{6.176}$$

$$= -\hat{K}^A(s)\hat{U}_\tau^A(s) + \tau\hat{\mathscr{H}}^A(s)\hat{U}_\tau^A(s) \tag{6.177}$$

$$= \left(\tau\hat{\mathscr{H}}^A(s) - \hat{K}^A(s)\right)\hat{U}_\tau^A(s). \tag{6.178}$$

Here Eqs. (6.125) and (6.137) were used. Moreover, the time-evolution operator at $s = 0$ is given by

$$\hat{U}_\tau^A(0) = 1. \tag{6.179}$$

Proof of the adiabatic theorem

Let us prove the adiabatic theorem. In the limit of $\tau \to \infty$, Eq. (6.178) is approximated as

$$i\frac{d}{ds}\hat{U}_\tau^A(s) \simeq \tau\hat{\mathscr{H}}^A(s)\hat{U}_\tau^A(s). \tag{6.180}$$

Thus, in the limit of $\tau \to \infty$, an operator $\hat{V}_\tau^A(s)$ which satisfies

$$i\frac{d}{ds}\hat{V}_\tau^A(s) = \tau\hat{\mathscr{H}}^A(s)\hat{V}_\tau^A(s) \tag{6.181}$$

is expected to approach $\hat{U}_\tau^A(s)$. Here, the equation can be integrated exactly, because $\hat{\mathscr{H}}^A(s)$ can be written by Eq. (6.172). Then, we obtain

$$\hat{V}_\tau^A(s) = \exp\left[-i\tau\int_0^s \hat{\mathscr{H}}^A(s')ds'\right] \tag{6.182}$$

$$= \exp\left[-i\tau\sum_{i=0}^{N_s-1}\int_0^s \lambda_i(s')ds'\hat{P}_i(0)\right] \tag{6.183}$$

$$= \exp\left[-i\tau\sum_{i=0}^{N_s-1}\phi_i(s)\hat{P}_i(0)\right] \tag{6.184}$$

$$= 1+\sum_{k=1}^{\infty}\frac{1}{k!}\left[-i\tau\sum_{i=0}^{N_s-1}\phi_i(s)\hat{P}_i(0)\right]^k \tag{6.185}$$

$$= \sum_{i=0}^{N_s-1}\left\{1+\sum_{k=1}^{\infty}\frac{1}{k!}[-i\tau\phi_i(s)]^k\right\}\hat{P}_i(0) \tag{6.186}$$

$$= \sum_{i=0}^{N_s-1} \mathrm{e}^{-i\tau\phi_i(s)}\hat{P}_i(0), \tag{6.187}$$

where Eqs. (6.127) and (6.128) were used. In addition, a new variable was introduced as

$$\phi_i(s) := \int_0^s \lambda_i(s')ds'. \tag{6.188}$$

By using the variable, we get

$$\hat{V}_\tau^{A\dagger}(s)\hat{V}_\tau^A(s) = \sum_{i=0}^{N_s-1}\hat{P}_i(0) = 1, \tag{6.189}$$

and it is confirmed that $\hat{V}_\tau^A(s)$ is a Hermitian operator. Furthermore, since $\phi_i(0)=0$ for all i, the following is satisfied:

$$\hat{V}_\tau^A(0) = \sum_{i=0}^{N_s-1}\hat{P}_i(0) = 1. \tag{6.190}$$

Next, to clarify the relation between $\hat{U}_\tau^A(s)$ and $\hat{V}_\tau^A(s)$, we define the overlap between two operators as follows:

$$\hat{W}_\tau^A(s) := \hat{V}_\tau^{A\dagger}(s)\hat{U}_\tau^A(s). \tag{6.191}$$

Because of Eqs. (6.178) and (6.181), the equation of $\hat{W}_\tau^A(s)$ is given by

$$i\frac{d}{ds}\hat{W}_\tau^A(s) = i\frac{d\hat{V}_\tau^{A\dagger}(s)}{ds}\hat{U}_\tau^A(s) + i\hat{V}_\tau^{A\dagger}(s)\frac{d\hat{U}_\tau^A(s)}{ds} \tag{6.192}$$

$$= -\tau\hat{V}_\tau^{A\dagger}(s)\hat{\mathscr{H}}^A(s)\hat{U}_\tau^A(s) + \hat{V}_\tau^{A\dagger}(s)\left[\tau\hat{\mathscr{H}}^A(s) - \hat{K}^A(s)\right]\hat{U}_\tau^A(s) \tag{6.193}$$

$$= -\hat{V}_\tau^{A\dagger}(s)\hat{K}^A(s)\hat{U}_\tau^A(s) \tag{6.194}$$

$$= -\hat{V}_\tau^{A\dagger}(s)\hat{K}^A(s)\hat{V}_\tau^A(s)\hat{V}_\tau^{A\dagger}(s)\hat{U}_\tau^A(s) \tag{6.195}$$

$$= -\sum_{0\le i,j\le N_s-1} \mathrm{e}^{i\tau(\phi_i(s)-\phi_j(s))}\hat{P}_i(0)\hat{K}^A(s)\hat{P}_j(0)\hat{W}_\tau^A(s) \tag{6.196}$$

$$= -\sum_{0\le i,j\le N_s-1} \mathrm{e}^{i\tau(\phi_i(s)-\phi_j(s))}\hat{K}_{ij}^A(s)\hat{W}_\tau^A(s), \tag{6.197}$$

where $\hat{K}_{ij}^A(s)$ is defined by

$$\hat{K}_{ij}^A(s) := \hat{P}_i(0)\hat{K}^A(s)\hat{P}_j(0) \tag{6.198}$$

$$= \hat{P}_i(0)\hat{A}^\dagger(s)\hat{K}(s)\hat{A}(s)\hat{P}_j(0) \tag{6.199}$$

$$= \hat{A}^\dagger(s)\hat{P}_i(s)\hat{K}(s)\hat{P}_j(s)\hat{A}(s). \tag{6.200}$$

Here, Eq. (6.150) was used. From Eq. (6.200), $\hat{W}_\tau^A(s)$ can be calculated by the integral equation given by

$$\hat{W}_\tau^A(s) = 1 + i\sum_{0\le i,j\le N_s-1}\int_0^s ds'\mathrm{e}^{i\tau(\phi_i(s')-\phi_j(s'))}\hat{K}_{ij}^A(s')\hat{W}_\tau^A(s'), \tag{6.201}$$

where we used the relation:

$$\hat{W}_\tau^A(0) = \hat{V}_\tau^{A\dagger}(0)\hat{U}_\tau^A(0) = 1. \tag{6.202}$$

In order to estimate $\hat{W}_\tau^A(s)$, we consider $\hat{F}_{ij}(s)$ defined by

$$\hat{F}_{ij}(s) := \int_0^s ds'\mathrm{e}^{i\tau(\phi_i(s')-\phi_j(s'))}\hat{K}_{ij}^A(s'). \tag{6.203}$$

If $i = j$, the following relation is satisfied:

$$\hat{F}_{ii}(s) = 0. \tag{6.204}$$

Next we consider the case when $i \neq j$. By taking the partial integral, we can obtain

$$\begin{aligned}\hat{F}_{ij}(s) &= \frac{1}{i\tau}\left\{\left[e^{i\tau(\phi_i(s')-\phi_j(s'))}\frac{\hat{K}^A_{ij}(s')}{\lambda_i(s')-\lambda_j(s')}\right]_0^s\right.\\ &\left.-\int_0^s ds' e^{i\tau(\phi_i(s')-\phi_j(s'))}\frac{d}{ds'}\left[\frac{\hat{K}^A_{ij}(s')}{\lambda_i(s')-\lambda_j(s')}\right]\right\}.\end{aligned} \tag{6.205}$$

The function inside the bracket in Eq. (6.205) does not diverge but has a finite value if the following two conditions are satisfied.

1. No degeneracy in eigenvalues. That is,

$$\forall i \neq j, \quad \lambda_i(s) \neq \lambda_j(s). \tag{6.206}$$

2. The derivatives of $\hat{K}^A_{ij}(s)$ and $\lambda_i(s)$ with respect to s are finite values. When the derivative of $\hat{K}^A_{ij}(s)$ is finite, $d\hat{P}_i(s)/ds$ and $d^2\hat{P}_i(s)/ds^2$ are finite values according to Eq. (6.157).

Thus, in the limit of $\tau \to \infty$, $\hat{F}_{ij}(s)$ asymptotically approaches zero:

$$\lim_{\tau\to\infty}\hat{F}_{ij}(s) = \mathcal{O}(1/\tau). \tag{6.207}$$

By taking the partial integral of Eq. (6.201), $\hat{W}^A_\tau(s)$ can be represented as,

$$\begin{aligned}&\hat{W}^A_\tau(s)\\ &= 1 + i\sum_{0\le i,j\le N_s-1}\left\{\left[\hat{F}_{ij}(s')\hat{W}^A_\tau(s')\right]_0^s - \int_0^s ds'\hat{F}_{ij}(s')\frac{d\hat{W}^A_\tau(s')}{ds'}\right\} &(6.208)\\ &\simeq 1 + \mathcal{O}\left(\frac{1}{\tau}\right), &(6.209)\end{aligned}$$

where the approximation is valid when τ is large. In this way, for large τ, the following relations are satisfied:

$$\hat{V}^{A\dagger}_\tau(s)\hat{U}^A_\tau(s) \simeq 1 + \mathcal{O}\left(\frac{1}{\tau}\right), \tag{6.210}$$

$$\hat{U}_\tau(s) \simeq \hat{A}(s)\hat{V}_\tau^A(s)\left[1+\mathscr{O}\left(\frac{1}{\tau}\right)\right], \tag{6.211}$$

from Eq. (6.174). Then, in the limit of $\tau \to \infty$, we obtain

$$\lim_{\tau\to\infty} \hat{U}_\tau(s) = \lim_{\tau\to\infty} \hat{A}(s)\hat{V}_\tau^A(s). \tag{6.212}$$

Here, from Eq. (6.133), the following relation is satisfied:

$$\hat{A}(s)\hat{P}_i(0) = \hat{P}_i(s)\hat{A}(s), \tag{6.213}$$

and by operating $\hat{V}_\tau^A(s)$ from the right-hand side, we obtain

$$\hat{A}(s)\hat{P}_i(0)\hat{V}_\tau^A(s) = \hat{P}_i(s)\hat{A}(s)\hat{V}_\tau^A(s), \tag{6.214}$$

and then the following relations are satisfied:

$$\hat{A}(s)\hat{V}_\tau^A(s)\hat{P}_i(0) = \hat{P}_i(s)\hat{A}(s)\hat{V}_\tau^A(s), \tag{6.215}$$

where we used the fact that $\hat{V}_\tau^A(s)$ and $\hat{P}_i(0)$ can commute because of the property given by Eq. (6.187). Then, in the limit of $\tau \to \infty$, we get the following relation:

$$\lim_{\tau\to\infty} \hat{A}(s)\hat{V}_\tau^A(s)\hat{P}_i(0) = \hat{P}_i(s) \lim_{\tau\to\infty} \hat{A}(s)\hat{V}_\tau^A(s), \tag{6.216}$$

and by substituting Eq. (6.212), we can prove the adiabatic theorem, that is,

$$\lim_{\tau\to\infty} \hat{U}_\tau(s)\hat{P}_i(0) = \hat{P}_i(s) \lim_{\tau\to\infty} \hat{U}_\tau(s). \tag{6.217}$$

Thus, if the initial state is an eigenstate of the Hamiltonian at $s = 0$, the state at each time is the same as the eigenstate of the instantaneous Hamiltonian.

6.8.3 Quantum adiabatic approximation

In this subsection, we consider the adiabatic condition where the adiabatic approximation is satisfied. For simplicity, here, we assume that the ground state is not degenerate. The adiabatic approximation is given by

$$\hat{U}_\tau(s) \simeq \hat{A}(s)\hat{V}_\tau^A(s), \quad (\tau \gg 1). \tag{6.218}$$

This means that the quantum adiabatic theorem given by Eq. (6.212) is approximately satisfied when τ is large enough (not necessary $\tau \to \infty$). Let $|\psi_\tau^i(s)\rangle$ $(i = 0, \cdots, N_\mathrm{s}-1)$ be the eigenstates of $\mathscr{H}(s)$, where $i = 0$ indicates the ground state. By using the eigenstates $|\psi_\tau^i(0)\rangle$ at $s = 0$, $|\psi_\tau^i(s)\rangle$ can be represented by

$$|\psi_\tau^i(s)\rangle = \hat{A}(s)\hat{V}_\tau^A(s)|\psi_\tau^i(0)\rangle. \tag{6.219}$$

The proof of the relation is as follows. Since $\hat{P}_i(s)$ is the projection operator defined by Eq. (6.126), we obtain the following equations according to Eq. (6.215):

$$\hat{P}_i(s)\hat{A}(s)\hat{V}_\tau^A(s)|\psi_\tau^i(0)\rangle = \hat{A}(s)\hat{V}_\tau^A(s)\hat{P}_i(0)|\psi_\tau^i(0)\rangle \tag{6.220}$$

$$= \hat{A}(s)\hat{V}_\tau^A(s)|\psi_\tau^i(0)\rangle. \tag{6.221}$$

Thus, Eq. (6.219) can be proved. The state at time s obtained by the time-evolution given by Eq. (6.122) is $\hat{U}_\tau(s)|\psi_\tau^i(0)\rangle$. Then, when the initial state is prepared as the jth state at $s=0$, the probability of the kth state at time s is given by

$$p_{j\to k} = |\langle\psi_\tau^k(s)|\hat{U}_\tau(s)|\psi_\tau^j(0)\rangle|^2 \tag{6.222}$$

$$= |\langle\psi_\tau^k(0)|\hat{V}_\tau^{A\dagger}(s)\hat{A}^\dagger(s)\hat{U}_\tau(s)|\psi_\tau^j(0)\rangle|^2 \tag{6.223}$$

$$= |\langle\psi_\tau^k(0)|\hat{V}_\tau^{A\dagger}(s)\hat{U}_\tau^A(s)|\psi_\tau^j(0)\rangle|^2 \tag{6.224}$$

$$= |\langle\psi_\tau^k(0)|\hat{W}_\tau^A(s)|\psi_\tau^j(0)\rangle|^2. \tag{6.225}$$

When the initial state is set to be the ground state at $s=0$, the probability of excited states η is given by

$$\eta := \sum_{0<k\leq N_s-1} p_{0\to k} \tag{6.226}$$

$$= \sum_{0<k\leq N_s-1} |\langle\psi_\tau^k(0)|\hat{W}_\tau^A(s)|\psi_\tau^0(0)\rangle|^2. \tag{6.227}$$

Here, when we consider the case that $\eta \ll 1$, we obtain

$$|\langle\psi_\tau^0(0)|\hat{V}_\tau^{A\dagger}(s)\hat{A}^\dagger(s)\hat{U}_\tau(s)|\psi_\tau^0(0)\rangle|^2 \simeq 1, \tag{6.228}$$

and Eq. (6.218) is satisfied. This fact means that the adiabatic approximation is guaranteed when $\eta \ll 1$. In other words, $\eta \ll 1$ is the condition of the adiabatic approximation. $\hat{W}_\tau^A(s)$ is given by the integral equation defined by Eq. (6.201) and can be obtained by the iterative method. For $\tau \gg 1$, $\hat{W}_\tau^A(s)$ can be approximated by the first degree of $\hat{F}_{ij}(s)$, because $\hat{F}_{ij}(s) \ll 1$ where $\hat{F}_{ij}(s)$ is defined by Eq. (6.203). That is, $\hat{W}_\tau^A(s)$ is approximated as

$$\hat{W}_\tau^A(s) \simeq 1 + i \sum_{0\leq i,j\leq N_s-1} \hat{F}_{ij}(s), \quad (\tau \gg 1). \tag{6.229}$$

Thus, $p_{0\to k}$ and η are approximated as

$$p_{0\to k} \simeq \left| \left\langle \psi_\tau^k(0) \middle| \sum_{0\le i,j\le N_s-1} \hat{F}_{ij}(s) \middle| \psi_\tau^0(0) \right\rangle \right|^2, \tag{6.230}$$

$$\eta \simeq \sum_{0<k\le N_s-1} \left| \left\langle \psi_\tau^k(0) \middle| \sum_{0\le i,j\le N_s-1} \hat{F}_{ij}(s) \middle| \psi_\tau^0(0) \right\rangle \right|^2. \tag{6.231}$$

In addition, we obtain

$$\langle \psi_\tau^k(0)|\hat{F}_{ij}(s)|\psi_\tau^0(0)\rangle$$

$$= \int_0^s ds' e^{i\tau(\phi_i(s')-\phi_j(s'))} \langle \psi_\tau^k(0)|\hat{K}_{ij}^A(s')|\psi_\tau^0(0)\rangle \tag{6.232}$$

$$= \int_0^s ds' e^{i\tau(\phi_i(s')-\phi_j(s'))} \langle \psi_\tau^k(0)|\hat{P}_i(0)\hat{A}^\dagger(s')\hat{K}(s')\hat{A}(s')\hat{P}_j(0)|\psi_\tau^0(0)\rangle \tag{6.233}$$

$$= \int_0^s ds' e^{i\tau(\phi_i(s')-\phi_j(s'))} \langle \psi_\tau^k(0)|\hat{A}^\dagger(s')\hat{K}(s')\hat{A}(s')|\psi_\tau^0(0)\rangle \delta_{ki}\delta_{j0}. \tag{6.234}$$

Here, according to the definition of the projection operator, we use the relations given by

$$\hat{P}_l(0)|\psi_\tau^m(0)\rangle = |\psi_\tau^m(0)\rangle \delta_{lm}. \tag{6.235}$$

Moreover, by using Eqs. (6.157) and (6.213), Eq. (6.234) is transformed as

$$\langle \psi_\tau^k(0)|\hat{F}_{ij}(s)|\psi_\tau^0(0)\rangle$$

$$= i \int_0^s ds' e^{i\tau(\phi_i(s')-\phi_j(s'))}$$

$$\times \left\langle \psi_\tau^k(0) \middle| \hat{A}^\dagger(s') \sum_{l=0}^{N_s-1} \frac{d\hat{P}_l(s')}{ds'} \hat{P}_l(s')\hat{A}(s') \middle| \psi_\tau^0(0) \right\rangle \delta_{ki}\delta_{j0} \tag{6.236}$$

$$= i \int_0^s ds' e^{i\tau(\phi_i(s')-\phi_j(s'))}$$

$$\times \left\langle \psi_\tau^k(0) \middle| \hat{A}^\dagger(s') \sum_{l=0}^{N_s-1} \frac{d\hat{P}_l(s')}{ds'} \hat{A}(s') \middle| \psi_\tau^0(0) \right\rangle \delta_{ki}\delta_{j0}\delta_{l0} \tag{6.237}$$

$$= i \int_0^s ds' e^{i\tau(\phi_i(s')-\phi_j(s'))}$$

$$\times \left\langle \psi_\tau^k(0) \left| \hat{A}^\dagger(s') \frac{d\hat{P}_0(s')}{ds'} \hat{A}(s') \right| \psi_\tau^0(0) \right\rangle \delta_{ki}\delta_{j0} \tag{6.238}$$

$$= i \int_0^s ds' \mathrm{e}^{i\tau(\phi_i(s') - \phi_j(s'))} \alpha_{k0}(s), \tag{6.239}$$

where a new variable was introduced by

$$\alpha_{k0}(s) := \left\langle \psi_\tau^k(0) \left| \hat{A}^\dagger(s) \frac{d\hat{P}_0(s)}{ds} \hat{A}(s) \right| \psi_\tau^0(0) \right\rangle. \tag{6.240}$$

According to Eq. (6.219), we obtain

$$\hat{A}(s)|\psi_\tau^i(0)\rangle = \hat{A}(s)\hat{V}_\tau^{A\dagger}(s)\hat{A}^\dagger(s)|\psi_\tau^i(s)\rangle \tag{6.241}$$

$$= \sum_{l=0}^{N_s-1} \mathrm{e}^{-i\tau\phi_l(s)} \hat{A}(s)\hat{P}_l(0)\hat{A}^\dagger(s)|\psi_\tau^i(s)\rangle \tag{6.242}$$

$$= \sum_{l=0}^{N_s-1} \mathrm{e}^{-i\tau\phi_l(s)} \hat{P}_l(s)\hat{A}(s)\hat{A}^\dagger(s)|\psi_\tau^i(s)\rangle \tag{6.243}$$

$$= \sum_{l=0}^{N_s-1} \mathrm{e}^{-i\tau\phi_l(s)} \hat{P}_l(s)|\psi_\tau^i(s)\rangle \tag{6.244}$$

$$= \mathrm{e}^{-i\tau\phi_i(s)} \hat{P}_i(s)|\psi_\tau^i(s)\rangle, \tag{6.245}$$

by using Eqs. (6.187) and (6.213). From the relation, we can obtain

$$\alpha_{k0}(s) = \left\langle \psi_\tau^k(s) \left| \mathrm{e}^{i\tau\phi_k(s)} \hat{P}_k^\dagger(s) \frac{d\hat{P}_0(s)}{ds} \hat{P}_0(s) \mathrm{e}^{-i\tau\phi_0(s)} \right| \psi_\tau^0(s) \right\rangle \tag{6.246}$$

$$=: \left\langle \bar{\psi}_\tau^k(s) \left| \hat{P}_k^\dagger(s) \frac{d\hat{P}_0(s)}{ds} \hat{P}_0(s) \right| \bar{\psi}_\tau^0(s) \right\rangle. \tag{6.247}$$

Here, we defined the eigenstates including the phase factor by

$$|\bar{\psi}_\tau^i(s)\rangle := \mathrm{e}^{-i\tau\phi_i(s)}|\psi_\tau^i(s)\rangle. \tag{6.248}$$

To rewrite this equation, we derive the relation between the Hamiltonian and the projection operator. According to Eq. (6.157), we get

$$\hat{P}_l(s)\hat{K}(s)\hat{P}_m(s) = i \sum_{i=0}^{N_s-1} \hat{P}_l(s) \frac{d\hat{P}_i(s)}{ds} \hat{P}_i(s)\hat{P}_m(s) \tag{6.249}$$

$$= i\hat{P}_l(s)\frac{d\hat{P}_m(s)}{ds}\hat{P}_m(s). \tag{6.250}$$

Furthermore, the following relation is satisfied as

$$\hat{P}_l(s)\hat{K}^\dagger(s)\hat{P}_m(s) = -i\sum_{i=0}^{N_s-1}\hat{P}_l(s)\hat{P}_i(s)\frac{d\hat{P}_i(s)}{ds}\hat{P}_m(s) \tag{6.251}$$

$$= -i\hat{P}_l(s)\frac{d\hat{P}_l(s)}{ds}\hat{P}_m(s). \tag{6.252}$$

Here, since $\hat{K}(s)$ is a Hermitian operator, Eqs. (6.250) and (6.252) should be the same, that is,

$$\hat{P}_l(s)\frac{d\hat{P}_m(s)}{ds}\hat{P}_m(s) = -\hat{P}_l(s)\frac{d\hat{P}_l(s)}{ds}\hat{P}_m(s). \tag{6.253}$$

Next, the Hamiltonian given by Eq. (6.129) is the same as

$$\hat{\mathscr{H}}(s) = \sum_{i=0}^{N_s-1}\lambda_i(s)\hat{P}_i^2(s). \tag{6.254}$$

By taking the derivative on both sides of Eq. (6.254) with respect to s, we obtain

$$\frac{d\hat{\mathscr{H}}(s)}{ds} = \sum_{i=0}^{N_s-1}\frac{d\lambda_i(s)}{ds}\hat{P}_i^2(s) + \sum_{i=0}^{N_s-1}\lambda_i(s)\frac{d\hat{P}_i(s)}{ds}\hat{P}_i(s) + \sum_{i=0}^{N_s-1}\lambda_i(s)\hat{P}_i(s)\frac{d\hat{P}_i(s)}{ds}. \tag{6.255}$$

Thus, the following relation is satisfied:

$$\hat{P}_l(s)\frac{d\hat{\mathscr{H}}(s)}{ds}\hat{P}_m(s)$$

$$= \sum_{i=0}^{N_s-1}\frac{d\lambda_i(s)}{ds}\delta_{li}\delta_{im}\hat{P}_l(s)\hat{P}_m(s) + \sum_{i=0}^{N_s-1}\lambda_i(s)\hat{P}_l(s)\frac{d\hat{P}_i(s)}{ds}\delta_{im}\hat{P}_m(s)$$

$$+ \sum_{i=0}^{N_s-1}\lambda_i(s)\hat{P}_l(s)\delta_{li}\frac{d\hat{P}_i(s)}{ds}\hat{P}_m(s) \tag{6.256}$$

$$= \frac{d\lambda_l(s)}{ds}\delta_{lm}\hat{P}_l(s)\hat{P}_m(s)$$

$$+\lambda_m(s)\hat{P}_l(s)\frac{d\hat{P}_m(s)}{ds}\hat{P}_m(s)+\lambda_l(s)\hat{P}_l(s)\frac{d\hat{P}_l(s)}{ds}\hat{P}_m(s) \tag{6.257}$$

$$=\frac{d\lambda_l(s)}{ds}\delta_{lm}\hat{P}_l(s)\hat{P}_m(s)+(\lambda_m(s)-\lambda_l(s))\hat{P}_l(s)\frac{d\hat{P}_m(s)}{ds}\hat{P}_m(s), \tag{6.258}$$

where we used Eq. (6.253) in the final line. From this equation, the relations:

$$\hat{P}_k(s)\frac{d\hat{P}_0(s)}{ds}\hat{P}_0(s)=-\frac{1}{\lambda_k(s)-\lambda_0(s)}\hat{P}_k(s)\frac{d\hat{\mathscr{H}}(s)}{ds}\hat{P}_0(s), \tag{6.259}$$

are obtained, and $\alpha_{k0}(s)$ can be written as

$$\alpha_{k0}(s) = -\frac{1}{\lambda_k(s)-\lambda_0(s)}\left\langle \bar{\psi}_\tau^k(s)\left|\hat{P}_k(s)\frac{d\hat{\mathscr{H}}(s)}{ds}\hat{P}_0(s)\right|\bar{\psi}_\tau^0(s)\right\rangle \tag{6.260}$$

$$= -\frac{1}{\Delta_{k0}(s)}\left\langle \bar{\psi}_\tau^k(s)\left|\frac{d\hat{\mathscr{H}}(s)}{ds}\right|\bar{\psi}_\tau^0(s)\right\rangle, \tag{6.261}$$

where the energy gap was defined as

$$\Delta_{ij}(s):=\lambda_i(s)-\lambda_j(s). \tag{6.262}$$

Then, we obtain

$$\langle\psi_\tau^k(0)|\hat{F}_{ij}(s)|\psi_\tau^0(0)\rangle = -i\int_0^s ds' e^{i\tau(\phi_i(s')-\phi_j(s'))}$$
$$\times\frac{1}{\Delta_{k0}(s')}\left\langle \bar{\psi}_\tau^k(s')\left|\frac{d\hat{\mathscr{H}}(s')}{ds'}\right|\bar{\psi}_\tau^0(s')\right\rangle\delta_{ki}\delta_{j0}. \tag{6.263}$$

Let us consider the upper bound of η. In order to take the integral of Eq. (6.263), we assume that $\Delta_{k0}(s')$ and $d\hat{\mathscr{H}}(s')/ds'$ do not depend on time. In this case, the following relation is satisfied:

$$\phi_i(s')-\phi_j(s')=\Delta_{ij}(s')s'. \tag{6.264}$$

We take the integral in Eq. (6.263) and obtain the following:

$$\langle\psi_\tau^k(0)|\hat{F}_{ij}(s)|\psi_\tau^0(0)\rangle$$

$$=-\frac{1}{\tau\Delta_{ij}(s)\Delta_{k0}(s)}\left(e^{i\tau\Delta_{ij}(s)s}-1\right)\left\langle \bar{\psi}_\tau^k(s)\left|\frac{d\hat{\mathscr{H}}(s)}{ds}\right|\bar{\psi}_\tau^0(s)\right\rangle\delta_{ki}\delta_{j0}. \tag{6.265}$$

Thus, $p_{0\to k}$ can be written as

$$p_{0\to k} = \left|\left\langle \psi_\tau^k(0) \left| \sum_{0\le i,j\le N_s-1} \hat{F}_{ij}(s) \right| \psi_\tau^0(0) \right\rangle\right|^2 \tag{6.266}$$

$$= \frac{1}{\tau^2}\frac{2}{\Delta_{k0}(s)^4}\left|\left\langle \bar{\psi}_\tau^k(s) \left| \frac{d\hat{\mathscr{H}}(s)}{ds} \right| \bar{\psi}_\tau^0(s) \right\rangle\right|^2 (1-\cos\tau\Delta_{k0}(s)s). \tag{6.267}$$

From this equation, when $\Delta_{k0}(s)$ and $d\hat{\mathscr{H}}(s)/ds$ change slow enough, we can understand the upper bound of $p_{0\to k}$ as follows:

$$p_{0\to k} \le \frac{1}{\tau^2}\frac{1}{\min\Delta_{k0}(s)^4}\max\left|\left\langle \bar{\psi}_\tau^k(s) \left| \frac{d\hat{\mathscr{H}}(s)}{ds} \right| \bar{\psi}_\tau^0(s) \right\rangle\right|^2. \tag{6.268}$$

From this equation, η is calculated as

$$\eta \le \sum_{0<k\le N_s-1} \frac{1}{\tau^2}\frac{1}{\min\Delta_{k0}(s)^4}\max\left|\left\langle \bar{\psi}_\tau^k(s) \left| \frac{d\hat{\mathscr{H}}(s)}{ds} \right| \bar{\psi}_\tau^0(s) \right\rangle\right|^2. \tag{6.269}$$

Here, since $\Delta_{k0}(s)$ is the excitation gap between the ground state and the kth state, $\Delta_{k0}(s)$ has the minimum value at $k = 1$, that is, the first excited state. Thus, we can estimate η as follows:

$$\eta \lesssim \frac{1}{\tau^2}\frac{1}{\min\Delta_{10}(s)^4}\max\left|\left\langle \bar{\psi}_\tau^1(s) \left| \frac{d\hat{\mathscr{H}}(s)}{ds} \right| \bar{\psi}_\tau^0(s) \right\rangle\right|^2. \tag{6.270}$$

From this equation, the condition of the adiabatic approximation, i.e., $\eta \ll 1$, becomes

$$\frac{1}{\min\Delta_{10}(s)^2}\max\left|\left\langle \bar{\psi}_\tau^1(s) \left| \frac{d\hat{\mathscr{H}}(s)}{ds} \right| \bar{\psi}_\tau^0(s) \right\rangle\right| \ll \tau. \tag{6.271}$$

That is, the adiabatic approximation is satisfied when the time evolution is slow enough as given by Eq. (6.271). Here, as the excitation gap of the first excited state $\Delta_{10}(s)$ becomes small, the left-hand side of Eq. (6.271) decreases. This fact indicates that the time evolution with longer time should be performed when the excitation gap is small. Moreover, for real time t, the adiabatic condition where the adiabatic approximation is satisfied is written as

$$\frac{1}{\min\Delta_{10}(t)^2}\max\left|\left\langle \bar{\psi}_1(t) \left| \frac{d\hat{\mathscr{H}}(t)}{dt} \right| \bar{\psi}_0(t) \right\rangle\right| \ll 1. \tag{6.272}$$

In this section, we considered the quantum adiabatic condition for systems with non-degenerated eigenenergies, i.e., systems which obey the no-level crossing theorem (Lax, 2007). Recently, the quantum adiabatic theorem for systems with degenerate energy spectra was studied by Rigolin and Ortiz (2012) using the technique developed by Rigolin et al. (2008) and Rigolin and Ortiz (2010).

6.9 Convergence Theorem for Quantum Annealing

In this section, the convergence theorem for quantum annealing is shown. Morita and Nishimori (2006, 2007, 2008) proved that when the transverse field $\Gamma(t)$ decreases by the schedule

$$\Gamma(t) \geq t^{-c/N}, \tag{6.273}$$

the convergence to the ground state at $\Gamma(t) = 0$ in the limit of $t \to \infty$ is guaranteed. Here, N is the number of spins, and c is a constant whose order is of $\mathscr{O}(N^0)$. The organization of this section is as follows. In Section 6.9.1, we show the convergence theorem for quantum annealing by using the Schrödinger equation. Section 6.9.2 is devoted to the derivation of the convergence theorem of quantum annealing via quantum-classical mapping. In Section 6.9.3, we show the case when the path-integral Monte Carlo method is used. Finally, we explain the convergence theorem of quantum annealing by using Green's function Monte Carlo method in Section 6.9.4.

6.9.1 Convergence theorem for quantum annealing by the Schrödinger equation

Consider the time evolution by the Schrödinger equation, which was explained in Section 6.2. The convergence theorem is given by

$$\Gamma(t) \geq a(\delta t - c)^{-\frac{1}{2N-1}}, \tag{6.274}$$

where a, b, c are constants with $\mathscr{O}(N^0)$ and δ is an infinitesimal value. Furthermore, N is the number of spins.

Adiabatic condition

Let us consider the time evolution of the system satisfying Eq. (6.1) by the Schrödinger equation. If the time evolution is adiabatic at each instantaneous time, the ground state of $\hat{\mathscr{H}}_0$ can be obtained definitely when the ground state of $\hat{\mathscr{H}}(0)$ is the initial state of the dynamics. This is the quantum adiabatic theorem introduced in Section 6.8.2. In order to realize that, the time evolution should obey the dynamics satisfying the adiabatic condition described in Section 6.8.3 where the adiabatic theorem is approximately satisfied. Let δ be a positive infinitesimal value. The time evolution guarantees the adiabatic condition when the following relation is satisfied:

$$\frac{\max \left| \left\langle \psi_1(t) \left| \frac{d\mathcal{H}(t)}{dt} \right| \psi_0(t) \right\rangle \right|}{\min \Delta(t)^2} = \delta \ll 1, \tag{6.275}$$

where $|\psi_0(t)\rangle$ and $|\psi_1(t)\rangle$ are the ground state and the first excited state at instantaneous time t respectively. In addition, the energy gap $\Delta(t)$ is defined by

$$\Delta(t) = \lambda_1(t) - \lambda_0(t), \tag{6.276}$$

where $\lambda_0(t)$ and $\lambda_1(t)$ are the energies at the ground state and the first-excited state, respectively.

Hopf's Theorem

In order to prove the convergence theorem, we use the Hopf theorem. Let $\hat{M}$ be the $N_s \times N_s$ positive matrix ($M_{ij} > 0$). The following inequality is satisfied:

$$|\varepsilon_i| \leq \frac{\kappa - 1}{\kappa + 1} \varepsilon_0, \tag{6.277}$$

where ε_i ($i = 0, 1, \cdots, N_s - 1$) are eigenvalues of $\hat{M}$ and ε_0 is the maximum eigenvalue. Note that the maximum eigenvalue does not degenerate according to the Perron–Frobenius theorem. Furthermore, κ in Eq. (6.277) is defined by

$$\kappa := \max_{i,j,k} \left(\frac{M_{ik}}{M_{jk}} \right). \tag{6.278}$$

Proof of the convergence theorem

Here we consider the system given by Eq. (6.1). We prove that when the transverse field decreases with the schedule given by Eq. (6.274), the time evolution satisfies the adiabatic condition (see Eq. (6.275)). Thus, the matrix elements of the time differential of the Hamiltonian $d\hat{\mathcal{H}}(t)/dt$ and the energy gap $\Delta(t)$ should be estimated. First we consider the matrix elements of $d\hat{\mathcal{H}}(t)/dt$. Since only the transverse field depends on time t in the Hamiltonian $\hat{\mathcal{H}}(t)$, all matrix elements for $j = 0, \cdots, N_s - 1$ where $N_s = 2^N$ are

$$\left| \left\langle \psi_j(t) \left| \frac{d\hat{\mathcal{H}}(t)}{dt} \right| \psi_0(t) \right\rangle \right| = -\left\langle \psi_j(t) \left| \sum_{i=1}^{N} \hat{\sigma}_i^x \frac{d\Gamma(t)}{dt} \right| \psi_0(t) \right\rangle \tag{6.279}$$

$$\leq -N \frac{d\Gamma(t)}{dt}. \tag{6.280}$$

Here, we used the relation:

$$\langle \psi_j(t) | \hat{\sigma}_i^x | \psi_0(t) \rangle \leq 1. \tag{6.281}$$

Next, to estimate the energy gap $\Delta(t)$, we use the Hopf theorem. The $N_s \times N_s$ square matrix $\hat{M}$ is introduced as

$$\hat{M} := \left(E_+\hat{I} - \mathscr{H}(t)\right)^N, \tag{6.282}$$

where $\hat{I}$ is the identity matrix and E_+ is a constant so that the following relation is satisfied:

$$E_+ \geq E_0^{\max} + \Gamma_0, \tag{6.283}$$

when $\Gamma(0) = \Gamma_0$. Here $E_0^{\max}$ is the maximum eigenvalue of $\hat{\mathscr{H}}_0$. We estimate the minimum element of the matrix $\hat{M}$. The lower bound of $E_+\hat{I} - \mathscr{H}(t)$ can be obtained as

$$E_+\hat{I} - \mathscr{H}(t) = E_+\hat{I} - \hat{\mathscr{H}}_0 + \Gamma(t)\sum_{i=1}^{N}\hat{\sigma}_i^x \tag{6.284}$$

$$\geq \underbrace{(E_0^{\max} + \Gamma_0)\hat{I} - \hat{\mathscr{H}}_0}_{\text{diagonal elements}} + \underbrace{\Gamma(t)\sum_{i=1}^{N}\hat{\sigma}_i^x}_{\text{off-diagonal elements}}. \tag{6.285}$$

The first, second, and third terms in the right-hand side of this equation are the diagonal elements, and the fourth term is the off-diagonal elements. The diagonal elements should be positive by the definition of $E_0^{\max}$. Since $(E_+\hat{I} - \mathscr{H}(t))_{\sigma\sigma'} = \Gamma(t)(>0)$ when σ' is generated by the single-spin flip, $(E_+\hat{I} - \mathscr{H}(t))^N_{\sigma\sigma'} > 0$ when $\sigma \neq \sigma'$ for arbitrary σ, σ'. Thus, all elements of $\hat{M}$ should be positive. In addition, since $\Gamma_0 > \Gamma(t)$ for $t > 0$, the diagonal elements of $\hat{M}$ are larger than the nonzero off-diagonal elements of $\hat{M}$. Thus, the minimum value of the elements of $\hat{M}$ appears in an off-diagonal element and is given by between two states where one of the states is given by the reversal operation to the other. Let $|\phi\rangle$ and $|\phi'\rangle$ be the completely opposite states, i.e., $|\phi'\rangle = \prod_{i=1}^{N}\hat{\sigma}_i^x|\phi\rangle$. The minimum elements of $\hat{M}$ is given by

$$\begin{aligned}\min_{i,j} M_{ij} &= \left\langle \phi' \left| \left[\Gamma(t)\sum_{i=1}^{N}\hat{\sigma}_i^x\right]^N \right| \phi \right\rangle \\ &= N!\Gamma(t)^N \left\langle \phi' \left| \prod_{i=1}^{N}\hat{\sigma}_i^x \right| \phi \right\rangle \\ &= N!\Gamma(t)^N.\end{aligned} \tag{6.286}$$

Next the maximum element of $\hat{M}$ is estimated. Let $E_0^{\min}$ be the minimum eigenvalue of $\hat{\mathscr{H}}_0$. It is only necessary to estimate the upper bound of Eq. (6.284), and the maximum elements of $\hat{M}$ can be obtained as

$$\max_{i,j} M_{ij} \le \left(E_+ - E_0^{\min} + N\Gamma_0\right)^N . \tag{6.287}$$

Here, we used the relation $\Gamma_0 > \Gamma(t)$ for $t > 0$. From this, since the maximum and minimum elements of $\hat{M}$ were estimated, the upper bound of κ defined by Eq. (6.278) can be obtained as

$$\kappa \le \frac{\left(E_+ - E_0^{\min} + N\Gamma_0\right)^N}{N!\Gamma(t)^N}. \tag{6.288}$$

Here, by using the Stirling formula which is valid for large N:

$$N! \simeq \sqrt{2\pi N}\frac{N^N}{\mathrm{e}^N}, \tag{6.289}$$

the following relation is obtained:

$$\kappa \le \frac{1}{\sqrt{2\pi N}}\left[\frac{E_+ - E_0^{\min} + N\Gamma_0}{N\Gamma(t)}\right]^N \mathrm{e}^N. \tag{6.290}$$

Thus, κ increases exponentially with N. Let $\lambda_i(t)$ $(i = 0, \cdots, N_s - 1)$ be the eigenvalues of $\mathscr{H}(t)$ and $\lambda_0(t)$ be the minimum eigenvalue of $\mathscr{H}(t)$, that is, the energy of the ground state. The Hopf theorem can be rewritten as

$$\left[E_+ - \lambda_i(t)\right]^N \le \frac{\kappa - 1}{\kappa + 1}\left[E_+ - \lambda_0(t)\right]^N . \tag{6.291}$$

Here, by using the relation:

$$\left(\frac{\kappa - 1}{\kappa + 1}\right)^{1/N} \le 1 - \frac{2}{N(\kappa + 1)}, \tag{6.292}$$

when $\kappa \ge 1$ and $N \ge 1$, the energy gaps between the ground state and the jth state is given by

$$\Delta_j(t) := \lambda_j(t) - \lambda_0(t) \tag{6.293}$$

$$\ge \frac{2}{N(\kappa + 1)}\left[E_+ - \lambda_0(t)\right] \tag{6.294}$$

$$\simeq \frac{2}{N\kappa}\left[E_+ - \lambda_0(t)\right]. \tag{6.295}$$

Here, since κ increases exponentially with N, the relation $\kappa \gg 1$ was used. Thus, the energy gaps are estimated as

$$\Delta_j(t) \geq \frac{2\sqrt{2\pi N}[E_+ - \lambda_0(t)]}{Ne^N} \left(\frac{N}{E_+ - E_0^{\min} + N\Gamma_0}\right)^N \Gamma(t)^N$$

$$=: A\Gamma(t)^N, \tag{6.296}$$

where A decreases exponentially with N.

From Eq. (6.296), by using Eqs. (6.280), (6.296), the adiabatic condition given by Eq. (6.275) is

$$-\frac{N}{A^2\Gamma(t)^{2N}}\frac{d\Gamma(t)}{dt} = \delta \ll 1. \tag{6.297}$$

By integrating both sides of the above equation, we obtain

$$\Gamma(t) = \left(\frac{2N-1}{N}A^2\right)^{-\frac{1}{2N-1}} (\delta t + c)^{-\frac{1}{2N-1}} \tag{6.298}$$

$$= a(\delta t + c)^{-\frac{1}{2N-1}}, \tag{6.299}$$

where

$$a := \left(\frac{2N-1}{N}A^2\right)^{-\frac{1}{2N-1}}. \tag{6.300}$$

Furthermore, c is a constant of integral. Thus, when $\Gamma(t)$ decreases by the schedule given by Eq. (6.298), the time evolution is adiabatic at each instantaneous time. As a result, in the limit of $t \to \infty$, the ground state of $\mathscr{H}_0$ can be obtained definitely in quantum annealing, if the initial state is prepared as the ground state at $t = 0$.

6.9.2 Correspondence between classical and quantum systems

In this subsection, we explain the fact that the convergence theorem of simulated annealing can be derived through the adiabatic conditions by using the quantum-classical mapping (Somma et al., 2007).

Quantum–classical correspondence

The Hamiltonian of classical Ising models is written as

$$\mathscr{H} = -\sum_{1\leq i,j\leq N} J_{ij}\sigma_i^z\sigma_j^z - \sum_{i=1}^{N} h_i\sigma_i^z, \quad (\sigma_i^z = \pm 1). \tag{6.301}$$

In this case, the thermal expectation value of a physical quantity $\mathscr{Q}$ can be obtained by

$$\langle \mathscr{Q} \rangle_T = \frac{\sum_{[\sigma]} \mathscr{Q} e^{-\beta \mathscr{H}}}{\sum_{[\sigma]} e^{-\beta \mathscr{H}}}, \tag{6.302}$$

where $\sum_{[\sigma]}$ represents the summation over all possible configurations. The number of all possible states is 2^N when the number of spins is N. Let $[\sigma] := [\sigma_1^z, \sigma_2^z, \cdots, \sigma_N^z]$ be a classical state of the system with N spins.

Next, we define the quantum model on the same lattice as Eq. (6.301). That is, the variables σ_i^z are transformed to the Pauli matrix $\hat{\sigma}_i^z$. The quantum Hamiltonian is written as

$$\hat{\mathscr{H}} = -\sum_{1 \le i,j \le N} J_{ij} \hat{\sigma}_i^z \hat{\sigma}_j^z - \sum_{i=1}^{N} h_i \hat{\sigma}_i^z. \tag{6.303}$$

Suppose the state depending on the temperature T is given by

$$|\psi(T)\rangle = e^{-\beta \hat{\mathscr{H}}/2} \sum_{[\sigma]} |[\sigma]\rangle. \tag{6.304}$$

As will be shown in the following, the quantum-mechanical expectation value of the physical quantity $\hat{\mathscr{Q}}$ by this state is represented by $\langle \hat{\mathscr{Q}} \rangle$ which is the same as the thermal expectation value $\langle \mathscr{Q} \rangle_T$. Here, $|[\sigma]\rangle$ is the state generated by mapping the classical state $[\sigma]$ into a quantum state, and $\hat{\mathscr{Q}}$ is the diagonal matrix with the physical quantity $\mathscr{Q}$. Thus, $\hat{\mathscr{Q}}$ and $\hat{\mathscr{H}}$ can commute, i.e., $[\hat{\mathscr{Q}}, \hat{\mathscr{H}}] = 0$. The relation $\langle \hat{\mathscr{Q}} \rangle = \langle \mathscr{Q} \rangle_T$ is shown as follows.

$$\langle \hat{\mathscr{Q}} \rangle = \frac{\langle \psi(T) | \hat{\mathscr{Q}} | \psi(T) \rangle}{\langle \psi(T) | \psi(T) \rangle} \tag{6.305}$$

$$= \frac{\sum_{[\sigma'],[\sigma]} \langle [\sigma'] | e^{-\beta \hat{\mathscr{H}}/2} \hat{\mathscr{Q}} e^{-\beta \hat{\mathscr{H}}/2} | [\sigma] \rangle}{\sum_{[\sigma'],[\sigma]} \langle [\sigma'] | e^{-\beta \hat{\mathscr{H}}} | [\sigma] \rangle} \tag{6.306}$$

$$= \frac{\sum_{[\sigma]} \langle [\sigma] | \hat{\mathscr{Q}} | [\sigma] \rangle e^{-\beta \hat{\mathscr{H}}}}{\sum_{[\sigma]} e^{-\beta \hat{\mathscr{H}}}} \tag{6.307}$$

$$= \frac{\sum_{[\sigma]} \mathscr{Q} e^{-\beta \hat{\mathscr{H}}}}{\sum_{[\sigma]} e^{-\beta \hat{\mathscr{H}}}} \tag{6.308}$$

$$= \langle \mathscr{Q} \rangle_T, \tag{6.309}$$

where $|\psi(T)\rangle$ is not normalized. From Eq. (6.304), the normalization factor of $|\psi(T)\rangle$ is $\sqrt{\mathscr{Z}(T)}$ which is the partition function of $\hat{\mathscr{H}}$ at T.

Furthermore, as will be proven in the following, the wave function defined by Eq. (6.304) is the ground state of

$$\hat{\mathcal{H}}_{\mathrm{q}}(T) := -\chi \sum_{i=1}^{N} \hat{\mathcal{H}}_{\mathrm{q}}^{i}(T), \tag{6.310}$$

$$\hat{\mathcal{H}}_{\mathrm{q}}^{i}(T) := \hat{\sigma}_i^x - \mathrm{e}^{\beta \hat{\mathcal{H}}_i}, \tag{6.311}$$

where $\hat{\mathcal{H}}_i$ represents the term including the ith site of Eq. (6.303) and is defined by

$$\hat{\mathcal{H}}_i := -\sum_{j=1}^{N} J_{ij} \hat{\sigma}_i^z \hat{\sigma}_j^z - h_i \hat{\sigma}_i^z, \tag{6.312}$$

$$\hat{\mathcal{H}} = \sum_{i=1}^{N} \hat{\mathcal{H}}_i. \tag{6.313}$$

Here, the following relation is satisfied:

$$\hat{\sigma}_i^x \sum_{[\sigma]} |[\sigma]\rangle = \sum_{[\sigma]} |[\sigma]\rangle. \tag{6.314}$$

This is because, $\sum_{[\sigma]}$ represents the summation over all possible states and $\hat{\sigma}_i^x$ is just the operation that the order of the summation is changed. In addition, since $\hat{\sigma}_i^z$ does not include $\hat{\mathcal{H}} - \hat{\mathcal{H}}_i$, $[\hat{\mathcal{H}} - \hat{\mathcal{H}}_i, \hat{\sigma}_i^z] = 0$. Moreover, anticommutation relation between $\hat{\mathcal{H}}_i$ and $\hat{\sigma}_i^x$ is satisfied, i.e., $\{\hat{\mathcal{H}}_i, \hat{\sigma}_i^x\} = 0$. From these facts, we obtain the following relation:

$$\hat{\sigma}_i^x \mathrm{e}^{-\beta \hat{\mathcal{H}}/2} \hat{\sigma}_i^x = \hat{\sigma}_i^x \mathrm{e}^{-\beta(\hat{\mathcal{H}} - \hat{\mathcal{H}}_i)/2} \mathrm{e}^{-\beta \hat{\mathcal{H}}_i/2} \hat{\sigma}_i^x \tag{6.315}$$

$$= \mathrm{e}^{-\beta(\hat{\mathcal{H}} - \hat{\mathcal{H}}_i)/2} \hat{\sigma}_i^x \mathrm{e}^{-\beta \hat{\mathcal{H}}_i/2} \hat{\sigma}_i^x \tag{6.316}$$

$$= \mathrm{e}^{-\beta(\hat{\mathcal{H}} - \hat{\mathcal{H}}_i)/2} \mathrm{e}^{\beta \hat{\mathcal{H}}_i/2} \tag{6.317}$$

$$= \mathrm{e}^{\beta \hat{\mathcal{H}}_i} \mathrm{e}^{-\beta \hat{\mathcal{H}}/2}. \tag{6.318}$$

From this equation, $\hat{\mathcal{H}}_{\mathrm{q}}^{i}(T)|\psi(T)\rangle$ can be calculated as

$$\hat{\mathcal{H}}_{\mathrm{q}}^{i}(T)|\psi(T)\rangle = \left(\hat{\sigma}_i^x - \mathrm{e}^{\beta \hat{\mathcal{H}}_i}\right) \mathrm{e}^{-\beta \hat{\mathcal{H}}/2} \sum_{[\sigma]} |[\sigma]\rangle \tag{6.319}$$

$$= \left(\hat{\sigma}_i^x \mathrm{e}^{-\beta \hat{\mathcal{H}}/2} \hat{\sigma}_i^x - \mathrm{e}^{\beta \hat{\mathcal{H}}_i} \mathrm{e}^{-\beta \hat{\mathcal{H}}/2}\right) \sum_{[\sigma]} |[\sigma]\rangle \tag{6.320}$$

$$= 0, \tag{6.321}$$

and we obtain

$$\hat{\mathscr{H}}_{\mathrm{q}}(T)|\psi(T)\rangle = 0. \tag{6.322}$$

Here, all off-diagonal elements are non-negative and the coefficients of $|\psi(T)\rangle$ are positive. Thus, according to the Perron–Frobenius theorem, it is clarified that the state $|\psi(T)\rangle$ is the ground state of $\hat{\mathscr{H}}_{\mathrm{q}}(T)$.

Next, we consider the physical meaning of $|\psi(T)\rangle$. In the quantum Hamiltonian defined by Eq. (6.310), the parameter exists, that is, there is a value of χ. This parameter is decided so that $|\hat{\mathscr{H}}_{\mathrm{q}}(T \to 0)| < \infty$ is satisfied in the low-temperature limit ($T \to 0$), and χ is chosen as

$$\chi = \mathrm{e}^{-\beta p}, \quad p = \max_{1 \le i \le N} |\hat{\mathscr{H}}_i|. \tag{6.323}$$

In fact, in the low-temperature limit, the following approximation is valid:

$$\hat{\mathscr{H}}_{\mathrm{q}}(T \to 0) \simeq \chi \sum_{i=1}^{N} \mathrm{e}^{\beta \hat{\mathscr{H}}_i} \tag{6.324}$$

$$= \sum_{i=1}^{N} \mathrm{e}^{\beta(\hat{\mathscr{H}}_i - p)}, \tag{6.325}$$

and $|\hat{\mathscr{H}}_{\mathrm{q}}(T \to 0)| < \infty$ is satisfied. Furthermore, from Eq. (6.304), $|\psi(T \to 0)\rangle$ can be written as

$$|\psi(T \to 0)\rangle \simeq |[\sigma]_{\mathrm{g}}\rangle, \tag{6.326}$$

where $[\sigma]_{\mathrm{g}}$ indicates the ground state of the classical Hamiltonian $\mathscr{H}$ defined by Eq. (6.301). Thus, $|\psi(T \to 0)\rangle$ is the quantum state corresponding to the ground state of $\mathscr{H}$. On the other hand, in the high temperature limit ($T \to \infty$),

$$\hat{\mathscr{H}}_{\mathrm{q}}(T \to \infty) \simeq -\sum_{i=1}^{N} (\hat{\sigma}_i^x - 1). \tag{6.327}$$

In this case, the quantum state $|\psi(T \to \infty)\rangle$ is obtained by the sum of all possible states which are generated by the classical state $[\sigma]$ into a quantum state, that is,

$$|\psi(T \to \infty)\rangle \simeq \sum_{[\sigma]} |[\sigma]\rangle. \tag{6.328}$$

This corresponds to the picture of the classical model at high temperatures. In this way, when we decide χ by Eq. (6.323), χ indicates the amplitude of the quantum fluctuation.

Furthermore, since χ corresponding to the quantum fluctuation increases as temperature T is increased, the thermal fluctuation can be mapped to the quantum fluctuation by the aforementioned operation.

From this consideration, the thermal expectation value of the classical model given by $\mathscr{H}$ can be obtained by the ground state $|\psi(T)\rangle$ of the quantum model represented by $\hat{\mathscr{H}}_{\mathrm{q}}(T)$. This is the quantum–classical correspondence.

Convergence theorem of simulated annealing by the adiabatic condition

Let us consider the case when simulated annealing is performed using the quantum Hamiltonian $\hat{\mathscr{H}}_{\mathrm{q}}(T)$. The temperature $T(t)$ depends on the time t and decreases gradually. Here we derive the convergence theorem of simulated annealing from the adiabatic condition of $\hat{\mathscr{H}}_{\mathrm{q}}(T(t))$. From Eq. (6.275), the adiabatic condition of this case can be written as

$$\frac{\max\left|\left\langle \psi_1(T(t)) \left| \frac{d\hat{\mathscr{H}}_{\mathrm{q}}(T(t))}{dT(t)} \frac{1}{\sqrt{\mathscr{Z}(T(t))}} \right| \psi(T(t)) \right\rangle\right|}{\min \Delta(T(t))^2} \left|\frac{dT(t)}{dt}\right| = \delta \ll 1, \quad (6.329)$$

where $|\psi_1(T(t))\rangle$ is the first excited state.

First we consider the lower bound of the energy gap $\Delta(T(t))$. The coefficient χ in $\hat{\mathscr{H}}_{\mathrm{q}}(T(t))$ corresponds to the quantum fluctuation, that is, the transverse field $\Gamma(t)$. Thus, from Eq. (6.296), the energy gap between the first excited state and the ground state $\Delta(T(t))$ is given by

$$\Delta(T(t)) \geq A\mathrm{e}^{-\beta pN}, \quad (6.330)$$

where since the ground state energy of $\hat{\mathscr{H}}_{\mathrm{q}}(T(t))$ is zero from Eq. (6.322), $\lambda_0(t) = 0$, and A is the coefficient that decreases exponentially with N. Next, from the temperature derivative of Eq. (6.322), we can obtain

$$\left[\frac{d}{dT(t)}\hat{\mathscr{H}}_{\mathrm{q}}(T(t))\right]|\psi(T(t))\rangle = -\hat{\mathscr{H}}_{\mathrm{q}}(T(t))\frac{d}{dT(t)}|\psi(T(t))\rangle \quad (6.331)$$

$$= -\hat{\mathscr{H}}_{\mathrm{q}}(T(t))\left[\frac{1}{2k_B T(t)^2}\mathscr{H}\right]|\psi(T(t))\rangle. \quad (6.332)$$

Furthermore, since the ground state energy of $\hat{\mathscr{H}}_{\mathrm{q}}(T(t))$ is zero, the eigenenergy of the first excited state $|\psi_1(T(t))\rangle$ is given by

$$\hat{\mathscr{H}}_{\mathrm{q}}(T(t))|\psi_1(T(t))\rangle = \Delta(T(t))|\psi_1(T(t))\rangle. \quad (6.333)$$

Thus, we can represent

$$\left\langle \psi_1(T(t)) \left| \frac{d\hat{\mathscr{H}}_{\mathrm{q}}(T(t))}{dT(t)} \right| \psi(T(t)) \right\rangle$$

$$= -\frac{\Delta(T(t)) \left\langle \psi_1(T(t)) \left| \hat{\mathscr{H}} \right| \psi(T(t)) \right\rangle}{2k_B T(t)^2}. \tag{6.334}$$

In addition, since $\hat{\mathscr{H}} = \sum_i \hat{\mathscr{H}}_i$ and $|\hat{\mathscr{H}}| \leq pN$ because of the definition of p (Eq. (6.323)), the following relation is satisfied:

$$\left| \left\langle \psi_1(T(t)) \left| \hat{\mathscr{H}} \right| \psi(T(t)) \right\rangle \right| \leq pN\sqrt{\mathscr{Z}(T(t))}, \tag{6.335}$$

where $\sqrt{\mathscr{Z}(T(t))}$ is the normalization factor of $|\psi(T(t))\rangle$.

By using these relations, the adiabatic condition of $\hat{\mathscr{H}}_{\mathrm{q}}(T(t))$ is written as

$$\frac{pN}{2k_B T(t)^2 \mathrm{e}^{-\beta pN}} \left| \frac{dT(t)}{dt} \right| = \delta' \ll 1. \tag{6.336}$$

By integrating this equation,

$$T(t) = \frac{pN}{\ln(\alpha t + 1)}. \tag{6.337}$$

When the temperature decreases with the schedule, $\hat{\mathscr{H}}_{\mathrm{q}}(T(t))$ evolves in keeping with the adiabatic approximation. Note that the schedule is the same as the convergence theorem of simulated annealing shown in Section 3.4. Thus, the convergence theorem of simulated annealing can be derived by the adiabatic condition of the corresponding quantum model.

6.9.3 Convergence theorem for quantum annealing by the path-integral Monte Carlo method

When we adopt the path-integral Monte Carlo method as the time evolution, the convergence theorem is given by

$$\Gamma(t) \geq \frac{P}{\beta}(t+2)^{-2/\Delta}, \tag{6.338}$$

where Δ is a positive constant which is proportional to the number of spins N. Here P and β are positive constants. Equation (6.338) was proved by Morita and Nishimori (2006). Let us prove this theorem in this subsection. In the time evolution of the path-integral Monte Carlo method, states are generated by the inhomogeneous Markov chain explained in Section 3.4.

Thus, in order to derive the convergence theorem for quantum annealing, we should prove that the inhomogeneous Markov chain is strong ergodicity in the path-integral Monte Carlo method.

Acceptance probability in quantum annealing

Herein we consider the Ising model in a transverse field. In the path-integral Monte Carlo method, the Hamiltonian given by Eq. (6.25) is used. The partition function of the Hamiltonian is expressed as

$$\mathscr{Z}(t) = \sum_{x \in \mathscr{S}} \exp\left[-\frac{E_0(x)}{T_0} - \frac{E_1(x)}{T_1(t)}\right], \tag{6.339}$$

where x is a state of all spin variables $\{\sigma_{i,k}^z\}$, $\mathscr{S}$ is the set of the all states, and

$$E_0(x) := -\sum_{1 \le i,j \le N} \sum_{k=1}^{P} \frac{J_{ij}}{P} \sigma_{i,k}^z \sigma_{j,k}^z - \sum_{i=1}^{N} \sum_{k=1}^{P} \frac{h_i}{P} \sigma_{i,k}^z, \tag{6.340}$$

$$E_1(x) := -\sum_{i-1}^{N} \sum_{k-1}^{P} \sigma_{i,k}^z \sigma_{i,k+1}^z, \tag{6.341}$$

$$T_0 := T, \tag{6.342}$$

$$T_1(t) := \frac{1}{2} \ln \coth\left[\frac{\beta \Gamma(t)}{P}\right]. \tag{6.343}$$

In the inhomogeneous Markov chain, the transition probability from the state x to the state y at time t is given by $G(y,x;t)$ which is defined by Eq. (3.91). In the path-integral Monte Carlo method, the acceptance probability $A(y,x;t)$ in $G(y,x;t)$ at time t is defined by

$$A(y,x;t) = f\left[\frac{g(y;t)}{g(x;t)}\right], \tag{6.344}$$

$$g(x;t) = \frac{1}{\mathscr{Z}(t)} \exp\left[-\frac{E_0(x)}{T_0} - \frac{E_1(x)}{T_1(t)}\right], \tag{6.345}$$

where $g(x;t)$ is the equilibrium Boltzmann distribution at the instantaneous temperature $T_1(t)$. $f(u)$ in this equation is defined for $u \geq 0$ and is given by Eqs. (3.110) and (3.111). Here, $f(u)$ is a monotonically increasing function and is defined such that the steady distribution generated by the transition probability $G(y,x;t)$ is $g(x;t)$. Thus, for a fixed time t,

$$g(y;t) = \sum_{x \in \mathscr{S}} G(y,x;t) g(x;t) \tag{6.346}$$

is satisfied. Quantum annealing based on the path-integral Monte Carlo method is performed by the inhomogeneous Markov chain, where the transition probability given by Eq. (3.91) with the acceptance probability [see Eq. (6.344)].

Confirmation for the ergodicity

The sufficient condition for the convergence of quantum annealing by using the path-integral Monte Carlo method is given by Eq. (6.338). Under the condition, since the inhomogeneous Markov chain satisfies the strong ergodicity, the probability distribution converges to the unique probability distribution that is proportional to $\exp[-E_0(x)/T_0]$.

Definitions of states and quantities

To prove the convergence theorem, we define the set of states. Let $\mathscr{S}_x$ be the set of the states that can be generated by the single step from the state x. $\mathscr{S}_x$ is written as

$$\mathscr{S}_x = \{y | y \in \mathscr{S}, P(y,x) > 0\}, \tag{6.347}$$

where $\mathscr{S}$ is the set of all the states. Let $\mathscr{S}_{\rm m}$ be the set of x such that $E_1(x) \geq E_1(y)$ for all states $y \in \mathscr{S}_x$. $\mathscr{S}_{\rm m}$ is written as

$$\mathscr{S}_{\rm m} = \{x | x \in \mathscr{S}, \forall y \in \mathscr{S}_x, E_1(x) \geq E_1(y)\}. \tag{6.348}$$

This is called the maximum state set of $E_1(x)$.

Next, we define some quantities. Let $d(y,x)$ be the minimum steps which are needed to transition from the state x to the state y. Let $\mathscr{S} \backslash \mathscr{S}_{\rm m}$ be the set obtained by removing $\mathscr{S}_{\rm m}$ from $\mathscr{S}$. $D(x)$ is the maximum value of $d(y,x)$ for the states $x \in \mathscr{S} \backslash \mathscr{S}_{\rm m}$. In other words, $D(x)$ is the necessary steps that are required to transition from the state x, except the maximum state set of $E_1(x)$, to an arbitrary state. Let R be the minimum value of $D(x)$, that is,

$$R = \min_{x \in \mathscr{S} \backslash \mathscr{S}_{\rm m}} \left\{ \max_{y \in \mathscr{S}} [d(y,x)] \right\}. \tag{6.349}$$

The state x^* is defined by

$$x^* = \arg \min_{x \in \mathscr{S} \backslash \mathscr{S}_{\rm m}} \left\{ \max_{y \in \mathscr{S}} [d(y,x)] \right\}. \tag{6.350}$$

Arbitrary states can be generated from the states $x^* \in \mathscr{S} \backslash \mathscr{S}_{\rm m}$ by at most R steps. Here, R increases with N linearly. In other words, the states $x^* \in \mathscr{S} \backslash \mathscr{S}_{\rm m}$ can be generated from an arbitrary state by at most R steps. R and x^* are used when the ergodic coefficient is estimated.

In order to represent the changes of $E_0(x)$ and $E_1(x)$ in a single step, L_0 and L_1 are defined by

$$L_0 := \max\{|E_0(y) - E_0(x)| \mid x, y \in \mathscr{S}, P(y,x) > 0\}, \tag{6.351}$$

$$L_1 := \max\{|E_1(y) - E_1(x)| \mid x, y \in \mathscr{S}, P(y,x) > 0\}. \tag{6.352}$$

The changes of $E_0(x)$ and $E_1(x)$ in the single step are smaller than or equal to L_0 and L_1, respectively. The non-zero minimum value of the generation probability $P(y,x)$ in $G(y,x;t)$ is defined by

$$w := \min\{P(y,x) \mid x, y \in \mathscr{S}, P(y,x) > 0\}. \tag{6.353}$$

Lower bound of the transition probability

We find the lower bound of the transition probability $G(y,x;t)$, which will be used in the proof of the convergence theorem. The lower bound of $G(y,x;t)$ is given by

- For $x, y \in \mathscr{S}$, where $P(y,x) > 0$,

$$G(y,x;t) \geq wf(1)\exp\left[-\frac{L_0}{T_0} - \frac{L_1}{T_1(t)}\right], \quad (t > 0). \tag{6.354}$$

- For $x \in \mathscr{S} \backslash \mathscr{S}_{\mathrm{m}}$,

$$G(x,x;t) \geq wf(1)\exp\left[-\frac{L_0}{T_0} - \frac{L_1}{T_1(t)}\right], \quad (t \geq t_1). \tag{6.355}$$

Here the initial time is set to $t = 0$, and t_1 will be introduced later.

Let us first consider Eq. (6.354). When $x \neq y$, the transition probability is written as

$$G(y,x;t) = P(y,x)A(y,x;t) \tag{6.356}$$

$$= P(y,x)f\left(\exp\left[-\frac{E_0(y) - E_0(x)}{T_0} - \frac{E_1(y) - E_1(x)}{T_1(t)}\right]\right) \tag{6.357}$$

$$= P(y,x)f\left(\mathrm{e}^{-a}\right), \tag{6.358}$$

where a is defined by

$$a := \frac{E_0(y) - E_0(x)}{T_0} + \frac{E_1(y) - E_1(x)}{T_1(t)}. \tag{6.359}$$

For $a \geq 0$, $G(y,x;t)$ becomes

$$G(y,x;t) \geq wf\left(\mathrm{e}^{-a}\right) \tag{6.360}$$

$$= wf\left(\mathrm{e}^{a}\right)\mathrm{e}^{-a} \tag{6.361}$$

$$\geq wf(1)\mathrm{e}^{-a} \tag{6.362}$$

$$= wf(1)\exp\left[-\frac{E_0(y)-E_0(x)}{T_0} - \frac{E_1(y)-E_1(x)}{T_1(t)}\right] \tag{6.363}$$

$$\geq wf(1)\exp\left[-\frac{L_0}{T_0} - \frac{L_1}{T_1(t)}\right]. \tag{6.364}$$

Here, we used Eqs. (6.351)–(6.353) and the property of $f(u)$ given by

$$f(u) = uf(1/u) \leq u. \tag{6.365}$$

On the other hand, for $a < 0$, $G(y,x;t)$ becomes

$$G(y,x;t) \geq wf\left(\mathrm{e}^{-a}\right) \tag{6.366}$$

$$\geq wf(1) \tag{6.367}$$

$$\geq wf(1)\exp\left[-\frac{L_0}{T_0} - \frac{L_1}{T_1(t)}\right]. \tag{6.368}$$

Thus, Eq. (6.354) is satisfied for arbitrary a.

Next we consider Eq. (6.355). For any $x \in \mathscr{S}\backslash\mathscr{S}_{\mathrm{m}}$, a state $y \in \mathscr{S}_x$ so that $E_1(y) - E_1(x) > 0$ exists. In this case, for the states x and y, we obtain

$$\lim_{t\to\infty} \exp\left[-\frac{E_0(y)-E_0(x)}{T_0} - \frac{E_1(y)-E_1(x)}{T_1(t)}\right] = 0. \tag{6.369}$$

Here, we used the fact that $T_1(t) \to 0$ in the limit of $t \to \infty$. Using the relation given by Eq. (6.365), we obtain

$$\lim_{t\to\infty} A(y,x;t) = \lim_{t\to\infty} f\left(\exp\left[-\frac{E_0(y)-E_0(x)}{T_0} - \frac{E_1(y)-E_1(x)}{T_1(t)}\right]\right) \tag{6.370}$$

$$\leq \lim_{t\to\infty} \exp\left[-\frac{E_0(y)-E_0(x)}{T_0} - \frac{E_1(y)-E_1(x)}{T_1(t)}\right] \tag{6.371}$$

$$= 0. \tag{6.372}$$

To satisfy the relation, it is necessary that the acceptance probability $A(y,x;t)$ becomes a small positive value δ at a certain time $t(\geq t_1)$. That is, the time t_1 exists satisfying

$$A(y,x;t) \leq \delta, \quad (t \geq t_1). \tag{6.373}$$

Here we consider $t \geq t_1$. According to the relation, for the states $x \in \mathscr{S} \backslash \mathscr{S}_{\mathrm{m}}$, $y \in \mathscr{S}_x$, the following relation is satisfied:

$$\sum_{z \in \mathscr{S}} P(z,x)A(z,x;t) = P(y,x)A(y,x;t) + \sum_{z \in \mathscr{S} \backslash y} P(z,x)A(z,x;t) \tag{6.374}$$

$$\leq P(y,x)\delta + \sum_{z \in \mathscr{S} \backslash y} P(z,x) \tag{6.375}$$

$$= -(1-\delta)P(y,x) + \sum_{z \in \mathscr{S} \backslash y} P(z,x) + P(y,x) \tag{6.376}$$

$$= -(1-\delta)P(y,x) + 1. \tag{6.377}$$

Here we used $A(z,x;t) \leq 1$ and $\sum_{z \in \mathscr{S}} P(z,x) = 1$. Thus, when $t \geq t_1$, $G(x,x;t)$ is written as

$$G(x,x;t) = 1 - \sum_{z \in \mathscr{S}} P(z,x)A(z,x;t) \tag{6.378}$$

$$\geq 1 - \{-(1-\delta)P(y,x) + 1\} \tag{6.379}$$

$$= (1-\delta)P(y,x) > 0. \tag{6.380}$$

In addition, in the limit of $t \to \infty$, since the following relation is satisfied:

$$wf(1)\exp\left[-\frac{L_0}{T_0} - \frac{L_1}{T_1(t)}\right] \to 0, \tag{6.381}$$

for large enough t_1, we obtain

$$G(x,x;t) \geq (1-\varepsilon)P(y,x) \tag{6.382}$$

$$\geq wf(1)\exp\left[-\frac{L_0}{T_0} - \frac{L_1}{T_1(t)}\right], \quad (t \geq t_1). \tag{6.383}$$

Thus, the lower bound of the transition probability was obtained by Eqs. (6.354) and (6.355).

Proof of weak ergodicity

In order to prove the weak ergodicity, we have to estimate the ergodic coefficient defined by Eq. (3.101). By the definition of x^*, the states $x^* \in \mathscr{S} \backslash \mathscr{S}_{\mathrm{m}}$ can be generated by at most R steps from arbitrary states $x \in \mathscr{S}$. Here, we consider the Markov chain that changes from the state x at time $t - R$ to the state x^* at t. In this case, at least one integer $l\,(0 \leq l \leq R)$ exists and the transition process from x to x^* is satisfied:

$$x := x'_0 \neq x'_1 \neq x'_2 \neq \cdots \neq x'_l = x'_{l+1} = \cdots = x'_R =: x^*. \tag{6.384}$$

When t is large enough and $t \geq t_1$, for $i\,(0 \leq i \leq R-1)$, the following relations are satisfied.

- For $i < l$,

$$G(x'_{i+1}, x'_i; t-R+i) \geq wf(1)\exp\left[-\frac{L_0}{T_0} - \frac{L_1}{T_1(t-R+i)}\right]. \tag{6.385}$$

- For $i \geq l$,

$$G(x'_{i+1}, x'_i; t-R+i) = G(x^*, x^*; t-R+i) \tag{6.386}$$

$$\geq wf(1)\exp\left[-\frac{L_0}{T_0} - \frac{L_1}{T_1(t-R+i)}\right]. \tag{6.387}$$

Here, we used Eqs. (6.354) and (6.355). Thus, when $t \geq t_1$, we obtain the (x^*, x)-element of the transition matrix $\mathcal{G}^{t,t-R}$ as

$$\mathcal{G}^{t,t-R}(x^*, x) = \sum_{x_1, x_2, \cdots, x_{R-1}} G(x^*, x_{R-1}; t-1) G(x_{R-1}, x_{R-2}; t-2) \cdots G(x_2, x_1; t-R+1) G(x_1, x; t-R) \tag{6.388}$$

$$\geq G(x^*, x'_{R-1}; t-1) G(x'_{R-1}, x'_{R-2}; t-2) \cdots G(x'_2, x'_1; t-R+1) G(x'_1, x; t-R) \tag{6.389}$$

$$\geq w^R f(1)^R \exp\left[-\frac{RL_0}{T_0} - \sum_{i=0}^{R-1} \frac{L_1}{T_1(t-R+i)}\right] \tag{6.390}$$

$$\geq w^R f(1)^R \exp\left[-\frac{RL_0}{T_0} - \frac{RL_1}{T_1(t-1)}\right]. \tag{6.391}$$

Here we used the fact that $T_1(t)$ is a monotonically decreasing function in the final inequality. By using the relation, we estimate the ergodic coefficient. The following is satisfied:

$$\sum_{z \in \mathcal{S}} \min\left\{\mathcal{G}^{t,t-R}(z,x), \mathcal{G}^{t,t-R}(z,y)\right\}$$

$$\geq \min\left\{\mathcal{G}^{t,t-R}(x^*,x), \mathcal{G}^{t,t-R}(x^*,y)\right\} \tag{6.392}$$

$$\geq w^R f(1)^R \exp\left[-\frac{RL_0}{T_0} - \frac{RL_1}{T_1(t-1)}\right]. \tag{6.393}$$

From this inequality, the following is satisfied:

$$\min_{x,y\in\mathscr{S}}\left\{\sum_{z\in\mathscr{S}}\min\left\{\mathscr{G}^{t,t-R}(z,x),\mathscr{G}^{t,t-R}(z,y)\right\}\right\}$$
$$\geq w^R f(1)^R \exp\left[-\frac{RL_0}{T_0}-\frac{RL_1}{T_1(t-1)}\right]. \tag{6.394}$$

Thus, the ergodic coefficient defined by Eq. (3.101) is given by

$$1-\alpha\left(\mathscr{G}^{t,t-R}\right)\geq w^R f(1)^R \exp\left[-\frac{RL_0}{T_0}-\frac{RL_1}{T_1(t-1)}\right]. \tag{6.395}$$

Here, we put: $t=kR\,(k=0,1,2,\cdots)$. k_1 exists satisfying $t_1=k_1R$, and for $k\geq k_1$, we obtain

$$1-\alpha\left(\mathscr{G}^{kR,kR-R}\right)\geq w^R f(1)^R \exp\left[-\frac{RL_0}{T_0}-\frac{RL_1}{T_1(kR-1)}\right]. \tag{6.396}$$

To satisfy Eq. (3.100), it is enough to use the schedule given by the condition as

$$T_1(t)\geq\frac{RL_1}{\ln(t+2)}. \tag{6.397}$$

By using $t=kR-1$, the condition is rewritten as

$$\exp\left[-\frac{RL_1}{T_1(kR-1)}\right]\geq\frac{1}{kR+1}. \tag{6.398}$$

Thus, we obtain

$$\sum_{k=0}^{\infty}\left[1-\alpha\left(\mathscr{G}^{kR,kR-R}\right)\right]\geq\sum_{k=k_1}^{\infty}\left[1-\alpha\left(\mathscr{G}^{kR,kR-R}\right)\right] \tag{6.399}$$
$$\geq w^R f(1)^R\exp\left(-\frac{RL_0}{T_0}\right)\sum_{k=k_1}^{\infty}\frac{1}{kR+1} \tag{6.400}$$
$$\to\infty. \tag{6.401}$$

As a result, it is proved that Eq. (3.100) is satisfied. In addition, we define Δ as

$$\Delta:=RL_1, \tag{6.402}$$

to obtain

$$\Gamma(t)\geq\frac{P}{\beta}\tanh^{-1}\left[\frac{1}{(t+2)^{2/\Delta}}\right]. \tag{6.403}$$

Moreover, for large enough t, the above Eq. (6.403) is written as

$$\Gamma(t) \geq \frac{P}{\beta}(t+2)^{-2/\Delta}. \tag{6.404}$$

Here, R is proportional to N, and thus Δ is a constant which is proportional to N.

Proof of strong ergodicity

In order to prove strong ergodicity, it is necessary to satisfy the three conditions explained in Section 3.4. The first condition is that the stochastic dynamics shows weak ergodicity when the annealing schedule represented by Eq. (6.338) is used, which has already been proved here. The second condition is satisfied since the Boltzmann distribution $g(x;t)$ is the steady distribution at t due to the definition of the acceptance probability $A(y,x;t)$. Then, the elements of the steady distribution $\mathscr{Q}_t$ are given by $g(x;t)$. Thus, in order to show strong ergodicity, all we have to do is to prove the third condition.

First, we focus on the monotonicity of $g(x;t)$. Let $E_1^{\min}$ be the minimum value of $E_1(x)$, and $\mathscr{S}_1^{\min}$ be the set of x such that $E_1(x) = E_1^{\min}$. In this case, $g(x;t)$ has the following properties:

- For $x \in \mathscr{S}_1^{\min}$,

$$g(x;t+1) \geq g(x;t), \quad (t > 0), \tag{6.405}$$

  where the initial time is $t = 0$.

- For $x \in \mathscr{S} \setminus \mathscr{S}_1^{\min}$,

$$g(x;t+1) \leq g(x;t), \quad (t > t_2), \tag{6.406}$$

  where t_2 will be introduced later.

In addition, we introduce

$$\delta E_1(x) = E_1(x) - E_1^{\min}. \tag{6.407}$$

Let us consider Eq. (6.405). For $x \in \mathscr{S}_1^{\min}$, the following is satisfied:

$$g(x;t)$$
$$= \frac{\exp\left[-\frac{E_0(x)}{T_0} - \frac{E_1^{\min}}{T_1(t)}\right]}{\sum_{y \in \mathscr{S}_1^{\min}} \exp\left[-\frac{E_0(y)}{T_0} - \frac{E_1^{\min}}{T_1(t)}\right] + \sum_{y \in \mathscr{S}\setminus\mathscr{S}_1^{\min}} \exp\left[-\frac{E_0(y)}{T_0} - \frac{E_1(y)}{T_1(t)}\right]} \tag{6.408}$$

$$= \frac{\exp\left[-\frac{E_0(x)}{T_0}\right]}{\sum_{y\in\mathscr{S}_1^{\min}} \exp\left[-\frac{E_0(y)}{T_0}\right] + \sum_{y\in\mathscr{S}\setminus\mathscr{S}_1^{\min}} \exp\left[-\frac{E_0(y)}{T_0} - \frac{\delta E_1(y)}{T_1(t)}\right]}. \tag{6.409}$$

Here, since $T_1(t+1) < T_1(t)$, $\delta E_1(y) > 0$. If $y \in \mathscr{S}\setminus\mathscr{S}_1^{\min}$, the following inequality is obtained:

$$\exp\left[-\frac{\delta E_1(y)}{T_1(t+1)}\right] < \exp\left[-\frac{\delta E_1(y)}{T_1(t)}\right]. \tag{6.410}$$

Thus, by using Eq. (6.409), we can obtain

$$g(x;t+1) > g(x;t). \tag{6.411}$$

Next let us consider Eq. (6.406). For $x \in \mathscr{S}\setminus\mathscr{S}_1^{\min}$, the following is satisfied:

$$g(x;t)$$

$$= \frac{\exp\left[-\frac{E_0(x)}{T_0} - \frac{E_1(x)}{T_1(t)}\right]}{\sum_{y\in\mathscr{S}_1^{\min}} \exp\left[-\frac{E_0(y)}{T_0} - \frac{E_1^{\min}}{T_1(t)}\right] + \sum_{y\in\mathscr{S}\setminus\mathscr{S}_1^{\min}} \exp\left[-\frac{E_0(y)}{T_0} - \frac{E_1(y)}{T_1(t)}\right]} \tag{6.412}$$

$$= \frac{\exp\left[-\frac{E_0(x)}{T_0} - \frac{\delta E_1(x)}{T_1(t)}\right]}{\sum_{y\in\mathscr{S}_1^{\min}} \exp\left[-\frac{E_0(y)}{T_0}\right] + \sum_{y\in\mathscr{S}\setminus\mathscr{S}_1^{\min}} \exp\left[-\frac{E_0(y)}{T_0} - \frac{\delta E_1(y)}{T_1(t)}\right]}. \tag{6.413}$$

We differentiate both sides of this equation with respect to t and obtain

$$\frac{d}{dt}g(x;t) = \frac{dT_1(t)}{dt}\frac{d}{dT_1(t)}g(x;t) \tag{6.414}$$

$$= \frac{dT_1(t)}{dt}\left[\delta E_1(x) - \sum_{y\in\mathscr{S}\setminus\mathscr{S}_1^{\min}} \delta E_1(y)g(y;t)\right]\frac{g(x;t)}{T_1(t)^2}. \tag{6.415}$$

Here for large enough t, the Boltzmann distribution $g(y;t)$ becomes very small at the states $y \in \mathscr{S}\setminus\mathscr{S}_1^{\min}$. Thus, time t_2, which is large enough, exists satisfying

$$\delta E_1(x) - \sum_{y\in\mathscr{S}\setminus\mathscr{S}_1^{\min}} \delta E_1(y)g(y;t) > 0, \quad (t \geq t_2). \tag{6.416}$$

Also, since $dT_1(t)/dt < 0$, the following inequality is satisfied:

$$\frac{d}{dt} g(x;t) < 0, \tag{6.417}$$

and we obtain

$$g(x;t+1) \leq g(x;t). \tag{6.418}$$

Thus, for $t \geq t_2$, $\|\mathscr{Q}_{t+1} - \mathscr{Q}_t\|$ is written as

$$\begin{aligned}
&\|\mathscr{Q}_{t+1} - \mathscr{Q}_t\| \\
&= \sum_{x \in \mathscr{S}} |g(x;t+1) - g(x;t)| && (6.419) \\
&= \sum_{x \in \mathscr{S}_1^{\min}} [g(x;t+1) - g(x;t)] - \sum_{x \in \mathscr{S} \setminus \mathscr{S}_1^{\min}} [g(x;t+1) - g(x;t)] && (6.420) \\
&= 2 \sum_{x \in \mathscr{S}_1^{\min}} [g(x;t+1) - g(x;t)]. && (6.421)
\end{aligned}$$

Here we used the relation given as

$$\|\mathscr{Q}_t\| = \sum_{x \in \mathscr{S}} g(x;t) = \sum_{x \in \mathscr{S}_1^{\min}} g(x;t) + \sum_{x \in \mathscr{S} \setminus \mathscr{S}_1^{\min}} g(x;t) = 1. \tag{6.422}$$

Thus, we obtain

$$\begin{aligned}
\sum_{t=t_2}^{\infty} \|\mathscr{Q}_{t+1} - \mathscr{Q}_t\| &= 2 \sum_{x \in \mathscr{S}_1^{\min}} [g(x;\infty) - g(x;t_2)] && (6.423) \\
&\leq 2. && (6.424)
\end{aligned}$$

In addition, by Eq. (6.422), for arbitrary t, the following inequality is satisfied:

$$\|\mathscr{Q}_{t+1} - \mathscr{Q}_t\| \leq 2. \tag{6.425}$$

Thus, we finally obtain

$$\begin{aligned}
\sum_{t=0}^{\infty} \|\mathscr{Q}_{t+1} - \mathscr{Q}_t\| &= \sum_{t=0}^{t_2-1} \|\mathscr{Q}_{t+1} - \mathscr{Q}_t\| + \sum_{t=t_2}^{\infty} \|\mathscr{Q}_{t+1} - \mathscr{Q}_t\| && (6.426) \\
&\leq 2t_2 + 2 && (6.427) \\
&< \infty. && (6.428)
\end{aligned}$$

Thus, the third condition for strong ergodicity is shown.

From the these facts, it is proved that the inhomogeneous Markov chain represented by Eq. (6.338) guarantees strong ergodicity. Thus, by using Eq. (6.338), the probability distribution converges to the unique probability distribution that is proportional to $\exp[-E_0(x)/T_0]$. That is, the ground state when $\Gamma(t) = 0$ can be obtained by quantum annealing by the path-integral Monte Carlo method.

6.9.4 Convergence theorem for quantum annealing by the Green's function Monte Carlo method

The convergence theorem for quantum annealing by using the Green's function Monte Carlo method explained in Section 6.4 is given by

$$\Gamma(t) \geq \frac{b}{(t+1)^c}, \tag{6.429}$$

where b and c are constants, and c satisfies the relation

$$0 < c \leq \frac{1}{N}. \tag{6.430}$$

This was proved by Morita and Nishimori (2006).

In the Green's function Monte Carlo method, we use the inhomogeneous Markov chain in which the transition probability is given by $G'_1(y,x;t)$ defined by Eq. (6.51). Here the transition probability $G'_1(y,x;t)$ is decomposed by the generation probability $P(y,x)$ and the acceptance probability $A(y,x;t)$. Each of them is given by

$$P(y,x) = \begin{cases} \dfrac{1}{N} & \text{(single-spin difference)} \\ 0 & \text{(otherwise)} \end{cases}, \tag{6.431}$$

$$A(y,x;t) = \frac{N\Delta t\Gamma(t)}{1 - \Delta t\,[E_0(x) - E_{\mathrm{T}}] + N\Delta t\Gamma(t)}, \tag{6.432}$$

where "single-spin difference" means the case when the state y is generated by flipping a single spin in the state x. Here, x and y are included in the state space $\mathscr{S}$.

The sufficient condition for the convergence of quantum annealing by using the Green's function Monte Carlo method is given by Eq. (6.429). Under the condition, the inhomogeneous Markov chain satisfies the strong ergodicity. We proved this fact as well in Section 6.9.3.

Lower bound of the transition probability

We find the lower bound of the transition probability $G'_1(y,x;t)$ which will be used in the proof of the convergence theorem. The lower bound of $G'_1(y,x;t)$ is given by

- When $x \neq y, (x, y \in \mathscr{S})$,

$$G_1'(y,x;t) \geq \frac{\Delta t \Gamma(t)}{1 - \Delta t\,(E_{\mathrm{min}} - E_{\mathrm{T}}) + N\Delta t \Gamma(t)}. \tag{6.433}$$

- When $x = y, (x, y \in \mathscr{S})$,

$$G_1'(x,x;t) \geq \frac{\Delta t \Gamma(t)}{1 - \Delta t\,(E_{\mathrm{min}} - E_{\mathrm{T}}) + N\Delta t \Gamma(t)}, \quad (t \geq t_1). \tag{6.434}$$

Here t_1 will be introduced later and E_{min} is the minimum value of $E_0(x)$.

Let us first consider Eq. (6.433). Since Eq. (6.432) is a monotonically increasing function of $E_0(x)$, $G_1'(y,x;t)$ takes the minimum value when $E_0(x)$ becomes the minimum value.

Next we consider Eq. (6.434). From Eq. (6.62), $G_1'(x,x;t)$ is defined as

$$G_1'(x,x;t) = 1 - \frac{N\Delta t \Gamma(t)}{1 - \Delta t\,[E_0(x) - E_{\mathrm{T}}] + N\Delta t \Gamma(t)}. \tag{6.435}$$

Since $\Gamma(t)$ is a monotonically decreasing function of t, the second term becomes a small positive value δ at a certain time $t (\geq t_1)$. That is, time t_1 exists satisfying

$$\frac{N\Delta t \Gamma(t)}{1 - \Delta t\,(E_0(x) - E_{\mathrm{T}}) + N\Delta t \Gamma(t)} \leq \delta, \quad (t \geq t_1). \tag{6.436}$$

Thus, the following inequality is satisfied:

$$G_1'(x,x;t) \geq 1 - \delta, \quad (t \geq t_1). \tag{6.437}$$

Since the right-hand side of Eq. (6.434) approaches zero in the limit of $t \to \infty$, we obtain

$$G_1'(x,x;t) \geq 1 - \delta \tag{6.438}$$

$$\geq \frac{\Delta t \Gamma(t)}{1 - \Delta t\,(E_{\mathrm{min}} - E_{\mathrm{T}}) + N\Delta t \Gamma(t)}, \quad (t \geq t_1), \tag{6.439}$$

and Eq. (6.433) is confirmed. Thus, the lower bound of the transition probability can be obtained.

Proof of weak ergodicity

When the number of spin is N, an arbitrary state y can be generated by another arbitrary state x by at most N steps. Suppose we consider the Markov chain that changes from the state x at time $t-N$ to the state y at t. In this case, at least one integer $l\,(0 \le l \le N)$ exists, and the transition process from x to y is satisfied:

$$x := x'_0 \neq x'_1 \neq x'_2 \neq \cdots \neq x'_l = x'_{l+1} = \cdots = x'_N =: y. \tag{6.440}$$

Thus, we obtain the (y,x)-element of transition matrix $\mathscr{G}^{t,t-N}$ as

$$\mathscr{G}^{t,t-N}(y,x) = \sum_{x_1,x_2,\cdots,x_{N-1}} G'_1(y,x_{N-1};t-1)G'_1(x_{N-1},x_{N-2};t-2)$$
$$\cdots G'_1(x_2,x_1;t-N+1)G'_1(x_1,x;t-N) \tag{6.441}$$

$$\geq G'_1(y,x'_{N-1};t-1)G'_1(x'_{N-1},x'_{N-2};t-2)$$
$$\cdots G'_1(x'_2,x'_1;t-N+1)G'_1(x'_1,x;t-N) \tag{6.442}$$

$$\geq \prod_{t'=t-N}^{t-1} \frac{\Delta t \Gamma(t')}{1-\Delta t(E_{\min}-E_{\mathrm{T}})+N\Delta t\Gamma(t')} \tag{6.443}$$

$$\geq \left[\frac{\Delta t\Gamma(t-1)}{1-\Delta t(E_{\min}-E_{\mathrm{T}})+N\Delta t\Gamma(t-1)}\right]^N. \tag{6.444}$$

Here we used the fact that $\Gamma(t)$ is a monotonically decreasing function of t in the final inequality. By using the relation, we obtain

$$\min_{x,y\in\mathscr{S}}\left\{\sum_{z\in\mathscr{S}} \min\left[\mathscr{G}^{t,t-N}(z,x),\mathscr{G}^{t,t-N}(z,y)\right]\right\}$$
$$\geq \left[\frac{\Delta t\Gamma(t-1)}{1-\Delta t(E_{\min}-E_{\mathrm{T}})+N\Delta t\Gamma(t-1)}\right]^N, \tag{6.445}$$

and the ergodic coefficient is given as

$$1-\alpha\left(\mathscr{G}^{t,t-N}\right) \geq \left[\frac{\Delta t\Gamma(t-1)}{1-\Delta t(E_{\min}-E_{\mathrm{T}})+N\Delta t\Gamma(t-1)}\right]^N. \tag{6.446}$$

Here, we put $t = kN\,(k=0,1,2,\cdots)$. k_1 exists satisfying $t_1 = k_1 N$. For $k \geq k_1$, we obtain

$$1-\alpha\left(\mathscr{G}^{kN,kN-N}\right) \geq \left[\frac{\Delta t\Gamma(kN-1)}{1-\Delta t(E_{\min}-E_{\mathrm{T}})+N\Delta t\Gamma(kN-1)}\right]^N. \tag{6.447}$$

Here, by substituting Eq. (6.429) for the right-hand side of Eq. (6.448), we obtain

$$\frac{\Delta t \Gamma(kN-1)}{1-\Delta t(E_{\min}-E_{\mathrm{T}})+N\Delta t\Gamma(kN-1)}$$

$$\geq \frac{\Delta t}{(1-\Delta t(E_{\min}-E_{\mathrm{T}}))\cdot(kN)^c/b+N\Delta t} \tag{6.448}$$

$$\underset{k\gg 1}{\simeq} \frac{1}{(kN)^c}\cdot\frac{b\Delta t}{1-\Delta t(E_{\min}-E_{\mathrm{T}})} \tag{6.449}$$

$$=: \frac{b'}{(kN)^c}. \tag{6.450}$$

Here b' is a constant. Thus, we obtain the following relation:

$$\sum_{k=0}^{\infty}\left[1-\alpha\left(\mathscr{G}^{kN,kN-N}\right)\right] \geq \sum_{k>k_1}^{\infty}\left[1-\alpha\left(\mathscr{G}^{kN,kN-N}\right)\right] \tag{6.451}$$

$$\geq \sum_{k>k_1}^{\infty}\frac{b'^N}{(kN)^{cN}} \tag{6.452}$$

$$\underset{0<c\leq 1/N}{\rightarrow} \infty. \tag{6.453}$$

That is, Eq. (3.100) is satisfied when the schedule of $\Gamma(t)$ is represented by Eq. (6.429). From these discussions, the time evolution is weak ergodic when $\Gamma(t)$ decreases with the annealing schedule given by Eq. (6.429).

Proof of strong ergodicity

In order to prove strong ergodicity, it is necessary to satisfy the three conditions explained in Section 3.4. The first condition is that the stochastic dynamics shows weak ergodicity when the annealing schedule represented by Eq. (6.429) is used, which has already been proved earlier.

Let us prove the second condition of strong ergodicity. It is enough to show that the steady distribution $q(x;t)$ exists so that the following equation is satisfied at each time t:

$$q(y;t) = \sum_{x\in\mathscr{S}} G_1'(y,x;t)q(x;t). \tag{6.454}$$

In the Green's function Monte Carlo method explained in Section 6.4, the steady distribution is given by

$$q(x;t) = \frac{w(x;t)}{\sum_{x\in\mathscr{S}} w(x;t)}, \tag{6.455}$$

where the weight $w(x;t)$ is defined by Eq. (6.59). When we consider the Ising model in a transverse field, the denominator is written as

$$\sum_{x\in\mathscr{S}} w(x;t)$$

$$= \mathrm{Tr}\left[1 - \Delta t\left(-\sum_{1\le i,j\le N} J_{ij}\hat{\sigma}_i^z\hat{\sigma}_j^z - \sum_{i=1}^{N} h_i\hat{\sigma}_i^z - E_{\mathrm{T}}\right) + N\Delta t\Gamma(t)\right] \tag{6.456}$$

$$= 2^N\left[1 + \Delta t E_{\mathrm{T}} + N\Delta t\Gamma(t)\right]. \tag{6.457}$$

Here we used the relation:

$$\mathrm{Tr}\left(-\sum_{1\le i,j\le N} J_{ij}\hat{\sigma}_i^z\hat{\sigma}_j^z - \sum_{i=1}^{N} h_i\hat{\sigma}_i^z\right) = 0. \tag{6.458}$$

The transition probability defined by Eq. (6.62) is rewritten as

$$G_1'(y,x;t) = \begin{cases} 1 - \dfrac{N\Delta t\Gamma(t)}{w(x;t)} & (x = y) \\ \dfrac{\Delta t\Gamma(t)}{w(x;t)} & (\text{single-spin difference}) \\ 0 & (\text{otherwise}) \end{cases} . \tag{6.459}$$

In this case, the following equation is satisfied:

$$\sum_{x\in\mathscr{S}} G_1'(y,x;t)q(x;t)$$

$$= \left[1 - \frac{N\Delta t\Gamma(t)}{w(y;t)}\right]\frac{w(y;t)}{\sum_{x\in\mathscr{S}} w(x;t)} + \sum_{x\in\mathscr{S}_y}\frac{\Delta t\Gamma(t)}{w(x;t)}\frac{w(x;t)}{\sum_{x\in\mathscr{S}} w(x;t)} \tag{6.460}$$

$$= q(y;t) - \frac{N\Delta t\Gamma(t)}{\sum_{x\in\mathscr{S}} w(x;t)} + \frac{\Delta t\Gamma(t)}{\sum_{x\in\mathscr{S}} w(x;t)}\sum_{x\in\mathscr{S}_y} 1 \tag{6.461}$$

$$= q(y;t). \tag{6.462}$$

Here $\mathscr{S}_y$ is the set of the states that can be generalized by a single-spin flip from the state y, where the number of states is N by the definition. From these discussions, it is shown that the distribution defined by Eq. (6.455) is a steady distribution at each time t.

Next, let us prove that the third condition is satisfied. The steady distribution given by Eq. (6.455) is rewritten as

$$q(x;t) = \frac{1}{2^N} - \frac{\Delta t E_0(x)}{1 + \Delta t E_{\mathrm{T}} + N\Delta t \Gamma(t)}. \tag{6.463}$$

Here $q(x;t)$ has the following properties:

- When $E_0(x) \geq 0$,

$$q(x;t+1) \leq q(x;t). \tag{6.464}$$

- When $E_0(x) < 0$,

$$q(x;t+1) > q(x;t). \tag{6.465}$$

These relations are directly understood from Eq. (6.463), since $\Gamma(t)$ is a monotonically decreasing function against t. Let $\mathscr{S}_-$ and $\mathscr{S}_+$ be the sets of states for $E_0(x) < 0$ and that of states for $E_0(x) \geq 0$, respectively. The following is obtained:

$$\|\mathscr{Q}_{t+1} - \mathscr{Q}_t\| = \sum_{x \in \mathscr{S}} |q(x;t+1) - q(x;t)| \tag{6.466}$$

$$= \sum_{x \in \mathscr{S}_-} [q(x;t+1) - q(x;t)] - \sum_{x \in \mathscr{S}_+} [q(x;t+1) - q(x;t)] \tag{6.467}$$

$$= 2 \sum_{x \in \mathscr{S}_-} [q(x;t+1) - q(x;t)]. \tag{6.468}$$

Here we used the relation given by

$$\|\mathscr{Q}_t\| = \sum_{x \in \mathscr{S}} q(x;t) = \sum_{x \in \mathscr{S}_-} q(x;t) + \sum_{x \in \mathscr{S}_+} q(x;t) = 1. \tag{6.469}$$

Thus, we obtain

$$\sum_{t=0}^{\infty} \|\mathscr{Q}_{t+1} - \mathscr{Q}_t\| = 2 \sum_{x \in \mathscr{S}_-} [q(x;\infty) - q(x;0)] \tag{6.470}$$

$$\leq 2 \tag{6.471}$$

$$< \infty, \tag{6.472}$$

that is, the third condition of strong ergodicity is shown. Thus, it is proved that when $\Gamma(t)$ decreases with the annealing schedule represented by Eq. (6.429), the stochastic dynamics shows strong ergodicity.

From these facts, it is proved that the inhomogeneous Markov chain represented by Eq. (6.429) guarantees strong ergodicity. Thus, by using Eq. (6.429), the probability distribution converges to the unique probability distribution that is proportional to $\exp[-E_0(x)/T_0]$. That is, the ground state when $\Gamma(t) = 0$ can be obtained by quantum annealing by the Green's function Monte Carlo method.

References

Amico, L., R. Fazio, A. Osterloh, and V. Vedral. 2008. 'Entanglement in many-body systems'. *Rev. Mod. Phys.* 80: 517.

Barahona, F. 1982. 'On the computational complexity of Ising spin glass models'. *J. Phys. A: Math. Gen.* 15: 3241.

Battaglia, D., M. Kolář, and R. Zecchina. 2004. 'Minimizing energy below the glass thresholds'. *Phys. Rev. E* 70: 036107.

Battaglia, D. A., G. E. Santoro, and E. Tosatti. 2005. 'Optimization by quantum annealing: Lessons from hard satisfiability problems *Phys. Rev. E* 71: 066707.

Bieche, L., J. P. Uhry, R. Maynard, and R. Rammal. 1980. 'On the ground states of the frustration model of a spin glass by a matching method of graph theory'. *J. Phys. A: Math. Gen.* 13: 2553.

Brooke, J., D. Bitko, T. F. Rosenbaum, and G. Aeppli. 1999. 'Quantum annealing of a disordered magnet'. *Science* 284: 779.

Cohen, E. and B. Tamir. 2014. 'D-Wave and predecessors: From simulated to quantum annealing'. *Int. J. Quantum Inform.* 12: 1430002.

Das, A., and B. K. Chakrabarti. 2008a. 'Colloquium: Quantum annealing and analog quantum computation'. *Rev. Mod. Phys.* 80: 1061.

Das, A., and B. K. Chakrabarti. 2008b. 'Reaching the ground state of a quantum spin glass using a zero-temperature quantum Monte Carlo method'. *Phys. Rev. E* 78: 061121.

Eisert, J., M. Cramer, and M. B. Plenio. 2010. 'Colloquium: Area laws for the entanglement entropy'. *Rev. Mod. Phys.* 82: 277.

Farhi, E., J. Goldstone, S. Gutmann, J. Lapan, A. Lundgren, and D. Preda. 2001. 'A quantum adiabatic evolution algorithm applied to random instances of an NP-complete problem'. *Science* 292: 472.

Grest, G. S., C. M. Soukoulis, and K. Levin. 1986. 'Cooling-rate dependence for the spin-glass ground-state energy: Implications for optimization by simulated annealing'. *Phys. Rev. Lett.* 56: 1148.

Hallberga, K. A. 2006. 'New trends in density matrix renormalization'. *Adv. Phys.* 55: 477.

Inoue, J.-I., Y. Saika, and M. Okada. 2009. 'Quantum mean-field decoding algorithm for error-correcting codes'. *J. Phys.: Conf. Ser.* 143: 012019.

Jarzynski, C. 1997a. 'Equilibrium free-energy differences from nonequilibrium measurements: A master-equation approach'. *Phys. Rev. E* 56: 5018.

Jarzynski, C. 1997b. 'Nonequilibrium equality for free energy differences'. *Phys. Rev. Lett.* 78: 2690.

Jeckelmann, E. 2002. 'Dynamical density-matrix renormalization-group method'. *Phys. Rev. B* 66: 045114.

Johnson, D. S., and L. A. McGeoch. 1997. *Local Search in Combinatorial Optimization*. Wiley.

Kadowaki, T., and H. Nishimori. 1998. 'Quantum annealing in the transverse Ising model'. *Phys. Rev. E* 58: 5355.

Kirkpatrick, S., C. D. Gelatt Jr., and M. P. Vecchi. 1983. 'Optimization by simulated annealing'. *Science* 220: 671.

Kurihara, K., S. Tanaka, and S. Miyashita. 2009. 'Quantum annealing for clustering'. In *Proceedings of the 25th Conference on Uncertainty in Artificial Intelligence (UAI2009)*.

Lax, P. D. 2007. *Linear Algebra and its Applications*. Wiley.

Martoňák, R., G. E. Santoro, and E. Tosatti. 2002. 'Quantum annealing by the path-integral Monte Carlo method: The two-dimensional random Ising model'. *Phys. Rev. B* 66: 094203.

Martoňák, R., G. E. Santoro, and E. Tosatti. 2004. 'Quantum annealing of the traveling-salesman problem'. *Phys. Rev. E* 70: 057701.

Matsuda, Y., H. Nishimori, and H. G. Katzgraber. 2009. 'Ground-state statistics from annealing algorithms: quantum versus classical approaches'. *New J. Phys.* 11: 073012.

Messiah, A. 1976. *Quantum Mechanics*. Wiley.

Montanari, A., G. Parisi, and F. Ricci-Tersenghi. 2004. 'Instability of one-step replica-symmetry-broken phase in satisfiability problems'. *J. Phys. A: Math. Gen.* 37: 2073.

Morita, S., and H. Nishimori. 2006. 'Convergence theorems for quantum annealing'. *J. Phys. A: Math. Gen.* 39: 13903.

Morita, S., and H. Nishimori. 2007. 'Convergence of quantum annealing with real-time Schrödinger dynamics'. *J. Phys. Soc. Jpn.* 76: 064002.

Morita, S., and H. Nishimori. 2008. 'Mathematical foundation of quantum annealing'. *J. Math. Phys.*, 49: 125210.

Nishimori, H., and Y. Nonomura. 1996. 'Quantum effects in neural networks'. *J. Phys. Soc. Jpn.* 65: 3780.

Nishino, T. 1995. 'Density matrix renormalization group method for 2D classical models'. *J. Phys. Soc. Jpn.* 64: 3598.

Nishino, T., and K. Okunishi. 1996. 'Corner transfer matrix renormalization group method'. *J. Phys. Soc. Jpn.* 65: 891.

Nishino, T., and K. Okunishi. 2007. 'Density matrix and renormalization for classical lattice models'. in. *Strongly Correlated Magnetic and Superconducting Systems*. P. 167. Springer.

Nishino, T., T. Hikihara, K. Okunishi, and Y. Hieida. 1999. 'Density matrix renormalization group: Introduction from a variational point of view'. *Int. J. Mod. Phys. B* 13: 1.

Nishino, T., K. Okunishi, Y. Hieida, T. Hikihara, and H. Takasaki. 2007. 'Transfer-matrix approach to classical systems'. in *Density-Matrix Renormalization*.

Ohzeki, M. 2010. 'Quantum annealing with the Jarzynski equality'. *Phys. Rev. Lett.* 105: 050401.

Ohzeki, M., and H. Nishimori. 2011. 'Quantum annealing: An introduction and new developments'. *J. Comput. Theor. Nanosci.* 8: 963.

Ohzeki, M., H. Katsuda, and H. Nishimori. 2011. 'Nonequilibrium work on spin glasses in longitudinal and transverse fields'. *J. Phys. Soc. Jpn.* 80: 084002.

Okuyama, M., Y. Yamanaka, H. Nishimori, and M. M. Rams. 2015. 'Anomalous behavior of the energy gap in the one-dimensional quantum XY model'. *Phys. Rev. E* 92: 052116.

Otsubo, Y., J.-I. Inoue, K. Nagata, and M. Okada. 2012. 'Effect of quantum fluctuation in error-correcting codes'. *Phys. Rev. E* 86: 051138.

Otsubo, Y., J.-I. Inoue, K. Nagata, and M. Okada. 2014. 'Code-division multiple-access multiuser demodulator by using quantum fluctuations'. *Phys. Rev. E* 90: 012126.

Reich, D. H., B. Ellman, J. Yang, T. F. Rosenbaum, G. Aeppli, and D. P. Belanger. 1990. 'Dipolar magnets and glasses: Neutron-scattering dynamical and calorimetric studies of randomly distributed Ising spins'. *Phys. Rev. B* 42: 4631.

Rigolin, G., and G. Ortiz. 2010. 'Adiabatic perturbation theory and geometric phases for degenerate systems'. *Phys. Rev. Lett.* 104: 170406.

Rigolin, G., and G. Ortiz. 2012. 'Adiabatic theorem for quantum systems with spectral degeneracy'. *Phys. Rev. A* 85: 062111.

Rigolin, G., G. Ortiz, and V. H. Ponce. 2008. 'Beyond the quantum adiabatic approximation: Adiabatic perturbation theory'. *Phys. Rev. A* 78: 052508.

Santoro, G. E., and E. Tosatti. 2006. 'Optimization using quantum mechanics: Quantum annealing through adiabatic evolution'. *J. Phys. A: Math. Gen.* 39: R393.

Santoro, G. E., R. Martoňák, E. Tosatti, and R. Car. 2002. 'Theory of quantum annealing of an Ising spin glass'. *Science* 295: 2427.

Sarjala, M., V. Petäjä, and M. Alava. 2006. 'Optimization in random field Ising models by quantum annealing'. *J. Stat. Mech.* P1008.

Sato, I., K. Kurihara, S. Tanaka, H. Nakagawa, and S. Miyashita. 2009. 'Quantum annealing for variational Bayes inference'. In *Proceedings of the 25th Conference on Uncertainty in Artificial Intelligence (UAI2009)*.

Sato, I., S. Tanaka, K. Kurihara, S. Miyashita, and H. Nakagawa. 2013. 'Quantum annealing for Dirichlet process mixture models with applications to network clustering'. *Neurocomputing* 121: 523.

Schollwöck, U. 2005. 'The density-matrix renormalization group'. *Rev. Mod. Phys.* 77: 259.

Seki, Y., and H. Nishimori. 2012. 'Quantum annealing with antiferromagnetic fluctuations'. *Phys. Rev. E* 85: 051112.

Seki, Y., and H. Nishimori. 2015. 'Quantum annealing with antiferromagnetic transverse interactions for the Hopfield model'. *J. Phys. A: Math. Gen.* 48: 335301.

Shibata, N., and H. Tsunetsugu. 1999. 'Thermodynamics of doped Kondo insulator in one dimension: Finite-temperature DMRG study'. *J. Phys. Soc. Jpn.* 68: 744.

Somma, R. D., C. D. Batista, and G. Ortiz. 2007. 'Quantum approach to classical statistical mechanics'. *Phys. Rev. Lett.* 99: 030603.

Stella, L., and G. E. Santoro. 2007. 'Quantum annealing of an Ising spin-glass by Green's function Monte Carlo'. *Phys. Rev. E* 75: 036703.

Suzuki, M. 1976a. 'Generalized Trotter's formula and systematic approximants of exponential operators and inner derivations with applications to many-body problems'. *Commun. Math. Phys.* 51: 183.

Suzuki, M. 1976b. 'Relationship between d-dimensional quantal spin systems and $(d+1)$-dimensional Ising systems'. *Prog. Theor. Phys.* 56: 1454.

Suzuki, M. 1977. 'On the convergence of exponential operators the Zassen-Haus formula, BCH formula and systematic approximants'. *Comm. Math. Phys.* 57: 193.

Suzuki, M. 1990. 'Fractal decomposition of exponential operators with applications to many-body theories and Monte Carlo simulations'. *Phys. Lett. A* 146: 319.

Suzuki, M. 1991. 'General theory of fractal path integrals with applications to many-body theories and statistical physics'. *J. Math. Phys.* 32: 400.

Suzuki, S., and H. Nishimori. 2007. 'Quantum annealing by ferromagnetic interaction with the mean-field scheme'. *Physica A* 384: 137.

Suzuki, S, and M. Okada. 2005. in *Quantum annealing and related optimization methods"*, Ed. by A. Das and B. K. Chakrabarti. Springer-Verlag. Chap. Simulated quantum annealing by the real-time evolution, page 207.

Suzuki, S., and M. Okada. 2007. 'Study on quantum annealing using the density matrix renormalization group'. *Interdisciplinary Information Sciences* 13: 49.

Suzuki, S., H. Nishimori, and M. Suzuki. 2007. 'Quantum annealing of the random-field Ising model by transverse ferromagnetic interactions'. *Phys. Rev. E* 75: 051112.

Suzuki, S., J. Inoue, and B. K. Chakrabarti. 2013. *Quantum Ising phases and Transitions in Transverse Ising Models*. Springer.

Tanaka, K., and T. Horiguchi. 2000. 'Quantum statistical-mechanical iterative method in image restoration'. *Electron Comm. Jpn.* 83: 84.

Tanaka, S., and R. Tamura. 2012. 'Quantum Annealing from the Viewpoint of Statistical Physics, Condensed Matter Physics, and Computational Physics'. in *Quantum Computing series/Lectures on Quantum Computing, Thermodynamics and Statistical Physics*. p. 1 World Scientific., *page 1*.

Trotter, H. F. 1959. 'On the product of semi-groups of operators'. *Proc. Am. Math. Soc.* 10: 545.

Tsuda, J., Y. Yamanaka, and H. Nishimori. 2013. 'Energy gap at first-order quantum phase transitions: An anomalous case'. *J. Phys. Soc. Jpn.* 82: 114004.

White, S. R. 1992. 'Density matrix formulation for quantum renormalization groups'. *Phys. Rev. Lett.* 69: 2863.

Wittek, P. 2014. *Quantum machine learning: What quantum computing means to data mining*. Elsevier.

Yoshikawa, S., K. Okunishi, M. Senda, and S. Miyashita. 2004. 'Quantum fluctuation-induced phase transition in S=1/2 XY-like Heisenberg antiferromagnets on the triangular lattice'. *J. Phys. Soc. Jpn.* 73: 1798.

PART TWO
ADDITIONAL NOTES

CHAPTER 7

Notes on Adiabatic Quantum Computers

Boaz Tamir[1] and Eliahu Cohen[2]

We discuss in this chapter the basics of adiabatic computation, as well as some physical implementations. After a short introduction to the quantum circuit model, we describe quantum adiabatic computation, quantum annealing, and the strong relations between the three. We conclude with a brief presentation of the D-Wave computer and some future challenges.

7.1 Introduction

During the last two decades, a great deal of attention has been focused on quantum computation following a sequence of results (Grover, 1996; Shor, 1997) suggesting that quantum computers are more powerful than classical probabilistic computers. Following Shor's result (1997), that factoring and extraction of discrete logarithms are both solvable on quantum computers in polynomial time, it is natural to ask whether other hard (consuming exponential resources) problems can be efficiently solved on quantum computers in polynomial time. It was Feynman's idea (1982) that quantum phenomena could not always be simulated by classical computers, and whenever there are such simulations there is an exponential growth in the required resources. Feynman also suggested the use of quantum computers and conjectured that quantum computers can be programmed to simulate any local quantum system. Since then, a vast literature has been written, addressing the theoretical and practical advantages of quantum computers, as well as some challenges in implementing them. In 1996, Lloyd supported Feynman's claim and concluded (Lloyd, 1996): "The wide variety of atomic, molecular and semiconductor quantum devices available suggests that quantum simulation may soon be reality". Just three years later, D-Wave systems were founded with the goal of making practical quantum computers (http://www.dwavesys.com). Indeed, quantum technology is maturing

[1]Faculty of Interdisciplinary Studies, Bar-Ilan University, Ramat-Gan, Israel
[2]H. H. Wills Physics Laboratory, University of Bristol, Tyndall Avenue, Bristol, UK

to the point where quantum devices, such as quantum communication systems, quantum random number generators and quantum simulators are built with capabilities exceeding classical computers. Quantum annealers (Das and Chakrabarti, 2008), in particular, solve hard optimization problems by evolving a known initial configuration towards the ground state of a Hamiltonian encoding a given problem. Quantum annealing is an advanced alternative to classical simulated annealing (van Laarhoven and Aarts, 1987), an approach to solve optimization problems based on the observation that the problem's cost function can be viewed as the energy of a physical system, and that energy barriers can be crossed by thermal hopping. However, to escape local minima it can be advantageous to explore low energy configurations quantum mechanically by exploiting superpositions and tunneling (see Fig. 7.1). Quantum annealing and adiabatic quantum computation are algorithms based on this idea, and programmable quantum annealers, such as D-Wave computers, are their physical realization. Quantum information processing offers dramatic speed ups, yet is famously susceptible to decoherence, the process whereby quantum superposition decays into mutually exclusive classical alternatives, a mixed state, thus robbing quantum computers of their power. For this reason, many researchers question the quantum features of D-Wave computers (Boixo et al., 2013, 2014, http://www.scottaaronson.com). In what follows, we shall refer to the controversy concerning the quantum properties of D-Wave computers.

In this short review work, we aim to present the crux of the subject matter. We shall focus on some fundamental results, leaving the small details outside. A strictly related, extensive work can be found in the paper by Suzuki and Das (2015).

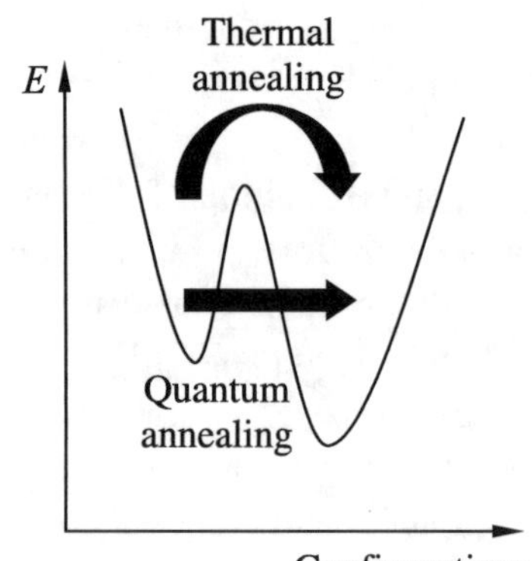

Fig. 7.1 Quantum annealing vs thermal annealing in a graph of energy as a function of configuration space.

7.2 The Circuit Model

As mentioned earlier, quantum computation was first suggested by Feynman as a way to overcome the problem of simulating quantum phenomena on a classical computer (Feynman, 1982). Feynman pointed out that a set of measurements on

Einstein–Podolsky–Rosen (EPR) entangled quantum particles could not be simulated in principle by classical means. Moreover, even when one can use classical computers to simulate quantum phenomena, the growth in resources needed is exponential. Therefore the natural solution would be to think of quantum computers. Soon after, Benioff (1982) and Deutsch (1985) presented a quantum version of a Turing machine (see also the paper by Albert (1983)). However, the quantum Turing machine model was not practical. In 1989, Deutsch suggested the idea of a quantum gate network computer (Deutsch, 1989). He also provided a strong argument showing that any finite dimensional unitary operator on a quantum state could be simulated by a simple universal gate. Deutsch's universal gate was a 3-qubit gate, a variant of the known Toffoli gate for reversible classical computations. This universal gate approximates any other quantum gate by using the well-known Kronecker (Hardy and Wright, 1979) approximation. Deutsch also presented the first known "quantum algorithm", later extended to the Deutsch–Josza algorithm (Deutsch and Jozsa, 1992). These algorithms can distinguish between a balanced function and a constant one by using a small number of measurements. They showed an exponential benefit over classical deterministic algorithms. In the scheme presented by Deutsch, quantum computers have no architecture and in that sense they resemble old, one purpose, analogue computers. Following the work of Deutsch, two main families of algorithms were introduced – Grover's search and Shor's factoring. In 1996, Grover (1996) presented a quantum search algorithm for an element in an unsorted array. The Grover algorithm has a speedup of a square root over the classical search algorithm (that is, if the size of the search space is 2^n, then the Grover complexity is $\sqrt{2^n}$). Although such a speedup does not cross a computational complexity class line (i.e., it does not turn a hard problem into a simple one), it shows a clear (and proven) gap between the quantum and classical computational complexity. We can easily demonstrate the algorithm for the two qubit case. In general, the algorithm consists of $O(\sqrt{2^n})$ iterations; in the 2-qubit case, one iteration is enough. Each such iteration consists of 2 substeps; the first marks the solution (without knowing its position, therefore using a black box) by a –1 phase, leaving all other elements unchanged, the second step is a reflection of each of the amplitudes over the (new) average of all amplitudes. In particular, for the 2-qubit case, following the first step, assuming that the 3rd element is the solution, we will get the amplitudes as in Fig. 7.2.

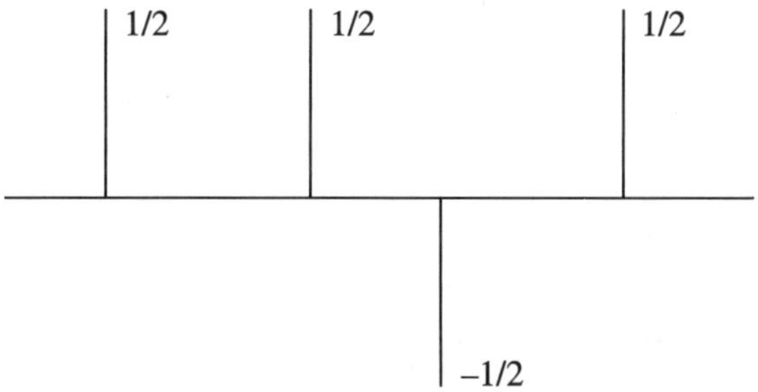

Fig. 7.2 The set of amplitudes following the first step of the Grover iteration in the 2-qubit case.

Now the average of all the amplitudes is 1/4. Reflecting the 1/2 amplitudes over the 1/4 line brings them to 0, while reflecting the –1/2 amplitudes over the same line brings it to 1 (see Fig. 7.3). Hence, one Grover iteration is enough.

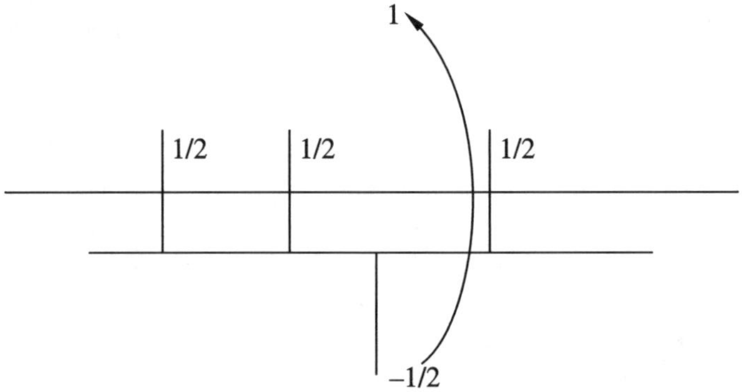

Fig. 7.3 The set of amplitudes following the second step of the Grover iteration in the 2-qubit case.

Note that the algorithm is a "black-box" (or oracle algorithm), and as such can be generalized and used to speed up many classical algorithms (Boyer et al., 1998). Later, Grover also used this algorithm to present a scheme for the construction of a superposition in a 2^n dimensional vector space using $\sqrt{2^n}$ steps at most (Grover, 2000). In 1994, Shor (1994) presented a polynomial algorithm for prime factorization (for a composite number N, having $\log N = n$ digits; a non-trivial factor of N could be found with $O((\log N)^3)$ operations) and discrete logarithms. So far all known classical algorithms for the factorization or discrete logarithm have exponential complexity. Therefore the reduction in complexity seems exponential. However, we have no proof for the claim that the complexity of such classical algorithms should be bounded from below by exponential functions. This is a manifestation of the well-known "P vs NP" problem (note however that factorization is not NP-complete). Shor's algorithm used a Fourier transform module that can identify the order of a modular function. This was the extension of a previous quantum algorithm by Simon (1994). The Fourier quantum module could serve also for phase estimation (Kitaev, 1995), for order finding (Shor, 1994, 1997), and in general, for the identification of other "hidden subgroups" of symmetries (Jozsa, 1998).

A severe drawback in quantum computation was and still is the problem of decoherence (Namiki et al., 1997). It is hard to construct a stable superposition of even a small number of qubits. It is even harder to apply unitary gates between the qubits. So far there are several suggestions as to the way to construct a quantum computer. Clearly, it is enough to construct the set of universal quantum gates (for the existence of such a set see the paper by DiVincenzo (1995)): a XOR or what is known in quantum computation as a $CNOT$ gate, and a one qubit rotation.

A major breakthrough came with the presentation of fault tolerant quantum gates (Shor, 1996). The basic ideas are: we first replace each qubit with a block of qubits, using some error correction code. The two physical states of the original qubit correspond to "logical" states of the block. Next, we write a "logic" gate between the blocks (a universal gate). Our aim is to control the propagation of noise, such that an error inside a block will not leak to far away blocks, by employing fault tolerant gates to control it. These fault tolerant quantum gates and error correction quantum codes sustain the hope that one day a large-scale quantum computer could indeed be realized.

Several criteria were suggested by DiVincenzo (2000) for the physical possibility of the realization of quantum computers:

(1) One clearly needs a physical representation of qubits.

(2) The coherence time of the qubits should be large enough to allow the computation.

(3) There should be a physical mechanism realizing the unitary evolution of the qubits. This mechanism must be controllable.

(4) Initial qubit states should be conveniently prepared.

(5) There should be a way of performing a projective measurement of the final qubit states.

In 2000, Farhi (Farhi et al., 2000) described a new model of quantum computation based on the quantum adiabatic theorem. It turned out that the quantum adiabatic model is equivalent to the quantum gate network model of Deutsch (Aharonov et al., 2008). The adiabatic model is discussed in the next section.

So far, small-scale quantum computers based on different kinds of physical qubits have been developed. To name a few: single photon quantum computers (Chuang and Yamamoto, 1995; Knill et al., 2001), nuclear spins (Gershenfeld and Chuang, 1997; Vandersypen et al., 2001), trapped ions (Cirac and Zoller, 1995), neutral atoms in optical lattices (Brennen et al., 1999), states of superconducting circuits (Mooij et al., 1999), quantum dots (Imamoglu et al., 1999) and electrons' spin on helium (Platzman and Dykman, 1999). We will focus our attention on quantum adiabatic computation as the "software" of quantum computers, while for some "hardware" details, we refer the reader to Cohen and Tamir (2014).

7.3 Adiabatic Computation

Adiabatic quantum computation (AQC) is a scheme of quantum computation that is theoretically predicted to be more robust against noise than other methods (Paz and Zurek, 1999; Childs et al., 2001; Lloyd, 2008). In this scheme a physical system is initially prepared in its known lowest energy configuration, or ground state. The computation involves gradually deforming the system's Hamiltonian, very slowly, to assure that the system remains in its ground state throughout the evolution process. One designs the evolution of the Hamiltonian such that the ground state of the final Hamiltonian is the solution to the optimization problem. AQC is based on the adiabatic theorem stated by Born and Fock (1928):

"A physical system remains in its instantaneous eigenstate if a given perturbation is acting on it slowly enough and if there is a gap between the eigenvalues and the rest of the Hamiltonian spectrum."

The Hamiltonian is therefore time dependent $H = H(t)$. The initial Hamiltonian $H(0) = H_0$ and its lowest energy eigenvector should be easy to construct. We assume the final Hamiltonian H_T is also easy to construct, however its ground state which is the solution to our optimization problem could be exponentially hard to find with a classical algorithm. The Hamiltonian of the AQC is therefore:

$$H(t) = (1 - \frac{t}{\tau})H_0 + \frac{t}{\tau} H_T \equiv (1-s)H_0 + sH_T, \tag{7.1}$$

where τ is the adiabatic time scale, and t goes from 0 to τ. The complexity of the adiabatic algorithm is manifested in the time it takes to evolve the computer from its initial to its final state. It can be shown (Farhi et al., 2000) that the adiabatic approximation is valid when the annealing time satisfies:

$$\tau \gg \frac{\max_{0\leq s\leq 1}[\langle 1(s)|\frac{dH(s)}{ds}|0(s)\rangle]}{\min_{0\leq s\leq 1}[\Delta_{10}(s)]^2}, \tag{7.2}$$

where $|i(s)\rangle$ for $i = 0,1$ are the ground and first excited states of $H(s)$, and $\Delta_{10}(s)$ is their energy difference. To understand this lower bound on the time complexity we go back to basic principles (Messiah, 1965). Given a time-dependent Hamiltonian, we can write:

$$H|\psi_n(t)\rangle = E_n(t)|\psi_n(t)\rangle, \tag{7.3}$$

where $|\psi_n(t)\rangle$ (resp. $E_n(t)$) is the n-th eigenvector (resp. eigenvalue). We can write a general solution to the Schrödinger equation as:

$$|\psi(t)\rangle = \sum c_n(t)|\psi_n(t)\rangle e^{i\theta_n(t)}, \tag{7.4}$$

where $\theta_n(t)$ is known as the dynamic phase. Using the Schrödinger equation, one can verify that:

$$\dot{c}_n(t) = -c_n(t)\langle\psi_n(t)|\dot{\psi}_n(t)\rangle - \sum_{m\neq n} c_m(t)\frac{\langle\psi_n(t)|\dot{H}|\psi_m(t)\rangle}{E_m - E_n}e^{i(\theta_m-\theta_n)}. \tag{7.5}$$

Now, to ensure that the evolution of the n-th eigenstate remains as the n-th eigenstate, we have to reduce the amplitude:

$$\frac{\langle\psi_n(t)|\dot{H}|\psi_m(t)\rangle}{E_m - E_n}. \tag{7.6}$$

This can be done by varying $H(t)$ very slowly with respect to $E_m - E_n$. This is the origin of the argument for the complexity of the adiabatic computer. In fact, this also shows that one can preserve all eigenstates if the evolution is slow enough. In particular if this first gap decreases very rapidly or exponentially as a function of the number of variables in our problem, then we should exponentially slow down the evolution. A more accurate estimation of the complexity time using perturbation theory was suggested by Jansen et al. (2007). Young et al. (2008), studied the computation of the minimal gap for the Exact Cover problem using quantum Monte Carlo simulations, with the adiabatic computation implementation in mind. It turned out that the time complexity for the computation of the minimal gap is so far exponential.

Rigolin and Ortiz (2012) stated an adiabatic theorem for degenerate states. A similar principle also applies there. If the evolution is slow enough, then the transition between eigenspaces, that is between different energy states, has low amplitude. A necessary and sufficient condition for the adiabaticity was computed.

AQC was first proposed by Farhi et al. (2000), where the 3-SAT problem was discussed. Farhi et al. (2000) also suggested that one way to attack the time complexity problem is by looking for a tensor decomposition of the total Hamiltonian to smaller-dimensional Hamiltonians.

Can an adiabatic computer solve an NP-complete problem efficiently? This question is not answered yet, however, the discussion by Farhi et al. (2001) is so far the closest we can get to an answer. There, the Exact Cover problem for a set of random instances was discussed. Each instance was constituted by several random iterations, where on each iteration, a random clause was picked and added to the set of previously picked clauses, thereby reducing the number of satisfying assignments, down to only one. In this generating process, the relation between the number of variables and the number of clauses is close to 1, this is believed to be the phase transition point between instances with several solutions (low number of clauses) and instances with no solutions (high number of clauses). Such instances are believed to be hard to solve on a classical computer. The quantum adiabatic algorithm was simulated on a classical computer up to $n = 20$ (where n is the number of variables). The time needed to get a success probability higher than some fixed value (1/8) was computed. It turns out that the time needed is quadratically related to n. Note that this quadratic relation is true only upto low values of n. Moreover, it is not clear if the aforementioned randomly generated set of instances are really hard to compute on a classical computer; nevertheless the results obtained by Farhi et al. (2001) are most challenging.

Farhi et al. (2002) gave a possible illustration of the exponential time complexity of the AQC in solving 3-SAT problems. Given a 2^n dimensional space, a cost function was defined by taking the sum of a 3-local cost function $h_3(z_i, z_j, z_k)$ on all sets of three variables. Such a cost function can be shown to be symmetric and a function of (only) the Hamming weight. The Hamiltonian therefore resembles a 3-SAT Hamiltonian. It turns out that the minimal eigenvector $|\theta, s\rangle$ can be parameterized by an angular parameter θ. At some point s^*, there is a degeneracy, where $|\theta_1, s\rangle$ and $|\theta_2, s\rangle$ are two distance vectors with almost 0 eigenvalue. At $s < s^*$, $|\theta_1, s\rangle$ is closer than $|\theta_2, s\rangle$to the 0 eigenvector and at $s > s^*$, $|\theta_2, s\rangle$ is closer to the 0-eigenvector. For that to happen, the system, while crossing s^* should tunnel through a barrier which might take an exponential time. One has to evolve the system slowly enough to allow the tunneling near s^*. Otherwise it will not stay in the global minimum. The time needed for such a tunneling can be computed by tunneling methods such as instantons. This emphasizes the importance of tunneling in quantum computing.

van Dam et al. (2001) suggested that AQC resembles a local search: when given a problem where the global optimum lies in a narrow basin, while there is a close local minimum with a much larger basin, the AQC might need an exponential time to reach the global minimum. This is as if the AQC is stuck in the large basin's local minimum. Note however, that this is not the case, since an AQC always stays at the global minimum, and therefore the existence of a large basin local minimum should be translated to an exponential contraction of the first gap. To demonstrate this behaviour of the AQC, the authors suggested an artificial example where a global function was defined by the Hamming weight of its input configuration, being equal to this weight if it is lower than $\frac{1+\varepsilon}{2}n$, and -1 (the global minimum) if the weight is higher.

van Dam et al. (2001) also proved that one can simulate any AQC by a series of unitary gates. This is one direction of the proof of the equivalence between AQC and the circuit model, below we shall describe the second direction (from unitary gates to adiabatic evolution). The proof by van Dam et al. (2001) relies on the discretization of the continuous time-dependent Hamiltonian into small intervals, where on each interval, the instantaneous Hamiltonian is time independent. It is easy to see that the norm difference between the continuous Hamiltonian and the discrete one is bounded above by an efficient function, and therefore the norm difference of the corresponding unitary gates (as proved in the paper) is also bounded by a similar bound. One has yet to show (as done in the paper) that each of the unitary gates could be efficiently constructed.

A major breakthrough was achieved when Aharonov et al. (2008) showed that adiabatic quantum computation is equivalent to the circuit model. Aharonov's theorem is based on the fact that for any circuit model algorithm, we can produce a "history vector" describing the whole process of concatenated unitary gates on the initial vector. This history vector could be written as the ground state of a certain Hamiltonian; hence, the circuit computation turns out to be an adiabatic one.

Here we shall present the main arguments given by Aharonov et al. (2008) since they have major significance for the whole subject.

Given a quantum circuit algorithm, we have to show that there exists an adiabatic algorithm that can produce the same output vector with high probability. A circuit algorithm is a concatenation of L unitary gates U_l for $l = 0$ to L such that $U_l\,|\alpha_l\rangle = |\alpha_{l+1}\rangle$. The trick is to look at the history vector:

$$|\eta\rangle = \frac{1}{\sqrt{L+1}} \sum_{l=0}^{L} |\alpha_l\rangle \otimes |l\rangle\,, \tag{7.7}$$

where $|l\rangle$ denotes a clock vector in the l-th state, and the $|l\rangle$s are orthogonal. In fact, such a vector was used by Kitaev et al. (2002) and was previously suggested by Feynman (1986). Consider the Hamiltonian H_P:

$$H_P = \frac{1}{2} \sum_{l=0}^{L} H_l \tag{7.8}$$

where

$$H_l = I \otimes |l\rangle\langle l| - U_l \otimes |l+1\rangle\langle l| - U_l^\dagger \otimes |l\rangle\langle l+1| + I \otimes |l+1\rangle\langle l+1|\,. \tag{7.9}$$

By the definition of H_l and the orthogonality of the $|l\rangle$s:

$$\begin{aligned} H_l(|\alpha_l\rangle \otimes |l\rangle) &= |\alpha_l\rangle \otimes |l\rangle - |\alpha_{l+1}\rangle \otimes |l+1\rangle \\ H_l(|\alpha_{l+1}\rangle \otimes |l+1\rangle) &= |\alpha_{l+1}\rangle \otimes |l+1\rangle - |\alpha_l\rangle \otimes |l\rangle\,, \end{aligned} \tag{7.10}$$

and therefore, $|\eta\rangle$ is the 0 eigenvector of the Hamiltonian H_P (one has to take care of the definitions of boundary conditions for H_l and α_l at 0 and L). Note that H_l is designed to verify that the history vector has the right concatenation at the lth point.

Now define

$$|\gamma_l\rangle = |\alpha_l\rangle \otimes |l\rangle\,. \tag{7.11}$$

We can then write:

$$\frac{1}{2}H_l + \frac{1}{2}H_{l-1} = |\gamma_l\rangle\langle\gamma_l| - \frac{1}{2}|\gamma_{l+1}\rangle\langle\gamma_{l+1}| - \frac{1}{2}|\gamma_{l-1}\rangle\langle\gamma_{l-1}|\,, \tag{7.12}$$

and hence, H_P is a Toeplitz matrix on the space spanned by $|\gamma_l\rangle$ for $l = 0$ to L :

$$H_P = \begin{pmatrix} \cdot & \cdot & \cdot & \cdot & \cdot & \cdot & \cdot \\ -\frac{1}{2} & 1 & -\frac{1}{2} & \cdot & \cdot & \cdot & \cdot \\ \cdot & -\frac{1}{2} & 1 & -\frac{1}{2} & \cdot & \cdot & \cdot \\ \cdot & \cdot & -\frac{1}{2} & 1 & -\frac{1}{2} & \cdot & \cdot \\ \cdot & \cdot & \cdot & -\frac{1}{2} & 1 & -\frac{1}{2} & \cdot \\ \cdot & \cdot & \cdot & \cdot & -\frac{1}{2} & 1 & -\frac{1}{2} \\ \cdot & \cdot & \cdot & \cdot & \cdot & \cdot & \cdot \end{pmatrix}.$$

It is left to compute the first gap of the $L+1$ dimensional matrix and show that it does not contract to 0 faster than some polynomial function in L.

For small s values, we can use the Gershgorin theorem (Bhatia, 1997). For s close to 1, we have to use stochastic methods. Consider $G(s) = I - H_P(s)$ and note that $G(s)$ is primitive. Therefore, by the Perron–Frobenius lemma (Horn and Johnson, 1985), it has a non-degenerate eigenvector for its highest eigenvalue $\mu(s)$ with positive entries $(\alpha_0, ..., \alpha_L)$. This highest weight vector is also the ground state of $H_P(s)$.

Using $G(s)$, we can define a stochastic matrix $P = P(s)$:

$$P_{i,j}(s) = \frac{\alpha_j}{\mu(s)\alpha_i} G_{i,j}(s). \tag{7.13}$$

The rest of the proof follows from the following two arguments (which are shown in the paper):

a) The limiting distribution of $P(s)$ is $(\frac{\alpha_0^2}{Z}, ..., \frac{\alpha_L^2}{Z})$, where $Z = \sum_i \alpha_i^2$ is a normalizing factor and the first gap of $P(s)$ is $\Delta H_P(s)/\mu(s)$.

Hence, the limiting distribution and the first gap of this stochastic matrix is directly related to the ground state and the first gap of $H_P(s)$. Therefore, we can use stochastic arguments to complete the proof. It is best to look at the conductance of the stochastic process. If B is a subset in $\{0, ..., L\}$, $P_{i,j}$ a stochastic process with limiting distribution π, then the flow out of B is defined by:

$$F(B) = \sum_{i \in B, j \notin B} \pi_i P_{i,j}, \tag{7.14}$$

and the "weight" of B is $\pi(B) = \sum_{i \in B} \pi_i$. The conductance of P is then defined as:

$$\phi(P) = \min_B \frac{F(B)}{\pi(B)}, \tag{7.15}$$

where we minimize over all the subsets B such that $\pi(B) < \frac{1}{2}$. This is the minimal normalized flow.

b) The conductance $\phi(P)(s)$ can be easily estimated to satisfy

$$\phi(P)(s) \geq \frac{1}{6L}. \tag{7.16}$$

This is shown by using the monotonicity of the coordinates of the ground state of $H_P(s)$. The theorem now follows from the known connection between the conductance and the first gap in stochastic processes. Low conductance means a low first gap. In particular, the first gap is at least $\frac{1}{2}\phi(P)^2$ (Sinclair and Jerrum, 1989). Hence, we achieve the desired inverse polynomial condition for the first gap.

The computation ends where we measure the output of the adiabatic computer, i.e., the history vector. We can first measure the clock register, and if the clock points to L, then the vector register holds $|\alpha_L\rangle$. To decrease the angle between the history vector and $|\alpha_L\rangle$ (in fact, the embedding of $|\alpha_L\rangle$ in the history vector), one can add several identity gates at the end of the circuit network; then the same vector $|\alpha_L\rangle$ will appear in the last few vectors $|\gamma_l\rangle$.

Since this proof, several adiabatic protocols were suggested to solve, for instance, the problems of: graph isomorphism, quantum counting, Grover's search problem, the Deutsch–Jozsa problem and Simon's problem (Farhi et al., 2000; van Dam et al., 2001; Hen, 2014a,b). However, in general, there is no direct and simple way to translate an algorithm in terms of the circuit model to an algorithm in terms of adiabatic computation, the proof by Aharonov et al. (2008) does not provide a simple way to go from one model to the other. This difficulty is mainly due to the fact that the exponentiation of a sum of Hamiltonians that do not commute is not the product of the exponentiation of each of the Hamiltonians. Therefore, in the circuit model we can present a simple set of universal gates, whereas for the adiabatic model it is much harder.

Albeit all the aforementioned difficulties, the equivalence between the models provides a new vantage point from which to tackle the central issues in quantum computation, namely designing new quantum algorithms and constructing fault tolerant quantum computers.

Childs et al. (2001) discusses the robustness of AQC using a master equation. A problem Hamiltonian solving an instance of a 3-bit Exact Cover problem was constructed using spin $\frac{1}{2}$ variables. There is an inherent robustness to the adiabatic evolution against dephasing in-between the eigenvectors since the computer stays in its ground state. Therefore, the problem is decoherence, due to the environment, into a higher eigenvector, or more generally, into a Gibbs state. Childs et al. (2001) uses a weak system-bath coupling. It was also assumed that the initial state of the system and bath is a tensor product of density matrices, and moreover, that the unitary evolution in the master equation is governed by a time-independent Hamiltonian. This last assumption is plausible since the evolution is very slow being adiabatic. The bath was assumed to be made of photons. An explicit master equation was computed in the energy eigenstates. The results were rather expected. The success probability increased with computation time (being more adiabatic), also being larger for a higher gap problem. For high temperature, the decoherence effect decreased the success probability. For very low temperatures and short computation times, the decoherence effect increased the probability by letting the system relax to the ground state from higher energy states, where it was placed due to the fast evolution. However, the results were non-conclusive being computed for very low dimensional instances.

Rigolin et al. (2008) introduced an adiabatic perturbation theory (APT) which expressed $|\psi(s)\rangle$ in terms of a power series in $v = \frac{1}{\tau}$ (where τ is the adiabatic time scale):

$$|\psi(s)\rangle = \sum_{p} v^p \, |\psi^p(s)\rangle \, . \tag{7.17}$$

Here $|\psi^p(s)\rangle$ is the pth perturbation term, and $|\psi^0(s)\rangle$ is the standard adiabatic term. For a very large τ, the only term in this sum will be the standard adiabatic one. We thus write $|\psi(s)\rangle$ in terms of the eigenvectors $|n(s)\rangle$ of $H(s)$:

$$|\psi(s)\rangle = \sum_{n,m}\sum_{p} v^p e^{\frac{-i}{v}\omega_m(s)} e^{i\gamma_m(s)} b^p_{n,m}(s)\,|n(s)\rangle\,, \tag{7.18}$$

where γ_m is a dynamic phase and ω_m is a geometric Berry phase. The main result of Rigolin et al.'s study (2008) is a recursion formula between $b^{p+1}_{n,m}(s)$, $\dot{b}^p_{n,m}(s)$, and $b^p_{k,m}(s)$ for all k. This enables the computation of the $p+1$ perturbation term using the p terms.

Such a perturbation theory is a major step in computing adiabatic sufficient and necessary conditions.

7.3.1 Simulated annealing and adiabatic computation

There is a deep analytic relation between simulated annealing and quantum adiabatic processes. In this section, we describe these relations (Somma et al., 2007; Morita and Nishimori, 2008), also known as the classical to quantum mapping. Suppose we are given a set of Boolean variables $\sigma_i = \pm 1$ defined on an n dimensional lattice. Let E be an energy (cost) function defined on the 2^n dimensional configuration space $\overline{\sigma} = (\sigma_1, .., \sigma_n)$, for example

$$E(\overline{\sigma}) = \sum_{i,j} J_{i,j}\sigma_i\sigma_j, \tag{7.19}$$

where $J_{i,j}$ are coupling coefficients. One can define a stochastic process S_β (for $\beta = \frac{1}{kT}$, an inverse temperature) such that its limiting distribution π_β satisfies:

$$\pi_\beta(\overline{\sigma}) = Z_\beta^{-1} e^{-\beta E(\overline{\sigma})} \tag{7.20}$$

$$Z_\beta = \sum_{\text{all } \overline{\sigma}} e^{-\beta E(\overline{\sigma})}. \tag{7.21}$$

If β is big enough, then sampling from π_β will result in a lowest energy configuration with high probability. Now for the quantum analogue; for each $\overline{\sigma}$, we can construct a corresponding quantum pure state $|\overline{\sigma}\rangle$, defined in the tensor space of Pauli z-spinors. Set S_β to be the stochastic matrix satisfying the detailed balance condition:

$$S_\beta(\overline{\sigma}_i|\overline{\sigma}_j)\pi_\beta(\overline{\sigma}_j) = S_\beta(\overline{\sigma}_j|\overline{\sigma}_i)\pi_\beta(\overline{\sigma}_i). \tag{7.22}$$

We can now define a Hamiltonian by setting its matrix coefficients to:

$$\langle\overline{\sigma}_i|\, H_\beta\, |\overline{\sigma}_j\rangle = \delta_{i,j} - \sqrt{S_\beta(\overline{\sigma}_i|\overline{\sigma}_j)S_\beta(\overline{\sigma}_j|\overline{\sigma}_i)}. \tag{7.23}$$

Define the state

$$|\psi_\beta\rangle = \frac{1}{\sqrt{Z_\beta}} \sum_{\text{all } \overline{\sigma}} e^{-\beta E(\overline{\sigma})/2}\, |\overline{\sigma}\rangle\,. \tag{7.24}$$

Then $|\psi_\beta\rangle$ corresponds to the Gibbs state for E and β. It turns out that $|\psi_\beta\rangle$ is the unique ground state of H_β (Boixo et al., 2015). In such a correspondence, one can also show that a transverse field component such as $\sum_j \sigma_j^x$, is only natural in this correspondence between the stochastic transition matrix and the Hamiltonian. In fact, the ground state of $I - \frac{1}{n}\sum_j \sigma_j^x$ which is the sum of all states in the computational z-basis with equal amplitudes, corresponds to the completely mixed state at infinite temperature.

We can now let $T = T(t)$ be a function of time. The question is how to choose the correct pace to decrease T. Somma et al. (2007) and Morita and Nishimori (2008) computed the evolution rate to guarantee that the whole process is adiabatic. The correct rate was found to be (no faster than)

$$T(t) \sim \frac{n}{\log(t)}, \tag{7.25}$$

which is very close to the known simulated annealing rate. Therefore, it seems that this correspondence between stochastic matrices and Hamiltonians embed simulated annealing into the set of quantum annealing, as if a special path of the quantum annealing process (in fact an adiabatic one) can produce a classical simulated annealing process.

In Somma et al. (2008) suggested a spectral gap amplification method for the classical annealing algorithm. This is done by quantum simulation of the classical annealing process using a version of the Grover algorithm suggested by Ambainis (2014). While solving the problem of element distinctness, a new version of the Grover algorithm was suggested. The algorithm alternates between two type of transformations, the first marks the target state (the ground state of the final Hamiltonian) using a –1 phase and leaves all orthogonal states untouched (see the discussion in the previous chapter), while the second transformation leaves the initial (start) state untouched while multiplying any of its perpendicular vectors by the same phase; this corresponds to the reflection transformation in the original Grover model. It was shown by Ambainis (2014) that such alternate iterations take the start state close to the target state. Somma et al. (2008) used this method to artificially amplify the spectral gap.

7.3.2 Different paths from an initial to a final eigenstate

An alternative way to the standard continuous path evolution of the AQC, was suggested by Boixo et al. (2009). Using the quantum Zeno effect (Misra and Sudarshan, 1977), one can discretely evolve the initial state to the final state or very close to it. In the quantum Zeno effect, we need $\frac{L^2}{1-p}$ steps to cross the angular distance L with fidelity p. We use a discrete set of Hamiltonians $H(l)$ $0 \le l \le L$, and we guarantee that the evolution by each of the Hamiltonians is close to the Zeno projection into $|\psi_l\rangle$, a non degenerate eigenvector of $H(l)$. Each Hamiltonian is evolved for a time t, which is randomly chosen from some distribution. Let R_l^t denote the corresponding evolution operator of $H(l)$ for a time t. We define a projective measurement operator by:

$$M_l(\rho) = P_l \rho P_l + (I - P_l)\rho(I - P_l), \tag{7.26}$$

where $P_l = |\psi(l)\rangle\langle\psi(l)|$. One can now estimate the difference between the projective measurement operator and the evolution operators:

$$||(M_l - R_l^t)(\rho)||_{tr} \leq \sup_{\omega_j} \Phi(\omega_j), \tag{7.27}$$

where ω_j are the energy difference to the other eigenvalues of $H(l)$, and Φ is the characteristic function of the distribution of t. We can now estimate the expectation value of the distribution of t, i.e., the average time we need to evolve any of the Hamiltonians, in terms of Φ (which lies in the frequency domain). With some knowledge of the random variable t, we can write

$$\langle t \rangle \gg \frac{1}{\min_s \Delta(s)} = \frac{1}{\Delta}. \tag{7.28}$$

Using the above condition on $\langle t \rangle$, we can guarantee that the random evolution operators are close to the projective measurement operators. It then follows that the complexity of the whole process is:

$$O\left(\frac{(L)^2 \log(L/(1-p))}{(1-p)\Delta}\right). \tag{7.29}$$

7.3.3 Imaginary time and simulations

An analysis of imaginary time was presented by Morita and Nishimori (2008). It was found that the asymptotic behavior of the imaginary time quantum annealing (IT-QA) is the same as the real time quantum annealing (RT-QA), also the error of the IT-QA is no larger than the error of RT-QA. However, the importance of the use of imaginary time lies in the fact that this new algorithm can be simulated on a classical computer and could be considered as a form of a quantum Monte Carlo method.

This is the essence of quantum MC techniques; we use imaginary time in the Schrödinger equation, turning the quantum equation into a classical one. Some of the common methods are the variational (McMillan, 1965), diffusion (Grimm and Storer, 1971), auxiliary field MC (Ceperley et al., 1977), path integral (Barker, 1979), Gaussian (Corney and Drummond, 2004), and stochastic Green function (Rousseau, 2008)

7.3.4 Complexity and universality

A few words on complexity class theory and universality are in order. The quantum parallel of the classical NP complexity class is known as QMA (Watrous, 2000). It is the class of all languages L that can be probabilistically verified by a quantum verifier in polynomial time. In particular, let B be the Hilbert space of some qubit, V a polynomial time quantum verifier and p a polynom. We say that L is in QMA if the following two conditions are satisfied:

a) If $x \in L$ then there exists a proof π in $B^{p(|x|)}$ such that:

$$pr(V(x,\pi) = 1) > \frac{2}{3}$$

b) If $x \notin L$ then for all π in $B^{p(|x|)}$;

$$pr(V(x,\pi) = 1) \leq \frac{1}{3}$$

The connection between the above definition of QMA and our discussion goes through the definition of the k-Local Hamiltonain Problem. Given a k-local Hamiltonian, that is, a Hamiltonian that acts on only k qubits, suppose it is promised that the ground state of the Hamiltonian is either below a or above b where $a < b \in [0,1]$ and $1/(b-a) = O(n^c)$ for some constant c. Then one has to distinguish between the two cases.

A k-local Hamiltonian can be thought of as a set of local constrains on the set of n qubits. Therefore, the problem of k-local Hamiltonian resembles the MAX-k-SAT problem. Kitaev showed that the $log(n)$-local Hamiltonian problem is in QMA (Kitaev, 2002), where $|x| = n$. Moreover, the 5-local Hamiltonian is QMA-complete. This fact could be interpreted as a map between the language of computational complexity theory and the language of condensed matter physics, hence its importance. Some refinments of Kitaev's theory quickly followed; in (Kempe and Regev, 2003) it was proven that the 3-local Hamiltonian problem is also QMA-complete, and in (Kempe et al., 2005) it was shown that the 2-local Hamiltonian problem is QMA-complete.

It was later proven by Biamonte (Biamonte and Love, 2008) that the 2-local Hamiltonian problem is QMA-complete even when restricted to real valued Hamiltonians (that is, represented by real matrices). Biamonte also showed that one can approximate the ground state of such Hamiltonian by a set of simple and realizable 2-local Hamiltonians; two such universal sets were introduced; in particular it was shown that:

a) The 2-local ZZXX Hamiltonians are QMA-complete, where:

$$H_{ZZXX} = \sum h_i \sigma_i^z + \sum \Delta_i \sigma_i^x + \sum J_{i,j} \sigma_i^z \sigma_j^z + \sum K_{i,j} \sigma_i^x \sigma_j^x$$

b) The 2-local ZX Hamiltonians are QMA-complete, where:

$$H_{ZX} = \sum h_i \sigma_i^z + \sum \Delta_i \sigma_i^x + \sum_{i<j} J_{i,j} \sigma_i^z \sigma_j^x + \sum_{i<j} K_{i,j} \sigma_i^x \sigma_j^z$$

In Ref. (Biamonte, 2008) a general scheme for writing the ground states of k-local Hamiltonians using 2-local Hamiltonians was presented. This resembles the use of simple Karnaugh maps in reduction of variables.

7.3.5 Additional methods

Boixo et al. (2015) also suggested a "partly adiabatic-partly diabatic" process. Suppose the two lowest eigenvalues are separated from the rest by a polynomially decreasing gap, while the first gap decreases exponentially with the size of the problem. One can compute the probability of jumping to the second eigenvector if the evolution is too fast (polynomial) near the (normalized) time s where the gap is minimal. In a similar way, one can compute the probability of going back to the ground state later on. This could be easy in case the problem is symmetric with respect to s. Such an example for the random glued tree problem was discussed by Boixo et al. (2015).

A few final remarks before we go on to discuss quantum annealing. The adiabatic model has several setbacks. The most important is the lack of a guaranteed fault tolerant method. In the circuit model, one can control the amount of noise passed on to neighboring qubits (see for instance the review by Nielsen and Chuang (2000)). We can concatenate circuits on which we control the amount of noise. This cannot be done in the adiabatic case. This is connected to the fact that the adiabatic computer model has no universal subset of computers. However, the adiabatic computer is robust against several types of noise as discussed earlier.

One last remark, it is best to do the adiabatic evolution in zero temperature. In practice, if kT is much smaller than the gap, then the adiabatic evolution will overcome thermal noise. In general, if kT is larger than the gap, it might be useful to describe the evolution of the Hamiltonian within the context of the master equation (Albash et al., 2012).

7.4 Quantum Annealing

Quantum annealing was suggested as an improvement to the simulated annealing technique which suffers a severe setback in cases where the system is "non-ergodic" (e.g., systems described by the spin glass model). In such cases, configurations of n spins corresponding to the minimum of the cost function could be separated by $O(n)$ sized barriers (Drossel and Moore, 2004), so that at any finite temperature, thermal fluctuations take practically infinite time to relax the system to the global minimum.

There are clear similarities between simulated and quantum annealing. In both methods, one has to strictly control the relevant parameters and change them slowly to tune the strengths of thermal or quantum fluctuations. In addition, the main idea behind both classical and quantum annealing is to keep the system close to its instantaneous ground state. Quantum annealing excels in tunneling through narrow (possibly cuspidal) barriers. Classical simulated annealing schedules might still have an advantage when the barrier is wide and shallow.

The basic scheme is as follows. First the computational problem has to be mapped to a corresponding physical problem, where the cost function is represented by some Hamiltonian H_0 of the Ising form:

$$H_0 = -\sum_{i<j} J_{ij}\sigma_i^z\sigma_j^z - \sum_i h_i\sigma_i^z. \tag{7.30}$$

Here J_{ij} denotes the coupling strength between spins i and j, and h_i describes the magnetic field at site i. Then a suitably chosen non-commuting quantum tunneling Hamiltonian H_1 is to be added,

$$H_1 = \sum_i \Delta_i\sigma_i^x$$

where Δ_i denotes the interaction strength with the "tunneling" term, so that the total Hamiltonian takes the form of:

$$H = H_0 - \Gamma(t)\sum_i \Delta_i\sigma_i^x \doteq H_0 + H_1(t), \tag{7.31}$$

where $\Gamma(t)$ describes H_1's time dependence. One can then solve the time-dependent Schrödinger equation for the wave function $|\psi(t)\rangle$:

$$i\hbar\frac{\partial}{\partial t}|\psi\rangle = [H_0 + H_1(t)]\,|\psi\rangle\,. \tag{7.32}$$

The solution approximately describes the tunneling dynamics of the system between different eigenstates of H_0. Like thermal fluctuations in classical simulated annealing, the quantum fluctuations owing to $H_1(t)$ help the system escape from the local "trapped" states. Eventually, $H_1(t) \to 0$ and the system settles in one of the eigenstates of H_0; hopefully, the ground state. This serves as a quantum analog of cooling the system. The introduction of such a quantum tunneling is supposed to make the high (but very narrow) barriers transparent to the system; it can make transitions to different configurations trapped between such barriers, in the course of annealing. In other words, it is expected that application of a quantum tunneling term will make the free energy landscape ergodic (see review by Chakrabarti et al. (1996)), and the system will consequently be able to visit any configuration with finite probability. Finally the quantum tunneling term is tuned to zero to get back the Ising Hamiltonian. It may be noted that the success of quantum annealing is directly connected to the replica symmetry restoration in quantum spin glass due to tunneling through barriers.

The fact that one can use the quantum tunneling effect produced by a transverse field to help evolve the system into its ground state was first suggested Ray et al. in 1989. It was initially contested by Altshuler et al. (2010) on the ground that the Anderson localization will not allow it.

7.4.1 Relation between simulated annealing, quantum annealing, and adiabatic computation

This relation between simulated and quantum annealing was discussed, for example, by Rose and Macready (2007). Assume first that a classical system at a temperature T is examined. We would like to minimize its free (Helmholtz) energy. If $p(x)$ is the probability of the state x, then eventually the distribution $p = p(x)$ will minimize the free energy:

$$F_T(p) = E_p(H) - TS(p),$$

where $E_p(H)$ is the expected value of the energy function,

$$E_p(H) = \sum_x p(x)H(x),$$

and $S(p)$ is the entropy:

$$S(p) = -\sum_x p(x)\log[p(x)].$$

There are two considerations in reaching the minimum value (distribution) of $F_T(p)$. First, the entropy should be maximized; this occurs when the probability of each state is the same. Second, $E_p(H)$ should be minimized. The shape of $E_p(H)$ is the landscape of the energy or cost function. When T is big enough, the entropy is the dominant part in the expression for the free energy, and hence the equipartition of the probabilities is most important. In other words, if the temperature is high enough, we can reach any point in the landscape of the cost function. If T is reduced, we discover the minima of the energy function; we can then hopefully descend into the global minimum. This is what we do in the simulated annealing algorithm, as well as in the physical process of annealing. For the quantum case, we look at a similar free energy function, this time over density matrices:

$$F_T(\rho) = \mathrm{tr}\rho(H) - TS(\rho).$$

When the temperature is 0, we are left with tr $\rho(H)$. We can write H as a sum of two terms, "kinetic" and "potential" :

$$H = K + V.$$

Now we can let K play the same role as the entropy obtained earlier. If, for example, we let

$$K = -\sum_i \sigma_i^x,$$

then the minimal eigenvector of K is an equal sum of all elements in the z computational basis. We can also add a coefficient Γ to control the amplitude of K. This is all done in the 0 temperature case and therefore

$$F_T(\rho) = \mathrm{tr}\rho(V) + \Gamma\mathrm{tr}\rho(K).$$

This resembles a classical annealing process: we start with high Γ and slowly reduce to $\Gamma = 0$, where we hope to end in a global minimum.

To sum up, we have a phase space of two variables T and Γ, (T,Γ). Now $(T,0)$ is the classical annealing path, and $(0,\Gamma)$ is the quantum annealing path. Can we find different paths where both coefficients are non-zero with a low computation time complexity?

The quantum annealing, as evident from the earlier discussion, does not have to be adiabatic. Being adiabatic restricts the evolution time to be slow enough.

Finnila et al. (1994) discusses a tunneling effect for a double well cost function, where one of the wells has a lower minimum. The authors used the diffusion Monte Carlo method, where an imaginary time is used in the Schrödinger equation, turning it into a classical diffusion function. Thereafter, one can use random walk agents to simulate the diffusion. This is a simple way to compute the ground state. The results were rather surprising and the tunneling effect was clearly shown; an initial function located in the higher well was able to leap over the barrier into the lower energy state.

What are the differences between simulated and quantum annealing from the point of view of computational complexity theory? A convergence criterion was proved by Morita and Nishimori (2006) which is similar to the well-known one in simulated annealing (Geman and Geman, 1984). If we let $\Gamma(t) = t^{-\gamma/n}$ (where γ is some positive constant), we are guaranteed to get a solution. This, however, could take an infinite amount of time. If we stop the relaxation at some final time t_f where the "temperature" $\Gamma(t)$ is small, $\Gamma(t) = \delta \ll 1$, then it is enough to wait until $t_f = e^{-n\ln(\delta)/2\gamma}$. Compare this to the relaxation time t_f for the simulated annealing protocol; there $T(t) \sim \frac{n}{k\log(t)}$ (see the discussion in the previous section), and if $T(t_f) = \delta$, then $t_f = e^{\frac{n}{\delta k}}$. Clearly, for very small δ, i.e., $\frac{1}{\delta} \gg -\text{In}(\delta)$, the quantum annealing scheme will be better than its simulated annealing counterpart. This is true in general, but could be hard to utilize, since both relaxation times are exponential.

In fact, for some specific problems the advantage of quantum annealing over simulated annealing is much more clear. Kadowaki and Nishimori (1998) tested it on a toy model of 8 qubits with a transverse Ising field. The authors showed that quantum annealing leads to the ground state with much larger probability than the classical scheme, where the same annealing schedule is used. Martoňák et al. (2004) used the path-integral Monte Carlo quantum annealing to show better results for the traveling salesman problem for 1002 cities. Here the algorithm was stopped after various number of steps and the results were compared to a simulated annealing algorithm. QA was shown to anneal more efficiently, and to decrease the solution residual error at a much steeper rate than SA. Farhi et al. (2002) constructed an example where the width between local minima is small and therefore the tunneling effect is strong. The simulated annealing counterpart of the example shows an exponential computational complexity.

Recent results suggest that for first order phase transitions, the adiabatic algorithm has exponential time complexity. For second order phase transition, the adiabatic algorithm has polynomial complexity. It was also suggested that by adding an annealing term, one can solve first order transition problems (Jörg et al., 2010; Seoane and Nishimori, 2012).

Brooke et al. (1999) applied the aforementioned model to a disordered ferromagnet. Their aim was to find the ground state for the ferromagnet with a certain proportion of randomly inserted antiferromagnetic bonds. Cooling it down to 30 mK and varying a transverse magnetic field, they were able to compare simulated and quantum annealing, concluding eventually that their experiment directly demonstrates the power of a quantum tunneling term in the Hamiltonian.

Another evidence for the existence of tunneling effects was presented by Boixo et al. (2014), who tested the D-Wave computer on a family of randomly generated Ising problems. It showed a clear distinction between easy and hard problems. In the hard cases where the success probability was low, the Hamming distance between the final vector and the ground state was high, whereas in the easy cases (high success probability), the Hamming distance was low. This could suggest a tunneling effect. Hard cases are those where the evolution is too fast for them; hence, the computer get stuck at an excited eigenvector. However the fact that these are exactly the states with high Hamming distance means that the tunneling was avoided there. For if the Hamming distance is d, then to cross that distance by tunneling is Γ^d (exponentially) hard (this is clear if we look at the transverse field as a perturbation).

There is also a deep connection between the number of free qubits in the ground and first excited eigenstates and the first gap (Boixo et al., 2014). It is easy to see that the transverse field breaks the degeneracy of a free qubit. Thus, if the first excited state has more free qubits than the ground state, the splitting of the energy states by the transverse field are such that the minimal gap reduces. This does not happen in classical simulated annealing. Therefore, such problems could be harder for quantum simulated annealing than for classical annealing.

7.5 D-Wave Computer

On May 2011, D-Wave Systems Inc. announced "D-Wave One", as "the world's first commercially available quantum computer". The D-Wave one contained 128 qubits. It provoked an immediate controversy about its true properties. Recently, Google has purchased "D-Wave Two" containing 512 qubits.

D-Wave computers utilize flux qubits of the persistent current flux qubit (PCQ) type (Wendin and Shumeiko, 2005). A set of 8 qubits are inter-coupled into a cell. In Fig. 7.4, qubit a is coupled to qubits A, B, C, and D. Similarly, qubit A is coupled to qubits a, b, c, and d. All 8 qubits and their interconnections can be described by the graph in Fig. 7.4 in D-Wave Two, 64 such cells constitute a two dimensional grid. Each cell is connected to its neighboring cells. The whole 512 qubits therefore implement a graph known as the chimera graph C_n.

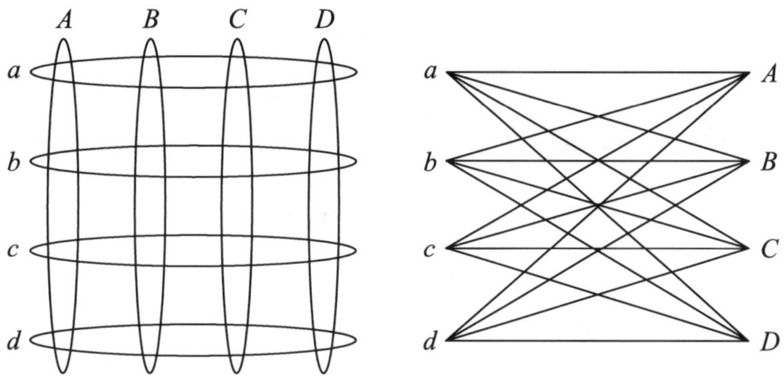

Fig. 7.4 8 qubit cell.

Since C_n is not a complete graph, it is not clear how to implement a general graph G into the hardware. One should distinguish between logical qubits and physical ones. For example, if C_n could only be connected to 4 of its neighbors, assume G has a vertex v_i with degree higher than 4, then to implement G inside such a C_n type machine, we first need to map v_i into a subtree of several such vertices. We will use the leaves of the tree to connect to other vertices (see Fig. 7.5). On the hardware, we will get a graph $\mathscr{G}$ of the physical qubits. The original graph G is called the minor of the expanded graph $\mathscr{G}$. Such a simple embedding demands that $|G|^2 \sim |\mathscr{G}|$, however there are more efficient embeddings (the task of finding such an embedding is in itself a hard computational problem, see also paper by Choi (2008)).

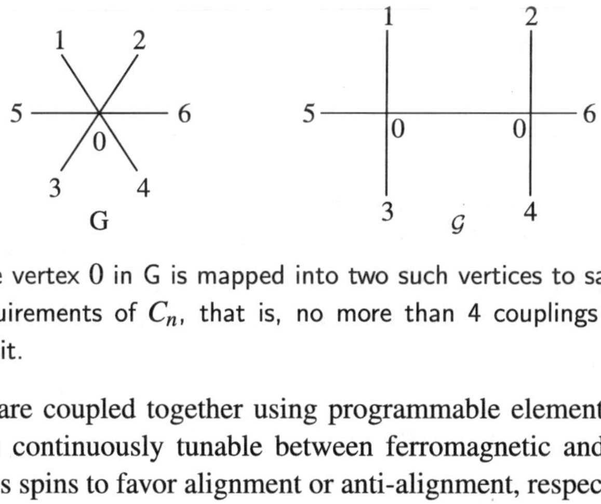

Fig. 7.5 The vertex 0 in G is mapped into two such vertices to satisfy the physical requirements of C_n, that is, no more than 4 couplings to each physical qubit.

The qubits (spins) are coupled together using programmable elements which provide an energy term that is continuously tunable between ferromagnetic and anti-ferromagnetic coupling; this allows spins to favor alignment or anti-alignment, respectively. The behavior of this system can be described with an Ising model Hamiltonian, similar to the ones in Eq. (7.30) and Eq. (7.31).

The computer is cooled to 20 mK by a dilution refrigerator, the pressure is set to 10^{-10} atmospheric pressure, and the computer is shielded magnetically to $5 \cdot 10^{-5}$ of earth magnetic field. The D-Wave computer has the size of a small chamber; however its core, a 512 qubits board, is much smaller.

The "Orion" prototype of the D-Wave contained only 16 qubits. Some of the problems it solved were pattern matching, Seating Arrangement and a Sodoku puzzle (Minkel, 2007). The "D-Wave One" (incorporating 128 qubits) was used by a team of Harvard University researchers to present some results of the largest protein folding problem solved to date (Ortiz et al., 2012). The current "D-Wave Two" consists of 512 qubits (not all active), and thus enables the solution of much more complex problems like network optimization and radiotherapy optimization which were demonstrated by the company (http://www.dwavesys.com).

McGeoch and Wang (2013) claimed that on quadratic unconstrained binary optimization problems, the D-Wave hardware returns results faster than the best known IBM CPLEX optimizer, by a factor of about 3600 for a problem of size $N = 439$. This was contested by CPLEX developers (Puget, 2013). Several similar claims were recently made by the D-Wave group, but so far for each such claim, there is a contesting one questioning the results.

The D-Wave computer is introduced as a quantum annealer, or as an adiabatic computer having a programmable transverse field for tunneling. In general, it is hard to guarantee that the time evolution will meet the requirements of the adiabatic theorem, and indeed it turns out that the D-Wave computer manifests a regime which is in-between adiabatic and thermic. The excitations to higher eigenvectors in the course of evolution is expected to be followed by a later relaxation into a ground state. It was claimed by the D-Wave group that such a regime could even improve the probability to get the correct result (Dickson et al., 2013).

7.5.1 Is the D-Wave a quantum computer? The future of quantum annealers

As a new apparatus it is only natural to ask how we can be sure this machine is indeed a quantum computer. If we have two parties having two computers, we can play the Clauser–Horne–Shimony–Holt (CHSH) (Dahl et. al., 2011) and if the success probability is higher than the classical computer (75%), we can assume the parties are using some quantum procedure with entanglements (see study by Reichardt et al. (2012)), however, for one computer the question is still open. In what follows, we present several criteria which we believe are important for the identification of the D-Wave (or any other computer) as a quantum computer. Some of the criteria can be attributed to the D-Wave computer, a few cannot, and the rest are still controversial.

7.5.2 Universality

For the adiabatic model, there is no natural set of universal gates generating the whole theoretic spectrum of the model. We can call this "inner universality". Since there is no such inner universality, the computer resembles the old analog "one purpose" computers (Das and Chakrabarti, 2008). Nevertheless we can ask what kind of problems can the

D-Wave computer solve? Currently, only particular optimization problems are considered. However, as was shown by Farhi et al. (2000), adiabatic computation can solve the 3-SAT problem (in exponential time) and therefore, in principle, any NP-hard problem could be solved. The (polynomial) equivalence of adiabatic computation and quantum circuit computation (Aharonov et al., 2008) suggests that, theoretically, the adiabatic method can be generalized towards solving any problem that can be solved by the circuit model. Indeed, several protocols for solving specific problems other than optimization were suggested: graph isomorphism, quantum counting, Grover's search problem, the Deutsch–Jozsa problem and Simon's problem (Farhi et al., 2000; van Dam et al., 2001; Hen, 2014a,b). However, in practice, there is no direct and simple way to translate an algorithm in terms of the circuit model to an algorithm in terms of adiabatic computation (then again, see the method suggested by Lloyd (2008)). Therefore, in the circuit model, we can present a simple set of universal gates, whereas for the adiabatic model, it is much harder.

7.5.3 Coherence time of the SQUIDs

The coherence time of the qubit should be larger than the time needed for the algorithm to compute. This is far from being achieved in the D-Wave computer. The coherence time of the SQUID is about 10 ns while the annealing time needed is 5–15 μs (Boixo et al., 2014). Indeed, how can one achieve quantum computation when the annealing time (depending on the first energy gap) of the computer is about 3 orders of magnitude longer than the predicted single-qubit coherence time? This seems to force a thermodynamic regime on the computer. The main reason for using such flux qubits is that they are relatively simple to manufacture, using common methods of lithography. A set of such SQUIDs, their coupling apparatus and measurement gates are concatenated as on a printed circuit board. Hence, we are still in need of a qubit "transistor", that is, a simple apparatus presenting a behavior of a two-state system, that can maintain coherence for a long time (in comparison to the operation time), and can be read off, coupled, and easily manipulated.

It seems that the D-Wave computer operates under a semi-adiabatic-semi-thermodynamic protocol. Near the anti-crossing, the evolution might be too fast for an adiabatic computer. Therefore, the eigenvectors are excited to higher energy states, later to be relaxed to the 0 eigenstate again. It was suggested by the D-Wave group that such relaxations could even help in the adiabatic evolution into the ground state (Dickson et al., 2013).

Recently, a promising progress was achieved regarding the decoherence time of flux qubits (Stern et al., 2014).

Another promising direction for increasing coherence time, is the research on anyons (Kitaev, 2003). These quasiparticles are topologically protected against decoherence and hence might be very valuable as the building blocks of a quantum annealer.

7.5.4 Scalability

How many qubits can a D-Wave have? D-Wave computers have made a major leap when incorporating the largest number of qubits ever seen on a single device. The question now is of scalability. It is possible that the complexity to construct such a computer with all its inter-couplings, grows exponentially. This will mean that the possible gain in algorithmic complexity is paid out in building a coherent circuit (see also (Kalai, 2011)). This question is deeply connected to the lack of a fault tolerant gate theory for adiabatic computation. A scalable architecture of adiabatic computing was suggested by Kaminsky and Lloyd (2004), by translating NP hard problems to the max independent set problem. For that problem, a highly robust Hamiltonian was suggested. A more fundamental research in this issue should be done in the context of the master equation (see also (Albash et al., 2012)).

7.5.5 Speed up

Are the D-Wave computers faster than other computers running different optimization algorithms? For which problems? Right now, it seems that the answer to the first question is "sometimes" and it is not clear yet what is the answer to the second problem (Boixo et al., 2014). In our opinion this is the most important indicator because of its practical significance, but currently it is a problematic issue (Boixo et al., 2014; Katzgraber et al., 2014). In 2013, it was indeed admitted by the D-Wave group (D-Wave group, 2013) that different software packages running on a single core desktop computer can solve those same problems as fast or faster than D-Wave's computers (at least 12,000 times faster than the D-Wave for quadratic assignment problems, and between 1 and 50 times faster for quadratic unconstrained binary).

Rϕnnow et al. (2014) discussed the question of defining and detecting quantum speedup. It was implicitly suggested that one should think of a new way to define computational complexity, at least for cases where the instances of the computational problem are randomly generated. In the following, we describe the general idea.

Consider the data presented by Boixo et al. (2014) and Albash et al. (2015), where 1000 different spin-glass instances (randomly picked) were investigated. Each instance was run 1000 times and the success probability s for finding the correct solution was computed. The parameter s could also describe the "hardness" of the problem.

Suppose one annealing run takes t_a time and has a success probability s. Therefore the total success probability of finding the solution at least once in R runs is $p = 1-(1-s)^R$. Set now $p = 0.99$ and write $R = R(s)$. Let $T_{DW}(N,s)$ be the time complexity of the D-Wave computer wired to a problem of size $\sqrt{N}$, (see the discussion earlier on the minor problem) and hardness s, let $T_C(N,s)$ be the corresponding classical time complexity. Clearly, $T_{DW}(N,s)$ would be proportional to Rt_a. One way to define a speedup would be to look at the quotient of quantiles:

$$\frac{T_{DW}(N,s)|_{s\leq s_0}}{T_C(N,s)|_{s\leq s_0}}. \tag{7.33}$$

This means that we average both complexities on a large set of instances (indexed by their hardness s) and only then compute the quotient. This suggests a new way to look at computational complexity theory, as a quotient of integrals or averages. Another way to define the speedup would be to look at the quantile of the quotient:

$$\left. \frac{T_{DW}(N,s)}{T_C(N,s)} \right|_{s \leq s_0} \tag{7.34}$$

which compares the complexity of both computers on the same instance and only then as a function of hardness s. Both methods presented inconclusive results of speedup, although the second showed a small advantage for using the D-Wave computer when N is large (Rϕnnow et al., 2014).

7.5.6 "Quantumness"

Since there is no clear evidence for a speedup, there is a possibility of comparing the behavior of the D-Wave computer to other models of computation with respect to a large family of computational problems. Consider again the data presented by Boixo et al. (2014) and Albash et al. (2015). A histogram describing the number of instances for each success probability s was presented. The D-Wave histogram was found to be strongly correlated with quantum annealers (in fact simulated quantum annealers) rather than classical annealers. Both the D-Wave and the quantum annealer had a bimodal histogram, a large set of problems which are very easy to solve (high success probability) and a large set that are hard to solve (low success probability). The classical simulated annealer had a normal distribution type of histogram with respect to success probability (hardness to solve). This was considered as a proof of the quantumness of the D-Wave machine.

Note that by these success probability distributions, a problem that is hard for one computer can be easy for the other, while the distribution for the whole "hardness" may look the same. This, in itself, questions the interpretation given to the aforementioned results.

These conclusions were also criticized by J. Smolin and others (Shin et al., 2014; Smolin and Smith, 2014). It was claimed that the difference between the histograms could be explained according to several grounds. Simulated annealing algorithms start from different initial points each time, while adiabatic algorithms start from the same point and evolve almost the same each time. Hence, different adiabatic trials naturally show more resemblance. This implies that time scales for the simulated annealing algorithm and for the adiabatic algorithm could not be compared as such. It would be of interest to increase the number of trials given to the SA. This way, one could probably find a good correlation between the simulated annealing and the D-Wave.

In fact, Shin et al. (2014) and Smolin and Smith (2014) presented a classical simulated annealing model on a set of 2-dimensional vectors, a compass $O(2)$ model (an SD model). Indeed the model showed a bimodal behavior with respect to success probabilities (Smolin and Smith, 2014). Being a classical model, this questions the quantum interpretation of the earlier results regarding the D-Wave computer.

Boixo et al. (2014) computed the correlation between the success probabilities of solving the same problem instance on any two computers in the set (SQA, DW, SD, SA). Note that this time each single instance was tested on two computers. High correlation between the DW and SQA (simulated quantum annealer) was shown. However, Shin et al. (2014) presented similar correlations (even slightly better) presented between the classical $O(2)$ model and the D-Wave, suggesting a classical behavior of the D-Wave.

Boixo et al. (2013) tested the D-Wave One on an artificial problem of a set of 8 spins: 4 core spin and 4 ancillae. The ground space for the particular wiring presented was highly degenerate and had two components, one was a cluster of 16 states, the second was a singular separate state. In the simulated annealing case writing a Lindblad set of equations shows that the separated state is enhanced, i.e., the probability to end in that state is higher than the average probability to end in one of the other ground states. In short, this could be explained by the fact that the separate state is close (in the Hamming distance) to a high number of (first) excited states, while any of the other ground states are close to a lower number of excited states. Since the evolution is thermic, the computer easily jumps into excited states and relaxes back with higher probability into the separated ground state. The overall result is an enhancement of the separated state. As for the quantum adiabatic computer, in the midst of the evolution, the separated state is no longer a ground state, due to the transverse field added. The state "joins" the ground space only at the end of the evolution; however, there the transverse field is too low to swap it with one of the other ground state. The overall result is an attenuation of the separated state. Hence we get a clear distinction between the two models. We can therefore use this toy problem as a test for the quantumness of the computer, and indeed the D-Wave One showed the expected adiabatic behavior. In response, it was shown by Shin et al. (2014) that the $O(2)$ classical model of the same problem exhibits a distribution of 0 eigenvectors similar to the one presented by the adiabatic computer, although it is a classical computer.

As another proof for quantumness, the response of the computer to a change in the properties of the flux qubits, was suggested by Johnson et al. (2011). For each qubit, the thermal fluctuations are proportional to $e^{-\varepsilon(U)/kT}$, where $\varepsilon(U)$ is the barrier height (see the description of PCQ SQUIDs by Wendin and Shumeiko (2005)). If we increase $\varepsilon(U)$, the thermal fluctuations gradually stop until they freeze out at some freezing time t_0^c, such that $\varepsilon(U)(t_0^c) = kT$. Similarly, the tunneling effects also freeze out when $\varepsilon(U)$ is increased above some value $\varepsilon(U)(t_0^q)$. We expect the freezing time t_0^c of the thermal fluctuation to be linearly dependent on temperature, and the tunneling freezing time t_0^q to be independent of temperature. The authors thereby apparently proved the existence of a tunneling quantum effect.

7.5.7 Does the computer exhibit entanglement?

Recently Smirnov and Amin (2013) introduced a theorem that connects the magnetic susceptibility of the adiabatic Hamiltonian with the existence of entangled states. Suppose we define the susceptibility of a qubit i to be:

$$\chi_i^\lambda = \frac{\partial \langle \sigma_i^z \rangle}{\partial \lambda}, \tag{7.35}$$

where λ controls the evolution of the Hamiltonian (such as a time parameter). Suppose χ_i^λ and χ_j^λ are both non-zero, $J_{ij} \neq 0$, and suppose the evolution is slow enough to reside on the ground state (there exists an anti-crossing), then the theorem states that at some far point in the evolution process the eigenstate is entangled.

Lanting et al. (2014) tested this theorem on the D-Wave computer, for two toy models: a two qubit circuit and a cell of 8 qubits. Both the ground and the first exited states turn out to be entangled vectors. The measurement was done using qubit tunneling spectroscopy (Berkley et al., 2013).

Note however, that the sole existence of entanglement in the process of computation does not guarantee the quantum properties we need from a quantum computer; therefore this criterion is weak.

7.5.8 Open questions and possible future routes

We believe that at this stage, where much work is still ongoing, we can only conclude this chapter with a few open questions. The D-Wave group has definitely made great progress in the field, both on theoretical and practical aspects. However, the D-Wave computer is now at the apex of a controversy.

To sum up, we wish to raise several questions and ideas concerning future research.

(1) Choosing the hardware or the gates of a quantum computer, there are two main factors to be considered: the coherence time and the operation time. There should be a high relation between the two. The superconducting flux circuits of the D-Wave are far from being the best in that point. For coherence and operation time scales of other qubits, see (Nielsen and Chuang, 2000; Ladd et al., 2010), for achieving long coherence time (0.1 ms) in superconducting qubits, see (Rigetti et al., 2012).

(2) The benefits of the flux qubits of the D-Wave are clear: they are easy to build using known techniques of lithography, the flux qubits are easy to couple, etc. However, with respect to other computational properties, they are only moderate (Ladd et al., 2010).

(3) If the D-Wave computer has quantum properties, and also thermic properties, then the best way to analyze its behavior is by Markovian master equations (see also (Albash et al., 2012)).

(4) Katzgraber et al. (2014) suggested that the glassy chimeras of the D-Wave might not be the right architecture for testing quantum annealing. It seems that its energy

landscape near zero temperature is too simple and does not have significant barriers to tunnel through. This attenuates the properties we want to use in the quantum computation.

(5) It could be that the chimera graph of the D-Wave makes the embedding of graphs into the computer hard. Different wiring of the computer could make it easy to test other problems (Choi, 2008).

(6) One can test the quantumness of the D-Wave computer by testing its performance on a specific problem having a large (known) computational complexity gap between its classical annealing and quantum adiabatic versions. A simple version of such a test function was suggested by Farhi et al. (2002). It was demonstrated there that the time complexity of a classical simulated annealing computer solving such a problem is exponential due to the height of some spikes, while an adiabatic computer could easily tunnel (that is, in polynomial time) through the spikes if these are narrow enough.

(7) It could be useful to simulate other quantum informational tasks besides optimization, and even to test the D-Wave with hard fundamental tasks such as area law behavior etc. (Wolf et al., 2008).

Acknowledgements

E.C. was supported by ERC-AD NLST.

References

Aharonov, D., W. van Dam, J. Kempe, Z. Landau, S. Lloyd, and O. Regev. 2008. 'Adiabatic quantum computation is equivalent to standard quantum computation'. *SIAM Rev.* 50: 755.

Albash, T., S. Boixo, D. A. Lidar, and P. Zanardi. 2012. 'Quantum adiabatic Markovian master equations'. *New J. Phys.* 14: 123016.

Albash, T., W. Vinci, A. Mishra, P. A Warburton, and D. A. Lidar. 2015. 'Consistency tests of classical and quantum models for a quantum annealer'. *Phys. Rev. A* 91: 042314.

Albert, D. Z. 1983. 'On quantum-mechanical automata'. *Phys. Lett. A* 98: 249.

Altshuler, B., H. Krovi, J. Roland, and B. I. Halperin. 2010. 'Anderson localization makes adiabatic quantum optimization fail'. *Proceedings of the National Academy of Sciences of the United States of America* 107: 12446.

Ambainis, A. 2014. 'Quantum walk algorithm for element distinctness'. arXiv:quant–ph/0311001.

Barker, J. A. 1979. 'A quantum-statistical Monte Carlo method; path integrals with boundary conditions'. *J. Chem. Phys.* 70: 2914.

Benioff, P. A. 1982. 'Quantum mechanical Hamiltonian models of discrete processes that erase their own histories: Application to Turing machines'. *Int. J. Theor. Phys.* 21: 177.

Berkley, A. J., A. J. Przybysz, T. Lanting, R. Harris, N. Dickson, F. Altomare, M. H. Amin, P. Bunyk, C. Enderud, E. Hoskinson, M. W. Johnson, E. Ladizinsky, R. Neufeld, C. Rich, A. Y. Smirnov, E. Tolkacheva, S. Uchaikin, and A. B. Wilson. 2013. 'Tunneling spectroscopy using a probe qubit'. *Phys. Rev. B* 87: 020502(R).

Bhatia, R. 1997. *Matrix Analysis (Graduate Texts in Mathematics)*. Springer.

Biamonte, J. D., P. J. Love. 2008. 'Realizable Hamiltonians for universal adiabatic quantum computer'. *Phys. Rev. A* 78: 012352.

Biamonte, J. D. 2008. 'Nonperturbative k-body to two-body commuting conversion Hamiltonians and embedding problem instances into Ising spins'. *Phys. Rev. A* 77: 052331.

Boixo, S., E. Knill, and R. D. Somma. 2009. 'Eigenpath traversal by phase randomization'. *Quant. Inf. and Comput.* 9: 833.

Boixo, S., T. Albash, F. Spedalieri, N. Chancellor, and D. A. Lidar. 2013. 'Experimental signature of programmable quantum annealing'. *Nat. Commun.* 4: 2067.

Boixo, S., T. F. Rønnow, S. V. Isakov, Z. Wang, D. Wecker, D. A. Lidar, J. M. Martinis, and M. Troyer. 2014. 'Evidence for quantum annealing with more than one hundred qubits'. *Nat. Phys.* 10: 218.

Boixo, S., G. Ortiz, and R. Somma. 2015. 'Fast quantum methods for optimization'. *Eur. Phys. J. Special Topics* 224: 35.

Born, M., and V. Fock. 1928. 'Beweis des adiabatensatzes'. *Z. Phys.* 51: 165.

Boyer, M., G. Brassard, P. Høyer, and A. Tapp. 1998. 'Tight bounds on quantum searching'. *Fortsch. Phys.* 46: 493.

Brennen, G. K., C. M. Caves, P. S. Jessen, and I. H. Deutsch. 1999. 'Quantum logic gates in optical lattices'. *Phys. Rev. Lett.* 82: 1060.

Brooke, D. J. Bitko, T. F. Rosenbaum, and G. Aeppli. 1999. 'Quantum annealing of a disordered magnet'. *Science* 284: 779.

Ceperley, D., G. V. Chester, and M. H. Kalos. 1977. 'Monte Carlo simulation of a many-fermion study'. *Phys. Rev. B* 16: 3081.

Chakrabarti, B. K., A. Dutta, and P. Sen. 1996. *Quantum Ising Phases and Transitions in Transverse Ising Models*. Springer.

Childs, A. M., E. Farhi, and J. Preskill. 2001. 'Robustness of adiabatic quantum computation'. *Phys. Rev. A* 65: 012322.

Choi, V. 2008. 'Minor-embedding in adiabatic quantum computation: I. The parameter setting problem'. *Quantum Inf. Process.* 7: 193.

Chuang, I. L., and Y. Yamamoto. 1995. 'Simple quantum computer'. *Phys. Rev. A* 52: 3489.

Cirac, J. I., and P. Zoller. 1995. 'Quantum computations with cold trapped ions'. *Phys. Rev. Lett.* 74: 4091.

Cohen, E., and B. Tamir. 2014. 'D-Wave and predecessors: From simulated to quantum annealing'. *Int. J. Quantum Inf.* 12: 1430002.

Corney, J. F., and P. D. Drummond. 2004. 'Gaussian quantum Monte Carlo methods for fermions and bosons'. *Phys. Rev. Lett.* 93: 260401.

D-Wave group. 2013. 'D-Wave: comment on comparison with classical computers'. http://www.archduke.org/stuff/d–wave–comment–on–comparison–with–classical– computers.

Das, A., and B. K. Chakrabarti. 2008. 'Colloquium: Quantum annealing and analog quantum computation'. *Rev. Mod. Phys.* 80: 1061.

Dahl, G. B. and S. E. Landsburg. 2011. 'Quantum Strategies'. https://arxiv.org/abs/1110.4678.

Deutsch, D. 1985. 'Quantum theory, the Church-Turing principle and the universal quantum computer'. *Proc. Roy. Soc. Lond. A* 400: 97.

Deutsch, D. 1989. 'Quantum computational networks'. *Proc. Roy. Soc. Lond. A* 425: 73.

Deutsch, D., and R. Jozsa. 1992. 'Rapid solution of problems by quantum computation'. *Proc. Roy. Soc. Lond. A* 439: 553.

Dickson, N. D., M. W. Johnson, M. H. Amin, R. Harris, F. Altomare, A. J. Berkley, P. Bunyk, J. Cai, E. M. Chapple, P. Chavez, F. Cioata, T. Cirip, P. deBuen, M. D. Brook, C. Enderud, S. Gildert, F. Hamze, J. P. Hilton, E. Hoskinson, K. Karimi, E. Ladizinsky, N. Ladizinsky, T. Lanting, T. Mahon, R. Neufeld, T. Oh, I. Perminov, C. Petroff, A. Przybysz, C. Rich, P. Spear, A. Tcaciuc, M. C. Thom, E. Tolkacheva, S. Uchaikin, J. Wang, A. B. Wilson, Z. Merali, and G. Rose. 2013. 'Thermally assisted quantum annealing of a 16-qubit problem'. *Nat. Commun.* 4: 1903. DiVincenzo, D. P. 1995. 'Two-bit gates are universal for quantum computation'. *Phys. Rev. A* 51: 1015.

DiVincenzo, D. P. 2000. 'The physical implementation of quantum computation'. *Fortschr. Phys.* 48: 771.

Drossel, B., and M. A. Moore. 2004. 'Energy barriers in spin glasses'. *Phys. Rev. B* 70: 064412.

Farhi, E., J. Goldstone, S. Gutmann, and M. Sipser. 2000. 'Quantum computation by adiabatic evolution'. arXiv:quant–ph/0001106.

Farhi, E., J. Goldstone, S. Gutmann, J. Lapan, A. Lundgren, and D. Preda. 2001. 'A quantum adiabatic evolution algorithm applied to random instances of an NP-complete problem'. *Science* 292: 472.

Farhi, E., J. Goldstone, and S. Gutmann. 2002. 'Quantum adiabatic evolution algorithms versus simulated annealing'. arXiv:quant–ph/0201031.

Feynman, R. P. 1982. 'Simulating physics with computers'. *Int. J. Theor. Phys.* 21: 467.

Feynman, R. P. 1986. 'Quantum mechanical computers'. *Found. Phys.* 16: 507.

Finnila, A. B., M. A. Gomez, C. Sebenik, C. Stenson, and J. D. Doll. 1994. 'Quantum annealing: A new method for minimizing multidimensional functions'. *Chem. Phys. Lett.* 219: 343.

Geman, S., and D. Geman. 1984. 'Stochastic relaxation, Gibbs distributions, and the Bayesian restoration of images'. *IEEE Trans. Patt. Anal. Machine Intel.* 6: 721.

Gershenfeld, N. A., and I. L. Chuang. 1997. 'Bulk spin-resonance quantum computation'. *Science* 275: 350.

Grimm, R. C., and R. G. Storer. 1971. 'Monte-Carlo solution of Schrödinger's equation'. *J. Comput. Phys.* 7: 134.

Grover, L. K. 1996. 'A fast quantum mechanical algorithm for database search'. *Proc. 28th Ann ACM Symp. Theory of Computing*. p. 212

Grover, L. K. 2000. 'Synthesis of quantum superpositions by quantum computation'. *Phys. Rev. Lett.* 85: 1334.

Hardy, G. H., and E. M. Wright. 1979. *An Introduction to the Theory of Numbers*. Clarendon Press.

Hen, I. 2014a. 'Continuous-time quantum algorithms for unstructured problems'. *J. Phys. A: Math. Theory.* 47: 045305.

Hen, I. 2014b. 'Period finding with adiabatic quantum computation'. *Europhys. Lett.* 105: 50005.

Horn, R. A., and C. R. Johnson. 1985. *Matrix Analysis*. Cambridge University Press.

Imamglu, A., D. D. Awschalom, G. Burkard, D. P. DiVincenzo, D. Loss, M. Sherwin, and A. Small. 1999. 'Quantum information processing using quantum dot spins and cavity QED'. *Phys. Rev. Lett.* 83: 4204.

Jansen, S., M.-B. Ruskai, and R. Seiler. 2007. 'Bounds for the adiabatic approximation with applications to quantum computation'. *J. Math. Phys.* 48: 102111.

Johnson, M. W., M. H. S. Amin, S. Gildert, T. Lanting, F. Hamze, N. Dickson, R. Harris, A. J. Berkley, J. Johansson, P. Bunyk, E. M. Chapple, C. Enderud, J. P. Hilton, K. Karimi, E. Ladizinsky, N. Ladizinsky, T. Oh, I. Perminov, C. Rich, M. C. Thom, E. Tolkacheva, C. J. S. Truncik, S. Uchaikin, J. Wang, B. Wilson, and G. Rose. 2011. 'Quantum annealing with manufactured spins'. *Nature* 473: 194.

Jörg, T., F. Krzakala, G. Semerjian, and F. Zamponi. 2010. 'First-order transitions and the performance of quantum algorithms in random optimization problems'. *Phys. Rev. Lett.* 104: 207206.

Jozsa, R. 1998. 'Quantum algorithms and the Fourier transform'. *Proc. Roy. Soc. Lond. A* 454: 323.

Kadowaki, T., and H. Nishimori. 1998. 'Quantum annealing in the transverse Ising model'. *Phys. Rev. E* 58: 5355.

Kalai, G. 2011. 'How quantum computers fail: Quantum codes, correlations in physical systems, and noise accumulation'. arXiv:1106.0485.

Kaminsky, W. M., and S. Lloyd. 2004. 'Scalable architecture for adiabatic quantum computing of NP-hard problems'. *Quantum Computing and Quantum Bits in Mesoscopic Systems*. Springer.

Katzgraber, H. G., F. Hamze, and R. S. Andrist. 2014. 'Glassy chimeras could be blind to quantum speedup: Designing better benchmarks for quantum annealing machines'. *Phys. Rev. X* 4: 021008.

Kempe, J., and O. Regev. 2003. '3-local Hamiltonian is QMA complete'. *Quantum Inform. Compu.* 3: 258.

Kempe, J., A. Y. Kitaev, O. Regev. 2005. 'The complexity of the local Hamiltonian problem'. *SIAM Journal of Computing* 35: 1070.

Kitaev, A. Y. 1995. 'Quantum measurements and the abelian stabilizer problem'. arXiv:quant–ph/9511026.

Kitaev, A. Y. 2003. 'Fault-tolerant quantum computation by anyons'. *Ann. Phys.* 303: 2.

Kitaev, A. Y., A. H. Shen, and M. N. Vyalyi. 2002. *Classical and Quantum Computation*. American Mathematical Society.

Knill, E., R. Laamme, and G. J. Milburn. 2001. 'A scheme for efficient quantum computation with linear optics'. *Nature* 409: 46.

Ladd, T. D., F. Jelezko, R. Laamme, Y. Nakamura, C. Monroe, and J. L. O'Brien, 2010. 'Quantum computers'. *Nature* 464: 45.

Lanting, T., A. Przybysz, A. Y. Smirnov, F. M. Spedalieri, M. H. Amin, A. J. Berkley, R. Harris, F. Altomare, S. Boixo, P. Bunyk, N. Dickson, C. Enderud, J. P. Hilton, E. Hoskinson, M. W. Johnson, E. Ladizinsky, N. Ladizinsky, R. Neufeld, T. Oh, I. Perminov, C. Rich, M. C. Thom, E. Tolkacheva, S. Uchaikin, A. B. Wilson, and G. Rose. 2014. 'Entanglement in a quantum annealing processor'. *Phys. Rev. X* 4: 021041.

Lloyd, S. 1996. 'Universal quantum simulators'. *Science* 273: 1073.

Lloyd, S. 2008. 'Robustness of adiabatic quantum computing'. arXiv:0805.2757.

Martoňák, R., G. E. Santoro, and E. Tosatti. 2004. 'Quantum annealing of the traveling-salesman problem'. *Phys. Rev. E* 70: 057701.

McGeoch, C. C., and C. Wang. 2013. 'Experimental evaluation of an adiabiatic quantum system for combinatorial optimization*CF '13 Proceedings of the ACM International Conference on Computing Frontiers*. 23

McMillan, W. L. 1965. 'Ground state of liquid He 4'. *Phys. Rev.* 138: A442.

Messiah, A. 1965. *Quantum Mechanics, Volume II*. North Holland Publishing Company.

Minkel, J. R. 2007. 'First "commercial" quantum computer solves sudoku puzzles'. http://www.scientificamerican.com/article/first–commercial–quantum–computer.

Misra, B., and E. C. G. Sudarshan. 1977. 'The Zeno's paradox in quantum theory'. *J. Math. Phys.* 18: 756.

Mooij, J. E., T. P. Orlando, L. Levitov, L. Tian, C. H. van der Wal, and S. Lloyd. 1999. 'Josephson persistent-current qubit'. *Science* 285: 1036.

Morita, S., and H. Nishimori. 2006. 'Convergence theorems for quantum annealing'. *J. Phys. A: Math. Gen.* 39: 13903.

Morita, S., and H. Nishimori. 2008. 'Mathematical foundation of quantum annealing'. *J. Math. Phys.* 49: 125210.

Namiki, M., H. Nakazato, and S. Pascazio. 1997. *Decoherence and Quantum Measurements*. World Scientific Press.

Nielsen, M. A., and I. L. Chuang. 2000. *Quantum Computation and Quantum Information*. Cambridge University Press.

Ortiz, A. P., N. Dickson, M. D. Brook, G. Rose, and A. A. Guzik. 2012. 'Finding low-energy conformations of lattice protein models by quantum annealing'. *Sci. Rep.* 2: 571.

Paz, J. P., and W. H. Zurek. 1999. 'Quantum limit of decoherence: Environment induced superselection of energy eigenstates'. *Phys. Rev. Lett.* 82: 5181.

Platzman, P. M., and M. I. Dykman. 1999. 'Quantum computing with electrons Coating on liquid helium'. *Science* 284: 1967.

Puget, J. F. 2013. 'D-Wave vs. CPLEX comparison, Part 1, Part 2, Part 3'. https://www.ibm.com/developerworks/community.

Ray, P., B. K. Chakrabarti, and A. Chakrabarti. 1989. 'Sherrington-Kirkpatrick model in a transverse field: Absence of replica symmetry breaking due to quantum fluctuations'. *Phys. Rev. B* 39: 11828.

Reichardt, B. W., F. Unger, and U. Vazirani. 2012. 'A classical leash for a quantum system: Command of quantum systems via rigidity of CHSH games'. arXiv:1209.0449.

Rigetti, C., J. M. Gambetta, S. Poletto, B. L. T. Plourde, J. M. Chow, A. D. Córcoles, J. A. Smolin, S. T. Merkel, J. R. Rozen, G. A. Keefe, M. B. Rothwell, M. B. Ketchen, and M. Steffen. 2012. Superconducting qubit in a waveguide cavity with a coherence time approaching 0.1 ms. *Phys. Rev. B* 86: 100506(R).

Rigolin, G., and G. Ortiz. 2012. 'Adiabatic theorem for quantum systems with spectral degeneracy'. *Phys. Rev. A* 85: 062111.

Rigolin, G., G. Ortiz, and V. H. Ponce. 2008. 'Beyond the quantum adiabatic approximation: Adiabatic perturbation theory'. *Phys. Rev. A* 78: 052508.

Rønnow, T. F., Z. Wang, J. Job, S. Boixo, S. V. Isakov, D. Wecker, J. M. Martinis, D. A. Lidar, and M. Troyer. 2014. 'Defining and detecting quantum speedup'. *Science* 345: 420.

Rose, G., and W. G. Macready. 2007. 'An introduction to quantum annealing'. DWAVE/Technical Document 0712.

Rousseau, V. G. 2008. 'Stochastic Green function algorithm'. *Phys. Rev. E* 77: 056705.

Seoane, B., and H. Nishimori. 2012. 'Many-body transverse interactions in the quantum annealing of the p-spin ferromagnet'. *J. Phys. A: Math. Gen.* 45: 435301.

Shin, S. W., G. Smith, J. A. Smolin, and U. Vazirani. 2014. 'How "quantum" is the D-Wave machine?' arXiv:1401.7087.

Shor, P. W. 1994. 'Algorithms for quantum computation: Discrete logarithms and factoring'. *Proceedings of the 35th Annual Symposium on Foundations of Computer Science*. p. 124.

Shor, P. W. 1996. 'Fault-tolerant quantum computation'. *Proceedings of the 37th Annual Symposium on Foundations of Computer Science*. p. 56.

Shor, P. W. 1997. 'Polynomial-time algorithms for prime factorization and discrete logarithms on a quantum computer'. *SIAM J. Sci. Statist. Comput.* 26: 1484.

Simon, D. R. 1994. 'On the power of quantum computation'. *Proceedings of the 35th Annual Symposium on Foundations of Computer Science*. p. 116.

Sinclair, A., and M. Jerrum. 1989. 'Approximate counting, uniform generation and rapidly mixing Markov chains'. *Information and Computation* 82: 93.

Smirnov, A. Y., and M. H. Amin. 2013. 'Ground-state entanglement in coupled qubits'. *Phys. Rev. A* 88: 022329.

Smolin, J. A., and G. Smith. 2014. 'Classical signature of quantum annealing'. *Front. Phys.* 2: 52.

Somma, R. D., C. D. Batista, and G. Ortiz. 2007. 'Quantum approach to classical statistical mechanics'. *Phys. Rev. Lett.* 99: 030603.

Somma, R. D., S. Boixo, H. Barnum, and E. Knill. 2008. 'Quantum simulations of classical annealing processes'. *Phys. Rev. Lett.* 101: 130504.

Stern, M., G. Catelani, Y. Kubo, C. Grezes, A. Bienfait, D. Vion, D. Esteve, and P. Bertet. 2014. 'Flux qubits with long coherence times for hybrid quantum circuits'. *Phys. Rev. Lett.* 113: 123601.

Suzuki, S., and A. Das. 2015. 'Quantum annealing: The fastest route to quantum computation?' *Eur. Phys. J. Special Topics* 224: 1.

van Dam, W., M. Mosca, and U. Vazirani. 2001. 'How powerful is adiabatic quantum computation?' *Proceedings of the 42nd Annual Symposium on Foundations of Computer Science*. p. 279.

van Laarhoven, P. J. M., and E. H. L. Aarts. 1987. *Simulated Annealing: Theory and Applications*. Reidel Publishing Company.

Vandersypen, L. M. K., M. Steffen, G. Breyta, C. S. Yannoni, M. H. Sherwood, and I. L. Chuang. 2001. 'Experimental realization of Shor's quantum factoring algorithm using nuclear magnetic resonance'. *Nature* 414: 883.

Watrous, J. 2000. 'Succinct quantum proofs for properties of finite groups.' *Proceedings 41st Annual Symposium on Foundations of Computer Science* 537.

Wendin, G., and V. S. Shumeiko. 2005. 'Superconducting quantum circuits, qubits and computing'. arXiv:cond–mat/0508729.

Wolf, M. M., F. Verstraete, M. B. Hastings, and J. I. Cirac. 2008. 'Area laws in quantum systems: Mutual information and correlations'. *Phys. Rev. Lett.* 100: 070502.

Young, A. P., S. Knysh, and V. N. Smelyanskiy. 2008. 'Size dependence of the minimum excitation gap in the quantum adiabatic algorithm'. *Phys. Rev. Lett.* 101: 170503.

CHAPTER 8

Quantum Information and Quenching Dynamics

Atanu Rajak[1] and Uma Divakaran[2]

This chapter deals with two important areas of quantum physics as indicated in the title. We first study quantum information theoretic measures like ground state fidelity, fidelity susceptibility, dynamical counterpart of the ground state fidelity (the Loschmidt echo) and the entanglement entropy for some spin models in the context of characterizing the associated quantum phase transitions. We then discuss the time evolution of the Majorana edge state of a one-dimensional p-wave superconductor after performing both the sudden and slow quenches in the Hamiltonian across quantum critical points.

Quantum phase transitions: Introduction

A quantum phase transition (QPT) is a zero-temperature phase transition of a quantum many-body system driven by a non-commuting term of the quantum Hamiltonian (Chakrabarti et al., 1996; Sachdev, 1999; Vojta, 2003; Suzuki et al., 2013; Dutta et al., 2015). This non-commutativity actually introduces quantum fluctuations resulting in various phase transitions at zero temperature in contrast to the thermal fluctuations in classical phase transitions. At a quantum critical point (QCP), the ground state energy of the system is a non-analytic function of a non-thermal parameter. There is a level-crossing between the ground state and the first excited state energies that is avoided for second order quantum phase transitions in finite systems; the energy levels actually cross at the first order transition point (Sachdev, 1999). On the other hand, in a second order transition point, the energy gap Δ_λ between the ground state and the first excited state vanishes for a system of infinite size. For a quantum system with the Hamiltonian $H(\lambda) = H_0 + \lambda H_1$

[1]Condensed Matter Physics Division, Saha Institute of Nuclear Physics, 1/AF Bidhan Nagar, Kolkata 700064, India
[2]UM-DAE Center for Excellence in Basic Sciences, Mumbai 400098, India

such that H_0 and H_1 do not commute, Δ_λ vanishes with the parameter λ in the following way, as it approaches λ_c

$$\Delta_\lambda \propto |\lambda - \lambda_c|^{\nu z}. \tag{8.1}$$

Here λ_c is the critical point, and ν and z are the critical exponents associated with the correlation length and dynamical exponent respectively. In a continuous phase transition, the system has two length scales: one is the length of the system and the other is called the correlation length ξ. The correlation length is a measure of length over which the system is spatially correlated. The correlation length governs the exponential decay of the equal time two-point connected correlation in the ground state as a function of distance between the two points. The equal time connected correlation function of the order parameter (O) is defined as

$$G(r) = \langle O(0,t)O(r,t)\rangle - \langle O(0,t)\rangle\langle O(r,t)\rangle \propto \frac{e^{-r/\xi}}{r^{d-2+\eta}}, \tag{8.2}$$

where d is the dimensionality of space and η denotes the Fisher exponent associated with the quantum critical point. This function determines the spatial correlation between two order parameters at two different points separated by a distance r and it contains important information about a phase transition. The correlation length diverges with the critical exponent ν as the system approaches the critical point, the form of which is given by

$$\xi \propto |\lambda - \lambda_c|^{-\nu}. \tag{8.3}$$

The correlation function $G(r)$ then depends only on the power law part (see Eq. (8.2)) which does not carry any characteristic length. In addition to equal time correlations in space, there are analogous equal space correlations in time that define a time scale ξ_τ. Similar to ξ, close to the critical point, the correlation time ξ_τ diverges as

$$\xi_\tau \sim \Delta_\lambda^{-1} \propto \xi^z \propto |\lambda - \lambda_c|^{-\nu z}. \tag{8.4}$$

As mentioned earlier, z is the dynamical exponent. This exponent does not come in the critical analysis of thermal phase transitions of classical systems because they have no intrinsic dynamics. One can see from Eq. (8.4) that in contrast to the classical phase transitions, space and time are interconnected to each other for QPTs. These characteristic length and time scales also scale in different ways for $z \neq 1$. This property of QPT makes it an interesting area of research.

8.1 Quantum Information Theoretic Measures: Indicator of a QCP

The studies of QPTs have been useful in connecting various fields like quantum information theory, quantum computation and condensed matter to each other. In recent years, the investigation of quantum information theoretic measures like quantum fidelity (Zanardi and Paunkovic, 2006; Zhou and Barjaktarevic, 2008; Gritsev and Polkovnikov, 2009; Gu, 2010; Gritsev and Polkovnikov, 2010; Rams and Damski, 2011), quantum discord (Sarandy, 2009; Dillenschneider, 2008; Nag et al., 2011), decoherence (Zurek, 1991; Haroche, 1998; Zurek, 2003; Joos et al., 2013; Suzuki et al., 2016) and entanglement in the context of characterizing QPTs has emerged as an important field of research (Nielsen and Chuang, 2000; Vedral, 2007; Polkovnikov et al., 2011; Dutta et al., 2015). The singularities associated with a QPT can be successfully captured by these information theoretic measures. They also show interesting scaling behaviors close to the QCP, attracting the attention of the scientific community towards it. This approach to quantum phase transitions is advantageous as compared to traditional methods like the Landau–Ginzburg theory (Chaikin and Lubensky, 1998), because here we do not need any knowledge about the order parameter and symmetry-breaking of the system. In this section, we introduce some quantum information theoretic measures.

8.1.1 Quantum fidelity

We briefly discuss here the concept of quantum ground state fidelity and its connection with quantum critical phenomena (Gritsev and Polkovnikov, 2009; Gu, 2010). Quantum fidelity is defined as the overlap of two ground state wavefunctions at two different points of the parameter space of a quantum Hamiltonian. A mathematical description of the same shall be provided later. Since fidelity describes the sensitivity to the dissimilarity between the states, it is related to the measure of loss of information encoded in quantum states. So, the fidelity is a very important measure in the context of quantum information theory which also successfully detects all the QCPs of a quantum many-body system. The structure of the ground state wavefunctions are very different across the two sides of the quantum critical point. Therefore the fidelity of two ground states which are very close to each other in the parameter space shows a significant drop at a QCP for a finite size system. At the same time, in the thermodynamic limit, two ground states are always orthogonal to each other even if they are separated by a very small amount of parameter space. As a result, the fidelity vanishes for any parameter value of the Hamiltonian even though one can anticipate a sharper drop in fidelity close to a QCP. This phenomenon has been realized in various quantum many-body systems and is known as the Anderson orthogonality catastrophe (Anderson, 1967).

Let us consider a general Hamiltonian of a d-dimensional quantum many-body system (Gu, 2010; Dutta et al., 2015)

$$H(\lambda) = H_0 + \lambda H_I, \tag{8.5}$$

where H_I is the term that produces quantum fluctuations in the system and λ denotes its strength. The eigenvalue equation for the system is given by $H(\lambda)|\psi_n(\lambda)\rangle = E_n|\psi_n(\lambda)\rangle$, where $n = 0,1,2\cdots$, with $n = 0$ representing the ground state of the Hamiltonian and $|\psi_n(\lambda)\rangle$ forms a complete basis set in the Hilbert space. Then the ground state fidelity between two ground states corresponding to the parameter values λ and $\lambda+\delta$ is defined as

$$F(\lambda,\delta) = |\langle\psi_0(\lambda)|\psi_0(\lambda+\delta)\rangle|. \tag{8.6}$$

In the limit of $\delta \to 0$, one can expand $|\psi_0(\lambda+\delta)\rangle$ in terms of $|\psi_0(\lambda)\rangle$ using the celebrated Taylor expansion technique so that the overlap is given by

$$\langle\psi_0(\lambda)|\psi_0(\lambda+\delta)\rangle = 1+\delta\langle\psi_0(\lambda)|\frac{\partial}{\partial\lambda}|\psi_0(\lambda)\rangle+\frac{\delta^2}{2}\langle\psi_0(\lambda)|\frac{\partial^2}{\partial\lambda^2}|\psi_0(\lambda)\rangle+\cdots. \tag{8.7}$$

Now following a trivial step, the absolute value of the overlap, i.e., the fidelity is calculated as

$$F(\lambda,\delta) = 1 - \frac{1}{2}\delta^2 L^d \chi_F(\lambda)+...,\tag{8.8}$$

where the linear term in δ becomes zero due to the normalization condition of the wavefunction and, L is the linear dimension of the system with $N = L^d$. Here, χ_F is defined as the fidelity susceptibility (FS) of the ground state and is given by Schwandt et al. (2009), Gu (2010) and Dutta et al. (2015),

$$\chi_F(\lambda) = \frac{1}{L^d}\left[\left\langle\frac{\partial}{\partial\lambda}\psi_0\middle|\frac{\partial}{\partial\lambda}\psi_0\right\rangle - \left\langle\frac{\partial}{\partial\lambda}\psi_0\middle|\psi_0\right\rangle\left\langle\psi_0\middle|\frac{\partial}{\partial\lambda}\psi_0\right\rangle\right]. \tag{8.9}$$

In case the ground state wavefunction is characterized by a large number of parameters, χ_F is calculated using metric tensor (Gu, 2010) which will not be discussed in this chapter.

The most relevant term in Eq. (8.8) in determining the fidelity is the coefficient of δ^2, i.e., the FS. It actually measures the rate at which the ground state changes with λ. Using Eq. (8.8) and considering $\delta^2 L^d \chi_F(\lambda) \ll 1$, the FS can be obtained as

$$\chi_F(\lambda) \equiv \lim_{\delta\to 0}\frac{1}{L^d}\frac{-2\ln F}{\delta^2} = -\frac{1}{L^d}\frac{\partial^2 F}{\partial\delta^2}. \tag{8.10}$$

One can observe that fidelity (see Eq. (8.8)) depends on δ, but χ_F depends only on λ and is insensitive to δ. As the fidelity shows a dip at a QCP, χ_F shows a large peak at that point (see Eq. (8.10)). Thus, χ_F is also able to detect the ground state singularities associated with QPTs without making reference to an order parameter. Close to a critical point, correlation length diverges, so there is only one length scale in the system which is L. Thus, for a d-dimensional system, the scaling of χ_F at the critical point is given by $\chi_F \sim L^{\frac{2}{\nu}-d}$, on the other hand, away from the critical point, the scaling is found to be $\chi_F \sim |\lambda - \lambda_c|^{\nu d-2}$ with $\nu d < 2$ (Schwandt et al., 2009; Grandi et al., 2010; Albuquerque et al., 2010; Mukherjee et al., 2011; Rams and Damski, 2011). For $\nu d > 2$, the high energy modes also contribute to the FS. Nevertheless, the quantum phase transition in some systems with $\nu d > 2$ can be detected by the FS if one uses twisted boundary conditions (Thakurathi et al., 2012). At the same time, the FS exhibits logarithmic scaling with L and λ in the case of marginality, i.e., when $\nu d = 2$ (Patel et al., 2013b).

8.1.2 Loschmidt echo

Decoherence which can be viewed as the loss of phase information of a system is a very important measure in the context of successful achievement of quantum computation (Nielsen and Chuang, 2000). It is a process through which the quantum-classical transition occurs by a reduction from a pure state to a mixed state (Zurek, 1991; Haroche, 1998; Zurek, 2003). In recent years, the Loschmidt echo (LE) which is closely related to decoherence, has been studied extensively for quantum critical systems (Quan et al., 2006; Rossini et al., 2007; Yuanet al., 2007; Zhang et al., 2009b; Damski et al., 2011; Mukherjee et al., 2012a; Nag et al., 2012; Sharma et al., 2012; Sharma and Rajak, 2012; Divakaran, 2013). The experimental studies on LE also have been done using NMR experiments (Buchkremer et al., 2000; Zhang et al., 2009b; Sanchez et al., 2009).

The LE is a measure of overlap of the two time-dependent wave functions evolving with two different Hamiltonians H_1 and H from the same initial state $|\psi_0\rangle$,

$$\mathscr{L}(t) = |\langle\psi_0|e^{iH_1t}e^{-iHt}|\psi_0\rangle|^2. \tag{8.11}$$

We now consider the state $|\psi_0\rangle$ as the ground state of H_1. Then, the LE is reduced to a simplified form given by

$$\mathscr{L}(t) = |\langle\psi_0|e^{-iHt}|\psi_0\rangle|^2. \tag{8.12}$$

The LE describes the hypersensitivity of the evolution of the system (with time) to the perturbation experienced by the environment which results in the decoherence of the system. Thus, LE is an important measure for quantum information processing.

According to Eq. (8.12), the quantification of the LE requires a Hamiltonian H that makes the evolution of the ground state $|\psi_0\rangle$ non-trivial. This Hamiltonian H can be generated considering a central spin model in which a central spin S is coupled globally to an environmental spin model E (Quan et al., 2006). In this regard, we will study the transition of the central spin S from a pure state to a mixed one induced by the criticality of the environmental spin system and show that the decay of the LE is best enhanced by the QPT of the surrounding environment. We will provide here a general formalism for calculating the LE considering a three-spin interacting transverse field Ising model as an environment (E) that is coupled to a central spin-$\frac{1}{2}$ (qubit) S. The ground state and the excited state of the central spin are denoted by $|g\rangle$ and $|e\rangle$ respectively. The composite Hamiltonian can be written in the form

$$H_T(h,\delta) = -\frac{1}{2}\sum_{n=1}^{N}\left[h\sigma_n^z + \delta\sigma_S^z\sigma_n^z + J_3\sigma_n^z\sigma_{n-1}^x\sigma_{n+1}^x + J_x\sigma_n^x\sigma_{n+1}^x\right], \tag{8.13}$$

where σ_n^x and σ_n^z are the standard Pauli spin matrices at each site n and σ_S^z denotes the Pauli matrix of the qubit. The parameters J_3, J_x and h are the three-spin interaction strength, the nearest neighbor ferromagnetic interaction in the x direction and the transverse field respectively. δ denotes the coupling strength of S to E. We shall consider here the $\delta \to 0$ limit or the weak coupling limit. With the chosen interaction term in the Hamiltonian (Eq. (8.13)), one can show the commutation relation as $[\sigma_S^z, H_T] = 0$ which signifies that σ_S^z is a constant of motion. As a result, the population of the ground and the excited states of the qubit do not change with time, i.e., there is no exchange of energy between the system and the bath (Quan et al., 2006; Rossini et al., 2007). This condition ensures that in this case, the qubit evolution leads to purely phase decoherence.

Let us consider that S is initially in a pure state $|\phi(0)\rangle_S = c_g|g\rangle + c_e|e\rangle$, where the coefficients satisfy the condition $|c_g|^2 + |c_e|^2 = 1$. On the other hand, the E is initially in the ground state $|\varphi(0)\rangle_E$. Then the initial composite wavefunction is given by

$$|\psi(0)\rangle = |\phi(0)\rangle_S \otimes |\varphi(0)\rangle_E. \tag{8.14}$$

It can be shown that the time evolution of the environmental spin model will split into two branches, given by $|\varphi_g(t)\rangle = \exp(-iH_gt)|\varphi(0)\rangle_E$ and $|\varphi_e(t)\rangle = \exp(-iH_et)|\varphi(0)\rangle_E$; here $|\varphi_g(t)\rangle$ evolves with the Hamiltonian $H_g = H_T(h,0) + V_g$ (when the S is in the ground state), whereas the evolution of $|\varphi_e(t)\rangle$ is determined by the Hamiltonian $H_e = H_T(h,0) + V_e$, where the effective potentials $V_g = \delta\sum\sigma_n^z$ and $V_e = -\delta\sum\sigma_n^z$ arise due to the interaction between S and E. Therefore, the coupling of a qubit with the transverse field term of the environmental spin chain reduces to only a change in h at each site by the amount δ or $-\delta$. The wave function of the composite system at a time t is given by

$$\begin{aligned}
|\psi(t)\rangle &= e^{-iH_T(h,\delta)}|\psi(0)\rangle \\
&= c_g|g\rangle \otimes e^{-iH_gt}|\varphi(0)\rangle_E + c_e|e\rangle \otimes e^{-iH_et}|\varphi(0)\rangle_E \\
&= c_g|g\rangle \otimes |\varphi_g(t)\rangle + c_e|e\rangle \otimes |\varphi_e(t)\rangle.
\end{aligned} \tag{8.15}$$

Although the initial composite wavefunction in Eq. (8.14) is unentangled, $|\psi(t)\rangle$ is an entangled state between S and E. At a general time t, the reduced density matrix of the qubit in the $\{|g\rangle, |e\rangle\}$ basis is given by

$$\rho_S(t) = \mathrm{Tr}_E|\psi(t)\rangle\langle\psi(t)| = \begin{pmatrix} |c_g|^2 & c_gc_e^*d^*(t) \\ c_ec_g^*d(t) & |c_e|^2 \end{pmatrix}, \tag{8.16}$$

where $d(t) = \langle \varphi_g(t)|\varphi_e(t)\rangle$ is called the decoherence factor. During its time evolution, the diagonal terms of $\rho_S(t)$ are constant in time; however, the off-diagonal terms evolve with time. Comparing Eq. (8.11) and Eq. (8.16) one can find the relation between $\mathscr{L}(t)$ and $d(t)$ given as

$$\mathscr{L}(t) = |d(t)|^2 = |\langle \varphi_g(t)|\varphi_e(t)\rangle|^2. \tag{8.17}$$

As the value of $\mathscr{L}(t)$ decreases from 1 to 0, the effective interaction between S and E increases; $\mathscr{L}(t) = 0$ corresponds to the maximum entanglement between the qubit and the environment. On the other hand, when $\mathscr{L}(t)$ is close to 1, the effective interaction between the qubit and the environment becomes almost zero. Therefore, close to a QCP of E at which $\mathscr{L}(t)$ takes minimum value, the central spin (qubit) transits from a pure state to a mixed state due to its maximum entanglement with the environment.

Recently, a lot of attention has be given to the study of the dynamics of a periodically driven closed quantum many-body system (Nag et al., 2014, 2015) using Floquet theory (Shirley, 1965). In this connection, the LE and work statistics, closely connected to the LE, have been studied in a transverse field Ising chain following a periodic driving of the transverse field (Sharma et al., 2014; Russomanno et al., 2015). The notion of dynamical phase transitions was also later introduced using the quantity LE. Recently, Heyl et al. (Heyl et al., 2013) indicated the close resemblance between canonical partition function of an equilibrium system given by $Z = \mathrm{Tr}e^{-\beta H}$, where H defines the Hamiltonian at an inverse temperature β, and the amplitude of the LE

$$G(t) = \langle \psi_0|e^{-iHt}|\psi_0\rangle. \tag{8.18}$$

Here $|\psi_0\rangle$, as before, is the ground state of the initial Hamiltonian evolving with final quenched Hamiltonian H. It is well known that the zeros of the partition function in a complex temperature plane defines a thermal phase transition. This analogy was extended to the non-analyticities in complex time domain z of the rate function defined as $r(z) = -\lim_{N\to\infty} \log G(z)/N$, where N is the number of degrees of freedom. The zeros of $G(z)$ in the complex time plane z at certain times t_n^* correspond to non-analyticities in the rate function. The existence of such a dynamical quantum phase transition (DQPT) is first shown by Heyl et al. in the transverse Ising model when the transverse field is suddenly quenched across the critical point. The study of DQPT has further been generalized to non-integrable quantum models suggesting the non-analytic dynamics as a generic feature of sudden quenches across quantum critical points (Karrasch and Schuricht, 2013). On the other hand, it has also been found in subsequent work that the DQPTs can occur in cases where the sudden quench takes place within the same phase, i.e., it does not cross the critical point (Andraschko and Sirker, 2014; Sharma et al., 2015). In the context of slow quenching, the occurrence of the DQPTs in a transverse Ising chain has already been reported by Pollmann et al. (2010), whereas the connection between these DQPTs and Fisher zeros is shown very recently (Sharma et al., 2016).

8.1.3 Entanglement

Entanglement is another important measure of quantum correlations between two systems. With the previous information theoretic measures, entanglement also has become a topic of immense research interest (Bose et al., 1998; Osborne and Nielsen, 2002; Osterloh et al., 2002; Vidal et al., 2003; Latorre et al., 2003; Hutton and Bose, 2004; Korepin, 2004; Calabrese and Cardy, 2004; Refael and Moore, 2004; Igloi and Lin, 2008; Song et al., 2011; Dubail and Stephan, 2011).

These studies lie at the interface of condensed matter physics and quantum information theory. Entanglement is a key ingredient in quantum information processing for sending informations in novel ways (Nielsen and Chuang, 2000). On the other hand, it also characterizes a QPT in a quantum many-body system with interesting scaling behaviors close to the QCP which are given in terms of the associated quantum critical exponents (Osborne and Nielsen, 2002; Osterloh et al., 2002). In the last few years, several measures of entanglement have been studied extensively for the quantum many-body systems using both analytical and numerical methods. One of the quantities to measure entanglement between two subsystems is the von Neumann entanglement entropy (Holzhey et al., 1994; Amico et al., 2008; Calabrese and Cardy, 2009; Eisert et al., 2010).

A quantum state $|\psi\rangle$ of a system consisting of two subsystems A and B is entangled if it cannot be written as the tensor product of individual subsystem states $|\psi_A\rangle$ and $|\psi_B\rangle$, i.e., the subsystems A and B are entangled if

$$|\psi\rangle \neq |\psi_A\rangle \otimes |\psi_B\rangle. \tag{8.19}$$

Consider a system which can be divided into two parts A and B of length L_A and L_B respectively. The density matrix for this system is defined as $\rho = |\psi\rangle\langle\psi|$, where $|\psi\rangle$ is assumed to be a pure state. The von Neumann entropy S of subsystem A with reduced density matrix $\rho_A = \mathrm{Tr}_B(\rho)$ is given by

$$S = -\mathrm{Tr}(\rho_A \ln \rho_A). \tag{8.20}$$

This von Neumann entropy is one form of measure of entanglement between two subsystems which will be discussed in this article, others being the Renyi entropy. As the quantum correlations between two subsystems increases, the entanglement entropy (EE) of the system increases. An unentangled combined state produces zero EE (S), whereas S is maximum for a maximally entangled state. The distinct scaling relations of EE at and close to a QCP are determined by the shortest length scale of the system. If we consider periodic boundary conditions in a critical one-dimensional system, i.e., if there are two boundary points between A and B, EE is found to scale as $S \sim \frac{c}{3} \ln L_A$, where c is a universal quantity and given by the central charge of the conformal field theory (Holzhey et al., 1994; Vidal et al., 2003; Latorre et al., 2003; Calabrese and Cardy, 2004). On the other hand, away from the critical point where the correlation length is small or $\xi \ll L_A$, EE is found to be $S \sim \frac{c}{3} \ln \xi$.

8.2 Dynamics in Quantum Many-body Systems

In Sec. 8.1, we discussed a few quantum information theoretic measures like fidelity, FS, LE and EE in the context of quantum phase transitions. We now discuss the non-equilibrium evolution of quantum many-body systems when some perturbation is applied to it. This perturbation to the Hamiltonian can be a sudden quench (Rossini et al., 2007; Calabrese and Cardy, 2007a; Rieger and Igloi, 2011) or a slow quench with a finite velocity (Zurek et al., 2005; Polkovnikov, 2005). In recent years, both types of quenches have been studied extensively in quantum many-body systems using different approaches resulting to interesting findings (Kadowaki and Nishimori, 1998; Igloi and Rieger, 2000; Calabrese and Cardy, 2006; Das et al., 2006; Calabrese and Cardy, 2007b; Cramer et al., 2008; Das and Chakrabarti, 2008; Iucci and Cazalilla, 2009; Divakaran et al., 2009; Dutta et al., 2010).

At the same time, the recent experimental realization of non-equilibrium dynamics on ultra cold atoms trapped in optical lattices (Greiner et al., 2002; Lewenstein et al., 2007; Bloch et al., 2008) inspires the theoretical work of quantum quenches in critical systems. Furthermore, the study of quenching dynamics is very important for addressing the questions on the thermalization of quantum systems (Patane et al., 2008; Iucci and Cazalilla, 2010; Canovi et al., 2011).

Following the non-equilibrium dynamics of quantum many-body systems across QCPs (for review articles see (Dziarmaga, 2010; Polkovnikov et al., 2011; Dutta et al., 2015)), there have been an increased interest in studies connecting the topology of condensed matter systems with the quenching dynamics (Tsomokos et al., 2009; Rahmani and Chamon, 2009; Halasz and Hamma, 2013).

Various topological systems have been investigated across a QCP in the context of quenching (Bermudez et al., 2009, 2010; Wang et al., 2014). In particular, the dynamics of an edge state of a topological insulator (Patel et al., 2013a) and also the p-wave superconducting chain (Rajak and Dutta, 2014; Sacramento, 2014; Rajak et al., 2014) have been explored following a quench in the Hamiltonian. At the same time, it has also been shown that the Majorana edge states can be generated dynamically in a non-topological phase of a p-wave superconducting chain (Thakurathi et al., 2013). In this connection, the formation and manipulation of Majorana edge states have been studied in a driven quantum system (Perfetto, 2013).

8.3 Fidelity Susceptibility and Loschmidt Echo for Generic Paths

In Sec. 8.1, we provided a general discussion on fidelity, FS and LE as a measure of quantum criticality of quantum systems at zero temperature (e.g., Suzuki et al., 2013; Dutta et al., 2015). We now consider a specific model: a three-spin interacting transverse field Ising model to study the FS and the LE along generic paths. In this connection, we also show the effect of different types of critical points such as ordinary critical points, multicritical points and quasicritical points on the aforementioned quantities (Rajak and

Divakaran, 2014). The method that we have used here is new and comparatively simpler than the standard method of the path dependent study which requires tensor analysis (Venuti and Zanardi, 2007; Mukherjee et al., 2011).

8.3.1 Model and phase diagram

We consider the Hamiltonian of a one-dimensional three-spin interacting transverse field Ising model (Kopp and Chakravarty, 2005; Divakaran and Dutta, 2007) which is given by Eq. (8.13) with $\delta = 0$

$$H = -\frac{1}{2}\sum_{n=1}^{L}\left[h\sigma_n^z + J_3\sigma_n^z\sigma_{n-1}^x\sigma_{n+1}^x + J_x\sigma_n^x\sigma_{n+1}^x\right]. \tag{8.21}$$

The different terms of the Hamiltonian have been defined around Eq. (8.13). The Hamiltonian in Eq. (8.21) can be diagonalized using the standard Jordan–Wigner (JW) transformation (Lieb et al., 1961; Pfeuty, 1970; Barouch et al., 1970; Barouch and McCoy, 1971; Kogut, 1979; Bunder and McKenzie, 1999; Kitaev, 2006; Feng et al., 2007; Chen and Nussinov, 2008) following a Fourier transformation. In the momentum (k) space, the Hamiltonian in Eq. (8.21) decouples as a direct sum of independent terms for each k, $H = \oplus_{k>0}H_k$, where H_k is a 2×2 matrix that is written in a basis $|0\rangle$ (with 0 c-fermions) and $|k,-k\rangle$ $(=c_k^\dagger c_{-k}^\dagger|0\rangle)$, given by (see (Rajak and Divakaran, 2014) for details)

$$H_k = \begin{pmatrix} h + J_x\cos k - J_3\cos 2k & J_x\sin k - J_3\sin 2k \\ J_x\sin k - J_3\sin 2k & -(h + J_x\cos k - J_3\cos 2k) \end{pmatrix}. \tag{8.22}$$

The Hamiltonian H_k for each k can be diagonalized by a rotation of an angle $\theta_k/2$, where θ_k is defined as

$$\tan\theta_k = \frac{J_x\sin k - J_3\sin 2k}{h + J_x\cos k - J_3\cos 2k}, \tag{8.23}$$

so that the corresponding energy spectrum is given by

$$\varepsilon_k = \left(h^2 + J_3^2 + J_x^2 + 2hJ_x\cos k - 2hJ_3\cos 2k - 2J_xJ_3\cos k\right)^{1/2}. \tag{8.24}$$

The phase diagram of the Hamiltonian in Eq. (8.21) with $J_3 = -1$ is shown in Fig. 8.1. It is easily observed from Eq. (8.24) that the energy gap becomes zero on the critical lines $h = J_3 + J_x$ and $h = J_3 - J_x$ for the momentum modes $k = \pi$ and 0, respectively. The system has a long-range ferromagnetic order only for a weak transverse field lying in the range $J_3 - J_x < h < J_3 + J_x$. These two critical lines separate the ferromagnetic phase from the paramagnetic one. The critical exponents associated with these phase transitions are given by $\nu = z = 1$ which are also the critical exponents for a one-dimensional transverse field Ising model (TFIM) (Pfeuty, 1970). So these transitions belong to the universality class of the one-dimensional TFIM and therefore are called "Ising transitions". In this model, there is another phase which can be called the three-spin dominated phase. The line $h = -J_3$ represents the phase transition line between this three-spin dominated phase and the quantum paramagnetic phase, also shown in Fig. 8.1. This phase transition is indeed analogous to the anisotropic phase transition observed in the one-dimensional transverse field XY model (Lieb et al., 1961; Barouch et al., 1970; Bunder and McKenzie, 1999; Mukherjee et al., 2007; Sharma et al., 2012). In fact, using a duality transformation (dual spins are located at the centres of the bonds of the original lattice), the system with the Hamiltonian in Eq. (8.21) can be reduced to the one-dimensional quantum XY model in a transverse field (Bunder and McKenzie, 1999; Kopp and Chakravarty, 2005). The point A in Fig. 8.1 corresponds to a multicritical point (MCP), since two critical lines $h = J_3 + J_x$ and $h = -J_3$ pass through it. There exists another MCP that appears at the intersection of $h = -J_3$ and $h = J_3 - J_x$. It is straightforward to show that the critical exponents associated with these MCPs are given by $z = 2$ and $\nu = 1/2$. The existence of some special points called quasicritical points can also be observed near the MCPs which, we will see, indeed determine the scaling of various quantities close to the MCPs. This is due to the fact that the energy ε_k has local minima at these quasicritical points shifted from the actual critical point.

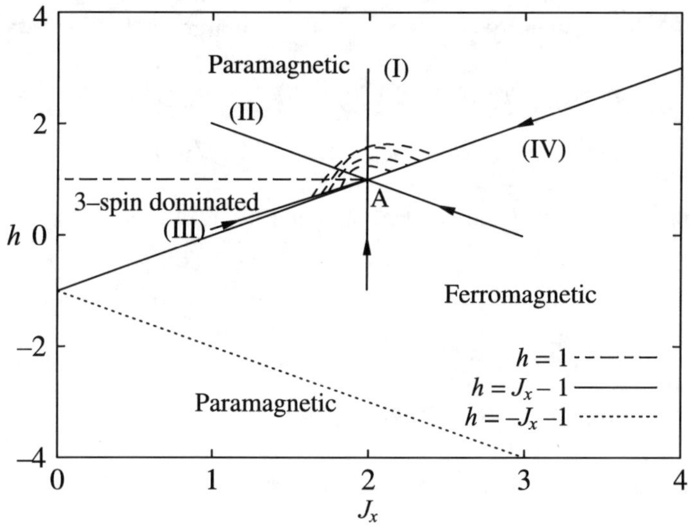

Fig. 8.1 The phase diagram of the one-dimensional three-spin interacting TFIM with different paths along which the MCP (point A) is approached. Three different critical lines are shown satisfying the relations $h = 1$, $h = J_x - 1$ and $h = -J_x - 1$, whereas, Paths I, II, III and IV are denoted by lines with arrows. A shaded region also has been shown in the figure where quasicritical points exist (from Rajak and Divakaran, 2014).

8.3.2 Fidelity susceptibility

As defined in Sec. 8.1.1, fidelity is the overlap of two ground state wave functions separated in the parameter space by a small amount, whereas χ_F measures the rate at which the ground state changes with a parameter of the Hamiltonian (Gritsev and Polkovnikov, 2009; Schwandt et al., 2009; Gu, 2010; Grandi et al., 2010; Albuquerque et al., 2010; Mukherjee et al., 2011; Rams and Damski, 2011; Thakurathi et al., 2012; Patel et al., 2013b; Damski, 2013; Damski and Rams, 2014). In recent years, FS has been studied using its various mathematical forms χ_F (Zanardi and Paunkovic, 2006; Yang et al., 2008; Gu et al., 2008; Gu, 2010).

We here focus on one particular form given by

$$\chi_F = \frac{1}{4L}\sum_k \left(\frac{\partial \theta_k}{\partial \lambda}\right)^2 \tag{8.25}$$

where $\theta_k/2$ is the angle of rotation that is needed to diagonalize the Hamiltonian (see Eq. (8.23)). We study the behavior of the FS close to the MCP by approaching the MCP along four different paths using a method which does not require the FS tensor as is done in earlier studies (Venuti and Zanardi, 2007; Mukherjee et al., 2011). We consider these paths for the following reasons. There exists quasicritical points on Paths I and II (See Fig. 8.1) when they approach the MCP, whereas Path IV is a gapless critical line which does not possess any quasicritical point. Path III is a special path having quasicritical points; but, it is very close to the critical line that might exhibit some interesting behavior in the FS. As the Hamiltonian in Eq. (8.21) has three parameters, we fix $J_3 = -1$ for our convenience and study the parameter space spanned by h and J_x.

In our method of calculating χ_F along a generic path, we redefine the Hamiltonian with a single variable employing the equation of the path. We then rotate H_k by an angle ϕ_k such that the parameter that is changed, say λ, is brought to the diagonal term. Then the χ_F can be found from Eq. (8.25) after evaluating the angle θ_k. We now briefly discuss the method to determine the angle, ϕ_k. Let R_k represent the rotation matrix for each k mode with the matrix elements $R_k(1,1) = \cos(\phi_k) = R_k(2,2)$ and $R_k(2,1) = -R_k(1,2) = \sin(\phi_k)$. We then rewrite H_k in terms of only one variable λ and rotate the Hamiltonian by an angle ϕ_k to get a matrix $H_k' = R^T H_k R$. The angle ϕ_k is determined using a condition that the off-diagonal term of the Hamiltonian H_k' is λ independent. After substituting the angle ϕ_k in H_k', the diagonal term in general will take the form $a_k\lambda + b_k$ and the off-diagonal term will have the form c_k. We assume $b_k \sim k^{z_1}$ and $c_k \sim k^{z_2}$ when expanded close to the critical mode. We first consider $z_1 < z_2$,; hence the scaling of the energy spectrum ε_k at the QCP $\lambda = 0$ is given by $\varepsilon_k \sim k^{z_1}$ that provides the dynamical exponent $z = z_1$. On the other hand, some situations can arise for which the path exhibits minima of the energy at $a_k\lambda_0 + b_k = 0$ with $\varepsilon_k \sim k^{z_2}$; these special points λ_0 are called quasicritical points (Deng et al., 2009; Mukherjee and Dutta, 2010). It has been established that the exponent z_2, different from the exponent z at the critical point, will determine the scaling of various quantities when there exist quasicritical points. If a path does not contain quasicritical points, then only at $\lambda = 0$ will the energy be minimum, and ε_k will scale as k^z where z is the minimum of z_1 and z_2, as is the case in Path IV. For $z_1 > z_2$, the dynamics will always be determined by the exponent z_2 that does not depend on the presence of the quasicritical point.

We now present our results on the FS along the four paths which cross the MCP and also discuss the effect of the presence of quasicritical points in each path. Considering the form of χ_F in Eq. (8.25), we get

$$\left(\frac{\partial \theta_k}{\partial \lambda}\right)^2 = \frac{a_k^2 c_k^2}{\varepsilon_k^4}$$

where $\theta_k = \tan^{-1}(c_k/(a_k\lambda + b_k))$. The expressions for a_k, b_k, c_k and ϕ_k are different for each path shown in Fig. 8.1.

Path I: We first consider Path I, $J_x = 2$ and approach the MCP ($h = 1$) by changing $\lambda = h - 1$. In the present case, the Hamiltonian H_k in Eq. (8.22) already has λ independent off-diagonal term, i.e., $H_k(1,1) = \lambda + 1 + 2\cos k + \cos 2k = -H_k(2,2)$ and $H_k(1,2) = H_k(2,1) = 2\sin k + \sin 2k$. We expand the different terms in the Hamiltonian close to the critical mode $k_c = \pi$ and get $a_k \sim 1, b_k \approx -k^2$ and $c_k \approx -k^3$. Finding θ_k as $\theta_k = \tan^{-1}$ $(H_k(1,2)/H_k(1,1))$, the fidelity susceptibility is given by

$$\chi_F = \frac{1}{4L}\sum_{k>0}\left(\frac{\partial \theta_k}{\partial \lambda}\right)^2 = \frac{1}{4L}\sum_{k>0}\frac{(2\sin k + \sin 2k)^2}{\varepsilon_k^4}$$

$$\approx \frac{1}{4L}\sum_{k>0}\frac{k^6}{\varepsilon_k^4}.$$

This relation provides L^5 scaling of the FS, since $\varepsilon_k \sim k^3$ at the quasicritical point $\lambda \sim k^2$ with redefining $(k - k_c)$ as k. Here, a_k, b_k and c_k have been expanded around $k \to 0$ which we will follow throughout the section. It can be noted that here $z_1 = 2$ and $z_2 = 3$ such that $z = z_1$, although the scaling of χ_F is determined by z_2.

We have shown χ_F as a function of h in Fig. 8.2. Close to the MCP $h = 1$, χ_F shows oscillations, which is shown in the inset of the same figure. These oscillations are explained using quasicritical points. As mentioned before, the quasicritical point occurs at $\lambda = -b_k/a_k$. Due to quantization of the momentum mode k in units of $2\pi/L$, for all discrete momentum modes close to the critical mode k_c, i.e., $k = k_c + 2\pi n/L$ with n as an integer, the energy spectrum will show $\varepsilon_k \sim k^3$ behavior. For each value of k, we can find different values of λ_0 close to $\lambda = 0$ which indicate the presence of more than one quasicritical point near the MCP (Mukherjee et al., 2011). The scaling of χ_F with L when the MCP is approached along the aforementioned path is shown in Fig. 8.3(a) for the first two peaks in χ_F as shown in the inset of Fig. 8.2.

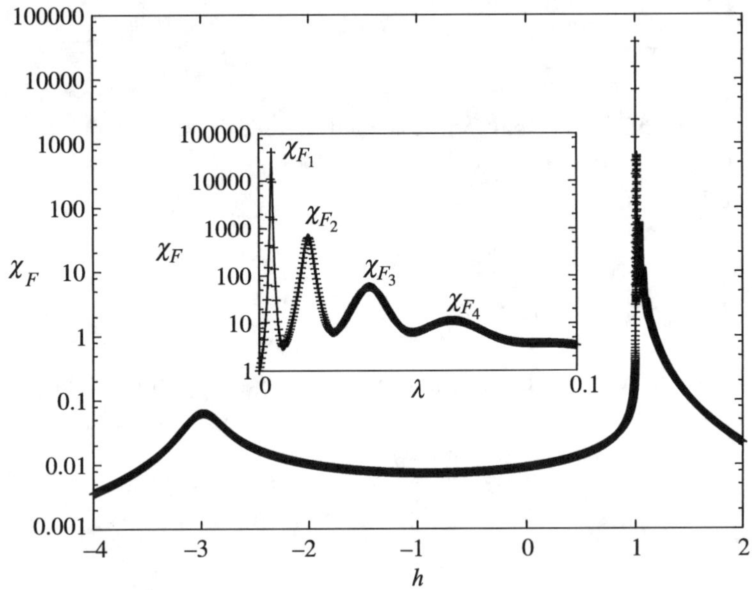

Fig. 8.2 The plot shows χ_F as a function of h at $J_x = 2, J_3 = -1$. Here the system size is $L = 100$. The first peak that occurs at $h = -3$ is an Ising critical point which shows linear scaling with L, and the second peak occuring at $h = 1$ corresponds to the MCP where L^5 scaling is found. The inset figure shows that the FS oscillates close to the MCP, indicating the presence of quasicritical points there (from Rajak and Divakaran, 2014).

Path II: Let us now focus on approaching the MCP along the path $h + J_x = 3$, Path II as shown in Fig. 8.1 by varying $\lambda = h - 1$. We provide a rotation of an angle ϕ_k in H_k of Eq. (8.22) to bring λ to the diagonal term and find that

$$H'_k(11) = \lambda\left[\cos 2\phi_k - \cos(k - 2\phi_k)\right]$$

$$+\left[\cos 2\phi_k + 2\cos(k - 2\phi_k) + \cos(2k - 2\phi_k)\right]$$

$$= \lambda a_k + b_k$$

$$H'_k(12) = -\sin 2\phi_k + 2\sin(k - 2\phi_k) + \sin(2k - 2\phi_k) = c_k$$

where,

$$a_k = \cos 2\phi_k - \cos(k - 2\phi_k)$$

$$b_k = \cos 2\phi_k + 2\cos(k - 2\phi_k) + \cos(2k - 2\phi_k)$$

$$c_k = -\sin 2\phi_k + 2\sin(k - 2\phi_k) + \sin(2k - 2\phi_k)$$

$$\text{with } \tan(2\phi_k) = \frac{\sin k}{-1+\cos k}. \tag{8.26}$$

After expanding around the critical mode, we get $a_k \approx 2$, $b_k \approx -k^2$, $c_k \approx -k^3/2$ and find the presence of a quasicritical point at $\lambda = k^2/2$ where $\varepsilon_k \sim k^3$. With $\theta_k = \tan^{-1}(H'_k(1,2)/H'_k(1,1))$, χ_F follows as

$$\chi_F = \frac{1}{4L}\sum_{k>0}\frac{a_k^2 c_k^2}{\varepsilon_k^4} \approx \frac{1}{4L}\sum_{k>0}\frac{k^6}{\varepsilon_k^4}. \tag{8.27}$$

In this case also we find L^5 scaling of χ_F which is once again confirmed by our numerical result in Fig. 8.3(b).

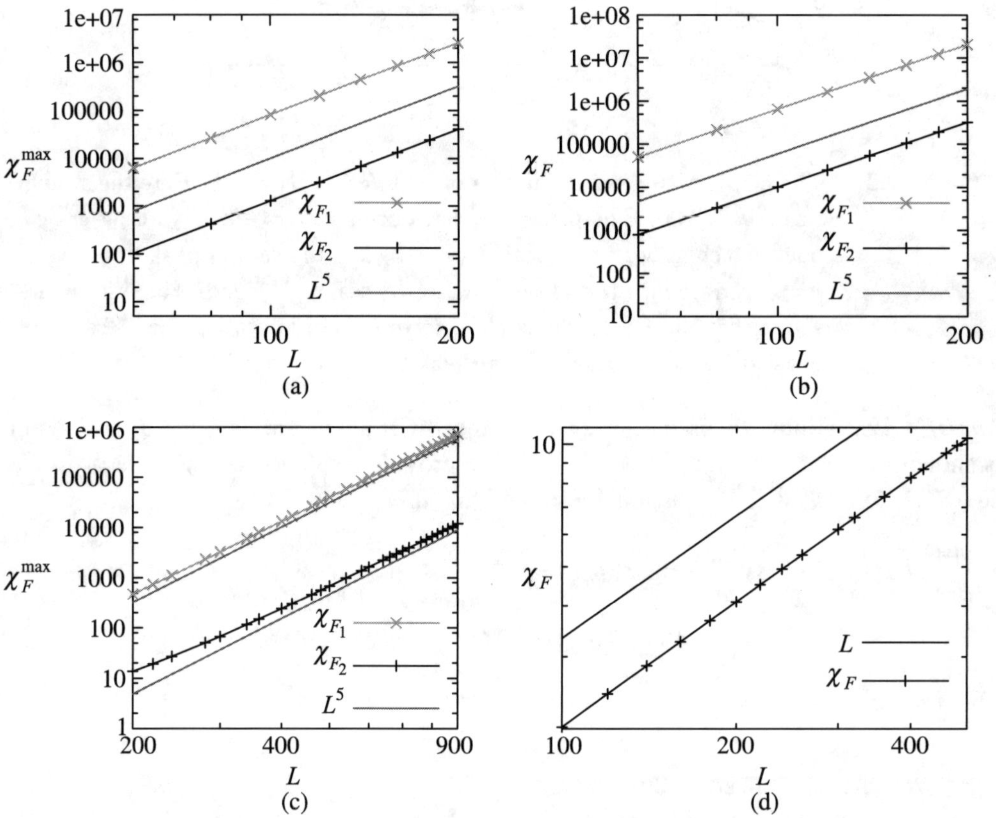

Fig. 8.3 The scaling of χ_F when the MCP is approached along four different paths. The plots shown in (a), (b) and (c) are for Paths I, II and III resulting in $\chi_F \propto L^5$ that shows the effect of quasicritical points; whereas it is linear in L for Path IV as shown in (d) due to the absence of quasicritical points along this path or line (from Rajak and Divakaran, 2014).

Path III: We now consider the path $h - 0.9J_x = -0.8$ and perform a rotation in H_k by an angle ϕ_k. In this process, we obtain the following elements:

$$a_k = \cos 2\phi_k + \frac{10}{9}\cos(k - 2\phi_k)$$

$$b_k = \cos 2\phi_k + 2\cos(k - 2\phi_k) + \cos(2k - 2\phi_k)$$

$$c_k = -\sin 2\phi_k + 2\sin(k - 2\phi_k) + \sin(2k - 2\phi_k)$$

$$\text{with } \tan(2\phi_k) = \frac{10/9\sin k}{1 + 10/9\cos k}. \tag{8.28}$$

Following an expansion around the critical mode $k_c = \pi$, we find $a_k \approx -1/9$, $b_k \approx -k^2$ and $c_k \approx 9k^3$ with a quasicritical point at $\lambda_0 = -9k^2$. In a similar fashion to the earlier cases, we get $\chi_F \sim L^5$ as $\varepsilon_k \sim k^3$ at the quasicritical point which is also shown in Fig. 8.3(c). It is to be noted that all the aforementioned three paths contain quasicritical points either in the paramagnetic phase or in the three-spin dominated phase of the model near the MCP. From Fig. 8.1, it is clear that Path III is very close to the critical line $h = J_x - 1$ which does not contain any quasicritical point. We consider this path to verify the consequence of this proximity to the critical line on the scaling behavior of the FS. Although we again obtain L^5 scaling, this is only for large L; the small departure for smaller L which is not observed in Paths I and II could be the effect of closeness to the critical line. Let us explore this path in detail. The quasicritical points in this path can be found at $\lambda_0 = -9k^2$, where k is inversely proportional to L. The factor of 9 in contrast to 1 and $1/2$ in Paths I and II respectively, shifts the position of the quasicritical point farther away with respect to the actual critical point. After expanding a_k, b_k and c_k close to the critical mode and the critical point which may not be accurate in this path for small L, we find a deviation in the scaling of the FS from L^5 scaling for small L.

Path IV: The final path considered here is a critical line $h - J_x = -1$. Following the same steps as in the previous cases, here also we find the functions a_k, b_k, c_k and ϕ_k with the following form

$$a_k = \left[\cos 2\phi_k + \cos(k - 2\phi_k)\right],$$

$$b_k = \left[\cos 2\phi_k + 2\cos(k - 2\phi_k) + \cos(2k - 2\phi_k)\right],$$

$$c_k = -\sin 2\phi_k + 2\sin(k - 2\phi_k) + \sin(2k - 2\phi_k),$$

$$\text{and } \tan(2\phi_k) = \frac{\sin k}{1 + \cos k}, \tag{8.29}$$

which when expanded close to the critical momentum $k_c = \pi$ provides $a_k \approx k$, $b_k \approx k^3$ and $c_k \approx -k^2$. It can be observed that for Paths I, II and III, we get quasicritical points where the a_ks are independent of k. However, a_k depends on k for Path IV and its exponent is less than z_1. We shall not therefore consider the term b_k for non-zero λ. Hence, ε_k goes as k for all non-zero λ, i.e, this path is the critical line, and a quasicritical point cannot be found along this path near the MCP which provides minimum energy. Because ε_k varies as k^2 at the MCP ($\lambda = 0$) where the energy becomes minimum, χ_F scales linearly with L, as can be seen from Fig. 8.3(d).

8.3.3 Loschmidt echo

As defined in Eq. (8.11), the LE is the overlap between two quantum states which evolve with two different Hamiltonians although both start from a state that may be considered as the ground state of one of the Hamiltonians (Quan et al., 2006; Sharma et al., 2012; Sharma and Rajak, 2012). In the momentum space, the expression of the LE for the Hamiltonian in Eq. (8.21) is given by (see Quan et al., 2006; Dutta et al., 2015)

$$\mathscr{L}(\lambda,t) = \prod_k \mathscr{L}_k(\lambda,t) = \prod_k \left(1 - \sin^2 2\gamma_k \sin^2(\varepsilon_k(\lambda+\delta)t)\right), \tag{8.30}$$

where $2\gamma_k = \theta_k^{\lambda} - \theta_k^{\lambda+\delta}$ and $\varepsilon_k(\lambda+\delta)$ is the energy spectrum corresponding to the Hamiltonian $H_k(\lambda+\delta)$. We shall calculate the LE using this expression where γ_k and ε_k will take care of the effect of paths. To study the behavior of the LE, we assume a partial sum $S = \ln \mathscr{L}$ along the lines similar to that done by Quan et al. (2006).

We first present the LE as a tool to detect a quantum critical point by considering a path parallel to Path I as in Fig. 8.1 at $J_x = 1$ consisting of three critical points as h is changed for a fixed time t. It can be seen from Fig. 8.4, LE is able to detect all the critical points in its path successfully. Quan et al. (2006) stated that at critical points, the LE exhibits collapse and revival with time; thus, LE can be used to indicate the presence of critical points. The time period of oscillations scales linearly with the system size L for an Ising critical point, whereas, it has also been shown that it can vary in a non-linear way for other types of QCPs (Sharma et al., 2012). We here show the difference in the behavior of LE for various critical points such as the anisotropic critical point (ACP), the MCP, and also the quasicritical point of the model discussed in Sec. 8.3.1 by analyzing its short time decay and the by analyzing its short time decay and the time period of LE oscillations (Quan et al., 2006). In the present case also, we follow the same method for considering the effect of generic path as discussed in Sec. 8.3.2.

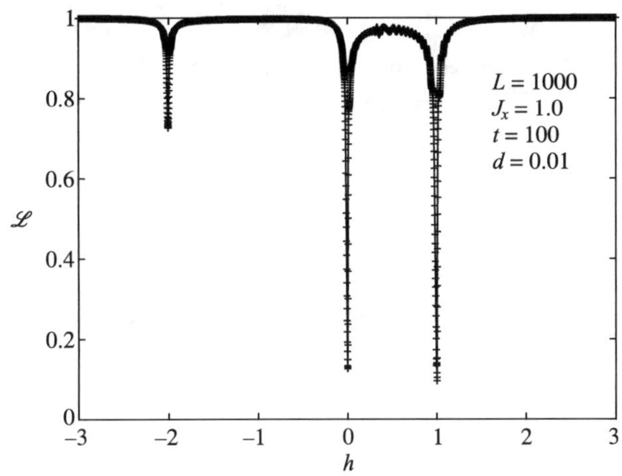

Fig. 8.4 Variation of the LE as a function of h. The sharp dips in the LE are seen at the Ising critical points ($h = -2, 0$) as well as at anisotropic critical point at $h = 1$ (from Rajak and Divakaran, 2014).

Short time decay: It can be shown that the energy gap on the ACP line vanishes for the critical mode $k_c = \cos^{-1}(\frac{-J_x}{2})$. Now, we consider $J_x = 1$ and vary h to study the short time behavior of the LE. In this case, the transverse field h is already in the diagonal term of the Hamiltonian H_k and therefore a rotation in H_k is not needed similar to path I in Sec. 8.3.2. Expanding Eq. (8.30) close to the critical mode $k_c = \cos^{-1}(-1/2)$, we obtain $\sin^2 \varepsilon_k(h+\delta)t \approx (h+\delta-1)^2 t^2$ and $\sin^2(2\alpha_k) \approx 9k^2\delta^2/4(h-1)^2(h+\delta-1)^2$. Using small time limit we get $S \propto -\Gamma t^2$, i.e.,

$$\mathscr{L}(h,t) \approx \exp\left(-\Gamma t^2\right) \tag{8.31}$$

where $\Gamma \propto \delta^2/(h-1)^2 L^2$ is the decay constant. Using Eq. (8.31), it can be easily shown that the LE is invariant under the transformation $L \to L\alpha$ and $t \to t\alpha$ for a fixed δ with α being some integer. The collapse and revival of the LE (see Fig. 8.5(a)) also verifies these scaling relations. A similar short time analysis of the LE has also been done close to a MCP and a quasicritical point. To approach a MCP ($J_x = 2$ and $h = 1$), we consider the critical line $h - J_x = -1$ and change $\lambda = h - 1$. In this case also, we get the same form of the LE as given in Eq. (8.31) with $\Gamma \propto \delta^2/\lambda^2 L^4$. We find that $\mathscr{L}(\lambda,t)$ remains invariant under the transformation $L \to L\alpha$ and $t \to t\alpha^2$ with a fixed δ. Again our numerical result (see Fig. 8.5(b)) confirms the aforementioned scaling relations. Finally, we consider Path I ($J_x = 2$) where quasicritical points exist in addition to a MCP as mentioned in Sec. 8.3.2. Once again we find Eq. (8.31) for a quasicritical point with a decay rate $\Gamma \propto \delta^2/(h-1)^2 L^6$, so that it is invariant under $L \to L\alpha$, $t \to t\alpha^3$ and fixed δ. The collapse and revival of the LE when fixing the system at a quasicritical point is shown in Fig. 8.5(c).

Time period analysis: On the ACP line, the collapse and revival of the LE has been found by choosing the parameter values $h = 1 - \delta$, $J_x = 1.0$ and $\delta = 0.01$. By expanding $\varepsilon_k(h+\delta)$

near the critical mode k_c, we get $\varepsilon_k(h+\delta) \approx \sqrt{4-J_x^2}(k-k_c)$. For the large L limit, the maximum contribution to $\mathscr{L}(h,t)$ comes from the momentum mode $k = k_c + 2\pi/L$. It is now easy to find that the time dependency in the LE in Eq. (8.30) comes from the term $\sin^2(\varepsilon_k(h+\delta)t)$. Hence, the quasi period of the oscillation in the LE is given by

$$T = \frac{L}{\sqrt{4-J_x^2}}. \tag{8.32}$$

This has been shown in Fig. 8.5(a) for three different system sizes. Going through similar steps of calculations, one can find that the quasi periods of the collapse and revival of the LE for the MCP and the quasicritical point are $T = L^2/2\pi$ and $T = L^3/4\pi^2$ respectively (see Figs. 8.5(b) and (c)).

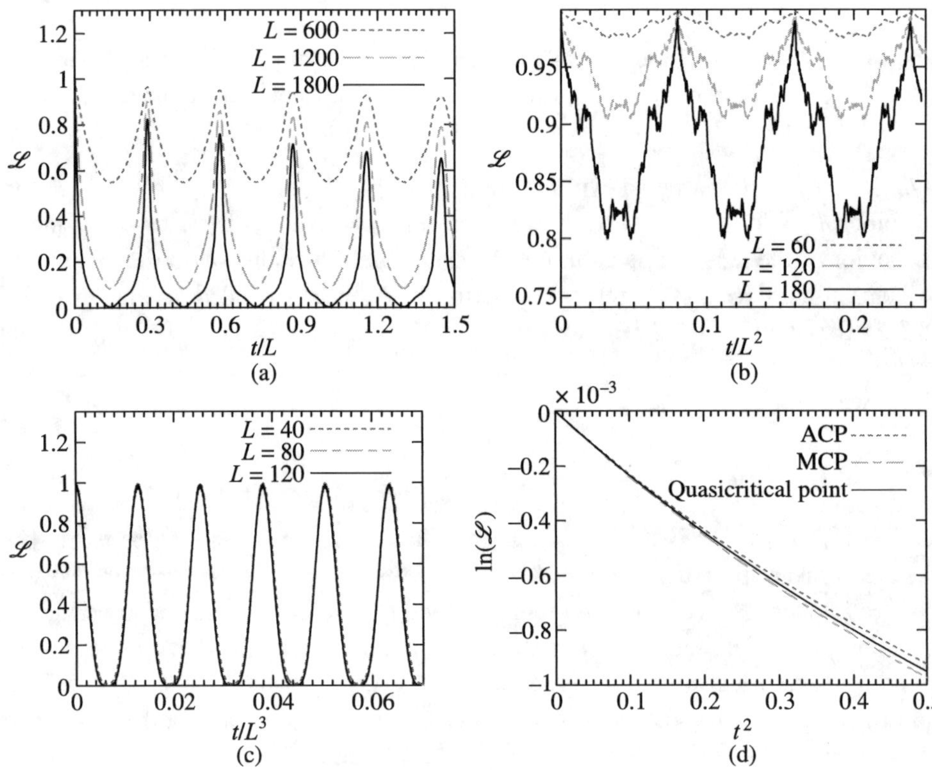

Fig. 8.5 Variation of LE as a function of scaled time t/L^α to highlight the scaling of time period with system size L, with α being the scaling exponent ($T \propto L^\alpha$) as obtained in the text for various types of critical points, i.e., (a) the anisotropic critical point where $T \propto L$, (b) multicritical point where $T \propto L^2$ and (c) quasicritical point with $T \propto L^3$. In (d), we present the almost linear variation of $\ln \mathscr{L}$ with t^2 for small times with $L = 100$ at the various critical points confirming the general small time behavior given by $\mathscr{L} \sim e^{-\Gamma t^2}$ (from Rajak and Divakaran, 2014).

8.4 Loschmidt Echo: Effect of Gapless Phase

In the previous section, we have discussed the scaling of the FS and the LE along different paths of a three-spin interacting transverse Ising model consisting of different types of isolated quantum critical points. We will now discuss the effect of gapless phase on the LE considering a central spin model where a two-dimensional Kitaev model on a honeycomb lattice has been chosen as an environment (Sharma and Rajak, 2012). In this case, the central spin or the qubit is coupled globally with the environment.

8.4.1 Model, phase diagram and anisotropic quantum critical point (AQCP)

Let us consider the Hamiltonian of the Kitaev model residing on a honeycomb lattice (Kitaev, 2006; Sengupta et al., 2008)

$$H = \sum_{j+l=even} \left(J_1 \sigma^x_{j,l} \sigma^x_{j+1,l} + J_2 \sigma^y_{j-1,l} \sigma^y_{j,l} + J_3 \sigma^z_{j,l} \sigma^z_{j,l+1} \right), \tag{8.33}$$

where the column and row indices of the honeycomb lattice are indicated by j and l respectively. The parameters J_1, J_2 and J_3 denote the couplings for three different bonds (see Fig. 8.6); $\sigma^a_{j,l}$ represent the standard Pauli spin matrices with $a(= x,\ y$ and $z)$, signifying the spin component.

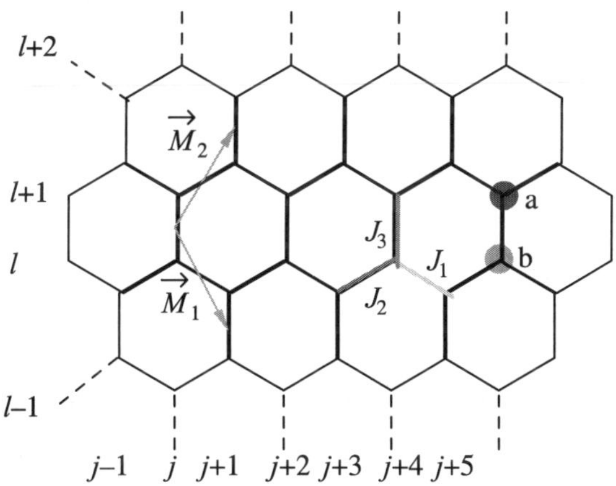

Fig. 8.6 The figure shows a schematic representation of the Kitaev model on a honeycomb lattice with the interactions denoted by J_1, J_2 and J_3; $\vec{M}_1 = \frac{\sqrt{3}}{2}\hat{i} - \frac{3}{2}\hat{j}$ and $\vec{M}_2 = \frac{\sqrt{3}}{2}\hat{i} + \frac{3}{2}\hat{j}$ are the spanning vectors of the lattice (from Sharma and Rajak, 2012).

This model can be solved exactly using JW transformation (Lieb et al., 1961). Although the model is two-dimensional, the special topology of the model actually makes it suitable for JW transformation (Kitaev, 2006; Sengupta et al., 2008; Mukherjee et al., 2012b). Using JW and Fourier transformations, the Hamiltonian in Eq. (8.33) can be represented in the momentum space in the following form (see (Sengupta et al., 2008; Mukherjee et al., 2012b; Sharma and Rajak, 2012) for details)

$$H = \sum_{\vec{k}} \begin{pmatrix} a^{\dagger}_{\vec{k}} & b^{\dagger}_{\vec{k}} \end{pmatrix} H_{\vec{k}} \begin{pmatrix} a_{\vec{k}} \\ b_{\vec{k}} \end{pmatrix}, \tag{8.34}$$

where the decoupled 2×2 Hamiltonian $H_{\vec{k}}$ is expressed in terms of Pauli matrices as

$$H_{\vec{k}} = \alpha_{\vec{k}} \sigma^1 + \beta_{\vec{k}} \sigma^2,$$

where $\quad \alpha_{\vec{k}} = 2[J_1 \sin(\vec{k}.\vec{M}_1) - J_2 \sin(\vec{k}.\vec{M}_2)],$

and $\beta_{\vec{k}} = 2[J_3 + J_1 \cos(\vec{k}.\vec{M}_1) + J_2 \cos(\vec{k}.\vec{M}_2)]. \qquad (8.35)$

In Eq. (8.34), $\vec{k}$ runs over half of the Brillouin zone of the hexagonal lattice, while the full Brillouin zone on the reciprocal lattice is represented by a rhombus with vertices $(k_x, k_y) = (\pm 2\pi\sqrt{3}, 0)$ and $(0, \pm 2\pi/3)$. By solving the 2×2 eigenvalue problem for $H_{\vec{k}}$ we obtain the eigenvalues

$$E^{\pm}_{\vec{k}} = \pm \varepsilon_{\vec{k}} \quad \text{with} \quad \varepsilon_{\vec{k}} = \sqrt{\alpha^2_{\vec{k}} + \beta^2_{\vec{k}}}. \tag{8.36}$$

The unitary matrix $U_{\vec{k}}$ that diagonalize the 2×2 matrix $H_{\vec{k}}$ for each $\vec{k}$ is given by

$$U_{\vec{k}} = \frac{1}{\sqrt{2}} \begin{pmatrix} 1 & 1 \\ -e^{i\theta_{\vec{k}}} & e^{i\theta_{\vec{k}}} \end{pmatrix}, \tag{8.37}$$

where

$$e^{i\theta_{\vec{k}}} = \frac{\alpha_{\vec{k}} + i\beta_{\vec{k}}}{\sqrt{\alpha^2_{\vec{k}} + \beta^2_{\vec{k}}}}.$$

Now, we can re-write the Hamiltonian in Eq. (8.35) in the diagonalized form

$$H = \sum_{\vec{k}} \begin{pmatrix} a^{\dagger}_{\vec{k}} & b^{\dagger}_{\vec{k}} \end{pmatrix} U_{\vec{k}} U^{\dagger}_{\vec{k}} H_{\vec{k}} U_{\vec{k}} U^{\dagger}_{\vec{k}} \begin{pmatrix} a_{\vec{k}} \\ b_{\vec{k}} \end{pmatrix} = \sum_{\vec{k}} [-\varepsilon_{\vec{k}} A^{\dagger}_{\vec{k}} A_{\vec{k}} + \varepsilon_{\vec{k}} B^{\dagger}_{\vec{k}} B_{\vec{k}}], \tag{8.38}$$

where the $A_{\vec{k}}$s and the $B_{\vec{k}}$s are Bogoliubov fermionic operators defined as

$$A_{\vec{k}} = \frac{1}{\sqrt{2}} [a_{\vec{k}} - e^{-i\theta_{\vec{k}}} b_{\vec{k}}], \quad B_{\vec{k}} = \frac{1}{\sqrt{2}} [a_{\vec{k}} + e^{-i\theta_{\vec{k}}} b_{\vec{k}}]. \tag{8.39}$$

There exists two energy bands corresponding to the positive and negative energy spectrums of the system as defined in Eq. (8.36). This model has a gapless phase for $|J_1 - J_2| \leq J_3 \leq$

(J_1+J_2), where the band gap $\Delta_{\vec{k}} = E^+_{\vec{k}} - E^-_{\vec{k}}$ becomes zero for some particular $\vec{k}$ modes. As shown in Fig. 8.7, the phase diagram of the model is represented by an equilateral triangle satisfying $J_1, J_2, J_3 > 0$ and the relation $J_1+J_2+J_3 = 4$ consisting of three gapped phases and one gapless phase. The boundaries between the gapped phases and the gapless phase are given by three gapless critical lines $J_1 = J_2+J_3$, $J_2 = J_3+J_1$ and $J_3 = J_1+J_2$.

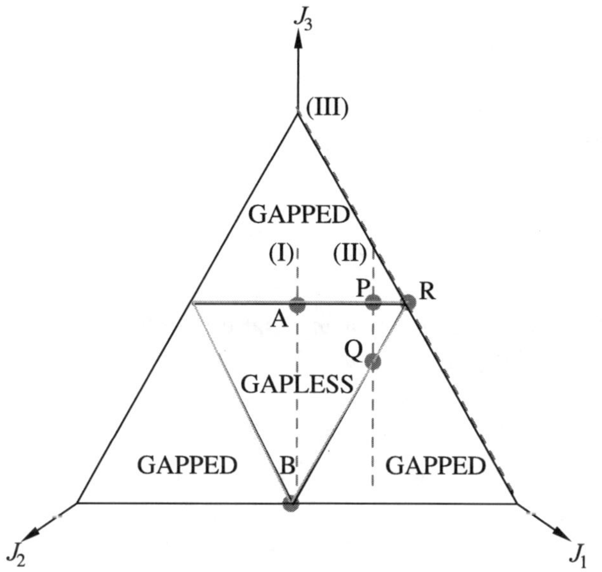

Fig. 8.7 The phase diagram of the Kitaev model in which all the points satisfy the relation $J_1+J_2+J_3 = 4$. The gapless phase of the model is given by the inner equilateral triangle satisfying the relations $J_1 \le J_2+J_3$, $J_2 \le J_3+J_1$ and $J_3 \le J_1+J_2$. The study of LE is carried out along Paths I, II and III where J_3 is varied (from Sharma and Rajak, 2012).

One can see that the band gap vanishes on the critical line $J_3 = J_1+J_2$ for the four $\vec{k}$ modes: (k_x, k_y)= $(\pm 2\pi/\sqrt{3}, 0)$ and $(\pm 2\pi/3, 0)$ which are indeed the four vertices of the Brillouin zone. Expanding $\alpha_{\vec{k}}$ and $\beta_{\vec{k}}$ close to the critical mode $\vec{k} = (2\pi/\sqrt{3}, 0)$ for $J_3 = J_1+J_2$, we get

$$\alpha_{\vec{k}} = 2\Big[J_1 \sin\big(\frac{\sqrt{3}}{2}k_x - \frac{3}{2}k_y\big) + J_1 \sin\big(\frac{\sqrt{3}}{2}k_x + \frac{3}{2}k_y\big)\Big],$$

$$\approx \sqrt{3}(J_2-J_1)k_x + 3(J_1+J_2)k_y,$$

$$\beta_{\vec{k}} = 2\Big[J_3 + J_1 \cos\big(\frac{\sqrt{3}}{2}k_x - \frac{3}{2}k_y\big) + J_2 \cos\big(\frac{\sqrt{3}}{2}k_x + \frac{3}{2}k_y\big)\Big],$$

$$\approx 2(J_3 - J_1 - J_2) + \frac{3}{4}(J_1 + J_2)k_x^2$$

$$+\frac{9}{4}(J_1 + J_2)k_y^2 + \frac{3\sqrt{3}}{2}(J_2 - J_1)k_x k_y,$$

$$= \frac{3}{4}(J_1 + J_2)k_x^2 + \frac{9}{4}(J_1 + J_2)k_y^2 + \frac{3\sqrt{3}}{2}(J_2 - J_1)k_x k_y, \tag{8.40}$$

where k_x and k_y correspond to the deviations from the critical modes. It can be noted that the variation of $\alpha_{\vec{k}}$ is linear, while $\beta_{\vec{k}}$ is quadratic in k_x and k_y. The point A ($J_{3c} = 2J_1$ with $J_1 = J_2$) in Fig. 8.7 is a special critical point that needs to be investigated carefully. A is found to be an AQCP with the energy relation $E_{\vec{k}} \sim k_x^2$ along k_x ($k_y = 0$) and $E_{\vec{k}} \sim k_y$ along k_y ($k_x = 0$). We can now get two different dynamical exponents, i.e., $z_\perp = 1$ and $z_\parallel = 2$ along $\hat{k_x}$ and $\hat{k_y}$ respectively. A general critical point with $J_1 \neq J_2$, $J_{3c} = J_1 + J_2$ is also an AQCP that can be exhibited considering a new coordinate system by giving a rotation to the previous one (Sharma and Rajak, 2012).

8.4.2 Loschmidt echo for the Kitaev model

We here assume a central spin model (see Sec. 8.1.2) where a two-dimensional Kitaev model on a honeycomb lattice has been considered as an environment and a central spin is coupled globally to the J_3 term of the environmental Hamiltonian. The coupling strength between the central spin and the environment is considered to be δ with $\delta \to 0$. Following similar steps of calculation as done by Quan et al. (2006) and Dutta et al. (2015), we obtain the expression of the LE, given by

$$\mathcal{L}(J_3, t) = \prod_{\vec{k}} \mathcal{L}_{\vec{k}} = \prod_{\vec{k}} [1 - \sin^2(2\gamma_{\vec{k}}) \sin^2(\varepsilon_{\vec{k}}(J_3 + \delta)t)], \tag{8.41}$$

where $\tan\theta_{\vec{k}}(J_3 + \delta) = \alpha_{\vec{k}}/\beta'_{\vec{k}}$ and $\gamma_{\vec{k}} = [\theta_{\vec{k}}(J_3) - \theta_{\vec{k}}(J_3 + \delta)]/2$. Here $\beta'_{\vec{k}}$ corresponds to the value with $J_3 + \delta$ instead of J_3. From Eq. (8.41), we find that there is a close resemblance in the expression of the LE with the case when the transverse Ising chain is considered to be the environment (Quan et al., 2006).

We first determine the LE numerically as a function of J_3 using Eq. (8.41). It detects all the critical points showing dip at those points. We illustrate this phenomenon of the LE by choosing three paths along which the parameter J_3 is changed. Along the path $J_1 = J_2$ (Path I in Fig. 8.7), J_3 is changed so that the system enters the gapless phase (i.e., the region $J_3 \in [0, 2]$) from the gapped phase crossing the point A at $J_3 = 2 - \delta$. A sharp dip in the LE is seen at point A; it then revives with a small magnitude which further shows a dip at point B, $J_3 = 0$. We observe the surprising result when the path is chosen (Path II in Fig. 8.7) such that the model enters the gapless phase crossing an AQCP with $J_1 \neq J_2$ indicated by P as shown in Fig. 8.7. The LE exhibits a sharp dip at the point P and remains close to a minimum value (with a revival of small magnitude as found in Path I) inside the gapless phase; a sharp rise is also observed when the system leaves the gapless phase crossing the point Q. In contrast to the previous results, the LE shows only a single dip at $J_3 = 2-\delta$ when J_3 is varied along Path III, $J_1 + J_3 = 4$ with $J_2 = 0$ (see Fig. 8.8, inset (b)). As throughout Path III, $J_2 = 0$, the aforementioned result is indeed related to the critical point of the one-dimensional Kitaev model. We shall now investigate the behavior of the LE close to these critical points considering short time and also study the time evolution of the LE fixing the environmental Hamiltonian at the quantum critical points.

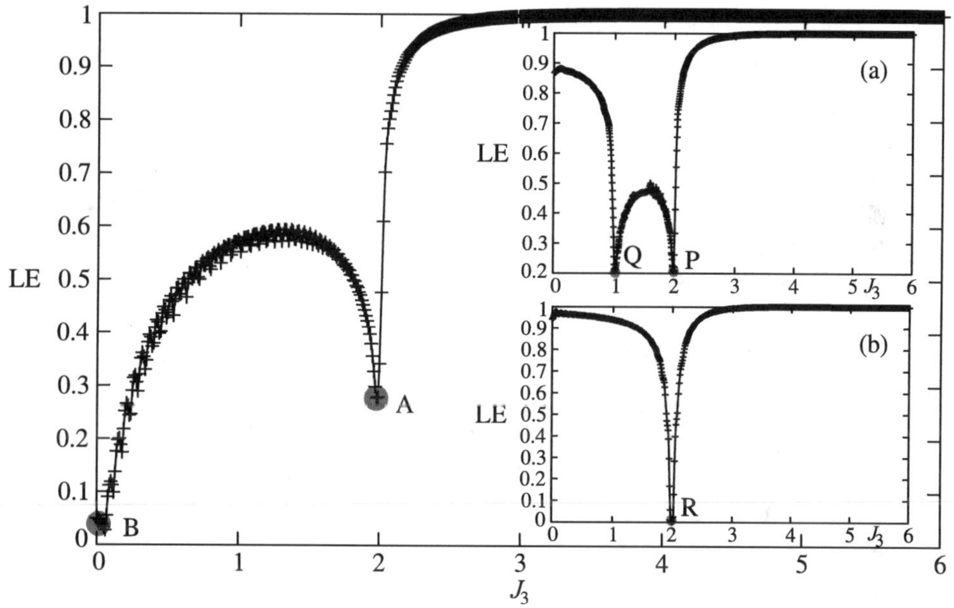

Fig. 8.8 The plots show variation of LE with the parameter J_3 (J_3 is changed Path I). The LE shows a sharp dip at $J_3 = 2-\delta$ (point A) and following a small but finite revival throughout the gapless phase, it further decays at point B, $J_3 = 0$ (see Fig. (8.7)). Here, $N_x = N_y = 200$, $\delta = 0.01$ and $t = 10$. Inset (a) shows LE as a function of J_3 when the change in J_3 is carried out along Path II ($J_1 = J_2 + 1$) with $N_x = N_y = 200$, $\delta = 0.01$ and $t = 10$ exhibiting a sharp dip at point P ($J_3 = 2-\delta$) and again there is a sharp increase at point Q ($J_3 = 1-\delta$). Inset (b) shows the dip in LE at $J_3 = 2-\delta$ when J_3 is changed along the line $J_1 + J_3 = 4$ (Path III). For this case, $N = 400$, $\delta = 0.01$ and $t = 10$ (from Sharma and Rajak, 2012).

Short time decay: As mentioned already, there is an AQCP $J_{3c} = 2J_1$ in this model with critical exponents given by $\nu_\perp = z_\perp = 1$ and $\nu_\parallel = 1/2, z_\parallel = 2$ along the directions $\hat{j}$ and $\hat{i}$ respectively. As in Sec. 8.3.3, here also we are interested in studying the short time behavior of the LE near the AQCP. We approach this critical point along Path I ($J_1 = J_2$) by varying the parameter J_3. The energy gap becomes zero at this point for the three momentum modes:

$(2\pi/\sqrt{3}, 0)$ and $(0, \pm 2\pi/3)$ that lie in half of the Brillouin zone. After expansion around one of the critical modes, we get $\sin^2 \varepsilon_k (J_3 + \delta)t \approx 4(J_3 + \delta - 2J_1)^2 t^2$ and $\sin^2(2\gamma_k) \approx 9J_1^2 k_y^2 \delta^2 / (J_3 - 2J_1)^2 (J_3 + \delta - 2J_1)^2$. Defining a cutoff frequency and a partial sum (see discussion around Eq. (8.30) and refer to (Quan et al., 2006; Sharma and Rajak, 2012)), we obtain the exponential decay of the LE in short time limit

$$\mathscr{L}(J_3, t) \approx \exp\left(-\Gamma t^2\right), \tag{8.42}$$

where the decay constant $\Gamma \propto \delta^2/(J_3 - 2J_1)^2 N_y^2$. It is interesting that the LE in short time limit can capture the anisotropic nature of the QCP since Γ scales as $1/N_y^2$ and does not depend on N_x. Again from the expression of $L(J_3, t)$, we can observe that it remains unaffected under the transformation $N_y \to N_y \alpha, \delta \to \delta/\alpha$ and $t \to t\alpha$, with α being some integer.

We have also shown the LE as a function of time t when the system is fixed at point A with $J_1 = J_2 = 1$ and $J_3 = 2 - \delta$ (see Fig. 8.9). The quasiperiod of the collapse and revival of the LE scales linearly with N_y, and remains unchanged in N_x. Therefore, the aforementioned short time scaling results are verified by this collapse and revival in the LE.

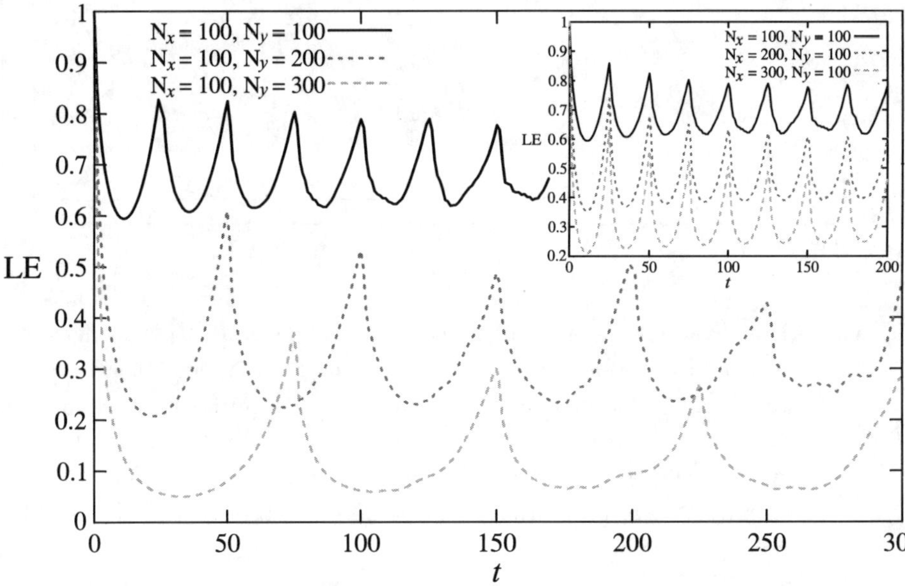

Fig. 8.9 The LE as a function of time t at the AQCP A (see Fig. 8.7) with $J_1 = J_2 = 1$, $\delta = 0.01$ and $J_3 = 2 - \delta$, considering $N_x (= 100)$ as fixed and N_y as changing confirms the scaling relations discussed in the text. The inset figure shows that the time period of the collapse and revival does not depend on N_x (from Sharma and Rajak, 2012).

8.5 Effect of Double Local Quenches on Loschmidt Echo and Entanglement Entropy

We have discussed the fidelity and its dynamical counterpart, i.e., the Loschmidt echo for two spin models in the previous two sections. In these cases, we have measured the time evolution of the LE in situations where the ground state of the Hamiltonian evolves with another Hamiltonian; the other Hamiltonian is effectively given by the first Hamiltonian with a very small change in tuning parameter. In this section, we study the time evolution of the LE and the EE in a situation when the system is out of equilibrium. In recent years, a considerable amount of focus is given to studying the EE in systems out of equilibrium (Eisler and Peschel, 2007; Calabrese and Cardy, 2007b; Eisler et al., 2008; Igloi et al., 2009; Hsu et al., 2009; Cardy, 2011; Stephan and Dubail, 2011; Igloi et al., 2012).

As mentioned already, one can now experimentally demonstrate such non-equilibrium studies using optical lattices (Bloch et al., 2008). One of the ways of generating such a non-equilibrium dynamics is a sudden quench. In global sudden quench, the EE generally shows a linear increase in time t up to some time t_0 (Calabrese and Cardy, 2005). On the other hand, in a local quench a parameter of the Hamiltonian is changed locally. For example, the entanglement entropy between two critical subsystems A and B of a homogeneous one-dimensional chain which are disconnected for $t < 0$ and connected at $t = 0$ increases as $S \sim \frac{2c}{3} \ln t$ for $t \ll L$ (Eisler and Peschel, 2007; Calabrese and Cardy, 2007b; Stephan and Dubail, 2011); here the final chain is periodic. On the other hand, if the final chain is open, the factor 2 in the expression of S is not present. These studies play an important role in the context of information propagation through a quantum many-body system. In this connection, the concept of light-cone-like behavior in condensed matter systems has been established in some recent theoretic studies (Bravyi et al., 2006; Eisert and Osborne, 2006; Naget al., 2014, 2015). The characteristic speed of the light-cone provides the maximum allowed speed with which information can propagate in the quantum system; it is known as the Lieb–Robinson limit (Lieb and Robinson, 1972). In the case of integrable quantum systems, this speed reduces to the maximum group velocity ($v_{\max}$) of the quasiparticles produced due to the quantum quench. In particular, it has been studied that the revival structure of the LE and the EE following a sudden quench is determined by the quasiparticles moving with speed $v_{\max}$ (Stephan and Dubail, 2011; Happola et al., 2012). The generalization of the Lieb–Robinson limit for a non-integrable quantum system has also been shown by Happola et al. (Happola et al., 2012). In Sec. 8.8.1, we also discuss the revival of the survival probability of an edge Majorana (which is in fact determined through the LE) under a sudden quench in the p-wave superconducting chain from the viewpoint of light-cone like propagation.

We here consider two decoupled transverse Ising spin chains which are either in the ferromagnetic phase or at critical point. These two spin chains are in their respective ground states for time $t < 0$; suddenly at $t = 0$, they are joined (J-quenching). Simultaneously, we change the value of the transverse field at a single site that belong to one of the spin chains (h-quenching). As a result the time evolution of the system becomes non-equilibrium. This process of double quench represents removal of a fragile bond with inclusion of a magnetic impurity in the system which locally increases the strength of the magnetic field at that lattice site. Although an analogous situation like removal of a weak bond can occur in the X-ray absorption problem (Eisler and Peschel, 2007; Mahan, 2013), the control of occurrence of such simultaneous double quenches is not very easy. At the same time, the numerical technique which will be used for this study has the limitation that the two quenches must occur simultaneously. However, in this study we shall show that the site of h-quench can be represented as a defect centre where the scattering of quasiparticles (wavepackets) has been observed. In short, we study the effect of sudden increase of the magnetic field in a lattice site which behaves like a scattering centre for the quasiparticles created due to the local J-quench. Our aim here is to study the non-trivial evolution of the LE and the EE using the theory of quasiparticles produced after the aforementioned local quenches (Rajak and Divakaran, 2016).

8.5.1 Model

We again consider a one-dimensional Ising chain in a transverse field, i.e., the model discussed in Sec. 8.3.1 with zero value for the three-spin interaction strength. The corresponding Hamiltonian is given by

$$H = -\sum_n (J_n \sigma_n^x \sigma_{n+1}^x + h_n \sigma_n^z), \tag{8.43}$$

where J_n and h_n are the site-dependent cooperative interactions and transverse magnetic fields respectively. If we consider the homogeneous chain ($h_n = h$ and $J_n = J$), the Hamiltonian in Eq. (8.43) has a QCP at $J = h$ which separates ferromagnetic and quantum paramagnetic phases. As discussed already, a homogeneous and periodic chain can be solved exactly in the momentum space using a JW transformation (Lieb et al., 1961; Pfeuty, 1970) followed by the Fourier transformation.

Nevertheless, one always find some defects in a real system which makes it inhomogeneous. The general procedure used to study systems which are not homogeneous is summarized as follows. Using JW transformation, the Hamiltonian in Eq. (8.43) can be expressed in a quadratic form of spinless fermions $c_n^\dagger$ and c_n (Lieb et al., 1961)

$$H = \sum_{m,n} \left[c_m^\dagger A_{m,n} c_n + \frac{1}{2}(c_m^\dagger B_{m,n} c_n^\dagger + \text{h.c.}) \right]. \tag{8.44}$$

Here, A is a $L \times L$ symmetric matrix since H is Hermitian and B is a $L \times L$ antisymmetric matrix which is the consequence of satisfying the anticommutation relations by c_n operators. In this case, the elements of A and B matrices are given by

$$A_{m,n} = -(J_m \delta_{n,m+1} + J_n \delta_{m,n+1}) - 2h_m \delta_{m,n},$$

$$B_{m,n} = -(J_m \delta_{n,m+1} - J_n \delta_{m,n+1}). \tag{8.45}$$

The Hamiltonian in Eq. (8.44) is diagonalized in terms of the fermionic operators η_q given by the relation (Lieb et al., 1961).

$$\eta_q = \sum_{n=1}^{N} (u_q(n)c_n + v_q(n)c_n^\dagger), \tag{8.46}$$

where $u_q(n)$ and $v_q(n)$ are some real numbers. Now, we can rewrite the aforementioned Hamiltonian in the diagonal form in terms of η_q fermions

$$H = \sum_{q=1}^{L} \varepsilon_q \left(\eta_q^\dagger \eta_q - \frac{1}{2} \right), \tag{8.47}$$

where ε_qs are the energies corresponding to different normal modes with index q. One can find that these $\varepsilon_q s$ can also be obtained from the solutions of the eigenvalue equations,

$$(A-B)(A+B)\Phi_q = \varepsilon_q^2 \Phi_q$$

$$(A+B)(A-B)\Psi_q = \varepsilon_q^2 \Psi_q\,. \tag{8.48}$$

It is easy to check that the elements of the eigenvectors of the eigenvalue equations in Eq. (8.48) are related to u and v matrices with the following forms: $\Phi_q(i) = u_q(i) + v_q(i)$ and $\Psi_q(i) = u_q(i) - v_q(i)$. The LE and the EE can be calculated using Φ, Ψ, u and v as given by Rossini et al. (2007), Igloi et al. (2009) and Rajak and Divakaran (2016).

8.5.2 Semiclassical theory of quasiparticles

We here present the semiclassical theory of quasiparticles (QPs) which shall be extensively used to explain the time evolution of the LE and the EE following two simultaneous local quenches. Let us consider a system at zero temperature which initially remains in its ground state and is suddenly perturbed. This forces the state of the system to go through a non-equilibrium evolution determined by the final Hamiltonian. The initial state which becomes an excited state after the perturbation, is a source of QPs with respect to the final Hamiltonian. In recent years, such studies involving non-equilibrium time evolution have been carried out in the light of the semiclassical theory of QP produced due to global (Rieger and Igloi, 2011; Blass et al., 2012) and local quenches (Divakaran et al., 2011), where a very good agreement between the numerics and the semiclassical theory were observed. We now outline the theory of QP generation that will be employed to explain the different time scales found in our numerical calculations. When a global quench is performed with a very small change in h from $h = 0$, QPs are generated in the system which has been discussed in detail by Rieger and Igloi (2011). The QPs associated with to a momentum are always created in pairs that follows from the conservation of momentum, the group velocity $v_g(q)(= |\partial \varepsilon_q/\partial q|)$ of them is equal and opposite to each other. As discussed by Rieger and Igloi (2011), these QPs in the limit of small h are considered as classical particles which simply flips the spin at a site when it crosses that. Although this picture is studied for a very specific quench (i.e., a small quench), it can be applied for stronger quenches and also for quenches in the paramagnetic region with some modifications. On the other hand, it is also claimed that these QPs are no longer point particles, but behave like extended objects for their large correlation length as the critical point is approached. In the next sections, our attempt will be to explain the numerical results at least qualitatively, using this point like a picture of QPs for spin chains residing in the critical or ferromagnetic regions.

8.5.3 Loschmidt echo and entanglement entropy for a critical chain

Let us first study double quenches as discussed earlier for a critical chain where already work has been done (Calabrese and Cardy, 2007b; Stephan and Dubail, 2011; Divakaran et al., 2011) for local J-quenches. As mentioned before, we consider the application of two simultaneous local quenches in the system and investigate the dynamics of the LE and the EE after such double quenches. For time $t < 0$, the whole system is constructed in the ground state of $H = H_1 + H_2$ with H_1 and H_2 being the Hamiltonians of two open homogeneous transverse Ising chains of lengths L_1 and L_2, respectively which are uncoupled. Two simultaneous quenches are made at $t = 0$: the two spin chains are suddenly joined together to give a chain of total length $L = L_1 + L_2$ and the transverse field at a specific site, say L', that belongs to either the chain 1 or 2 is changed from h to $h+\delta$. The evolution of the system is then determined by the final Hamiltonian

$$H_f = H_1 + H_2 + H_{12}^I - \delta\sigma_{L'}^z, \tag{8.49}$$

where H_{12}^I denotes the interaction Hamiltonian between the two spin chains of length L_1 and L_2, which has the form $J\sigma_{L_1}^x\sigma_{L_1+1}^x$. On the other hand, the term $-\delta\sigma_{L'}^z$ in the Hamiltonian determines the h-quenching in the system. These quenches can be incorporated in our numerics by considering a spin chain of total length $L(= L_1 + L_2)$, where the first L_1 spin represents System 1 and the remaining part is System 2. Initially, we consider $J_{L_1} = 0$, i.e., the spin chains are disconnected at $t < 0$; then, at $t = 0$, it is changed to J, equal to the interaction strength for all the other sites. We have set $J = 1$ for the entire calculation. Numerical calculations for LE and EE have been done by Rossini et al. (2007) and Igloi et al. (2009).

When the two local perturbations as discussed earlier are turned on, there will be a local increase in energy of the system at the sites of local perturbations (Calabrese and Cardy, 2007b; Divakaran et al., 2011). These two sites then indeed become the source of QP production at time $t = 0$. We shall now denote the QPs produced for the h-quenching at L' as QP^1; the QPs that move to left and right at $t = 0$ are denoted by QP_L^1 and QP_R^1, respectively. In a similar fashion, the left and right moving QPs produced at the site L_1 of J-quenching will be called QP_L^2 and QP_R^2. In the following, we qualitatively explain the various time scales associated with evolution of the LE and the EE using the picture of QP propagation.

Time scales: We here define a few time scales using the propagation of the QPs generated at the sites of local quenches to explain the time evolution of the LE and the EE. We mainly consider $L' < L_1 \leq L_A$, where L_A is explained in the context of EE in Sec. 8.1.3. The case where $L' > L_1$ is discussed in the caption of the respective figures. More importantly, the different time scales are determined by the fastest moving QPs with maximum group velocity $v_{\max} = \max_k v_g(k)$. These time scales are given as $t_1 = (2L')/v_{\max}$ (time of come back of QP_L^1), $t_2 = 2(L-L')/v_{\max}$ (time of come back of QP_R^1), $t_3 = (2L_1)/v_{\max}$ (time of come back of QP_L^2) and $t_4 = (2L_2)/v_{\max}$ (time of come back of QP_R^2). In addition to these obvious time scales, we find two more time scales in our numerical results: one is given by $t' = 2|L_1 - L'|/v_{\max}$, which provides the time taken by QP_L^2 (for $L_1 > L'$) to reach L' and come back to L_1 after getting partially reflected at L' where it observes a difference in potential from h to $h+\delta$. The second time scale that we

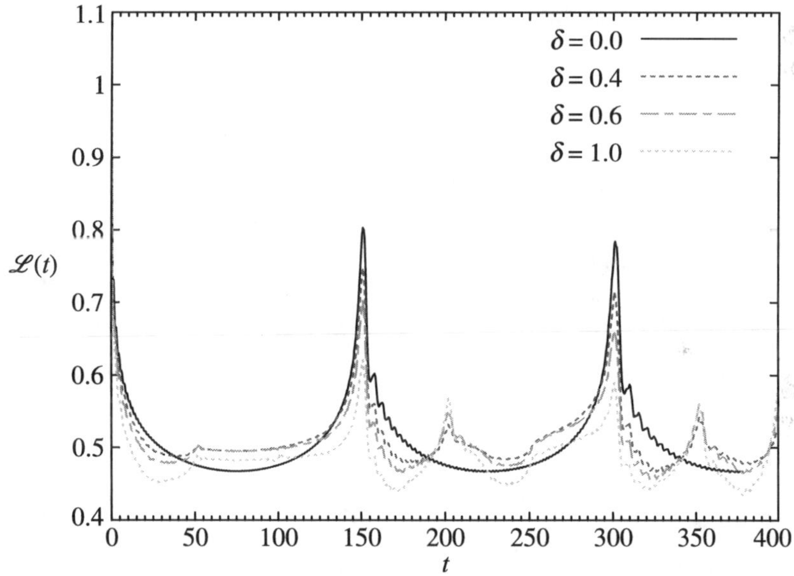

Fig. 8.10 The variation of LE with time for different values of δ when J-quench and h-quench are performed at $L_1 = L/2$ and $L' = L/3$ respectively. The h-quenching here is made by changing its value from 1 to $1+\delta$ at L'. In the case of only J-quenching (i.e., $\delta = 0$ case), the LE marks a peak at $t_3 = L/v_{\max} = 150$ and $T = 2t_3$. Here, $v_{\max} = 2$ and $L = 300$. By applying both the local quenches simultaneously at time $t = 0$, we find a small peak at $t' = 50$ and a relatively stronger peak at $t'' = 200$. It can also be noted that there are small fluctuations close to $t_1 = 100$ which is more clear for the $\delta = 1$ curve (from Rajak and Divakaran, 2016).

observe, is equal in magnitude to t_2, though this is due to QP^2 and can be found as $t'' = 2(L-L')/v_{\max}$, which is the time that the QP_L^2 (QP_R^2) takes to get reflected at L' (right boundary) and return to L_1 after being fully (partially) reflected at the right boundary (L'). In the absence of partial reflection, all the QPs come back to their origin at time period $T = 2L/v_{\max}$. The maximum group velocity $v_{\max}$ can be calculated for the homogeneous transverse Ising model using the elements provided in Eq. (8.45) which is given by 2 at the critical point and also in the paramagnetic phase. In the present case also, the value of $v_{\max}$ is 2, which can be verified numerically by differentiating the eigenvalues.

Time evolution of the LE and the EE: We now present our numerical results of the time evolution of LE and EE as a result of double local quenches in the spin chain. Let us first consider the evolution of the LE. In Fig. 8.10, the evolution of LE has been shown for different values of δ with $L_1 = L_2$ and $L' = L/3$. As the value of δ increases (i.e., the higher strength of h-quench), the decay of the LE has become faster as expected. At the same time, we have shown the effect of L' on the time scales in Fig. 8.11. If we consider only J-quenching (Stephan and Dubail, 2011), the LE exhibits three time scales given by t_3, t_4 and the time period T as discussed earlier (see $L' = 0$ plot of Fig. 8.11). As mentioned

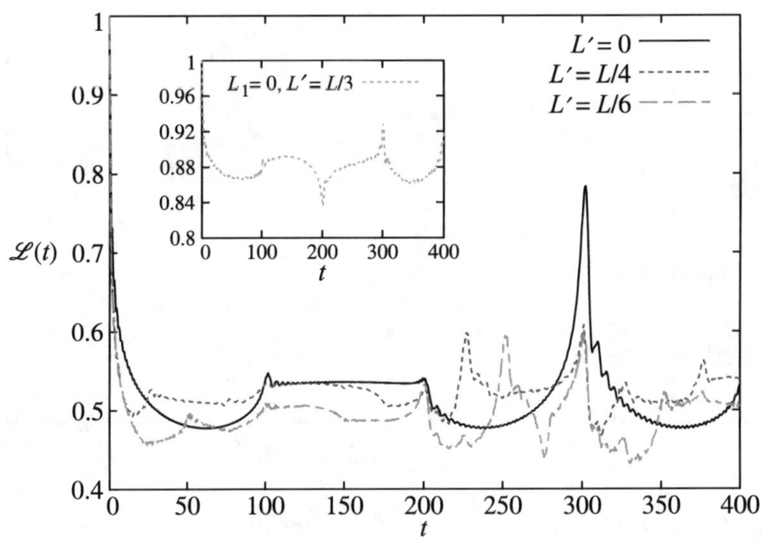

Fig. 8.11 Main: The time evolution of the LE for two local quenches having $L_1 = 100$, $L_A = 100$ and $L = 300$ and different L'. The first peak in the LE is seen at time $t' = 2(L_1 - L')/v_{\max}$; $t' = 25$ and 50 for $L' = L/4$ and $L/6$ respectively. The other time scales are given by $t_3 = 100$, $t_4 = 200$ and t''. It can be noted that $t'' = 225$ and 250 for $L' = 75$ and 50, respectively. Inset: The LE as a function of time t for only h-quench at $L/3$ with $L = 300$. Time scales $t_1 = 100$ and $t_2 = 200$ both are observed in this case. It is easy to find that the magnitude of decay in this case is very small (from Rajak and Divakaran, 2016).

before, the LE is indeed the overlap of two wavefunctions that initially had unity overlap and starts to decrease from 1 at $t = 0$ till one of the QPs reach the boundary. Intuitively, we can argue that during the time of coming back after reflection, the QPs will undo the effect of its dynamics, resulting in an increase in the LE. Therefore, one observes peaks at t_3 and t_4 when both the QPs of the pair come back to their origin. In addition to J-quenching if we change the transverse field from h to $h+\delta$ at L', it is expected to show some more time scales as discussed in the previous paragraph: $t_1 = (2L')/v_{\max}$, $t' = 2(L_1 - L')/v_{\max}$, $t_2 = t'' = 2(L-L')/v_{\max}$. These time scales are clearly observed in Fig. 8.11 (main) except t_1. We shall try to give an argument supporting this fact later. In a case when only h-quenching is performed by changing h from 1 to 2, it has been found that the decay in the LE is very less as compared to the case of J-quench, though we do observe t_1 and t_2 as shown in the inset of Fig. 8.11.

Let us now focus on the time evolution of the EE considering the scenario described here. We define another length of size L_A for subsystem A and calculate the entanglement between the subsystem A and the remaining system of size $L - L_A$ (subsystem B) using the EE. We shall assume the simplest case with $L_1 = L_A$, i.e., the position of J-quenching also corresponding to the right end of subsystem A of size L_A. It is interesting that the bipartite EE of two critical Ising chains in transverse field can also identify the effect of h-quenching. We find that for the single quench (J quenching), the EE exhibits complete periodic oscillations with a sharp decay at $t_3 = 2L_1/v_{\max}$ and a rise at t_4 (see Fig. 8.12) which also has been discussed by Stephan and Dubail (2011) using conformal field theory. Importantly, it is also possible to explain this result using the picture of QPs. The main idea here is that the EE will increase with time if one of the QPs of a pair remains in subsystem A and the other is in B. After the creation of a pair of QPs at $L_1 = L_A$, the QPs start travelling in opposite directions and an immediate increase in $S(t)$ occurs for $t > 0$. both the QPs reach the source point and a dip in S is seen once again; after that, there is a repetition of the same pattern. Additionally if we now accomplish local h-quenching at an arbitrary site L' of the spin chain, we find that the dynamics of the EE after double quenches corresponds to the J-quenching case with some deviations at certain times which also can be explained using the picture of QP generation. It has been observed that the case of double quench follows the single quench ($L' = 0.0$) till $t = t'/2$ and after that the EE shows a sudden deviation or increase from the single quench case. This is due to the fact that one of the QPs (QP_R^1) created at L' enters the subsystem B at this time while the other QP of the same pair is in subsystem A. This provides an additional increase in S. On the other hand, the QP_L^2 returns to $L_1 = L_A$ at time $t = t'$ after reflection at L' where a sharp decrease is observed in S since both QP_L^2 and QP_R^2 are in subsystem B. The decrease in the EE continues (following a small rise) till QP_R^2 enters subsystem A at $t = 2(L - L_1)/v_{\max}$ after the reflection from the right boundary. After that a sharp increase has been found in the EE. We observe one more time scale $t'' = 2(L-L')/v_{\max}$ which was also present in the LE. The QP_R^2 and QP_L^2 come back at this time where they exchange the subsystems, as also discussed earlier in the context of the LE. All these times scales can be found in Fig. 8.12.

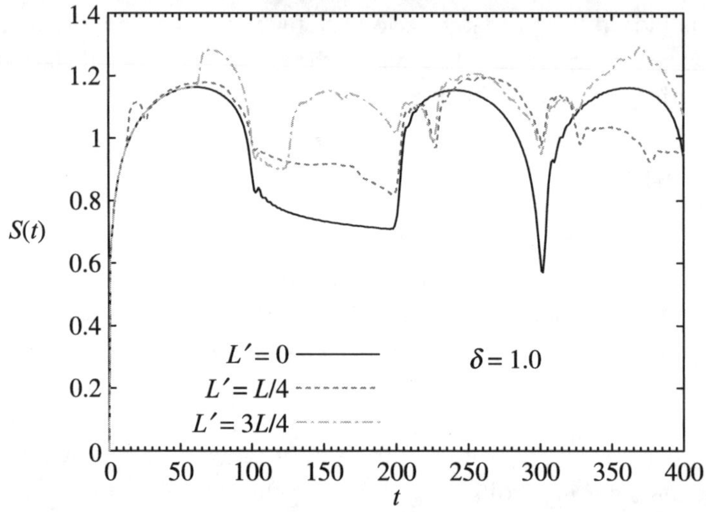

Fig. 8.12 The plot shows time evolution of the entanglement entropy having $L_1 = L_A = 100$ and different values of L'. For $L' = L/4$, the departure from the only J-quench case is seen at $t'/2 = 12.5$, whereas at $t' = 25$, QP_L^2 get in the system B after reflection at L' in which QP_R^2, the other partner, is already present. The EE keeps on decreasing till QP_R^2 enters system A at time $t = L_2 = 200$. A dip is also seen at $t = t'' = 225$ where QP_L^2 and QP_R^2 interchanges their systems. In a similar fashion, we can make an argument for the time evolution of EE when $L' = 3L/4$. The deviation from the case of J-quench alone starts at $t = 62.5$. There is a sharp decay in EE at $t = 100$ when QP_L^2 enters system B. On the other hand, QP_R^2 goes into system A at $t = 125$ that results in an increase in the EE since its other partner remains in system B. The normal rise at $t = 200$ which is clearly present for the single quenching (J-quenching) case can also be found here. This may be an artefact of getting a transmitted component of the QP wavepacket at L' which indicates that the reflection of the QPs at L' is not perfect, as already discussed in Sec. 8.5.4. We can also observe the time scales $t'' = 225$ and $T = 300$ (from Rajak and Divakaran, 2016).

From this discussion we can understand the main difference between the study of evolution of the LE and the EE. For the LE, the important time scales are when the generated QPs return back to their origin. On the other hand, the aforementioned time scales in the EE are observed when one of the QPs that belong to a pair enters from one system to the other. If due to this crossing both the QPs are in the same system, then one finds a decrease in the EE, else it increases. It can be also noted that the increase in the EE for small time with the application of double quenches is similar to only J-quenching, i.e., $S \sim \ln t$ for $t \ll L/v_{\max}$ (Eisler and Peschel, 2007; Calabrese and Cardy, 2007b; Stephan and Dubail, 2011). The effect of h-quenching in the evolution starts at $t'/2$.

Finally, it is found that for the quenches performed here, the main effects are due to QP^2 at times t', t'', t_3 and t_4. We also observe that the effect of quenches at t'' is stronger; this may be due to three different QPs returning to their origin after getting reflected from various points as discussed here: (i) QP_R^1 after getting reflected from the right boundary, (ii) QP_R^2 after two reflections, i.e., at the right boundary and then at L' so that it returns to L_1, (iii) QP_L^2 after two reflections, i.e., at L' and then at the right boundary so that it returns to L_1. A peak or dip is also present at $t = L$ which is the time when of all the fastest QPs return to their source point if the reflection does not occur at L'; thus providing us an indication of the presence of a transmitted wave of the QPs. We shall make some more observations about it after studying the results in the ferromagnetic region.

It can be observed that the existence of time scale t_1 for QP^1 is not very clear in these figures, but some perturbation can be seen at this time. On the other hand, we have clearly observed the effect of QP^1 at $t = t'/2$, pointing strongly to its presence. In Sec. 8.5.5, we shall try to give an argument about the absence of time scale t_1 in the dynamics of the LE.

8.5.4 Entanglement entropy: Ferromagnetic phase

In this section, we study the EE as a function of time when the total spin chain lies in the ferromagnetic phase. The motivation behind this study is to observe the two distinct natures of the QPs in the critical and the ferromagnetic regions (see Sec. 8.5.2) through the evolution of the EE. Here, we assume the asymmetric case, i.e., $(L_1 \neq L_2)$ with $L_1 = L_A$. We now focus on two different cases to determine the dynamics of the EE. In the first case, the total spin chain is considered to be deep inside the ferromagnetic phase (see Fig. 8.13 (a)) and for the second one, the spin chain is near the critical point (see Fig. 8.13 (b)). As mentioned before, we first consider two open spin chains having $h = 0.5$ at all sites. When there is only J-quenching ($L' = 0$), the EE marks t_3 and t_4 successfully (see Fig. 8.13(a)). Interestingly, one can find the difference between the critical and ferromagnetic region within the time t_3. As found in the critical case, the EE increases at $t = 0$ with a decay that begins around $t = t_3/2$, i.e., the time at which QP_L^2 reflects from the boundary; however the decrease is sharper at t_3. Whereas, in the ferromagnetic phase, we observe a sudden increase in the EE around $t = 0$ and after that, it remains almost at a constant value until time t_3 where the decrease starts suddenly. This indicates that the QPs in the ferromagnetic phase are point-like particles whose location can be determined exactly. On the other hand, for the critical case, these QPs are extended objects as also discussed by Blass et al. (2012), and as a consequence, the effect of reflection of a QP at the end of the chain is felt also at L_1. As also found in the critical case (see Sec. 8.5.3), there is a decay in the EE between times t_3 to t_4. In this time span, the QPs with maximum group velocity do not contribute in the EE. We now concentrate on the time dynamics of the EE following double quenches. It is easy to observe the time scales $t'/2$ and t' from Fig. 8.13. Surprisingly, the EE starts decreasing at t' and it keeps on decreasing up to t_4. The EE then shows a sharp increase at t_4 which is the consequence of QP_R^2 entering the system A at t_4; whereas QP_L^2 is already in system B. It i important to mention that the time scale t'' is not clearly observed here, which can also

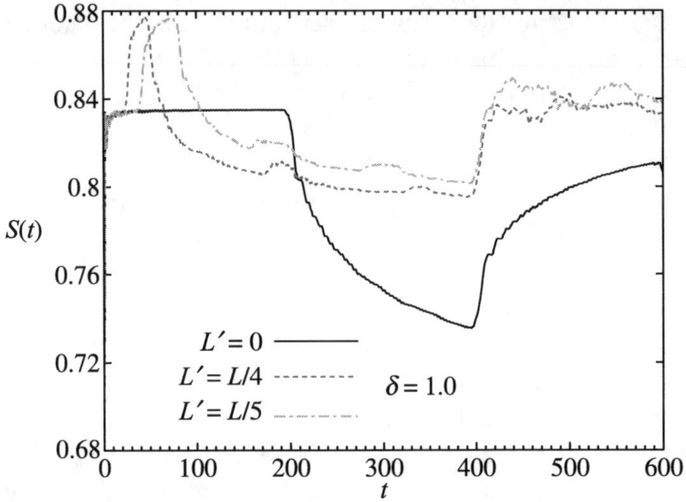

(a)

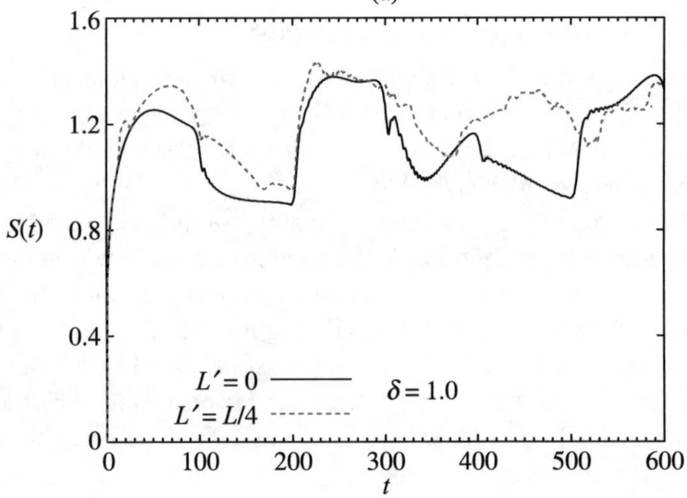

(b)

Fig. 8.13 (a) Variation of the EE as a function of time following single and double quenches when the total spin chain is in the ferromagnetic region at $h = 0.5$. Here, we consider $L_1 = 100$, $L_A = 100$ and $L = 300$ and different L'. For $L' = L/4$, the EE deviates from the only J-quenching case, starting at $t = t'/2$ with $t' = 50$ and $v_{\max} = 2h = 1$. For time $t > 50$, both the quasiparticles QP_L^2 and QP_R^2 exist in subsystem B that leads to the decrease in the EE continuing up to $t_4 = 400$. Again, for $L' = L/5$, we find $t' = 80$ and observe the expected behavior. The explanation of the absence of t'' is given in the text. The plot (b) is the same with $h = 0.99$. The observed time scales are given by $t' = 25.25$, $t'/2$, $t_3(\sim 101)$ $t_4(\sim 202)$ and $t''(\sim 227)$ calculated with $v_{\max} = 1.98$ (from Rajak and Divakaran, 2016).

be explained using the point-like behavior of QPs in the ferromagnetic phase. We find that for $t < t''$, QP_L^2 exists in system B; whereas QP_R^2 is in system A. At $t = t''$, these two QPs exchange their systems; therefore, their contribution in the EE is equal for $t < t''$ and $t > t''$. In contrast, due to the wavepacket-like behavior of the QPs, the critical case distinguishes the two cases, i.e., whether a QP approaches L_1 or moves away from L_1. The time evolution of the EE with $h = 0.99$ has been shown in Fig. 8.13(b); this gives an indication towards a more extended behavior of QPs similar to the critical case signaling a partial reflection at L'.

8.5.5 QP^1 and QP^2: a comparative study

This section provides an explanation regarding the dominance of QP^2 over QP^1 to determine the evolution of LE and EE. The energies of these QPs that are generated at the sites of local quenches are given by the eigenvalues of the Hamiltonian at the final parameter values. Let us consider the QP η_q with the corresponding energy ε_q, where q denotes different modes. The QPs η_q are created at the site of perturbation with probability f_q. The expression for f_q is given by

$$f_q = \langle \psi_i | \eta_q^\dagger \eta_q | \psi_i \rangle. \tag{8.50}$$

This indeed has a proportionality relation with the number of QPs η_q (created in the initial state $|\psi_i\rangle$) corresponding to the final Hamiltonian H. This relation of f_q can be calculated numerically using the elements of both initial and final Hamiltonians (see Ref. (Rajak and Divakaran, 2016)). We have plotted f_q as a function of q in Fig. 8.14 for three different quenching cases: h-quench alone (from 1 to 2 at L'), J-quench alone (from 0 to 1 at L_1) and also both the quenches together. Figure 8.14 clearly indicates that the probability of QP creation is higher as an order of magnitude in case of only J-quenching in comparison to the h-quench. It provides an indication that the site of J-quenching is the main source for the generation of QPs; thus the evolution of the LE and the EE are mainly determined by the dynamics of QP^2 for the quenching schemes which we have adopted here. Furthermore, we also compare the f_q for J-quenching from 0 to 1 at a specific site with the h-quenching from 0 to 1. It has been found that for such quenches of equal magnitude, f_q is of the same order and thus, the effect of QP^1 and QP^2 should be sensed equally.

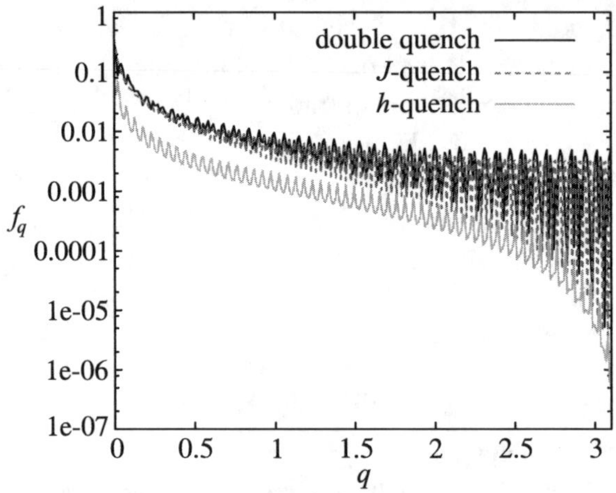

Fig. 8.14 The plot shows f_q as a function of q for a critical chain with $L_1 = 100$, $L' = 60$ and $L = 300$. The dashed line represents the case of the h-quench alone which is clearly smaller than the J-quench alone (the dotted line) by an order of magnitude. The solid line is for the double quench, which almost overlaps with the J-quenching case. A similar type of behavior can also be observed for a spin chain residing in the ferromagnetic region (from Rajak and Divakaran, 2016).

8.6 Topology in Condensed Matter Systems

In this section, we provide a brief introduction to the emergence of topology in condensed matter systems and then connect it to quenches later. Topology is a branch of mathematics concerned with the geometrical properties of a system which are preserved under small changes in the system. Two systems are topologically equivalent if they can be transformed continuously into each other. It is now an important task to define a quantity for a system which represents the topology of that system. This quantity is called topological invariant when it takes integer values and remains fixed under small changes in the system. Topologically distinct systems must have different topological invariant values. For example, in a two-dimensional parameter space, one can define a winding number (W) as a measure of topological invariance which counts the number of times a closed curve winds around the origin in the anticlockwise direction. Clearly winding numbers do not change continuously with some small changes in the system. To change the integer value of W, therefore, the curve must have to pass through the origin where W is ill-defined.

This concept of topology from mathematics has been used in condensed matter physics to define some special states of matter. In condensed matter physics, two quantum many-body systems are topologically equivalent if we can continuously transform their Hamiltonians into each other. Now if we consider this as a definition for topological invariance, we can face a problem since then all quantum systems would be topologically equivalent. Therefore, we have to apply some constraint in the Hamiltonians of the quantum systems. Drastic change occurs when we consider systems with an energy gap. We can now say that two gapped quantum systems are topologically equivalent if their Hamiltonians can be continuously transformed into each other without ever closing the energy gap.

In general, for condensed matter physics, a phase transition corresponds to some symmetries being broken. In the last few decades, studies of new states of matter, namely, topological states (Thouless et al., 1982; Wen, 1995) which are not associated with broken symmetries have emerged as a rapidly growing field of research (Hasan and Kane, 2010; Qi and Zhang, 2010, 2011). It is to be noted that distinct topological phases of a topological system cannot be defined by the usual local order parameters similar to the magnetic and superconducting systems. As defined earlier, those states of matters are represented by some topological invariants which change discontinuously across the transition. A topological condensed matter system has insulating bulk, i.e., there is a finite gap in spectrum between the ground state and the first excited state at low temperature, with gapless states at the boundaries. These boundary states actually contribute to electric conduction. The number of these edge states is determined by the non-zero topological invariant which is calculated from the bulk band structure of a system (Hasan and Kane, 2010). This is called bulk-boundary correspondence that allows the creation of gapless boundary modes because of the non-trivial topology of the insulating bulk (Hasan and Kane, 2010; Qi and Zhang, 2010, 2011).

In particular, the topological classification of condensed matter states has been first introduced in the study of quantum Hall systems (Thouless et al., 1982; Wen, 1995). For such systems, only the gapless edge states contribute in quantized Hall conductance. Here, the number of edge states and the quantized Hall conductance are the topological invariants which do not change under smooth deformation of the system parameters and remains constant until the system passes through quantum phase transitions. Another important example of a topological system is a topological insulator (Kane and Mele, 2005a,b; Fu et al., 2007); topological insulators have been investigated extensively in the last few years with the observation of these states in real materials (Bernevig et al., 2006; Fu and Kane, 2007; Zhang et al., 2009a). A topological insulator has a finite bulk energy gap between valance and conduction bands, i.e., its bulk is an insulator. On the other hand, the surface or edge states for such systems are gapless and are governed by two-dimensional or one-dimensional Dirac equations respectively. The gapless states of the topological insulators are protected by the time-reversal symmetry of the system (Hasan and Kane, 2010). A one-dimensional p-wave superconductor, introduced by Kitaev (Kitaev, 2001) is also a topological system that hosts zero energy modes at the two ends of the chain whose number is a topological invariant. These end modes can be represented in terms of the second quantized operators which are Hermitian, i.e., they are real; these modes are called Majorana modes. In two dimensions, a spinless $p_x + ip_y$ superconductor exhibits Majorana edge modes at the core of vortices (Volovik, 1999; Read and Green, 2000) which has a close resemblance with the chiral modes in the quantum Hall effect. In Secs. 8.8 and 8.9, we shall consider the dynamics of an edge Majorana mode following quantum quenches in the p-wave superconducting chain. We therefore provide a brief discussion about the appearance of Majorana fermions in condensed matter systems and the different topological phases of the one-dimensional p-wave superconductor in the next sections.

8.6.1 Majorana fermions and condensed matter systems

In his 1937 paper (Majorana, 1937), Ettore Majorana introduced the theoretical concept of real fermions in the context of the existence of a real solution of the Dirac equation; these fermions are now known as "Majorana fermions". The anti-particle of an electrically charged particle is different from the particle due to their opposite charges. But the Majorana fermions are their own anti-particles which signify that they are electrically neutral. As a usual fermionic operator, f is complex; it can be split into two real operators which are Hermitian

$$f = \frac{1}{2}(a + ib), \tag{8.51}$$

where $a^\dagger = a$ and $b^\dagger = b$. The operators a and b correspond to two Majorana fermions. Therefore, one complex fermion provides two Majorana fermions and thus a system with $N/2$ fermions can be described by N Majorana fermions. These Majorana operators satisfy the following relations $a_j^2 = 1 = b_j^2 = 1$, $\{a_j, a_l\} = \{b_j, b_l\} = 2\delta_{jl}$ and $\{a_j, b_l\} = 0$; these imply that to one fermionic site j, there are two Majorana sites or equivalently, a Majorana fermion can be viewed as occupying half of a state.

Although Ettore Majorana proposed his idea in the context of high-energy physics, in parallel, condensed matter physicists recently are giving more attention to realize Majorana fermions in a large variety of condensed matter systems (Wilczek, 2009; Alicea, 2012). For conventional solids made of electrons, it is difficult to observe Majorana fermions because electrons are charged and therefore different from their anti-matter counterparts, i.e., the holes with opposite charges. In second quantized operator language, if $f^\dagger$ represents the creation of an electron, then f creates a hole in the system. Superconductivity, on the other hand, may be an appropriate platform for realization of such exotic Majorana excitations because it violates charge conservation and as a result quasiparticles in a superconductor are mixtures of electrons and holes (Alicea, 2012). In such materials, electrons form Cooper pairs and one such pair can be added or subtracted from the condensate without significant changes of its properties. As a consequence, the electron number and therefore the electronic charge are no longer conserved quantities for superconducting systems. In the conventional s-wave superconductors, the overall requirement of antisymmetry of a Cooper pair means that the paired electrons must have opposite spins. The quasiparticle operators of the form $\eta = uc_\uparrow^\dagger + vc_\downarrow$ for s-wave superconductors then are not Hermitian for any situation. Thus, spin of the electrons prevents the emergence of Majorana fermions in such superconductors. This discussion suggests that spinless superconductors are ideal candidates for the emergence of exotic Majorana excitations. The Cooper pairing of two spinless (or spin polarized) electrons requires non-zero relative orbital angular momentum that provides an odd parity state resulting in a p-wave superconductor in one dimension (Kitaev, 2001) and also a $p+ip$ superconductor in two dimensions (Kopnin and Salomaa, 1991). The one-dimensional spinless p-wave superconductor is a topological system which has zero energy edge Majorana modes as midgap excitations that are guaranteed by the particle–hole symmetry of the system. The zero energy Majorana modes are indeed a mixture of particles and holes in equal measure; one can call these modes as "partiholes" (Wilczek, 2009). The superconducting pairing term of such a Hamiltonian actually induces zero energy excitations in the system. Whereas, for $p+ip$ superconductors, zero modes exist in the vortices. These p-wave superconductors are referred to as "topological superconductors" which host zero modes. But, in nature there are very few candidates which show p-wave superconductivity, e.g., ^{3}He superfluid phase (Vollhardt and Wolfle, 2013). On the other hand, a two-dimensional p-wave superconductor can be generated from the proximity effect between the surface states of a strong topological insulator and an s-wave superconductor (Fu and Kane, 2008; Das et al., 2012).

In recent experiments, the signature of Majorana modes has been observed through the zero-bias transport properties of nanowires coupled to superconductors (Mourik et al., 2012; Deng et al., 2012; Das et al., 2012; Chang et al., 2013; Lee et al., 2014); although there are some inconsistencies in experimental results with theoretical predictions (Rainis et al., 2013). Additionally, the hybridization of Majorana fermions has also been observed in experiment through the zero-bias anomalies in the differential conductance in an InAs nanowire coupled to a superconductor (Finck et al., 2013).

8.6.2 Kitaev chain: model and topological phases

The topological properties of a p-wave spinless superconductor have been studied from different points of views in recent years (Fulga et al., 2011; Lutchyn and Fisher, 2011; DeGottardi et al., 2011, 2013a; Sau and Sarma, 2012; Thakurathi et al., 2013; DeGottardi et al., 2013b) (For a review see (Alicea, 2012)). These studies lie at the interface of condensed matter physics, quantum information processing, decoherence and quantum computation (Kitaev and Laumann, 2009; Budich et al., 2012; Schmidt et al., 2012). The one-dimensional p-wave superconducting chain, introduced by Kitaev (Kitaev, 2001), is the simplest model in which we can realize zero energy Majorana modes. In the following sections, the Kitaev chain will be studied in the context of quantum quenches. We therefore provide here a brief discussion on the bulk and end properties of the p-wave superconducting chain in equilibrium.

The p-wave superconductor is a one-dimensional fermionic lattice model which consists of a nearest-neighbor hopping amplitude w, a superconducting p-wave pairing term Δ that couples two nearest-neighbor electrons and an on-site chemical potential μ.

The Hamiltonian of such a system of spinless (or spin polarized) fermions with system size N is given by

$$H = \sum_{j=1}^{N} [-w(f_j^\dagger f_{j+1} + f_{j+1}^\dagger f_j) + \Delta(f_j f_{j+1} + f_{j+1}^\dagger f_j^\dagger)] - \sum_{j=1}^{N} \mu(f_j^\dagger f_j - 1/2), \quad (8.52)$$

where we consider that all these parameters are real and the spacing of the lattice is chosen to be unity for simplicity. The fermionic annihilation and creation operators $f_j(f_j^\dagger)$ satisfy the relations $\{f_j, f_l\} = 0$ and $\{f_j, f_l^\dagger\} = \delta_{jl}$. We assume the periodic boundary condition that makes the Hamiltonian in Eq. (8.52) translationally invariant. The Hamiltonian can be therefore diagonalized in momentum basis, $f_k = \frac{1}{\sqrt{N}} \sum_{j=1}^{N} f_j e^{-ikj}$. In the momentum space, the Hamiltonian in Eq. (8.52) can be written in the standard Bogoliubov–de Gennes form

$$H = \sum_{k>0} \begin{pmatrix} f_k^\dagger & f_{-k} \end{pmatrix} H_k \begin{pmatrix} f_k \\ f_{-k}^\dagger \end{pmatrix},$$

$$H_k = -(2w\cos k + \mu)\, \tau^z \; + \; 2\Delta \sin k \; \tau^y, \quad (8.53)$$

where τ^y and τ^z are Pauli matrices in the particle–hole subspace. This provides a particle–hole symmetric energy dispersion given by

$$E_k = \pm\sqrt{(\mu + 2w\cos k)^2 + 4\Delta^2 \sin^2 k}. \quad (8.54)$$

The energy gap ($2E_k$) of the bulk becomes zero at certain values of Δ and μ for some particular k modes. The phase diagram of the model consisting of three different phases (represented by I, II and III) is shown in Fig. 8.15. It can be observed that $\mu = \pm 2w$ define two quantum critical lines with critical momentum modes $k_c = \pi$ and 0 (for which the bulk gap vanishes), respectively, while for the critical line $\Delta = 0$ (with μ lying between $-2w$ and $2w$), $k_c = \cos^{-1}(-\mu/2w)$.

Symmetries of the model: Symmetries have a strong influence on topology. In fact, discrete symmetries (e.g., particle–hole symmetry, time-reversal symmetry etc.) play an important role in classification of different topological systems (Altland and Zirnbauer, 1997; Fidkowski and Kitaev, 2011). The particle-hole symmetry is represented by an anti-unitary operator, $P = \tau^x K$, where K is complex conjugation which is an antilinear operator satisfying $P^2 = 1$ (Hasan and Kane, 2010; Qi and Zhang, 2010). The Hamiltonian H_k in Eq. (8.53) has an intrinsic particle-hole symmetry described by

$$P\, H_k\, P^{-1} \;=\; -H_{-k}. \quad (8.55)$$

One can easily prove this relation for the Hamiltonian H_k. The minus sign in Eq. (8.55) signifies that the spectrum (see Eq. (8.54)) of the system must be symmetric around zero energy. Indeed, Eq. (8.55) shows that if $\psi_k = (u_k, v_k)^T$ is an eigenvector of H_k with energy

E_k and momentum k, then $P\psi_k = (v_k^*, u_k^*)^T$ is also an eigenvector of the same Hamiltonian with energy $-E_k$ and momentum $-k$. But these two energy states do not correspond to two distinct quantum states because the Bogoliubov quasiparticles associated with them follow $b_{E_k}^\dagger = b_{-E_k}$. Therefore, creating a quasiparticle at the energy state E_k is the same as creating a hole at $-E_k$. Then for $E_k = 0$, $b_0^\dagger = b_0$, i.e., b_0 is its own anti-particle. As mentioned earlier, these quasiparticles are called Majorana zero modes which are protected by the bulk gap of the system. Although Majorana zero modes do not exist in periodic boundary conditions, its prediction can always be done from the bulk properties of the system. Another important symmetry of the system is time-reversal symmetry. As our model is spinless, the time-reversal symmetry of the system is given by the relation, $H_{-k}^* = H_k$ for all k.

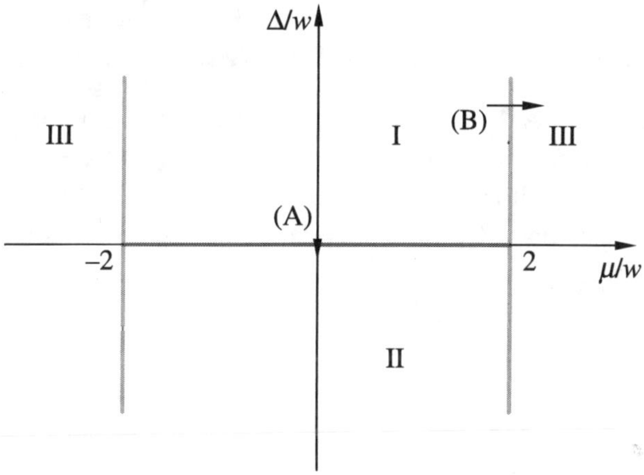

Fig. 8.15 Phase diagram of the p-wave superconducting model in one dimension (see Eq. (8.52)). In this diagram, Phases I and II are topologically non-trivial, whereas, Phase III is topologically trivial. Sudden quenching in this system is performed along Paths A and B (see Sec. 8.8).

In order to explore the non-trivial topology in the weak pairing phase of the model, we represent the Hamiltonian in Eq. (8.52) in terms of the two Majorana operators a_j and b_j at each site defined as

$$f_j = \frac{1}{2}(a_j + ib_j), \quad f_j^\dagger = \frac{1}{2}(a_j - ib_j). \tag{8.56}$$

We can now rewrite Eq. (8.52) with an open boundary condition in the form

$$H = \frac{i}{2}\sum_{j=1}^{N-1}[(-w+\Delta)a_j b_{j+1} + (w+\Delta)b_j a_{j+1}] - \frac{i}{2}\sum_{j=1}^{N}\mu a_j b_j. \tag{8.57}$$

The pictorial representation of various couplings of the Hamiltonian in Eq. (8.57) is shown in Fig. 8.16.

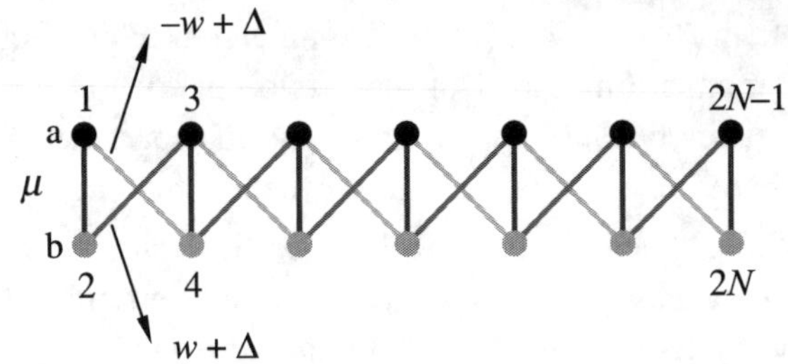

Fig. 8.16 Schematic representation of the Majorana chain with the Hamiltonian as defined in Eq. (8.57), considered a virtual ladder. It may be thought that the middle of each vertical bond here represents a fermionic site j which supports two Majorana fermions a_j and b_j represented by black and gray filled circles respectively with their intra-site interaction as μ. In this representation, other two couplings are symbolized by diagonal links.

The different phases (I, II and III) of the model (see Fig. 8.15) are characterized by the zero energy modes of the Hamiltonian in Eq. (8.57). Due to the existence of two zero energy Majorana modes at the two ends of an open and long chain, Phase I ($\Delta > 0$ and $|\mu| < 2w$) in Fig. 8.15 is one of the topological phases. One can show it using a certain condition, i.e, for $\Delta = w$ and $\mu = 0$. This condition reduces the Hamiltonian in Eq. (8.57) to the form

$$H = iw \sum_{j=1}^{N-1} b_j a_{j+1}, \tag{8.58}$$

which incorporates interaction between Majorana fermions only at adjacent lattice sites (see Fig. 8.17(b)). One can then express the aforementioned Hamiltonian in terms of the fermions combining two Majorana fermions at j and $j+1$ which is given by

$$H = 2w \sum_{j=1}^{N-1} \left(d_j^\dagger d_j - \frac{1}{2}\right), \tag{8.59}$$

where $d_j = \frac{1}{2}(b_j + ia_{j+1})$. From Eq. (8.59), we can observe that adding a d_j fermion in the chain has cost $2w$ amount of energy. This indicates the existence of a bulk gap in the system which is also consistent with our findings using periodic boundary conditions (see discussion around Eq. (8.54)). On the other hand, we can find from Eq. (8.58) that the Majorana operators a_1 and b_N do not enter in the Hamiltonian leading to two isolated zero energy Majorana modes a_1 and b_N at the left and right edge of the chain, respectively (see Fig. 8.17(b)). These two isolated Majorana modes can be combined to produce a highly non-local complex fermion, $f_m = \frac{1}{2}(a_1 + ib_N)$ which is the basic building block of fault-tolerant topological quantum computation (Freedman, 1998; Kitaev, 2003; Freedman et al., 2003). This non-local fermion costs zero energy and thus the ground state of the system becomes two-fold degenerate with opposite fermionic parity, i.e., if $|\psi_0\rangle$ is a ground state of the system satisfying $d_j|\psi_0\rangle = 0$ for $j = 1, \cdots, N-1$ and $f_m|\psi_0\rangle = 0$, then $|\psi_1\rangle = f_m^\dagger|\psi_0\rangle$

with $d_j|\psi_1\rangle = 0$ for $j = 1, \cdots, N-1$ is also a ground state. In a similar fashion, it can be shown that Phase II ($\Delta < 0$ and $|\mu| < 2w$) contains two unpaired Majorana modes, b_1 and a_N at the two ends of the chain (see Fig. 8.17(c)). Therefore, Phase II is also a topologically non-trivial phase with two-fold ground state degeneracy. The distinction of the topological phases I and II is made by the nature of Majorana modes in two ends of the chain. Phase I is a topological phase hosting a_1 (b_N) Majorana edge modes at the left (right) of the open chain, while for Phase II, it is the other way round. The Majorana end modes at the critical points are free to propagate through the chain since there is no bulk gap. As a result, they are no longer localized at the ends but become uniformly delocalized throughout the chain. Phase III ($|\mu| > 2w$) is non-topological with no Majorana edge modes. This can be shown using the special limit $\Delta = w = 0$, $\mu \neq 0$, when all the Majorana fermions are coupled to each other for each fermionic lattice site. As a consequence, there is no unpaired edge modes (see Fig. 8.17(a)) in Phase III. Introducing this special limit, one can note that adding a f_j fermion in the system requires a finite energy $|\mu|$; this confirms that in this limit, the system is gapped which also has been found from periodic boundary condition results. We thus find here that the system has very less boundary effects, making Phase III topologically trivial.

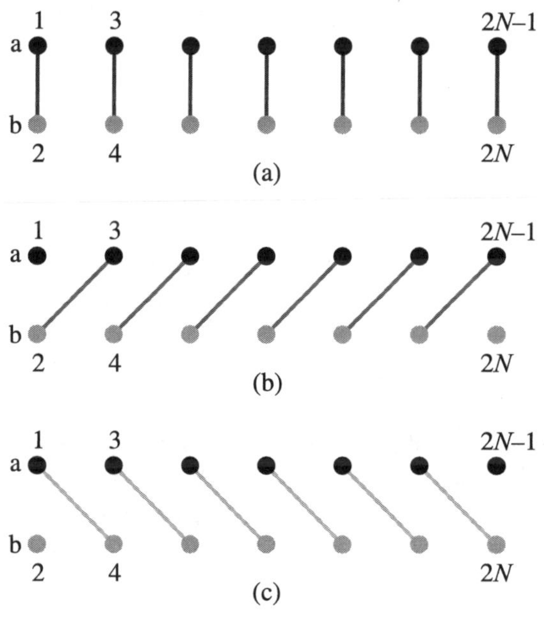

Fig. 8.17 Schematic representation of the Hamiltonian in Eq. (8.57) with different special conditions as defined in the text. (a) $\mu \neq 0$, $w = \Delta = 0$, (b) $\mu = 0$, $w = \Delta$ and (c) $\mu = 0$, $w = -\Delta$. In (a), the pair of Majoranas at each site is connected to each other by the strength μ resulting in a topologically trivial phase. (b) In this regime, there are two isolated Majorana modes a_1 and b_N at the left end and right end respectively. (c) This regime also represents a topologically non-trivial phase with two isolated Majorana modes b_1 and a_N.

Although we define different phases of the system in the context of zero energy Majorana modes using some special cases, it can be shown that those modes persist even if we move from those limiting cases (Kitaev, 2001). But then the zero energy Majorana modes cannot be simply represented by isolated modes a_1 (b_1) and b_N (a_N) for Phase I and Phase II respectively. In this case, the end Majorana modes will not be perfectly localized at the ends of the chain; rather their wavefunctions decay exponentially into the bulk of the chain (see Fig. 8.18) with the maximum probability of getting the Majoranas at the edges. The overlap of the wavefunctions gives rise to an interaction between these two Majorana end modes and they move from zero energy with energy scale $e^{-N/\xi}$, where ξ is the superconducting coherence length for this system (Kitaev, 2001; Alicea, 2012). As a result, except When $N \gg \xi$, the degeneracy between the ground states $|\psi_0\rangle$ and $|\psi_1\rangle$ breaks by the same energy scale defined earlier.

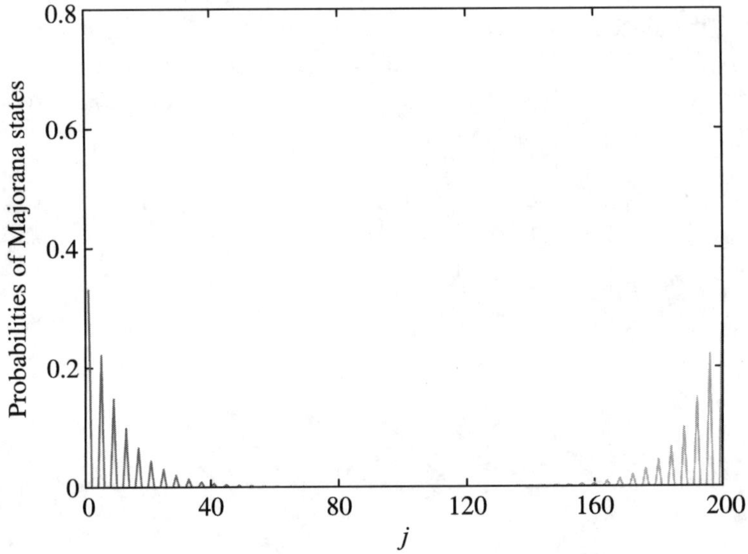

Fig. 8.18 Two Majorana edge states are localized at two ends of a 100-site open Majorana chain in Phase I ($\Delta = 0.1$ and $\mu = 0.0$) with an exponential decay into the bulk of the chain. j labels the Majorana sites $1, 2, \ldots 200$ (from Rajak and Dutta, 2014).

Topological Invariant: The system has a topological invariant through which we can distinguish different phases; this is known as the winding number (Niu et al., 2012; Tong et al., 2013) (see Sec. 8.6). The Hamiltonian in Eq. (8.53) can be defined in the form $H_k = \vec{d}_k.\vec{\tau}$, where $\vec{d}_k = (2\Delta \sin k)\ \hat{y} - (\mu + 2w\cos k)\ \hat{z}$ is the Anderson pseudospin vector (Anderson, 1958) in the $y-z$ plane. We can easily find the angle θ_k made by the vector $\vec{d}_k$ with respect to the $\hat{z}$-axis; that is given by

$$\theta_k = \tan^{-1}\left(\frac{-\mu - 2w\cos k}{2\Delta \sin k}\right). \tag{8.60}$$

Now, the winding number W that characterizes distinct phases of the model is defined as

$$W = \oint \frac{d\theta_k}{2\pi} = \int_0^{2\pi} \frac{dk}{2\pi}\frac{d\theta_k}{dk}, \tag{8.61}$$

where the integral is performed around the one-dimensional Brillouin zone. W, which takes only integer values, defines the number of rotations of the vector $\vec{d}_k$ in the $y-z$ plane around the one-dimensional Brillouin zone. The number of rotations becomes positive if the vector rotates in the anticlockwise direction. It can be shown with a few steps that W takes values -1, 1 and 0 in Phases I, II and III, respectively (see Fig. 8.15). This quantity does not change until the system passes through a quantum phase transition for some value of k, where W is ill-defined (DeGottardi et al., 2013b). It is now straightforward to find a relation between the bulk invariant W and the number of edge modes N_a (N_b) of a (b) type Majorana modes at the left end of the open chain. The relation is given by $W = N_b - N_a$ which correctly defines the bulk–edge correspondence as already mentioned in Sec. 8.6.

We can also describe these distinct phases in the spin-1/2 language. One can easily transform the Hamiltonian in Eq. (8.57) in to a spin-1/2 XY model Hamiltonian in a transverse field (Lieb et al., 1961; Barouch et al., 1970; Bunder and McKenzie, 1999) by using the inverse JW transformation. In terms of spin operators, Eq. (8.57) takes the form

$$H = -\sum_{j=1}^{N-1} (J_x\sigma_j^x\sigma_{j+1}^x + J_y\sigma_j^y\sigma_{j+1}^y) - h\sum_{j=1}^{N} \sigma_j^z, \tag{8.62}$$

where σ_j^x and σ_j^y are the Pauli spin matrices with x and y spin components and the interaction terms follow the relations $J_x = (w+\Delta)/2$, $J_y = (w-\Delta)/2$ and $\mu = -2h$. This Hamiltonian corresponds to a one-dimensional spin model with anisotropic interactions in x and y directions subjected to a transverse magnetic field. So, in the context of the spin system Hamiltonian in Eq. (8.62) Phases I and II correspond to ferromagnetically ordered phases with long range order in the x and y directions respectively. Phase III is paramagnetic with no long-range order and zero spontaneous magnetization.

8.7 One-dimensional Generalized Cluster Model

The presence of longer range hopping and pairing in a spinless superconducting chain provides a richer phase diagram consisting of phases with multiple zero energy Majorana modes in each end of an open chain. In this connection, a one-dimensional generalized cluster model has been numerically studied to investigate its topological properties considering both equilibrium and non-equilibrium situations (Ohta et al., 2015). The Hamiltonian of the afore mentioned model is given by

$$H_{\rm GC} = \sum_{i=1}^{N}(-J^{XZX}\sigma_i^x\sigma_{i+1}^z\sigma_{i+2}^x + J^{YY}\sigma_i^y\sigma_{i+1}^y + J^{YZY}\sigma_i^y\sigma_{i+1}^z\sigma_{i+2}^y), \tag{8.63}$$

where N is the system size. The three-spin interaction in Eq. (8.63) is indeed called the cluster interaction. As discussed by Ohta et al. (2015), the different phases of this model are characterized by string order parameters and entanglement spectrum.

The phase diagram of the model comprising a variety of phases is shown in Fig. 8.19(d). The boundaries between two different phases are determined using the condition of vanishing bulk gap. The calculation of various order parameters have been made using the time-evolving block decimation method for infinite systems (iTEBD). It can be easily realized that the different phases of the model occurs depending upon the dominancy of the parameters in Eq. (8.63). The cluster phase C is characterized by the non-zero string order parameter defined as $O_{XZX} = \lim_{L\to\infty} O_{XZX}(L)$ (see Fig. 8.19(e)). The quantity $O_{XZX}(L)$ is called the string correlation function of length L, given by

$$O_{XZX}(L) = (-1)^L \left\langle \sigma_1^x \sigma_2^y \left(\prod_{i=3}^{L-2} \sigma_i^z \right) \sigma_{L-1}^y \sigma_L^x \right\rangle . \qquad (8.64)$$

Similarly, the order parameter O_{YZY} defines another cluster phase which can be called dual cluster (C*) phase where the term J^{YZY} dominates. On the other hand, the string order parameters in ferromagnetic phases are given by $O_{XX} = \lim_{L\to\infty} \langle \sigma_1^x \sigma_L^x \rangle$ and $O_{YY} = \lim_{L\to\infty} \langle \sigma_1^y \sigma_L^y \rangle$ with ordering in the x and y directions respectively. As shown in Fig. 8.19, the order parameters O_{XX} and O_{YY} have finite values in their respective phases, thus characterizing those phases. Furthermore, using the JW transformation, the Hamiltonian in Eq. (8.64) can be represented in terms of Majorana fermions. Now, for an open chain, one can find that the system has two zero energy Majorana modes in the each end when J^{XZX} or J^{YZY} is dominant (see Figs. 8.19 (a), (c)). On the other hand, if the term J^{YY} dominates, then there exists one Majorana mode in each end of the chain, as shown in Fig. 8.19(b). The phase diagram in Fig. 8.19(d) can also be explained using the concept of entanglement spectrum (ES) (Lou et al., 2011; Tanaka et al., 2012). The ES for an open chain has been calculated using exact diagonalization. Interestingly, it has been shown that degeneracy in the lowest values of the ES in each phase provides the number of zero energy end Majorana modes in that phase (see (Ohta et al., 2015) for details). At the same time, the dynamics in this model has also been studied following a slow quench (or sweep) of an interaction term across the quantum critical point between the phases C and C*. In this context, the dual string correlation function and the entanglement entropy as the function of length have been investigated for different sweep times. A periodic structure in the length dependence of these quantities has been observed as a result of sweeping across a critical point.

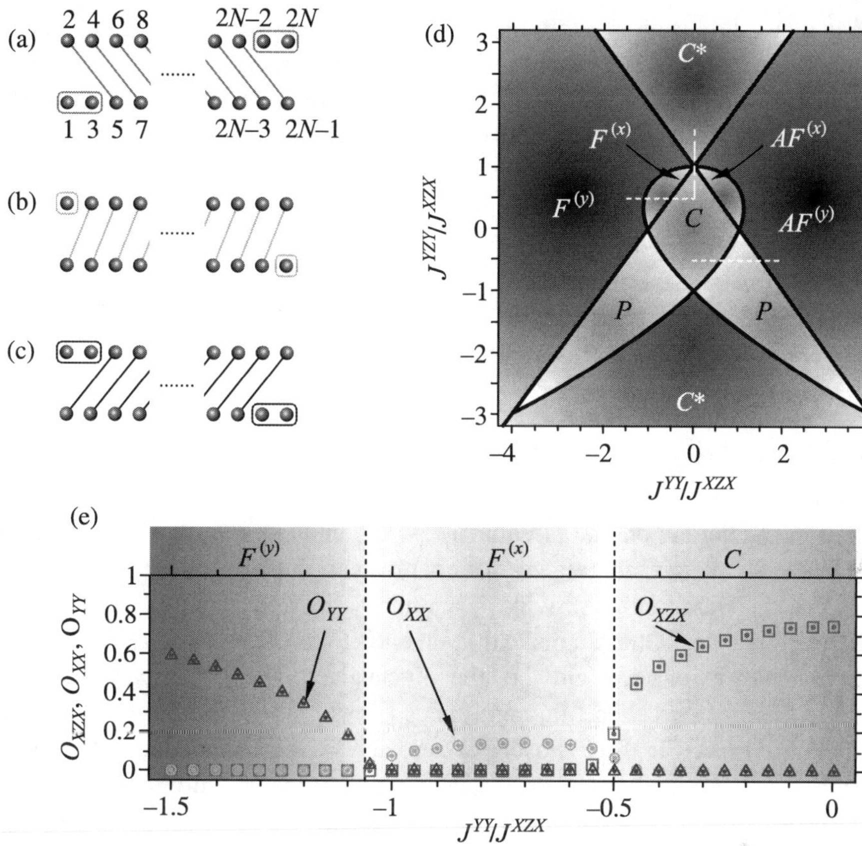

Fig. 8.19 (a)–(c) Schematic representation of three terms in Eq. (8.63) using the language of Majorana fermions where (a), (b), and (c) describe the first, second, and third terms respectively. The unpaired Majorana modes enclosed in dotted lines exist at two ends of the chain. (d) The plot shows the phase diagram of the generalized cluster model of Eq. (8.63) at zero temperature for $J^{XZX} > 0$. The thick solid curves are the phase boundaries where the energy gap vanishes at a particular k. The different phases like ferromagnetic, antiferromagnetic, paramagnetic, cluster and dual cluster are represented by F, AF, P, C and C* respectively. The superscripts in F and AF denote the direction of the order. (e) These phases are determined by string order parameters. The string order parameters O_{XZX}, O_{XX} and O_{YY} have been calculated along the thick dotted line $J^{YZY}/J^{XZX} = 0.5$ (see Fig. 8.19(d)) using the iTEBD (from Ohta et al., 2015).

8.8 Sudden Quench: Dynamics of a Majorana Edge State

Let us consider an open and long chain of spinless one-dimensional p-wave superconductor and investigate the time evolution of a zero energy Majorana edge state after a sudden quenching in the system from one topological phase to the topologically trivial phase (or to the other topological phase) and also to the QCP separating these phases (Rajak and Dutta, 2014). Following the quench, the Majorana edge mode interacts with the bulk modes and therefore is assumed to decohere. We consider different quenching schemes and address the question whether this situation occurs for all quenching schemes. Studying the Majorana survival probability (MSP), we show here that in some cases, the edge Majorana survives and, travels back and forth between two ends of the chain.

We now demonstrate the method to calculate the MSP following a quantum quench. We first describe the Hamiltonian in Eq. (8.57) by a generic quadratic form in terms of Majorana operators

$$H = \frac{i}{4}\sum_{j,k=1}^{2N} A_{jk} c_j c_k, \tag{8.65}$$

where c_j are the Majorana operators with $c_{2j-1} = a_j$ and $c_{2j} = b_j$, and A is a real skew-symmetric $2N \times 2N$ matrix; the eigenvalues of A come in pairs as $\pm i\varepsilon_n$ (where $\varepsilon_n > 0$). It indicates that the Hamiltonian provides even number of zero eigenvalues, and these are called zero energy modes. One can make the wavefunctions of zero energy modes real; they are well separated by a finite gap with the other eigenvalues. These characteristics of the end modes lead to them being called Majorana states (Thakurathi et al., 2013). As mentioned already, the Phase I (see Fig. 8.15) hosts two zero energy Majorana edge modes which are shown in Fig. 8.18. We get these states by diagonalizing the Hamiltonian in Eq. (8.65) to obtain the eigenvectors corresponding to zero eigenvalues and finding the appropriate real wavefunctions.

We now focus in studying the energy levels for the Majorana chain with open and periodic boundary conditions as functions of $\xi (= \Delta/w)$ with $\mu = 0$ (see Fig. 8.20). We have assumed here that $w = 1$ which also will be followed for the rest of the section. For open boundary condition, it can be clearly observed that the Hamiltonian in Eq. (8.65) supports two zero energy Majorana modes in both Phases I and II, while a periodic chain does not contain any zero energy states.

Let us now consider a sudden quench of a parameter of the Hamiltonian in Eq. (8.65) along the different paths on the phase diagram (Fig. 8.15) at time $t = 0$ and investigate the time evolution of a Majorana end mode. Immediately after a sudden quench, the Hamiltonian of a system changes accordingly, but the state cannot follow these changes. As a result the evolution of the initial state is determined by the final Hamiltonian. After a rapid quench at time $t = 0$, the time-evolved state of a Majorana mode is given by

$$|\psi_m(\Delta,\mu,t)\rangle = \sum_{n=1}^{2N} e^{-iE_n t} |\Phi_n(\Delta',\mu')\rangle\langle\Phi_n(\Delta',\mu')|\psi_m(\Delta,\mu)\rangle, \tag{8.66}$$

where $|\psi_m(\Delta,\mu)\rangle$ is a Majorana edge state for the initial parameter values Δ and μ, and $|\Phi_n(\Delta',\mu')\rangle$ correspond to the eigenstates of the final Hamiltonian with parameters Δ', μ'

after quenching. E_ns are the eigenvalues of the final Hamiltonian. One can get these states by a numerical diagonalization of the Hamiltonian in Eq. (8.65) for the initial and final parameter values. To study the time-evolution of the Majorana zero mode following a sudden quench, we consider the MSP $P_m(t)$ (which is indeed the LE (Quan et al., 2006; Sharma et al., 2012)) given by

$$P_m(t) = \left| \sum_{n=1}^{2N} |\langle \psi_m(\Delta,\mu)|\Phi_n(\Delta',\mu')\rangle|^2 e^{-iE_nt} \right|^2. \tag{8.67}$$

In the following subsections, the MSP will be studied for different quenching paths.

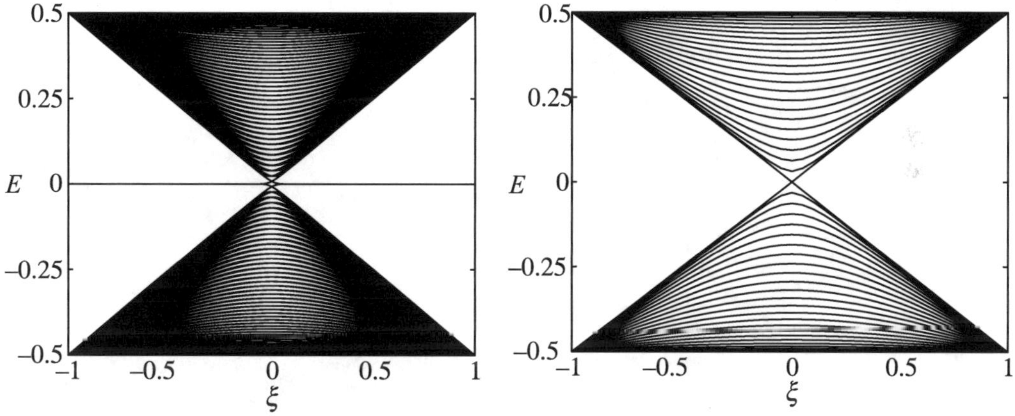

Fig. 8.20 The diagram for the energy levels of the Majorana chain of system size $N = 100$ as a function of $\xi = \Delta/w$ with $\mu = 0$ and $w = 1$ considering (a) open boundary condition and (b) periodic boundary condition, respectively. One can observe that two zero energy Majorana edge modes exist in case (a) but not in case (b). It is mentioned that the energy spectrum is scaled by a constant factor $1/4$ in Eq. (8.65) compared to Eq. (8.54) (from Rajak and Dutta, 2014).

8.8.1 Quenching schemes and results

Quenching along different paths: We first consider a path $\mu = 0$ denoted by A in Fig. 8.15. Along this path, the parameter Δ is suddenly changed so that the system moves from Phase I to II. We observe a rapid decay of the MSP as a function of time following a minimum value with some noisy fluctuation (Fig. 8.21(a)) for the aforementioned quenching scheme (Phase I to II). This indicates that the Majorana edge mode decoheres with time, but not completely since the evolution of the system is unitary.

Let us now concentrate on the most interesting situation emerging due to the quench of the Majorana chain to the QCP ($\Delta = 0$, $\mu = 0$). For this case, we find that the MSP of a Majorana edge mode shows collapse and revival as a function of time (Fig. 8.21(b)).

This can be attributed to the fact that the low-lying energy levels of the bulk at the QCP are nearly equispaced (see Fig. 8.20) and also there exists a bulk energy gap $E_g \sim 1/N$ as the dynamical critical exponent for this QCP is unity.

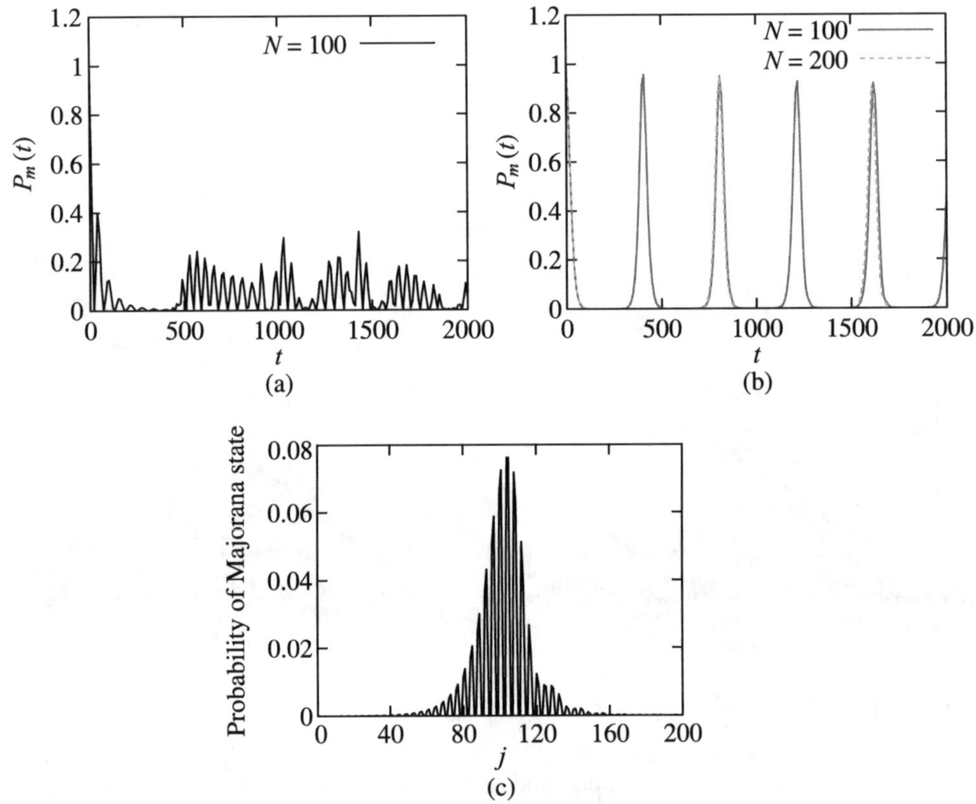

Fig. 8.21 Majorana survival probability of a Majorana edge state for various quench protocols along Path A and probability of Majorana at certain time t after quench. (a) There is a sudden decay in the MSP due to a quenching from Phase I ($\Delta = 0.1$) to Phase II ($\Delta = -0.1$) and after that, it remains minimum without any revival. (b) On the other hand, following a quench from Phase I ($\Delta = 0.1$) to the QCP ($\Delta = 0.0$), MSP exhibits nearly perfect oscillation with a time period scaling linearly with the system size N. (c) Probability of the Majorana end mode at time $t = 100$ for quenching at the QCP with the Majorana site j. At this time, the probability of the Majorana mode is maximum, close to the center of the chain (from Rajak and Dutta, 2014).

Following the quench, the Majorana edge state interacts with low-energy states of the bulk which are nearly equispaced. As a result, the main contribution to the summation of Eq. (8.67) comes from the overlap of the Majorana edge state with these low-energy states of the bulk. Equation (8.67) therefore can be represented as a Fourier series of a periodic function with fundamental frequency given by the energy difference of two consecutive equispaced energy levels which are proportional to $1/N$. This indicates that the quasiperiod of collapse and revival must have a scaling relation with system size N. This prediction is confirmed from Fig. 8.21(b) where the time period of oscillation becomes double if the system size is made twice the previous one. We can now conclude that for quenching at the QCP, separating two topologically non-trivial phases, the Majorana edge mode of Fig. 8.18 oscillates between two edges of the chain (see Fig. 8.21(c)); the time period of this oscillation is proportional to the length of the spin chain. However, there is a slow decrease in the amplitude of revival with time as a result of the coupling of the Majorana mode with interior bulk modes of higher energy as it oscillates between the two ends of the chain.

Furthermore, we analyze this oscillation of the Majorana modes considering the case with the periodic boundary condition; although in this situation, the system does not host an edge Majorana, the structure of the bulk energy band remains unchanged. With periodic boundary condition, the Hamiltonian in Eq. (8.52) can be written as the direct sum of 2×2 independent matrices for each momentum mode k (see Eq. (8.53)). The energy dispersion of the system is given by Eq. (8.54). We find that the bulk energy gap becomes zero for the critical mode $k_c = \pi/2$ at the critical point $(\Delta = 0, \mu = 0)$. Expanding E_k around the critical mode $k_c = \pi/2$, it can be shown that the first non-zero term in E_k varies linearly with k and the second term is proportional to k^3, where $(k - k_c)$ has been rescaled as k. If the ground state of the system in one of the phases evolves with the critical Hamiltonian, the square of modulus of the overlap between the initial and the time-evolved states (or the LE) is given by $\mathscr{L}(t) = \prod_{k>0} \prod_{k>0} \mathscr{L}_k(t) \left(1 - A_k \sin^2(E_k t)\right)$ (see Eq. (8.30)), where the function A_k changes very slowly with k and each $\mathscr{L}_k(t)$ has a periodicity π/E_k. Let us now consider the modes around $k \to 0$ separated by $\Delta k = 2\pi/N$; there is constructive interference between these modes at time instants given by $\Delta E_k t = p\pi$ with p being an integer; the resulting revival time for each mode $t_k = \frac{1}{2} pN |\partial E_k/\partial k|^{-1}$. If we have a linear energy dispersion (true for momentum with $k \to 0$), many modes interfere constructively and the first revival of the overlap will occur for the maximum of the group velocity $v_g(k) = |\partial E_k/\partial k|$ (see the discussion in Sec. 8.5). It can be found that this revival of the overlap occurs at intervals which is proportional to the length N of the chain. One can note that close to $\xi = 0$, the difference between two consecutive low-energy levels is $\Delta E = E_n - E_{n-1} \approx 2\pi/N$ (see Fig. 8.20). This leads to a fact that the time period of revival is given by

$$T = \frac{2\pi}{\Delta E} \propto N. \tag{8.68}$$

We also observe that the amplitude of the revival decays with time. The next higher order term in the expansion of E_k is non-linear, i.e., k^3; these modes do not add up constructively and an eventual decay in the amplitude of revival is found as the Majorana mode oscillates between two edges of the chain. In the asymptotic limit, the probability saturates at a finite value as the size of the system is finite. We have also found similar results for quenching along Path B as shown in Fig. 8.21.

Quenching within the same phase: In this case, we study the MSP when the parameters of the Hamiltonian in Eq. (8.65) are changed within a single phase (Phase I). As a consequence of this quenching, there is a decay in the MSP, but it does not become zero as the bulk of the system is gapped and the Majorana cannot propagate through the bulk. In addition, for this case, the energy levels are spaced unequally and hence there is no perfect periodic structure of MSP. As shown in Fig. 8.22(a), the MSP here fluctuates haphazardly having a mean value that is close to unity. In this context, we can make a conclusion that the Majorana edge mode is mainly localized at the end of the chain even after this quenching (see Fig. 8.22(b)).

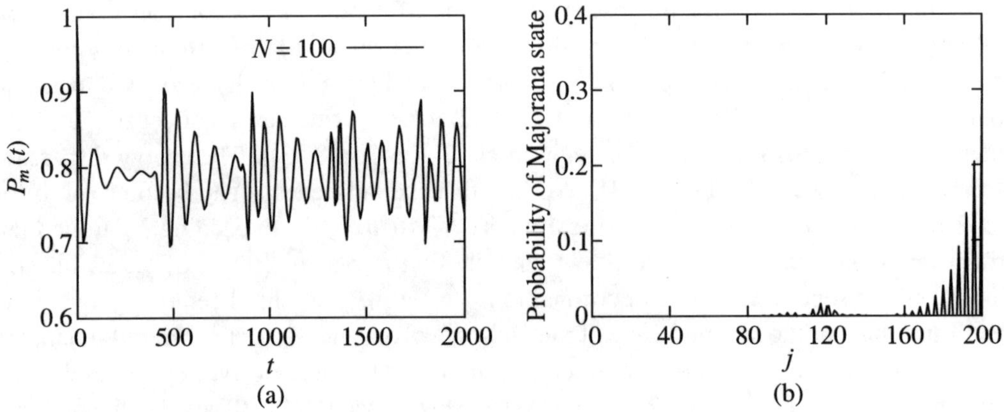

Fig. 8.22 A sudden quench is performed within Phase I from ($\Delta = 0.2$ and $\mu = 0.0$) to ($\Delta = 0.1$ and $\mu = 0.0$) and the MSP becomes (a) a rapidly oscillating function of time t having nearly 0.8 mean value and (b) probability of the Majorana end mode at $t = 100$, with the Majorana sites j following the quench (from Rajak and Dutta, 2014).

8.9 Slow Quench: Dynamics of a Majorana Edge State

Recently, a one-dimensional p-wave superconductor has been studied in the context of slow quenching dynamics and it has been established that a zero energy Majorana mode cannot tunnel adiabatically from one topologically non-trivial phase to the other separated by a quantum critical line (Bermudez et al., 2010). Now, if we consider the hopping term as a complex quantity, an extended gapless phase is generated in this model between two topological phases (DeGottardi et al., 2013b). This in fact breaks the effective time reversal symmetry (ETRS) of the system. Our aim here is to find out the possibility of an adiabatic transportation of a Majorana edge state from one topological phase to the other when the superconducting chain is driven across the extended gapless region (Rajak et al., 2014).

8.9.1 Model, phase diagram and energy spectrum

We here consider the Hamiltonian in Eq. (8.52) with a complex hopping term $w_0 e^{i\phi}$, where w_0 and ϕ are the nearest-neighbor hopping amplitude and the phase of the hopping term, respectively (DeGottardi et al., 2013b).

The Hamiltonian of the model defined here is diagonalized in the momentum space under periodic boundary condition and given by

$$H_k = (2w_0 \sin\phi \sin k)\, I \;-\; (2w_0 \cos\phi \cos k \,+\, \mu)\, \sigma^z + (2\Delta \sin k)\, \sigma^y, \tag{8.69}$$

where the Pauli matrices σ^y and σ^z are defined in the particle-hole subspace. It can be noted that for this model, $H^*_{-k} \neq H_k$ except $\phi = 0$ or π implying the breaking of ETRS. The QCPs for the Hamiltonian in Eq. (8.69) with the critical modes k_c can be located in the parameter space using the condition

$$(2w_0 \cos\phi \cos k_c \,+\, \mu)^2 \,+\, 4\Delta^2 \sin^2 k_c \;=\; 4w_0^2 \sin^2\phi \sin^2 k_c. \tag{8.70}$$

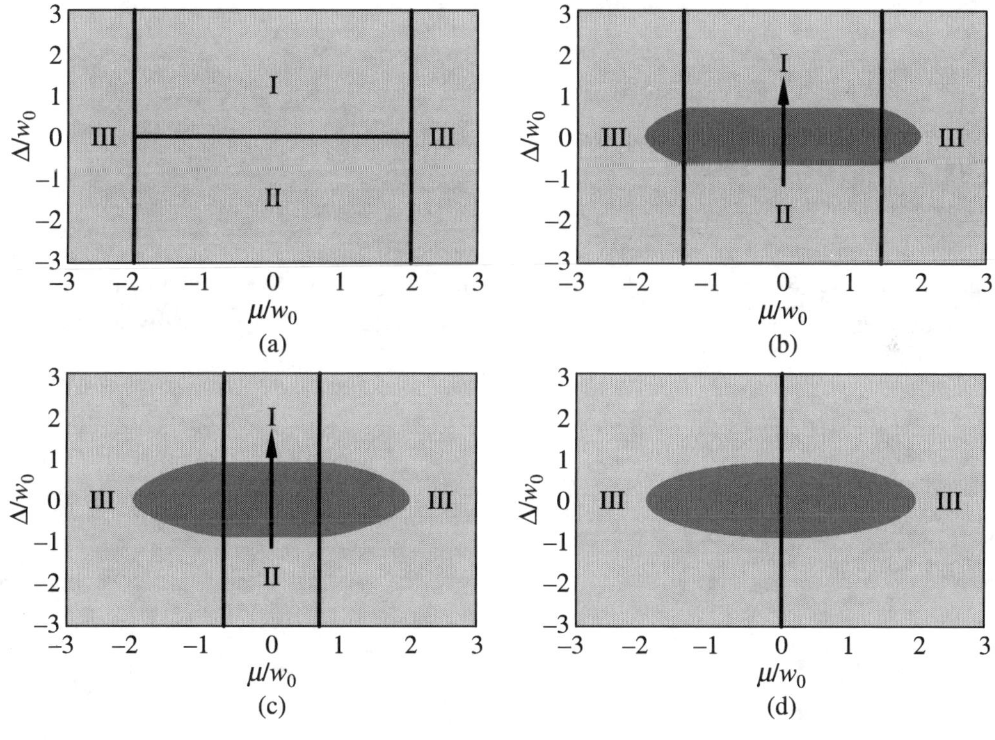

Fig. 8.23 Phase diagram of the system with Hamiltonian in Eq. (8.52) for various phases of complex hopping term (a) $\phi = 0$, (b) $\phi = \pi/4$, (c) $\phi = 2\pi/5$ and (d) $\phi = \pi/2$. In this plot, I and II represent two different topological phases, whereas III is the topologically trivial phase. The path $\mu = 0$ is represented by the vertical arrow along which quenching is performed (from Rajak et al., 2014).

We have shown the phase diagrams of the model for different $\phi \in [0, \pi/2]$ in Fig. 8.23 which are obtained by analyzing the energy spectrum. In the phase diagram, the shaded region represents the gapless phase which is bounded by a rectangle whose left and right sides are capped by two elliptical arcs (except the cases with $\phi = 0$ and $\pi/2$). The two sides of the rectangular region in the phase diagram are bounded by the horizontal lines $\Delta/w_0 = \pm\sin\phi$ and the vertical lines given by $\mu/w_0 = \pm 2\cos\phi$. If we increase ϕ, the two vertical phase boundaries approach one another and collapse at $\mu = 0$ for $\phi = \pi/2$. In this case, there are no topological phases and the gapless region is given by an ellipse described by $\mu^2 + 4\Delta^2 = 4w_0^2$.

We now represent the Hamiltonian in Eq. (8.52) in terms of the two Majorana operators a_j and b_j at each site where the hopping term is complex. As a result, the Hamiltonian with an open boundary condition can be written as

$$H = -\frac{i}{2}\sum_{j=1}^{N-1}\Big[w_0\cos\phi(a_jb_{j+1} + a_jb_{j-1})$$

$$-\Delta(a_jb_{j+1} - a_jb_{j-1}) + w_0\sin\phi(a_ja_{j+1} + b_jb_{j+1})\Big]$$

$$-\frac{i}{2}\sum_{j=1}^{N}\mu a_jb_j. \tag{8.71}$$

We can observe that the zero energy Majorana fermions (a and b) for such a system with complex hopping terms are coupled to each other. The Heisenberg equations of motion corresponding to a_j and b_j zero energy Majorana modes are then given by (DeGottardi et al., 2013b)

$$i\,\dot{a}_j = -[H, a_j] = 0, \quad i\,\dot{b}_j = -[H, b_j] = 0,$$

$$w_0\cos\phi\;(\beta_{j+1} + \beta_{j-1}) - \Delta\,(\beta_{j+1} - \beta_{j-1})$$

$$+\mu\,\beta_j + w_0\sin\phi\,(\alpha_{j+1} - \alpha_{j-1}) = 0, \tag{8.72}$$

$$w_0\cos\phi\;(\alpha_{j+1} + \alpha_{j-1}) - \Delta\,(\alpha_{j-1} - \alpha_{j+1})$$

$$+\,\mu\,\alpha_j + w_0\sin\phi\;(\beta_{j-1} - \beta_{j+1}) = 0. \tag{8.73}$$

To get these equations, we first consider the time-dependent Majorana operators $a_j = \alpha_j e^{-iE_jt}$ and $b_j = \beta_j e^{-iE_jt}$. We then find the solutions of the Heisenberg equations of motion for these Majorana operators with $E_j = 0$. It is easy to observe that Eqs. (8.72) and (8.73) consist of two coupled equations involving both a and b zero energy Majorana modes. We can numerically diagonalize the Hamiltonian in Eq. (8.71) to show that a_1 and b_1 are not isolated, though the probability of a_1 is much higher than b_1 at the left end of the chain in Phase I. Similarly, one can observe the existence of both a_N and b_N modes at the right end of the chain where the probability of b_N is much higher than a_N. A topological invariant for such systems with broken ETRS has been introduced in by DeGottardi et al. (2013b).

We have shown the energy spectrum of the Hamiltonian in Eq. (8.71) considering both periodic and open boundary conditions in Fig. 8.24(a) and Fig. 8.24(b), respectively. The existence of two zero energy lines in Fig. 8.24(b) indicates that Phase I and II contain two zero energy Majorana states at each edge of the chain. In contrast, the Majorana chain with periodic boundary conditions does not host any Majorana end mode. Interestingly, there is an inversion of few bulk energy levels near zero energy inside the gapless region of the model. Let us consider an energy level close to zero energy and study its behavior with the variation of ξ. It becomes minimum at $\xi = -\sin\phi$, when the system enters a gapless phase and then bends upward up to $\xi = 0$, where it starts decreasing following a minimum value at $\xi = \sin\phi$ and after that it increases. It also can be observed that the number of inverted energy levels increases as we increase N or ϕ (see Ref. (Rajak et al., 2014)).

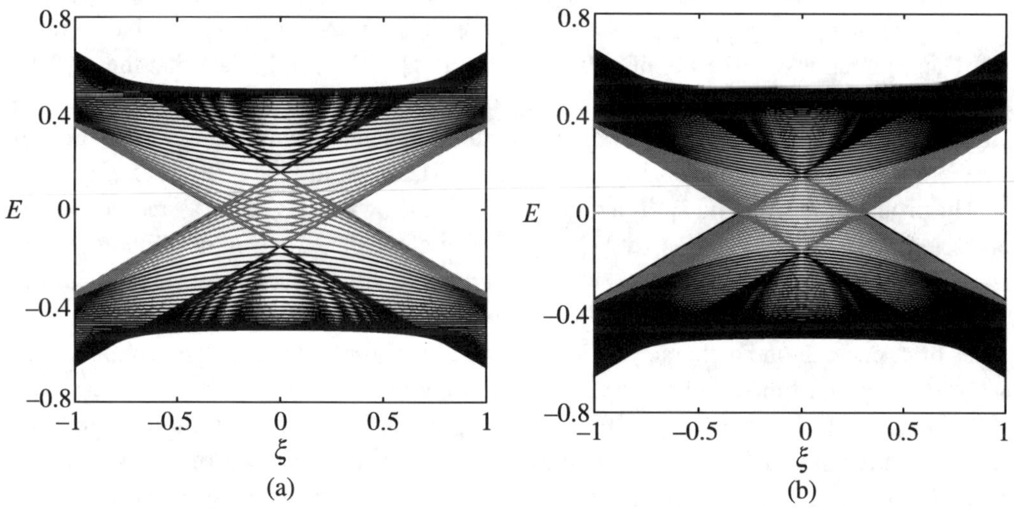

Fig. 8.24 The energy levels of the Hamiltonian in Eq. (8.52) as a function of parameter $\xi = \frac{\Delta}{w_0}$ (where $w_0 = 1$) for (a) periodic and (b) open boundary conditions with $N = 100$ and $\phi = \pi/10$. The inverted energy levels inside the gapless phase of the system is indicated by the red color (from Rajak et al., 2014).

8.9.2 Quenching dynamics of a Majorana edge state

We now focus on the dynamics of a Majorana edge mode in the Majorana chain shown in Eq. (8.71) following the quenching protocol $\Delta(t) = -1 + 2t/\tau$ along the path $\mu = 0$, where τ is the quenching time with t running from 0 to τ. We have also considered quenching for the model in Eq. (8.71) when the superconducting term is varied in a non-linear fashion (see (Rajak et al., 2014)). Starting from an initial zero energy Majorana edge mode at $\Delta(0) = -1$ (i.e., in Phase II) having a real wavefunction $|\Psi(0)\rangle$, we investigate whether this edge mode can be transported to the other topological phase. Following numerical integration of the time-dependent Schrödinger equation, we shall determine some probabilities at the final instant $t = \tau$: the probability of a Majorana mode getting excited to the positive (negative) energy band P_{def} (P_{neg})

$$P_{\text{def(neg)}} = \sum_{\varepsilon^+>0(\varepsilon^-<0)} |\langle \varepsilon^{+(-)}|\Psi(\tau)\rangle|^2, \tag{8.74}$$

where $|\varepsilon^{+(-)}\rangle$ are eigenstates of the final Hamiltonian within Phase I at $\Delta = 1$ corresponding to the positive and negative energy respectively, and $|\Psi(\tau)\rangle$ represents the time-evolved Majorana edge state at $t = \tau$. We also consider another probability which is given by the modulus square of the overlap between the equilibrium Majorana edge state at $\Delta = 1$ (denoted by $|\varepsilon_0\rangle$) and the final time-evolved state at $t = \tau$. This is defined as $P_{\text{m}} = |\langle \varepsilon_0|\Psi(\tau)\rangle|^2$. Therefore, if the aforementioned probability becomes non-zero, we can say that there is a finite probability of getting a Majorana edge mode in Phase I. Now, one can ask what happens to P_{m} after $t = \tau$, i.e., when the final state $|\Psi(\tau)\rangle$ evolves with the final Hamiltonian at $\Delta = 1$ as $\exp(-iH(\Delta = 1)t)|\Psi(\tau)\rangle$. Noting the fact that the state $|\varepsilon_0\rangle$ is an eigenstate of the final Hamiltonian, it can be easily shown that $P_m(t > \tau)$ evaluated through $|\langle \varepsilon_0|\exp(-iH(\Delta = 1)t)|\Psi(\tau)\rangle|^2$ remains unchanged with time which indicates that there exists a non-zero probability of the edge Majorana in Phase I even after $t = \tau$.

The probabilities P_{def} and P_{m} as a function of τ are shown in Fig. 8.25(a) and Fig. 8.25(b) for various ϕs with a system size $N = 100$. It is clear that there is no obvious scaling relation with τ in contrast to the periodic boundary condition scenario (see (Rajak et al., 2014)). Interestingly, it is observed that although P_{def} remains constant at 0.5 for small values of τ, one can find a characteristic τ denoted by τ_c for which the first prominent drop in P_{def} can be seen followed by a few drops. On the other hand, the first significant peak in P_{m} is also found at $\tau = \tau_c$ (see Fig. 8.25(b)), which at implies a finite probability of an adiabatic transport of the initial edge Majorana under a quenching of the system from Phase II to Phase I. This does not happen for $\phi = 0$: P_{m} remains fixed at zero for all values of τ which indicates that adiabatic tunneling of the Majorana edge state is not possible and P_{def} always becomes $1/2$ which implies the complete delocalization of the edge Majorana within the bulk energy modes as indicated by Bermudez et al. (2010). Furthermore, we find that there is a complete overlap between P_{def} and P_{neg} as a function of τ. This signifies that the time-evolved Majorana edge modes get delocalized within the equal number of positive and negative energy bulk states for all values of τ (see Fig. 8.26(a)).

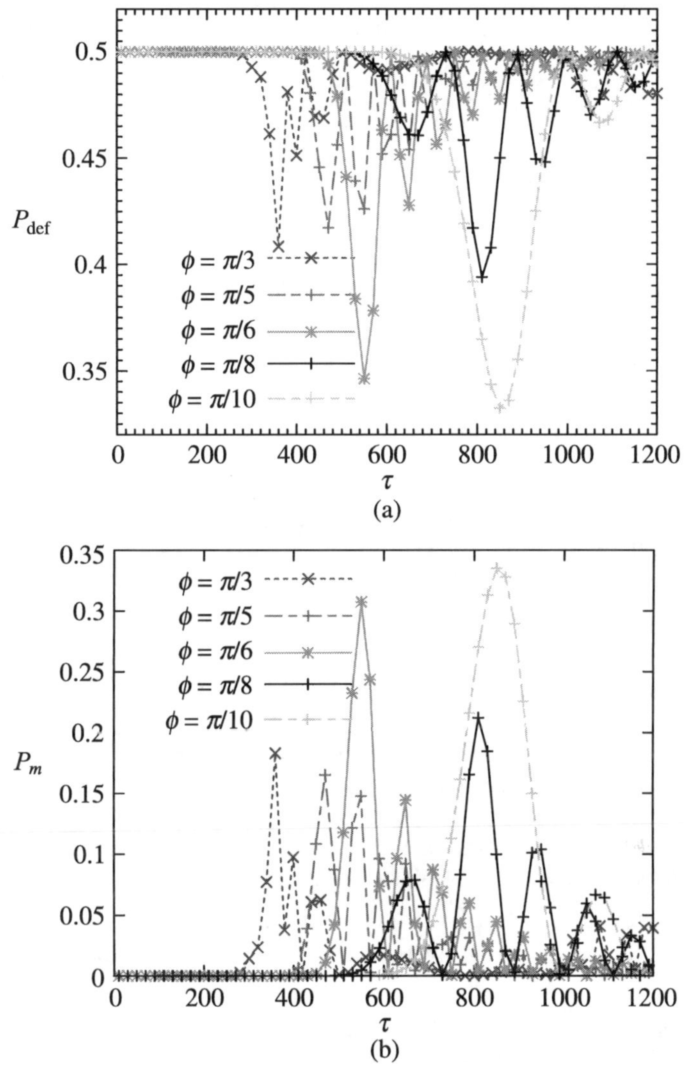

Fig. 8.25 (a) Variation of the probability of defect (P_{def}) with τ for different values of ϕ exhibiting dip at different values of $\tau \geq \tau_c$ where the value of τ_c increases as ϕ decreases. (b) The probability of Majorana (P_m) exhibits a peak precisely at those values of τ where P_{def} shows dips. Here, $N = 100$ (from Rajak et al., 2014).

Moreover, by studying the results shown in Figs. 8.25(a) and (b), one can find a relation between ϕ and τ_c which determines the positions of the first prominent peak (dip) in P_{m} (P_{def}). It is observed that the value of τ_c increases with decreasing ϕ. Let us define two times t_e and t_o when the system enters and leaves the gapless region respectively. Then the transit time through the gapless region is obtained as $\Delta t = t_0 - t_e = \tau \sin\phi$. We find that there is an optimal value of Δt that does not depend on ϕ (for a fixed N) when P_{m} becomes non-zero; investigating our numerical results, we can infer that this optimal value has a relation with τ_c given by

$$\Delta t_{\text{op}} = \tau_c \sin\phi. \tag{8.75}$$

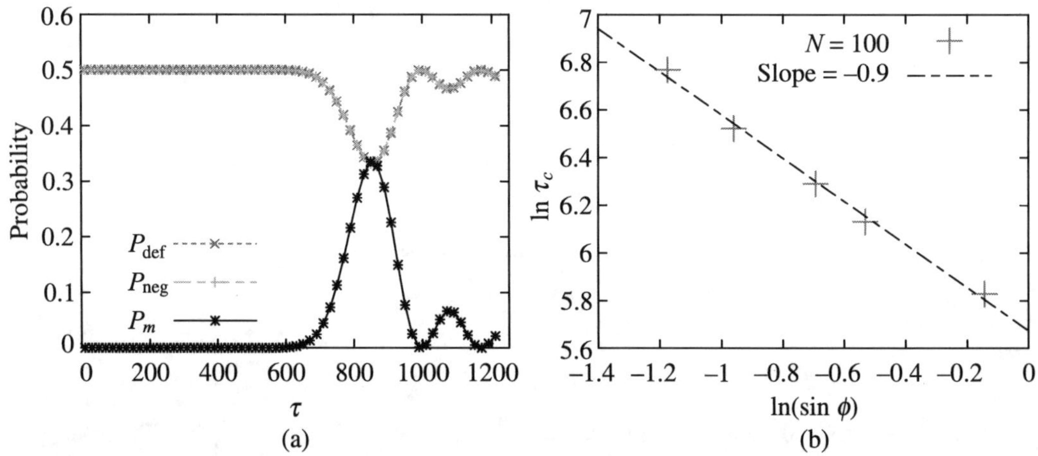

Fig. 8.26 (a) P_{def}, P_{neg} and P_m as a function of τ with $\phi = \pi/10$ exhibit that all of them add up to unity for any τ. (b) The plot shows that ln(τ_c) varies almost linearly with ln(sin ϕ) having slope ($= -0.9$) nearly equal to -1 (from Rajak et al., 2014).

It is then mentioned that τ_c itself must depend on ϕ and diverges for $\phi \to 0$. Then P_{m} becomes zero for any τ that suggests the impossibility of an adiabatic tunneling of the Majorana edge state in that case (Bermudez et al., 2010). Plotting $\ln \tau_c$ vs $\ln \sin\phi$ in Fig. 8.26(b), we establish the relation between τ_c and $\sin\phi$ as found in Eq. (8.75) when Δt_{op} remains fixed.

By a close observation of Fig. 8.25(a) and (b) one can suggest that there are n number of relations identical to Eq. (8.75) for the transit times $\Delta t_{\text{op1}}, \Delta t_{\text{op2}}, \cdots, \Delta t_{\text{opn}}$ ($\Delta t_{\text{op(n-1)}} < \Delta t_{\text{opn}}$) associated with the peaks (dips) in P_{m} (P_{def}); where n is an integer and given by the number of peaks (dips) in P_{m} (P_{def}). The disappearance of adiabatic transition probability P_m for $\tau < \tau_c$ implies that there exists a threshold value of passage time $\Delta t_{\text{th}} = \Delta t_{\text{op1}}$, which depends only on N, below which the adiabatic transition of edge Majorana is prohibited. We have also shown the variation of P_m as a function of τ with $\phi = \pi/5$ for different N in Fig. 8.27(a). Figure 8.27(b) exhibits the linear increase of τ_c with N for a given ϕ.

We have the following observations: the optimum passage time Δt_{op} associated with the adiabatic transition is dependent essentially only on the system size N, while it marginally depends on the quenching length along the path $\mu = 0$, and also the phase ϕ of the complex hopping term. Depending on these observations, we make a conjecture for this adiabatic transport of Majorana edge mode. We propose that for transit times

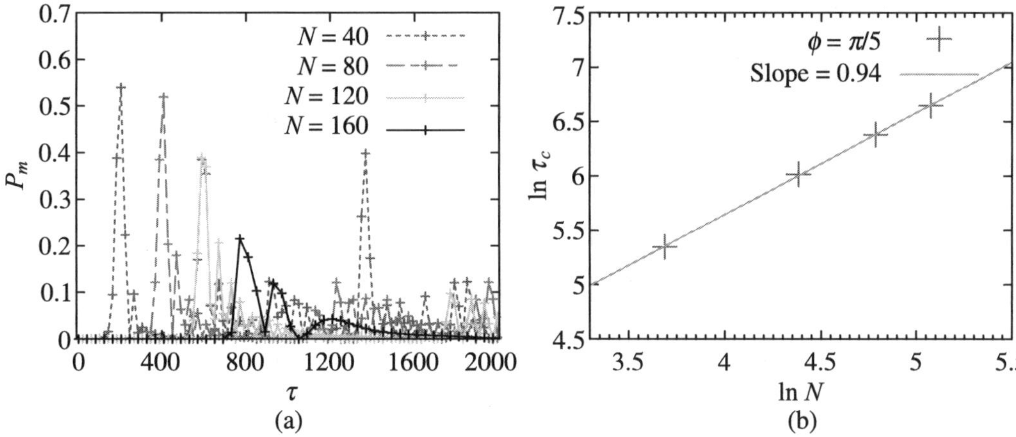

Fig. 8.27 (a) Variation of P_{m} as a function of τ for different system sizes N with $\phi = \pi/5$. P_{m} exhibits peak at different values of $\tau \geq \tau_c$. (b) The figure shows a plot between $\ln \tau_c$ and $\ln N$ for ϕ with slope $(= 0.94)$ nearly equal to 1 suggesting $\tau_c \sim N$ (from Rajak et al., 2014).

$\Delta t \geq \Delta t_{\mathrm{th}}$, the time-evolved Majorana delocalizes only within the inverted bulk energy levels (close to zero energy) present inside the gapless phase (see Fig. 8.24). As a result, there is a finite probability of recombinations of the Majorana mode from the inverted bands and we get an adiabatic tunneling to the other phase. On the other hand, for $\Delta t < \Delta t_{\mathrm{th}}$, the initial Majorana delocalizes over all the positive and negative bulk energy levels (see Fig. 8.28) prohibiting the possibility of adiabatic transport. After quenching, the probability of finding a Majorana edge mode at the other phase becomes maximum when the evolved Majorana state interacts with the minimum number of inverted energy levels and as a consequence, the final time-evolved Majorana state is closely connected to the equilibrium Majorana state at the other phase. The efficiency of the adiabatic transfer is indicated in Fig. 8.25(b) and Fig. 8.27(a) where the value of the first significant peak in P_{m} decreases as the number of inverted levels increases for larger N as well as ϕ.

Finally, one can ask the question why do τ_c and hence Δt_{op} increase with the system size (see Fig. 8.27(b)). As we increase N, the difference between two consecutive energy levels inside the "inverted" region decreases leading to an increase in the characteristic relaxation time of the system and hence a higher value of τ_c is required for an adiabatic passage.

8.10 Summary and Conclusion

We have investigated the quantum information theoretic measures like the ground state fidelity, the ground state fidelity susceptibility (FS), the Loschmidt echo (LE) and the entanglement entropy (EE) in the context of characterizing quantum phase transitions of some spin models. These measures capture the singularities associated with the ground state of the system at the quantum critical point (QCP) and successfully detect all the QCPs. They also show distinct scaling relations when the system is close to a QCP.

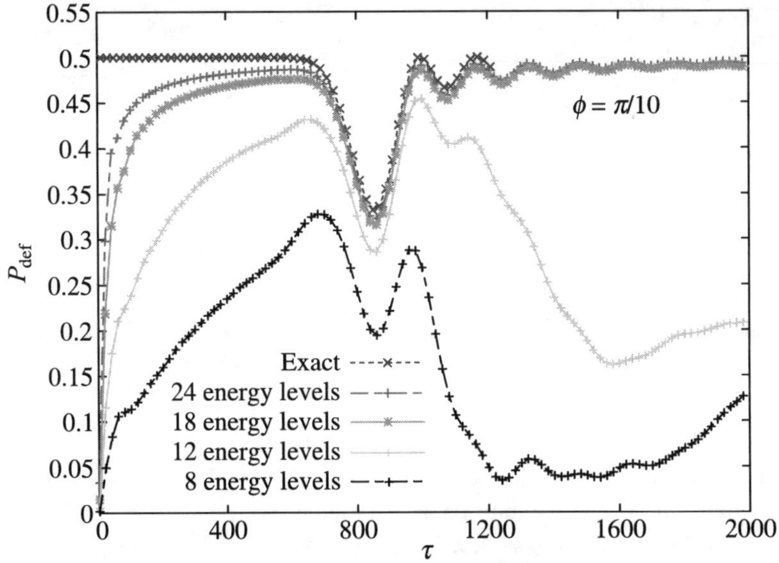

Fig. 8.28 Variation of P_{def} as a function of τ determining overlaps between the time-evolved Majorana state at final time and different number of positive energy states (near zero energy) at the final parameter values. Let us consider the case with $\phi = \pi/10$ and $N = 100$, where the number of inverted energy levels in both positive and negative sides of the zero energy is given by 18 (see Ref. (Rajak et al., 2014)). The plot indicates that for $\tau \leq \tau_c$ ($\Delta t < \Delta t_{\text{th}}$), the initial edge Majorana mixes with all the bulk energy levels. For $\tau \geq \tau_c$, on the other hand, the P_{def} determined by using only the inverted positive bulk energy levels nearly coincides with the exact P_{def} procured considering all positive energy levels. This clearly verifies that quenched Majorana state interacts only with the inverted energy levels for a passage time $\Delta t \geq \Delta t_{\text{th}}$ (from Rajak et al., 2014).

We introduced a new method to determine the FS and the LE for a generic path which can be used for models that are reducible to decoupled 2×2 matrix form (Rajak and Divakaran, 2014). The method used for the aforementioned study is simpler than the conventional method that requires tensor analysis. The verification of our method has been done considering a one-dimensional three-spin interacting transverse field Ising model which has a rich phase diagram consisting of isolated critical points, multicritical points (MCPs) and quasicritical points. We have also discussed the effect of gapless phase on the LE considering a central spin model where a two-dimensional Kitaev model on a honeycomb lattice has been chosen as an environment (Sharma and Rajak, 2012). In this case, the central spin or the qubit is coupled globally with the environment. The LE here successfully detects the gapless phase of the Kitaev model on a honeycomb lattice. It is interesting to note that the LE can also capture the anisotropic nature of the QCPs (lying on the boundary between the gapless and gapped phases) through its short time behavior.

We confirmed this behavior numerically by studying the collapse and revival of the LE when the system is fixed at a QCP. We also presented a numerical study of the time evolution of the LE and the EE for a one-dimensional transverse field Ising model with an open boundary condition following two simultaneous local quenches (Rajak and Divakaran, 2016). Here, one of the local quenches is given by the connection of two spin-$1/2$ open chains at a fixed time (J-quenching) and the other involves a rapid change in the magnitude of the transverse field term at a single site in one of two spin chains (h-quenching). We numerically evaluated the time-evolution of the LE and the EE after performing such double quenches, and explain various time scales associated with this dynamics using the picture of quasiparticles (QPs) produced due to such quenches. The shape of these curves is determined by the QPs propagating with the maximum velocity. In this process of double quenches, we have interestingly found the manifestation of partial or full reflections of QPs at the site of h-quench during the time evolution of the LE and the EE. We have also established that for the quenches considered here, the QPs produced in one of the quenches (i.e., the J-quench) is dominant in comparison to those generated in the other process (h-quench). This scenario is true for the specific quench that we have considered here. On the other hand, for two quenches of similar magnitude, we observe that both the QPs determine different time scales in the LE and the EE. As discussed by Rieger and Igloi (2011), the QPs behave like classical particles in the ferromagnetic region, whereas near the critical point, they are wavepackets of low lying excitations. Remarkably, our numerical results on double quenches also established this fact in a very transparent manner.

Recently, it has also been found that the studies involving quenching dynamics of a topological system across a QCP provide some interesting results not observed in the non-topological models, especially in the context of the Kibble–Zurek scaling of the defect density (Bermudez et al., 2009, 2010). In this regard, we first studied the survival probability of an edge Majorana when the superconducting pairing term of the p-wave superconducting chain is suddenly changed so that the system is driven from one topological phase to the other or to the QCPs and also to the non-topological phases. Interestingly, we found that for the quenching up to the QCP, dividing the two topological phases (or separating one topological phase from the topologically trivial one), the MSP shows a collapse and revival as a function of time signifying that the edge Majorana oscillates between the two physical edges of the chain. The time period of this oscillation has been found to be proportional to the system size. This phenomenon can be explained using the nearly equispaced nature of the bulk energy levels close to zero energy at the QCP. Nevertheless, the amplitude of collapse and revival eventually decreases with time when the bulk energy modes which are not equally spaced start contributing to the MSP. These results have also been analyzed in the context of the light cone like propagation of quantum correlations in a quenched quantum many-body system which also explains the collapse and revival of the LE in a generic situation. On the other hand, when the system is quenched from a topological phase to an other or to the trivial phase, the edge Majorana rapidly decoheres with time and fluctuates around a vanishingly small mean value. We have also studied slow quenching dynamics of the p-wave superconducting chain with a complex hopping term that breaks the effective time-reversal symmetry of the system and generates an extended gapless region. In this context, we have shown that for a finite open chain, one can find a nonzero probability of transporting an edge Majorana from one topological phase to the other through this intermediate gapless region under an adiabatic quench of the superconducting pair; this occurs for an optimum passage time across this gapless phase, $\Delta t_{\rm op} = \tau_c \sin\phi$ which leads to the condition $\tau_c = \Delta t_{\rm op}/\sin\phi$. Our numerical study has established that $\Delta t_{\rm op}$ is independent of ϕ (and depends only on the system size) while τ_c indeed depends on ϕ and diverges when $\phi \to 0$. This surprising observation has been explained noting the fact that the energy levels inside the gapless region of this system gets inverted. We have attributed the phenomenon of this adiabatic transfer to the delocalization of the time-evolved Majorana states only within the inverted energy levels which is only possible above an optimum transit time that the system requires to traverse the gapless region. Therefore, there exists a finite probability of tunneling of a Majorana edge state for some optimum values of transit time. We have also argued that the final time-evolved edge Majorana state is not instantaneous; rather it persists perpetually even when the quenching is stopped.

Acknowledgements

This chapter is based on a part of the thesis by AR, submitted to the Department of Physics of Calcutta University. AR is grateful to Amit Dutta, Bikas K. Chakrabarti, Tanay Nag and Shraddha Sharma for their active collaborations and very useful discussions. UD thanks Amit Dutta for related discussions and acknowledges DST, India for financial support through INSPIRE fellowship.

References

Albuquerque, A. F., F. Alet, C. Sire, and S. Capponi. 2010. 'Quantum critical scaling of fidelity susceptibility'. *Phys. Rev. B* 81: 064418.

Alicea, J. 2012. 'New directions in the pursuit of Majorana fermions in solid state systems'. *Rep. Prog. Phys.* 75: 076501.

Altland, A., and M. R. Zirnbauer. 1997. 'Nonstandard symmetry classes in mesoscopic normal-superconducting hybrid structures'. *Phys. Rev. B* 55: 1142.

Amico, L., R. Fazio, A. Osterloh, and V. Vedral. 2008. 'Entanglement in many-body systems'. *Rev. Mod. Phys.* 80: 517.

Anderson, P. W. 1958. 'Coherent excited states in the theory of superconductivity: Gauge invariance and the Meissner effect'. *Phys. Rev.* 110: 827.

Anderson, P. W. 1967. 'Infrared catastrophe in Fermi gases with local scattering potentials'. *Phys. Rev. Lett.* 18: 1049.

Andraschko, F., and J. Sirker. 2014. 'Dynamical quantum phase transitions and the Loschmidt echo: A transfer matrix approach'. *Phys. Rev. B* 89: 125120.

Barouch, E., and B. M. McCoy. 1971. 'Statistical mechanics of the XY model. II. Spin-correlation functions'. *Phys. Rev. A* 3: 786.

Barouch, E., B. M. McCoy, and M. Dresden. 1970. 'Statistical mechanics of the XY model. I'. *Phys. Rev. A* 2: 1075.

Bermudez, A., D. Patane, L. Amico, and M. A. Martin-Delgado. 2009. 'Topology-induced anomalous defect production by crossing a quantum critical point'. *Phys. Rev. Lett.* 102: 135702.

Bermudez, A., L. Amico, and M. A. Martin-Delgado. 2010. 'Dynamical delocalization of Majorana edge states by sweeping across a quantum critical point'. *New J. Phys.* 12: 055014.

Bernevig, B. A., T. L. Hughes, and S.-C. Zhang. 2006. 'Quantum spin Hall effect and topological phase transition in HgTe quantum wells'. *Science* 314: 1757.

Blass, B., H. Rieger, and F. Iglói. 2012. 'Quantum relaxation and finite-size effects in the XY chain in a transverse field after global quenches'. *Europhys. Lett.* 99: 30004.

Bloch, I., J. Dalibard, and W. Zwerger. 2008. 'Many-body physics with ultra-cold gases'. *Rev. Mod. Phys.* 80: 885.

Bose, S., V. Vedral, and P. L. Knight. 1998. 'Multiparticle generalization of entanglement swapping'. *Phys. Rev. A* 57: 822.

Bravyi, S., M. B. Hastings, and F. Verstraete. 2006. 'Lieb-Robinson bounds and the generation of correlations and topological quantum order'. *Phys. Rev. Lett.* 97: 050401.

Buchkremer, F. B. J., R. Dumke, H. Levsen, G. Birkl, and W. Ertmer. 2000. 'Wave packet echoes in the motion of trapped atoms'. *Phys. Rev. Lett.* 85: 3121.

Budich, J. C., S. Walter, and B. Trauzettel. 2012. 'Failure of protection of Majorana based qubits against decoherence'. *Phys. Rev. B* 85: 121405(R).

Bunder, J. E., and R. H. McKenzie. 1999. 'Effect of disorder on quantum phase transitions in anisotropic XY spin chains in a transverse field'. *Phys. Rev. B* 60: 344.

Calabrese, P., and J. Cardy. 2004. 'Entanglement entropy and quantum field theory'. *J. Stat. Mech.* P06002.

Calabrese, P., and J. Cardy. 2005. 'Evolution of entanglement entropy in one-dimensional systems'. *J. Stat. Mech.* P04010.

Calabrese, P., and J. Cardy. 2006. 'Time dependence of correlation functions following a quantum quench'. *Phys. Rev. Lett.* 96: 136801.

Calabrese, P., and J. Cardy. 2007a. 'Entanglement and correlation functions following a local quench: A conformal field theory approach'. *J. Stat. Mech.* P10004.

Calabrese, P., and J. Cardy. 2007b. 'Quantum quenches in extended systems'. *J. Stat. Mech.* P06008.

Calabrese, P., and J. Cardy. 2009. 'Entanglement entropy and conformal field theory'. *J. Phys. A: Math. Theor.* 42: 504005.

Canovi, E., D. Rossini, R. Fazio, G. E. Santoro, and A. Silva. 2011. 'Quantum quenches, thermalization, and many-body localization'. *Phys. Rev. B* 83: 094431.

Cardy, J. 2011. ' Measuring entanglement using quantum quenches'. *Phys. Rev. Lett.* 106: 150404.

Chaikin, P. M., and T. C. Lubensky. 1998. *Principles of condensed matter physics*. Cambridge University Press.

Chakrabarti, B. K., A. Dutta, and P. Sen. 1996. *Quantum Ising Phases and Transitions in Transverse Ising Model*. Springer-Verlag.

Chang, W., V. E. Manucharyan, T. S. Jespersen, J. Nygård, and C. M. Marcus. 2013. 'Tunneling spectroscopy of quasiparticle bound states in a spinful Josephson junction'. *Phys. Rev. Lett.* 110: 217005.

Chen, H.-D., and Z. Nussinov. 2008. 'Exact results of the Kitaev model on a hexagonal lattice: Spin states, string and brane correlators, and anyonic excitations'. *J. Phys. A: Math. Theor.* 41: 075001.

Cramer, M., C. M. Dawson, J. Eisert, and T. J. Osborne. 2008. 'Exact relaxation in a class of nonequilibrium quantum lattice systems'. *Phys. Rev. Lett.* 100: 030602.

Damski, B. 2013. 'Fidelity susceptibility of the quantum Ising model in a transverse field: The exact solution'. *Phys. Rev. E* 87: 052131.

Damski, B., and M. M. Rams. 2014. 'Exact results for fidelity susceptibility of the quantum Ising model: The interplay between parity, system size, and magnetic field'. *J. Phys. A: Math. Theor.* 47: 025303.

Damski, B., H. T. Quan, and W. H. Zurek. 2011. 'Critical dynamics of decoherence'. *Phys. Rev. A* 83: 062104.

Das, A., and B. K. Chakrabarti 2008. 'Colloquium: Quantum annealing and analog quantum computation'. *Rev. Mod. Phys.* 80: 1061.

Das, A., K. Sengupta, D. Sen, and B. K. Chakrabarti. 2006. 'Infinite-range Ising ferromagnet in a time-dependent transverse magnetic field: Quench and ac dynamics near the quantum critical point'. *Phys. Rev. B* 74: 144423.

Das, A., Y. Ronen, Y. Most, Y. Oreg, M. Heiblum, and H. Shtrikman. 2012. 'Zero-bias peaks and splitting in an Al-InAs nanowire topological superconductor as a signature of Majorana fermions'. *Nat. Phys.* 8: 887.

DeGottardi, W., D. Sen, and S. Vishveshwara. 2011. 'Topological phases, Majorana modes and quench dynamics in a spin ladder system'. *New J. Phys.* 13: 065028.

DeGottardi, W., D. Sen, and S. Vishveshwara. 2013a. 'Majorana fermions in superconducting 1D systems having periodic, quasiperiodic, and disordered potentials'. *Phys. Rev. Lett.* 110: 146404.

DeGottardi, W., M. Thakurathi, S. Vishveshwara, and D. Sen. 2013b. 'Majorana fermions in superconducting wires: Effects of long-range hopping, broken time-reversal symmetry, and potential landscapes'. *Phys. Rev. B* 88: 165111.

Deng, M. T., C. L. Yu, G. Y. Huang, M. Larsson, P. Caroff, and H. Q. Xu. 2012. 'Anomalous zero-bias conductance peak in a Nb–InSb nanowire–Nb hybrid device'. *Nano Lett.* 12: 6414.

Deng, S., G. Ortiz, and L. Viola. 2009. 'Anomalous nonergodic scaling in adiabatic multicritical quantum quenches'. *Phys. Rev. B* 80: 241109.

Dillenschneider, R. 2008. 'Quantum discord and quantum phase transition in spin chains'. *Phys. Rev. B* 78: 224413.

Divakaran, U. 2013. 'Three-site interacting spin chain in a staggered field: Fidelity versus Loschmidt echo'. *Phys. Rev. E* 88: 052122.

Divakaran, U., and A. Dutta. 2007. 'The effect of the three-spin interaction and the next nearest neighbor interaction on the quenching dynamics of a transverse Ising model'. *J. Stat. Mech.* P11001.

Divakaran, U., V. Mukherjee, A. Dutta, and D. Sen. 2009. 'Defect production due to quenching through a multicritical point'. *J. Stat. Mech.*, P02007.

Divakaran, U., F. Iglói, and H. Rieger. 2011. 'Non-equilibrium quantum dynamics after local quenches'. *J. Stat. Mech.* P10027.

Dubail, J., and J.-M. Stéphan. 2011. 'Universal behavior of a bipartite fidelity at quantum criticality'. *J. Stat. Mech.* L03002.

Dutta, A., R. R. P. Singh, and U. Divakaran. 2010. 'Quenching through Dirac and semi-Dirac points in optical lattices: Kibble-Zurek scaling for anisotropic quantum critical systems'. *Europhys. Lett.* 89: 67001.

Dutta, A., G. Aeppli, B. K. Chakrabarti, U. Divakaran, T. F. Rosenbaum, and D. Sen. 2015. *Quantum Phase Transitions in Transverse Field Models*. Cambridge University Press.

Dziarmaga, J. 2010. 'Dynamics of a quantum phase transition and relaxation to a steady state'. *Adv. Phys.* 59: 1063.

Eisert, J., and T. J. Osborne. 2006. 'General entanglement scaling laws from time evolution'. *Phys. Rev. Lett.* 97: 150404.

Eisert, J., M. Cramer, and M. B. Plenio. 2010. 'Colloquium: Area laws for the entanglement entropy'. *Rev. Mod. Phys.* 82: 277.

Eisler, V., and I. Peschel. 2007. 'Evolution of entanglement after a local quench'. *J. Stat. Mech.* P06005.

Eisler, V., D. Karevski, T. Platini, and I. Peschel. 2008. 'Entanglement evolution after connecting finite to infinite quantum chains'. *J. Stat. Mech.* P01023.

Feng, X.-Y., G.-M. Zhang, and T. Xiang. 2007. 'Topological characterization of quantum phase transitions in a spin-1/2 model'. *Phys. Rev. Lett.* 98: 087204.

Fidkowski, L., and A. Kitaev. 2011. 'Topological phases of fermions in one dimension'. *Phys. Rev. B* 83: 075103.

Finck, A. D. K.,D. J. Van Harlingen, P. K. Mohseni, K. Jung, and X. Li. 2013. 'Anomalous modulation of a zero-bias peak in a hybrid nanowire-superconductor device'. *Phys. Rev. Lett.* 110: 126406.

Freedman, M., A. Kitaev, M. Larsen, and Z. Wang. 2003. 'Topological quantum computation'. *Bull. Am. Math. Soc.* 40: 31.

Freedman, M. H. 1998. 'P/NP, and the quantum field computer'. *Proc. Natl. Acad. Sci.* 95: 98.

Fu, L., and C. L. Kane. 2007. 'Topological insulators with inversion symmetry'. *Phys. Rev. B* 76: 045302.

Fu, L., and C. L. Kane. 2008. 'Superconducting proximity effect and Majorana fermions at the surface of a topological insulator'. *Phys. Rev. Lett.* 100: 096407.

Fu, L., C. L. Kane, and E. J. Mele. 2007. 'Topological insulators in three dimensions'. *Phys. Rev. Lett.* 98: 106803.

Fulga, I. C., F. Hassler, A. R. Akhmerov, and C. W. J. Beenakker. 2011. 'Scattering formula for the topological quantum number of a disordered multimode wire'. *Phys. Rev. B* 83: 155429.

Grandi, C. D., V. Gritsev, and A. Polkovnikov. 2010. 'Quench dynamics near a quantum critical point'. *Phys. Rev. B* 81: 012303.

Greiner, M., O. Mandel, T. Esslinger, T. W. Hänsch, and I. Bloch. 2002. 'Quantum phase transition from a superfluid to a Mott insulator in a gas of ultracold atoms'. *Nature* 415: 39.

Gritsev, V., and A. Polkovnikov. 2009. 'Universal dynamics near quantum critical points'. arXiv:0910.3692.

Gritsev, V., and A. Polkovnikov. 2010. *Developments in Quantum Phase Transitions*. Taylor and Francis.

Gu, S. -J. 2010. 'Fidelity approach to quantum phase transitions'. *Int. J. Mod. Phys. B* 24: 4371.

Gu, S. -J., H. -M. Kwok, W. -Q. Ning, and H.-Q. Lin. 2008. 'Fidelity susceptibility, scaling, and universality in quantum critical phenomena'. *Phys. Rev. B* 77: 245109.

Halász, G. B., and A. Hamma. 2013. 'Topological Rényi entropy after a quantum quench'. *Phys. Rev. Lett.* 110: 170605.

Häppölä, J., G. B. Halász, and A. Hamma. 2012. 'Universality and robustness of revivals in the transverse field XY model'. *Phys. Rev. A* 85: 032114.

Haroche, S. 1998. 'Entanglement, decoherence and the quantum/classical boundary'. *Phys. Today* 51: 36.

Hasan, M. Z., and C. L. Kane. 2010. 'Colloquium: Topological insulators'. *Rev. Mod. Phys.* 82: 3045.

Heyl, M., A. Polkovnikov, and S. Kehrein. 2013. 'Dynamical quantum phase transitions in the transverse-field Ising model'. *Phys. Rev. Lett.* 110: 135704.

Holzhey, C., F. Larsen, and F. Wilczek. 1994. 'Geometric and renormalized entropy in conformal field theory'. *Nucl. Phys. B* 424: 443.

Hsu, B., E. Grosfeld, and E. Fradkin. 2009. 'Quantum noise and entanglement generated by a local quantum quench'. *Phys. Rev. B* 80: 235412.

Hutton, A., and S. Bose. 2004. 'Mediated entanglement and correlations in a star network of interacting spins'. *Phys. Rev. A* 69: 042312.

Iglói, F., and Y. -C. Lin. 2008. 'Finite-size scaling of the entanglement entropy of the quantum Ising chain with homogeneous, periodically modulated and random couplings'. *J. Stat. Mech.* P06004.

Iglói, F., and H. Rieger. 2000. 'Long-range correlations in the nonequilibrium quantum relaxation of a spin chain'. *Phys. Rev. Lett.* 85: 3233.

Iglói, F., Z. Szatmári, and Y. -C. Lin. 2009. 'Entanglement entropy with localized and extended interface defects'. *Phys. Rev. B* 80: 024405.

Iglói, F., Z. Szatmári, and Y. -C. Lin. 2012. 'Entanglement entropy dynamics of disordered quantum spin chains'. *Phys. Rev. B* 85: 094417.

Iucci, A., and M. A. Cazalilla. 2009. 'Quantum quench dynamics of the Luttinger model'. *Phys. Rev. A* 80: 063619.

Iucci, A., and M. A. Cazalilla. 2010. 'Quantum quench dynamics of the sine-Gordon model in some solvable limits'. *New J. Phys.* 12: 055019.

Joos, E., H. D. Zeh, C. Kiefer, D. J. W. Giulini, J. Kupsch, and I. -O. Stamatescu. 2013. *Decoherence and the Appearance of a Classical World in Quantum Theory*. Springer-Verlag.

Kadowaki, T., and H. Nishimori. 1998. 'Quantum annealing in the transverse Ising model'. *Phys. Rev. E* 58: 5355.

Kane, C. L., and E. J. Mele. 2005a. 'Quantum spin Hall effect in graphene'. *Phys. Rev. Lett.* 95: 226801.

Kane, C. L., and E. J. Mele. 2005b. 'Z_2 topological order and the quantum spin Hall effect'. *Phys. Rev. Lett.* 95: 146802.

Karrasch, C., and D. Schuricht. 2013. 'Dynamical phase transitions after quenches in nonintegrable models'. *Phys. Rev. B* 87: 195104.

Kitaev, A. 2006. 'Anyons in an exactly solved model and beyond'. *Ann. Phys.* 321: 2.

Kitaev, A., and C. Laumann. 2009. 'Topological phases and quantum computation'. arXiv:0904.2771.

Kitaev, A. Y. 2001. 'Unpaired Majorana fermions in quantum wires'. *Physics-Uspekhi* 44: 131.

Kitaev, A. Y. 2003. 'Fault-tolerant quantum computation by anyons'. *Ann. Phys.* 303: 2.

Kogut, J. B. 1979. 'An introduction to lattice gauge theory and spin systems'. *Rev. Mod. Phys.* 51: 659.

Kopnin, N. B., and M. M. Salomaa. 1991. 'Mutual friction in superfluid He_3: Effects of bound states in the vortex core'. *Phys. Rev. B* 44: 9667.

Kopp, A., and S. Chakravarty. 2005. 'Criticality in correlated quantum matter'. *Nat. Phys.* 1: 53.

Korepin, V. E. 2004. 'Universality of entropy scaling in one dimensional gapless models'. *Phys. Rev. Lett.* 92: 096402.

Latorre, J. I., E. Rico, and G. Vidal. 2003. 'Ground state entanglement in quantum spin chains'. arXiv: quant–ph/0304098.

Lee, E. J. H., X. Jiang, M. Houzet, R. Aguado, C. M. Lieber, and S. De Franceschi. 2014. 'Spin-resolved Andreev levels and parity crossings in hybrid superconductor-semiconductor nanostructures'. *Nat. Nanotechnol.* 9: 79.

Lewenstein, M., A. Sanpera, V. Ahufinger, B. Damski, A. Sen, and U. Sen. 2007. 'Ultracold atomic gases in optical lattices: Mimicking condensed matter physics and beyond'. *Adv. Phys.* 56: 243.

Lieb, E., and D. Robinson. 1972. 'The finite group velocity of quantum spin systems'. *Commun. Math. Phys.* 28: 251.

Lieb, E., T. Schultz, and D. Mattis. 1961. 'Two soluble models of an antiferromagnetic chain'. *Ann. Phys.* 16: 407.

Lou, J., S. Tanaka, H. Katsura, and N. Kawashima. 2011. 'Entanglement spectra of the two-dimensional Affleck-Kennedy-Lieb-Tasaki model: Correspondence between the valence-bond-solid state and conformal field theory'. *Phys. Rev. B* 84: 245128.

Lutchyn, R. M., and M. P. A. Fisher. 2011. 'Interacting topological phases in multiband nanowires'. *Phys. Rev. B* 84: 214528.

Mahan, G. D. 2013. *Many-particle Physics*. Springer-Verlag.

Majorana, E. 1937. 'Teoria simmetrica dell'elettrone e del positrone'. *Nuovo Cimento* 5: 171.

Mourik, V., K. Zuo, S. M. Frolov, S. R. Plissard, E. P. A. M. Bakkers, and L. P. Kouwenhoven. 2012. 'Signatures of Majorana fermions in hybrid superconductor-semiconductor nanowire devices'. *Science* 336: 1003.

Mukherjee, V., and A. Dutta. 2010. 'Adiabatic multicritical quantum quenches: Continuously varying exponents depending on the direction of quenching'. *Europhys. Lett.* 92: 37004.

Mukherjee, V., U. Divakaran, A. Dutta, and D. Sen. 2007. 'Quenching dynamics of a quantum XY spin-1/2 chain in a transverse field'. *Phys. Rev. B* 76: 174303.

Mukherjee, V., A. Polkovnikov, and A. Dutta. 2011. 'Oscillating fidelity susceptibility near a quantum multicritical point'. *Phys. Rev. B* 83: 075118.

Mukherjee, V., S. Sharma, and A. Dutta. 2012a. 'Loschmidt echo with a nonequilibrium initial state: Early-time scaling and enhanced decoherence'. *Phys. Rev. B* 86: 020301(R).

Mukherjee, V., A. Dutta, and D. Sen. 2012b. 'Quantum fidelity for one-dimensional Dirac fermions and two-dimensional Kitaev model in the thermodynamic limit'. *Phys. Rev. B* 85: 024301.

Nag, T., A. Patra, and A. Dutta. 2011. 'Quantum discord in a spin-1/2 transverse XY chain following a quench'. *J. Stat. Mech.* P08026.

Nag, T., U. Divakaran, and A. Dutta. 2012. 'Scaling of the decoherence factor of a qubit coupled to a spin chain driven across quantum critical points'. *Phys. Rev. B* 86: 020401(R).

Nag, T., S. Roy, A. Dutta, and D. Sen. 2014. 'Dynamical localization in a chain of hard core bosons under periodic driving'. *Phys. Rev. B* 89: 165425.

Nag, T., D. Sen, and A. Dutta. 2015. 'Maximum group velocity in a one-dimensional model with a sinusoidally varying staggered potential'. *Phys. Rev. A* 91: 063607.

Nielsen, M. A., and I. L. Chuang. 2000. *Quantum Computation and Quantum Information*. Cambridge University Press.

Niu, Y., S. B. Chung, C. -H. Hsu, I. Mandal, S. Raghu, and S. Chakravarty. 2012. 'Majorana zero modes in a quantum Ising chain with longer-ranged interactions'. *Phys. Rev. B* 85: 035110.

Ohta, T., S. Tanaka, I. Danshita, and K. Totsuka. 2015. 'Phase diagram and sweep dynamics of a one-dimensional generalized cluster model'. *J. Phys. Soc. Jpn.* 84: 063001.

Osborne, T. J., and M. A. Nielsen. 2002. 'Entanglement in a simple quantum phase transition'. *Phys. Rev. A* 66: 032110.

Osterloh, A., L. Amico, G. Falci, and R. Fazio. 2002. 'Scaling of entanglement close to a quantum phase transition'. *Nature* 416: 608.

Patanè, D., A. Silva, L. Amico, R. Fazio, and G. E. Santoro. 2008. 'Adiabatic dynamics in open quantum critical many-body systems'. *Phys. Rev. Lett.* 101: 175701.

Patel, A. A., S. Sharma, and A. Dutta. 2013a. 'Quench dynamics of edge states in 2D topological insulator ribbons'. *Euro. Phys. J. B* 86: 1.

Patel, A. A., S. Sharma, and A. Dutta. 2013b. 'Role of marginality in quantum fidelity and Loschmidt echo: Dirac points in 2-D'. *Europhys. Lett.* 102: 46001.

Perfetto, E. 2013. 'Dynamical formation and manipulation of Majorana fermions in driven quantum wires in contact with a superconductor'. *Phys. Rev. Lett.* 110: 087001.

Pfeuty, P. 1970. 'The one-dimensional Ising model with a transverse field'. *Ann. Phys.* 57: 79.

Polkovnikov, A. 2005. 'Universal adiabatic dynamics in the vicinity of a quantum critical point'. *Phys. Rev. B* 72: 161201(R).

Polkovnikov, A., K. Sengupta, A. Silva, and M. Vengalattore. 2011. 'Colloquium: Nonequilibrium dynamics of closed interacting quantum systems'. *Rev. Mod. Phys.* 83: 863.

Pollmann, F., S. Mukerjee, A. G. Green, and J. E. Moore. 2010. 'Dynamics after a sweep through a quantum critical point'. *Phys. Rev. E* 81: 020101.

Qi, X. -L., and S. -C. Zhang. 2010. 'The quantum spin Hall effect and topological insulators'. *Phys. Today* 63: 33.

Qi, X. -L., and S. -C. Zhang. 2011. 'Topological insulators and superconductors'. *Rev. Mod. Phys.* 83: 1057.

Quan, H. T., Z. Song, X. F. Liu, P. Zanardi, and C. P. Sun. 2006. 'Decay of Loschmidt echo enhanced by quantum criticality'. *Phys. Rev. Lett.* 96: 140604.

Rahmani, A., and C. Chamon. 2009. 'Exact results on the quench dynamics of the entanglement entropy in the toric code'. *Phys. Rev. B* 82: 134303.

Rainis, D., L. Trifunovic, J. Klinovaja, and D. Loss. 2013. 'Towards a realistic transport modeling in a superconducting nanowire with Majorana fermions'. *Phys. Rev. B* 87: 024515.

Rajak, A., and U. Divakaran. 2014. 'Fidelity susceptibility and Loschmidt echo for generic paths in a three-spin interacting transverse Ising model'. *J. Stat. Mech.* P04023.

Rajak, A., and U. Divakaran. 2016. 'Effect of double local quenches on Loschmidt echo and entanglement entropy of a one-dimensional quantum system'. *J. Stat. Mech.* 043107.

Rajak, A., and A. Dutta. 2014. 'Survival probability of an edge Majorana in a one-dimensional p-wave superconducting chain under sudden quenching of parameters'. *Phys. Rev. E* 89: 042125.

Rajak, A., T. Nag, and A. Dutta. 2014. 'Possibility of adiabatic transport of a Majorana edge state through an extended gapless region'. *Phys. Rev. E* 90: 042107.

Rams, M. M., and B. Damski. 2011. 'Quantum fidelity in the thermodynamic limit'. *Phys. Rev. Lett.* 106: 055701.

Read, N., and D. Green. 2000. 'Paired states of fermions in two dimensions with breaking of parity and time-reversal symmetries and the fractional quantum Hall effect'. *Phys. Rev. B* 61: 10267.

Refael, G., and J. E. Moore. 2004. 'Entanglement entropy of random quantum critical points in one dimension'. *Phys. Rev. Lett.* 93: 260602.

Rieger, H., and F. Iglói. 2011. 'Semiclassical theory for quantum quenches in finite transverse Ising chains'. *Phys. Rev. B* 84: 165117.

Rossini, D., T. Calarco, V. Giovannetti, S. Montangero, and R. Fazio. 2007. 'Decoherence induced by interacting quantum spin baths'. *Phys. Rev. A* 75: 032333.

Russomanno, A., S. Sharma, A. Dutta, and G. E. Santoro. 2015. 'Asymptotic work statistics of periodically driven Ising chains'. *J. Stat. Mech.* P08030.

Sachdev, S. 1999. *Quantum Phase Transitions*. Cambridge University Press.

Sacramento, P. D. 2014. 'Fate of Majorana fermions and Chern numbers after a quantum quench'. *Phys. Rev. E* 90: 032138.

Sánchez, C. M., P. R. Levstein, R. H. Acosta, and A. K. Chattah. 2009. 'NMR Loschmidt echoes as quantifiers of decoherence in interacting spin systems'. *Phys. Rev. A* 80: 012328.

Sarandy, M. S. 2009. 'Classical correlation and quantum discord in critical systems'. *Phys. Rev. A* 80: 022108.

Sau, J. D., and S. D. Sarma. 2012. 'Realizing a robust practical Majorana chain in a quantum-dot-superconductor linear array'. *Nat. Commun.* 3: 964.

Schmidt, M. J., D. Rainis, and D. Loss. 2012. 'Decoherence of Majorana qubits by noisy gates'. *Phys. Rev. B* 86: 085414.

Schwandt, D., F. Alet, and S. Capponi. 2009. 'Quantum Monte Carlo simulations of fidelity at magnetic quantum phase transitions'. *Phys. Rev. Lett.* 103: 170501.

Sengupta, K., D. Sen, and S. Mondal. 2008. 'Exact results for quench dynamics and defect production in a two-dimensional model'. *Phys. Rev. Lett.* 100: 077204.

Sharma, S., and A. Rajak. 2012. 'Study of Loschmidt echo for two-dimensional Kitaev model'. *J. Stat. Mech.* P08005.

Sharma, S., V. Mukherjee, and A. Dutta. 2012. 'Study of Loschmidt echo for a qubit coupled to an XY-spin chain environment'. *Eur. Phys. J. B* 85: 1.

Sharma, S., A. Russomanno, G. E. Santoro, and A. Dutta. 2014. 'Loschmidt echo and dynamical fidelity in periodically driven quantum systems'. *Europhys. Lett.* 106: 67003.

Sharma, S., S. Suzuki, and A. Dutta. 2015. 'Quenches and dynamical phase transitions in a nonintegrable quantum Ising model'. *Phys. Rev. B* 92: 104306.

Sharma, S., U. Divakaran, A. Polkovnikov, and A. Dutta. 2016. 'Slow quenches in a quantum Ising chain: Dynamical phase transitions and topology'. *Phys. Rev. B* 93: 144306.

Shirley, J. H. 1965. 'Solution of the Schrödinger equation with a Hamiltonian periodic in time'. *Phys. Rev.* 138: B979.

Song, H. F., C. Flindt, S. Rachel, I. Klich, and K. Le Hur. 2011. 'Entanglement entropy from charge statistics: Exact relations for noninteracting many-body systems'. *Phys. Rev. B* 83: 161408(R).

Stéphan, J.-M., and J. Dubail. 2011. 'Local quantum quenches in critical one-dimensional systems: Entanglement, the Loschmidt echo, and light-cone effects'. *J. Stat. Mech.* P08019.

Suzuki, S., J. -I. Inoue, and B. K. Chakrabarti. 2013. *Quantum Ising Phases and Transitions in Transverse Ising Models*. Heidelberg, Springer.

Suzuki, S., T. Nag, and A. Dutta. 2016. 'Dynamics of decoherence: Universal scaling of the decoherence factor'. *Phys. Rev. A* 93: 012112.

Tanaka, S., R. Tamura, and H. Katsura. 2012. 'Entanglement spectra of the quantum hard-square model: Holographic minimal models'. *Phys. Rev. A* 86: 032326.

Thakurathi, M., D. Sen, and A. Dutta. 2012. 'Fidelity susceptibility of one-dimensional models with twisted boundary conditions'. *Phys. Rev. B* 86: 245424.

Thakurathi, M., A. A. Patel, D. Sen, and A. Dutta. 2013. 'Floquet generation of Majorana end modes and topological invariants'. *Phys. Rev. B* 88: 155133.

Thouless, D. J., M. Kohmoto, M. P. Nightingale, and M. den Nijs. 1982. 'Quantized Hall conductance in a two-dimensional periodic potential'. *Phys. Rev. Lett.* 49: 405.

Tong, Q. -J., J. -H. An, J. Gong, H. -G. Luo, and C. H. Oh. 2013. 'Generating many Majorana modes via periodic driving: A superconductor model'. *Phys. Rev. B* 87: 201109(R).

Tsomokos, D. I., A. Hamma, W. Zhang, S. Haas, and R. Fazio. 2009. 'Topological order following a quantum quench'. *Phys. Rev. A* 80: 060302(R).

Vedral, V. 2007. *Introduction to Quantum Information Science*. Oxford University Press.

Venuti, L. C., and P. Zanardi. 2007. 'Quantum critical scaling of the geometric tensors'. *Phys. Rev. Lett.* 99: 095701.

Vidal, G., J. I. Latorre, E. Rico, and A. Kitaev. 2003. 'Entanglement in quantum critical phenomena'. *Phys. Rev. Lett.* 90: 227902.

Vojta, M. 2003. 'Quantum phase transitions'. *Rep. Prog. Phys.* 66: 2069.

Vollhardt, D., and P. Wole. 2013. *The Superfluid Phases of Helium 3*. Dover Publications, Inc.

Volovik, G. E. 1999. 'Fermion zero modes on vortices in chiral superconductors'. *JETP Lett.* 70: 609.

Wang, P., W. Yi, and G. Xianlong. 2014. 'Topological phase transition in the quench dynamics of a one-dimensional Fermi gas'. arXiv:1404.6848.

Wen, X.-G. 1995. 'Topological orders and edge excitations in fractional quantum Hall states'. *Adv. Phys.* 44: 405.

Wilczek, F. 2009. 'Majorana returns'. *Nat. Phys.* 5: 614.

Yang, S., S. -J. Gu, C. -P. Sun, and H.-Q. Lin. 2008. 'Fidelity susceptibility and long-range correlation in the Kitaev honeycomb model'. *Phys. Rev. A* 78: 012304.

Yuan, Z. -G., P. Zhang, and S.-S. Li. 2007. 'Loschmidt echo and Berry phase of a quantum system coupled to an XY spin chain: Proximity to a quantum phase transition'. *Phys. Rev. A* 75: 012102.

Zanardi, P., and N. Paunković. 2006. 'Ground state overlap and quantum phase transitions'. *Phys. Rev. E* 74: 031123.

Zhang, H., C. -X. Liu, X. -L. Qi, X. Dai, Z. Fang, and S. -C. Zhang. 2009a. 'Topological insulators in Bi_2Se_3, Bi_2Te_3 and Sb_2Te_3 with a single Dirac cone on the surface'. *Nat. Phys.* 5: 438.

Zhang, J., F. M. Cucchietti, C. M. Chandrashekar, M. Laforest, C. A. Ryan, M. Ditty, A. Hubbard, J. K. Gamble, and R. Laamme. 2009b. 'Direct observation of quantum criticality in Ising spin chains'. *Phys. Rev. A* 79: 012305.

Zhou, H. -Q., and J. P. Barjaktarevič. 2008. 'Fidelity and quantum phase transitions'. *J. Phys. A: Math. Theor.* 41: 412001.

Zurek, W. H. 1991. 'Decoherence and the transition from quantum to classical'. *Phys. Today* 44: 36.

Zurek, W. H. 2003. 'Decoherence, einselection, and the quantum origins of the classical'. *Rev. Mod. Phys.* 75: 715.

Zurek, W. H., U. Dorner, and P. Zoller. 2005. 'Dynamics of a quantum phase transition'. *Phys. Rev. Lett.* 95: 105701.

CHAPTER 9

A Brief Historical Note on the Studies of Quantum Spin Glasses, Annealing and Computation

Sudip Mukherjee[1]

This brief historical and documentary note traces the development of ideas and technology related to quantum spin glasses and annealing. We also discuss some very recent publications and arxiv preprints to highlight the present scenarios in quantum annealing.

9.1 Introduction

Optimization of a multivariable cost function is still a challenging computational problem. Such minimization becomes extremely difficult when we are negotiating with NP-hard types of problem, where the search time is not bounded by any polynomial in the total number of variables of the function. Kirkpatrick et al. (1983) introduced the idea of simulated annealing (SA) to handle such problems but the technique fails to work in problems like the determination of the ground state of an Ising spin glass of N spins or of the minimum travel path connecting N cities on a plane. In these cases, the system gets localized into some local minima of the free energy or cost function, which is separated from the global minima by macroscopically high ($O(N)$) free energy or cost function barriers. It was then suggested that quantum tunneling across such barriers could play an effective role in finding the global minima of the cost function. Such a possibility was first proposed by Ray et al. (1989); this is the key idea behind quantum annealing (QA). QA has been developed with the aid of theoretical (Kadowaki and Nishimori, 1998; Farhi et al., 2001; Santoro et al., 2002) and experimental (Brooke et al., 1999) works over the last two decades. In the year 2011, we witnessed the first successful hardware implementation of QA by D-Wave Systems (Johnson et al., 2011). Development of such machines provide an opportunity to check the validity of the idea of QA physically. Several tests have been

[1]Department of Physics, Barasat Govt. College, North 24 Parganas, West Bengal 700124, India and CMP Division, Saha Institute of Nuclear Physics, Kolkata 700064, India

made by some renowned research groups through this quantum computing device to scrutinize the proclaimed fastness of the quantum algorithm. Although the performance of the D-Wave machine (often called quantum annealer) has not become superior to the SA in the context of speed enhancement, the quantum nature in computation seems to be clear in the respective field.

PHYSICAL REVIEW B VOLUME 39, NUMBER 16 1 JUNE 1989

Sherrington-Kirkpatrick model in a transverse field:
Absence of replica symmetry breaking due to quantum fluctuations

P. Ray, B. K. Chakrabarti, and Arunava Chakrabarti
Saha Institute of Nuclear Physics, 92 Acharya Praffula Chandra Road, Calcutta 700 009, India
(Received 14 October 1988)

The Sherrington-Kirkpatrick model under a transverse field is studied here employing the Suzuki-Trotter formula to map the model to an equivalent classical one. The effective Thouless-Anderson-Palmer free energy is used to study the stability of the system, and Monte Carlo computer simulations of the effective classical model are performed to obtain the phase diagram and the magnetization overlap distribution. Our results indicate a trivial overlap distribution due to quantum fluctuations. The phase diagram shows a slight initial increase in the glass transition temperature T_g as the transverse field is switched on, confirming that obtained by Yokota.

Fig. 9.1 Title and abstract from the first published paper proposing the idea that quantum tunneling across the free energy barriers in the Sherrington–Kirkpatrick spin glass model can lead to an efficient way of searching for its ground state(s). It was argued that "quantum tunneling between the classical 'trap' states, separated by infinite (but narrow) barriers in the free energy surface, is possible, as quantum tunneling probability is proportional to the barrier area which is finite." They suggested that any amount of transverse field would lead to the collapse of the overlap distribution to a delta function. It may be noted that computationally hard problems can often be mapped into such long-range spin glass models; the advantage of quantum tunneling in such quantum spin glass models has lead ultimately to the development of the quantum annealer. A related reference is Chakrabarti (1981). (Permission to use title and abstract from the paper is given by American Physical Society)

9.2 A short History of the Development

In frustrated magnetic systems, like classical Ising spin glasses, the free energy landscape is highly rugged. Due to the uneven nature of the free energy landscape, the system cannot search the entire configurational space. Ray et al. (1989) suggested that the system can avoid the trapping in some local free energy minima by pursuing quantum tunneling across the tall ($O(N)$) but narrow free energy barriers, which essentially leads the system to flow towards its global free energy minima. Later (both in receiving and publishing dates), Thirumalai et al. (1989) suggested the same advantage of quantum tunneling through high

VOLUME 67, NUMBER 15 PHYSICAL REVIEW LETTERS 7 OCTOBER 1991

From Classical to Quantum Glass

Wenhao Wu, B. Ellman, and T. F. Rosenbaum
The James Franck Institute and Department of Physics, The University of Chicago, Chicago, Illinois 60637

G. Aeppli and D. H. Reich[(a)]
AT&T Bell Laboratories, Murray Hill, New Jersey 07974
(Received 11 March 1991)

We study the effects of a transverse magnetic field on the dynamics of the randomly diluted, dipolar coupled, Ising magnet $LiHo_{0.167}Y_{0.833}F_4$. The transverse field mixes the eigenfunctions of the ground-state Ising doublet with the otherwise inaccessible excited-state levels. We observe a rapid decrease in the characteristic relaxation times, large changes in the spectral form of the relaxation, and a depression of the spin-glass transition temperature with the introduction of quantum fluctuations.

Fig. 9.2 Title and abstract from the first paper reporting on the experimental studies of an Ising spin glass sample under the influence of a tunable transverse field. The observed nature of the tunneling induced phase diagram compared well with that predicted by Ray et al. (1989). (Permission to use title and abstract from the paper is given by American Physical Society)

ELSEVIER 18 March 1994 CHEMICAL PHYSICS LETTERS
Chemical Physics Letters 219 (1994) 343–348

Quantum annealing: a new method for minimizing multidimensional functions

A.B. Finnila, M.A. Gomez, C. Sebenik, C. Stenson, J.D. Doll
Department of Chemistry, Brown University, Providence, RI 02912, USA

Received 29 November 1993

Abstract

Quantum annealing is a new method for finding extrema of multidimensional functions. Based on an extension of classical, simulated annealing, this approach appears robust with respect to avoiding local minima. Further, unlike some of its predecessors, it does not require an approximation to a wavefunction. We apply the technique to the problem of finding the lowest energy configurations of Lennard-Jones clusters of up to 19 particles (roughly 10^5 local minima). This early success suggests that this method may complement the widely implemented technique of simulated annealing.

Fig. 9.3 Title and abstract of the first published paper demonstrating the search of the ground state of a Lennard–Jones system using 'quantum annealing' (term appearing for the first time in paper title). (Permission to use title and abstract from the paper is given by Elsevier)

PHYSICAL REVIEW E VOLUME 58, NUMBER 5 NOVEMBER 1998

Quantum annealing in the transverse Ising model

Tadashi Kadowaki and Hidetoshi Nishimori
Department of Physics, Tokyo Institute of Technology, Oh-okayama, Meguro-ku, Tokyo 152-8551, Japan
(Received 30 April 1998)

We introduce quantum fluctuations into the simulated annealing process of optimization problems, aiming at faster convergence to the optimal state. Quantum fluctuations cause transitions between states and thus play the same role as thermal fluctuations in the conventional approach. The idea is tested by the transverse Ising model, in which the transverse field is a function of time similar to the temperature in the conventional method. The goal is to find the ground state of the diagonal part of the Hamiltonian with high accuracy as quickly as possible. We have solved the time-dependent Schrödinger equation numerically for small size systems with various exchange interactions. Comparison with the results of the corresponding classical (thermal) method reveals that the quantum annealing leads to the ground state with much larger probability in almost all cases if we use the same annealing schedule. [S1063-651X(98)02910-9]

Fig. 9.4 Title and abstract from a paper demonstrating the clear advantage of quantum annealing in Ising models with frustrating interactions. It was found that in case of quantum annealing not only there was a higher probability of the initial state to converge in the ground state of such models but also the convergence time was comparatively less than that of simulated annealing. A related reference is Chakrabarti et al. (1996). (Permission to use title and abstract from the paper is given by American Physical Society)

and narrow barriers for the Sherrington–Kirkpatrick model (in almost the same language as in the earlier paper). Indeed, an earlier indication for quantum dynamical formulation for diffusion across random barriers by Ebeling et al. (1984) was taken forward by Rujan (1988) to demonstrate numerically that an effective tunneling algorithm can help in searching for optimal tours better than the classical algorithms for the traveling salesman problem. Such a simple idea has had its share of controversy; it was severely criticized by the literature (Altshuler et al., 2010; Hen and Young, 2011). They argued that the incoherent mixture of the tunneling waves (which are completely random in phase) coming from different energy barriers make the system localized. Immediately after the proposal (see Fig. 9.1) of Ray et al. (1989), some experimental results supporting the idea of such quantum tunneling was reported by Wu et al. (1991) (see Fig. 9.2). The possibility of annealing down to the ground state with tunable quantum fluctuations was mentioned by Finnila et al. (1994) (see Fig. 9.3). However the first numerical implementation of QA in the search of the ground state of frustrated Ising spin glass with a tunable tunneling field was successfully made by Kadowaki and Nishimori (1998) (see Fig. 9.4). This paper was instrumental in further growths in QA. Brooke et al. (1999) further developed their previous experimental works and reported the advantages of QA (see Fig. 9.5) in some disordered magnetic sample. Such experimental realizations of the advantage of QA promoted the field to exciting levels. Soon, many experimental and theoretical groups from all over the world started paying attention to studying QA exhaustively. Farhi et al. (2001) showed the potential benefit of zero temperature quantum adiabatic algorithm

Science
AAAS

Science **284**, 779 (1999)

Quantum Annealing of a Disordered Magnet

J. Brooke,[1] D. Bitko,[1] T. F. Rosenbaum,[1*] G. Aeppli[2]

Traditional simulated annealing uses thermal fluctuations for convergence in optimization problems. Quantum tunneling provides a different mechanism for moving between states, with the potential for reduced time scales. Thermal and quantum annealing are compared in a model disordered magnet, where the effects of quantum mechanics can be tuned by varying an applied magnetic field. The results indicate that quantum annealing hastens convergence to the optimum state.

Fig. 9.5 Title and abstract of the first published experimental paper reporting on the advantage of quantum annealing in finding the ground state of disordered magnets. Such experimental observations have put quantum annealing on firm physical ground. Some related references are Wu et al. (1991), Wu et al. (1993), Bitko et al. (1996), Kadowaki and Nishimori (1998). (Permission to use title and abstract from the paper is given by The American Association for the Advancement of Science)

in solving NP-hard computational problems (see Fig. 9.6). The decrease of residual energy (i.e., the energy difference between the final annealed and the original ground state energy) with annealing time was estimated by Santoro et al. (2002) (see Fig. 9.7). Between the years 2004 and 2006, many compact reviews on QA had come to the literature, which are really helpful in understanding the QA techniques in a more meaningful way. Santoro and Tosatti (2006) have illustrated the applications of the QA scheme in many toy models as well as in NP-hard disordered systems (see Fig. 9.8). They have also discussed the numerical implementation of QA through both deterministic and stochastic algorithms. The physical structure of QA and its applications in the computational field have been reviewed by Das and Chakrabarti (2008) (see Fig. 9.9). The mathematical framework behind QA have been discussed in the review of Morita and Nishimori (2008) (see Fig. 9.10). They have examined the convergence conditions of QA algorithms in finding the optimal cost of a multivariable function. Ancona-Torres et al. (2008) experimentally demonstrated that the dynamics of spin glass systems in the presence of quantum fluctuation is quite different from thermally driven dynamics (see Fig. 9.11).

Science
AAAS

Science **292**, 472 (2001)

A Quantum Adiabatic Evolution Algorithm Applied to Random Instances of an NP-Complete Problem

Edward Farhi,[1]* Jeffrey Goldstone,[1] Sam Gutmann,[2] Joshua Lapan,[3] Andrew Lundgren,[3] Daniel Preda[3]

A quantum system will stay near its instantaneous ground state if the Hamiltonian that governs its evolution varies slowly enough. This quantum adiabatic behavior is the basis of a new class of algorithms for quantum computing. We tested one such algorithm by applying it to randomly generated hard instances of an NP-complete problem. For the small examples that we could simulate, the quantum adiabatic algorithm worked well, providing evidence that quantum computers (if large ones can be built) may be able to outperform ordinary computers on hard sets of instances of NP-complete problems.

Fig. 9.6 Title and abstract of a paper reporting on the zero temperature quantum adiabatic algorithm for NP-hard problems. A related reference here is Kadowaki and Nishimori (1998). (Permission to use title and abstract from the paper is given by The American Association for the Advancement of Science)

The studies on QA have been able to draw the attention even at the popular level when D-Wave (Johnson et al., 2011) (see Fig. 9.12) systems announced the news of building the first quantum computer in 2011. That machine operated with 128 qubits based on the basic principle of QA. Such exotic news created an upsurge of interest in the scientific community. Some well-known research groups have shown their interest in checking the performance of such machines and their investigations have resulted in positive outcomes which are probably a good answer to the criticisms made on the QA as well as on the quantum annealer. In May 2013, the BBC news blog indicated the purchase deal between NASA group of consortium and D-Wave for a 512-qubit quantum annealer computer. Bapst et al. (2013) have reviewed both thermal and quantum annealing dynamics in the context of quantum spin glasses. Boixo et al. (2014) (see Fig. 9.13) as well as many newspapers have reported the successful tests of the quantum annealer machine with about 100 qubits. The reasons behind the absence of quantum speed-up in the D-Wave machine have been discussed by Katzgraber et al. (2014). They have conjectured that in the quantum annealer machine, the search for the ground state of spin

Science
AAAS

Science **295**, 2427 (2002)

Theory of Quantum Annealing of an Ising Spin Glass

Giuseppe E. Santoro,[1] Roman Martoňák,[2,3] Erio Tosatti,[1,4]* Roberto Car[5]

Probing the lowest energy configuration of a complex system by quantum annealing was recently found to be more effective than its classical, thermal counterpart. By comparing classical and quantum Monte Carlo annealing protocols on the two-dimensional random Ising model (a prototype spin glass), we confirm the superiority of quantum annealing relative to classical annealing. We also propose a theory of quantum annealing based on a cascade of Landau-Zener tunneling events. For both classical and quantum annealing, the residual energy after annealing is inversely proportional to a power of the logarithm of the annealing time, but the quantum case has a larger power that makes it faster.

Fig. 9.7 Title and abstract of a paper on the application of quantum annealing in estimating the remaining fraction of undesired solutions in some optimization searches in Ising spin glasses. They also indicated the relative fastness of quantum annealing with respect to simulated annealing. Some related references here are Wu et al. (1993), Finnila et al. (1994), Kadowaki and Nishimori (1998), Brooke et al. (1999), Brooke et al. (2001) and Farhi et al. (2001). (Permission to use title and abstract from the paper is given by The American Association for the Advancement of Science)

glass defined on the Chimera graph, is not the appropriate way to check the performance of QA. In order to test the efficiency of a quantum algorithm, Katzgraber et al. (2014) suggested to deal with problems (embedded in Chimera topology) that have spin glass phase at finite temperature. Cohen and Tamir (2014) discussed and compared the quantum gate-based computer with the quantum annealer of the D-Wave computer. Heim et al. (2015) deeply probed into the issue of quantum speed-up (see Fig. 9.14). Although some observations have been reported about the ascendancy of QA over SA in searching for the ground state of two-dimensional Ising spin glass studied by the quantum Monte Carlo technique, they suggest that such advantage is not available in a physical quantum device in continuous time limit. Liu et al. (2015) examined the dynamics of a Ising spin glass where each spin interacts with three other spins through antiferromagnetic couplings

INSTITUTE OF PHYSICS PUBLISHING — JOURNAL OF PHYSICS A: MATHEMATICAL AND GENERAL

J. Phys. A: Math. Gen. 39 (2006) R393–R431 — doi:10.1088/0305-4470/39/36/R01

TOPICAL REVIEW

Optimization using quantum mechanics: quantum annealing through adiabatic evolution

Giuseppe E Santoro[1,2] and Erio Tosatti[1,2]

[1] International School for Advanced Studies (SISSA) and CNR-INFM Democritos National Simulation Center, Via Beirut 2-4, I-34014 Trieste, Italy
[2] International Center for Theoretical Physics (ICTP), Trieste, Italy

Abstract We review here some recent work in the field of quantum annealing, alias adiabatic quantum computation. The idea of quantum annealing is to perform optimization by a quantum adiabatic evolution which tracks the ground state of a suitable time-dependent Hamiltonian, where 'h' is slowly switched off. We illustrate several applications of quantum annealing strategies, starting from textbook toy-models—double-well potentials and other one-dimensional examples, with and without disorder. These examples display in a clear way the crucial differences between classical and quantum annealing. We then discuss applications of quantum annealing to challenging hard optimization problems, such as the random Ising model, the travelling salesman problem and Boolean satisfiability problems. The techniques used to implement quantum annealing are either deterministic Schr¨odinger's evolutions, for the toy models, or pathintegral MonteCarlo and Green's functionMonteCarlo approaches, for the hard optimization problems. The crucial role played by disorder and the associated non-trivial Landau–Zener tunnelling phenomena is discussed and emphasized.

Fig. 9.8 Two important early reviews helped the subsequent development of quantum annealing significantly. Title and abstract for the first review on adiabatic quantum computation and annealing, proclaiming that the idea of quantum tunneling through the infinitely high energy barriers in long-range frustrated spin glasses was introduced in Ray et al. (1989). Other related references here are Finnila et al. (1994), Chakrabarti et al. (1996), Kadowaki and Nishimori (1998), Brooke et al. (1999) and Das et al. (2005). (Permission to use title and abstract from the paper is given by Institute of Physics)

(see Fig. 9.15). Their studies suggest that SA outperforms QA in bringing the system to the spin glass phase. Albash and Lidar (2015) addressed (see Fig. 9.16) the highly debated issue of decoherance in quantum systems during the course of QA, compared to that in SA. They conclude that, as such, decoherencence does not affect adiabatic quantum computers, while they affect gate-based quantum computers. Hashizume et al. (2015) applied a singular-value decomposition scheme through QA to analyze two-dimensional data sets and image data (see Fig. 9.17). Authors constructed a variance–covariance matrix from a given set of tabulated data. Hashizume et al. (2015) used QA to extract the eigen vector of the variance–covariance matrix corresponding to the largest eigenvalue but their studies do not indicate any quantum speed-up over the classical algorithm in this case. Fastness of quantum algorithm depends upon the energy gap between the ground state and the next excited state of the problem Hamiltonian. The dimension of Hilbert space grows

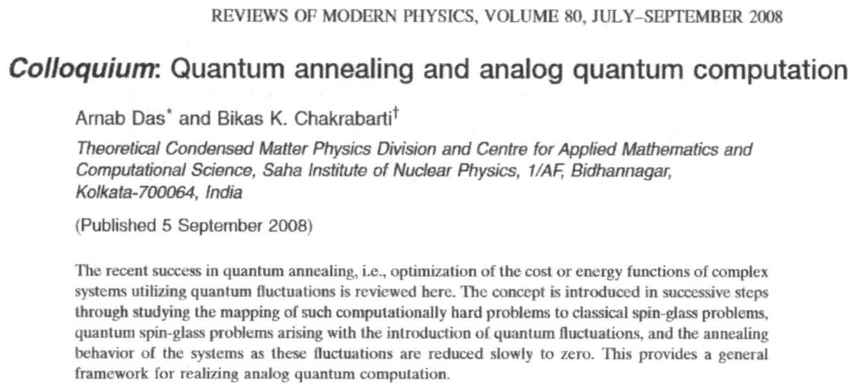
REVIEWS OF MODERN PHYSICS, VOLUME 80, JULY–SEPTEMBER 2008

***Colloquium*: Quantum annealing and analog quantum computation**

Arnab Das* and Bikas K. Chakrabarti†

Theoretical Condensed Matter Physics Division and Centre for Applied Mathematics and Computational Science, Saha Institute of Nuclear Physics, 1/AF, Bidhannagar, Kolkata-700064, India

(Published 5 September 2008)

The recent success in quantum annealing, i.e., optimization of the cost or energy functions of complex systems utilizing quantum fluctuations is reviewed here. The concept is introduced in successive steps through studying the mapping of such computationally hard problems to classical spin-glass problems, quantum spin-glass problems arising with the introduction of quantum fluctuations, and the annealing behavior of the systems as these fluctuations are reduced slowly to zero. This provides a general framework for realizing analog quantum computation.

Fig. 9.9 Title and abstract from the second (early) review on quantum annealing and quantum computation. Authors summarized that following the indication of Ray et al. (1989), both theoretical and experimental successful studies on QA made this technique extremely useful in solving hard optimization problems, in binary to analog quantum computers. Some related references here are Finnila et al. (1994), Kadowaki and Nishimori (1998), Brooke et al. (1999), Farhi et al. (2001) and Santoro et al. (2002). (Permission to use title and abstract from the paper is given by American Physical Society)

exponentially with the system size. Such thing puts a strong restriction on the analytic evaluation of the energy gap. Mandra et al. (2015) proposed a method to reduce the dimension of the Hilbert space without exploiting any symmetry of the problem Hamiltonian (see Fig. 9.18). Using this method, they found that QA works better than the classical algorithm in the Grover problem in which one needs to search for a specific target in an unstructured database.

9.3 Review of Some Recent Discussion Papers and Arxiv Preprints

Very recently a collection of reviews and discussion papers have been published in the European Physical Journal (Special Topics) covering the present scenario and many open questions related to QA. This follows an earlier collection by Das and Chakrabarti (2005). We briefly survey a few issues of this recent (2015) collection along with some other recent preprints or publications associated with QA.

With antiferromagnetic fluctuations, Seki and Nishimori (2015) performed QA on a Hopfield model with k-spin interactions. They observed that such a trick works well for $3 \leq k \leq 21$. They however observed that there is no such advantage for $k = 2$. An approximate calculation is shown in Mukherjee and Chakrabarti (2015) to demonstrate the effectiveness of QA in context of search time. They revisited the work of Das et al. (2005)

JOURNAL OF MATHEMATICAL PHYSICS **49**, 125210 (2008)

Mathematical foundation of quantum annealing

Satoshi Morita[1,a)] and Hidetoshi Nishimori[2]
[1]*International School for Advanced Studies (SISSA), Via Beirut 2-4, I-34014 Trieste, Italy*
[2]*Department of Physics, Tokyo Institute of Technology, Oh-okayama, Meguro-ku, Tokyo 152-8551, Japan*

(Received 4 June 2008; accepted 10 September 2008; published online 15 December 2008)

Quantum annealing is a generic name of quantum algorithms that use quantum-mechanical fluctuations to search for the solution of an optimization problem. It shares the basic idea with quantum adiabatic evolution studied actively in quantum computation. The present paper reviews the mathematical and theoretical foundations of quantum annealing. In particular, theorems are presented for convergence conditions of quantum annealing to the target optimal state after an infinite-time evolution following the Schrödinger or stochastic (Monte Carlo) dynamics. It is proved that the same asymptotic behavior of the control parameter guarantees convergence for both the Schrödinger dynamics and the stochastic dynamics in spite of the essential difference of these two types of dynamics. Also described are the prescriptions to reduce errors in the final approximate solution obtained after a long but finite dynamical evolution of quantum annealing. It is shown there that we can reduce errors significantly by an ingenious choice of annealing schedule (time dependence of the control parameter) without compromising computational complexity qualitatively. A review is given on the derivation of the convergence condition for classical simulated annealing from the view point of quantum adiabaticity using a classical-quantum mapping. © *2008 American Institute of Physics.*
[DOI: 10.1063/1.2995837]

Fig. 9.10 Title and abstract from the review discussing the detailed mathematical structure and theorems related to QA. Some related references here are Finnila et al. (1994), Kadowaki and Nishimori (1998), Santoro et al. (2002), Santoro and Tosatti (2006), Das et al. (2005), Das and Chakrabarti (2005) and Das and Chakrabarti (2008). (Permission to use title and abstract from the paper is given by American Institute of Physics)

to indicate the benefit of QA in evaluating the ground state of kinetically constrained systems. Authors also focused on the potential advantage of using tunable longitudinal fields in the QA of Ising spin glass indicated by Rajak and Chakrabarti (2014). Hen and Young (2015) discussed the several computational algorithms related to classical and quantum annealing. Authors also gave the physical picture of zero temperature adiabatic evolution of a quantum system in connection with QA. Finally, they gave suggestions on how to get sustained coherence of qubits in D-Wave machine; they also indicated some probable ways of improving the hardware of such machines. Boixo et al. (2015) showed the superiority of QA over SA in some specific problems. They provided a mapping of classical stochastic matrix into a quantum Hamiltonian. The adiabatic evolution of such a quantum Hamiltonian to reach the ground state, is restricted by the vanishingly small spectral gap (the energy gap between two lowest energy states) for large system sizes.

PRL **101**, 057201 (2008) PHYSICAL REVIEW LETTERS week ending 1 AUGUST 2008

Quantum and Classical Glass Transitions in $LiHo_xY_{1-x}F_4$

C. Ancona-Torres,[1] D. M. Silevitch,[1] G. Aeppli,[2] and T. F. Rosenbaum[1,*]

[1]*The James Franck Institute and Department of Physics, The University of Chicago, Chicago, Illinois 60637, USA*

[2]*London Centre for Nanotechnology and Department of Physics and Astronomy, UCL, London, WC1E 6BT, United Kingdom*

(Received 14 January 2008; revised manuscript received 2 April 2008; published 30 July 2008)

When performed in the proper low-field, low-frequency limits, measurements of the dynamics and the nonlinear susceptibility in the model Ising magnet in a transverse field $LiHo_xY_{1-x}F_4$ prove the existence of a spin-glass transition for $x = 0.167$ and 0.198. The classical behavior tracks for the two concentrations, but the behavior in the quantum regime at large transverse fields differs because of the competing effects of quantum entanglement and random fields.

Fig. 9.11 Title and abstract from a paper extending and clarifying the quantum annealing method used by Brooke et al. (1999) for quantum glasses. Authors argued that at low enough temperatures, and "in cases where barriers to relaxation are tall and narrow, quantum mechanics can enhance the ability to traverse the free energy surface (Ray et al. 1989)". Other related references cited are Wu et al. (1991), Wu et al. (1993). (Permission to use title and abstract from the paper is given by American Physical Society)

Boixo et al. (2015) devised a technique to amplify such spectral gaps for frustration-free Hamiltonians, which can make QA faster than SA. They also gave some idea on diabatic evolution of quantum states to enhance speed of QA. Fault detection in electrical power distribution system for given observations through QA was made by Perdomo-Ortiz et al. (2015). Their work provided a mapping of the optimization problem into quadratic unconstrained binary optimization (QUBO) and subsequent implementation in a quantum annealer. They cooked up an Ising Hamiltonian associated with the cost function which has gauge symmetry. Their analysis indicated that a proper choice of gauge can significantly enhance the performance of QA. Kechedzhi and Smelyanskiy (2015) proposed an algorithm to simulate the open system QA, which is naturally close to the study of the realistic quantum dynamics in quantum devices. Below some optimal limit of noise, they claimed such algorithms can utilize the thermal fluctuation in course of the optimization. They implemented the algorithm in studying the QA on p-spin models. They found that the quantum fluctuation breaks the degeneracy of metastable states, which essentially indicates the advantage of QA in this case. Very recently, Boixo et al. (2016) showed that the quantum computations in analog quantum devices are anchored by the phenomena of multiqubit tunneling across the energy barriers (see Fig. 9.19). Their experimental work with 200 qubits indicate that quantum tunneling becomes superior to thermal hopping in the search of global minima.

Silevitch et al. (2015) discussed their experimental works on QA. They performed both SA and QA on the family of magnetic samples $Li(Ho,Y)F_4$. Their observations

LETTER

194 | NATURE | VOL 473 | 12 MAY 2011
doi:10.1038/nature10012

Quantum annealing with manufactured spins

M. W. Johnson[1], M. H. S. Amin[1], S. Gildert[1], T. Lanting[1], F. Hamze[1], N. Dickson[1], R.Harris[1], A. J. Berkley[1], J. Johansson[$], P. Bunyk[1], E. M. Chapple[1], C. Enderud[1], J. P. Hilton[1], K. Karimi[1], E. Ladizinsky[1], N. Ladizinsky[1], T. Oh[1], I. Perminov[1], C. Rich[1], M. C. Thom[1], E. Tolkacheva[1], C. J. S. Truncik[·], S. Uchaikin[1], J.Wang[1], B.Wilson[1] & G. Rose[1]

Many interesting but practically intractable problemscan be reduced to that of finding the ground state of a system of interacting spins; however, finding such a ground state remains computationally difficult1. It is believed that the ground state of somenaturally occurring spin systems can be effectively attained through a process called quantum annealing [2,3]. If it could be harnessed, quantum annealing might improve on known methods for solving certain types of problem [4,5]. However, physical investigation of quantum annealing has been largely confined tomicroscopic spins in condensed-matter systems [6-12]. Here we use quantum annealing to find the ground state of an artificial Ising spin system comprising an array of eight superconducting flux quantum bits with programmable spin-spin couplings. We observe a clear signature of quantum annealing, distinguishable from classical thermal annealing through the temperature dependence of the time at which the system dynamics freezes. Our implementation can be configured in situ to realize a wide variety of different spin networks, each of which can be monitored as it moves towards a low-energy configuration [13,14].
This programmable artificial spin network bridges the gap between the theoretical study of ideal isolated spin networks and the experimental investigation of bulk magnetic samples. Moreover, with an increased number of spins, such a system may provide a practical physical means to implement a quantum algorithm, possibly allowing more-effective approaches to solving certain classes of hard combinatorial optimization problems.

[1]D-Wave Systems Inc., 100-4401 Still Creek Drive, Burnaby, British Columbia V5C 6G9, Canada.
[$]Department of Natural Sciences, University of Agder, Post Box 422, NO-4604 Kristiansand, Norway.
[·]Department of Physics, Simon Fraser University, Burnaby, British Columbia V5A 1S6, Canada.

Fig. 9.12 Title and abstract of the first paper by the D-Wave group giving the basic architecture of their quantum annealing processor. Some related references here are Finnila et al. (1994), Kadowaki and Nishimori (1998), Brooke et al. (1999) and Farhi et al. (2001). (Permission to use title and abstract from the paper is given by Nature Publishing Group)

asserted that the QA settles the system into the final state quicker than SA and the final annealed states of SA and QA are not identical. Chancellor et al. (2015) implemented the maximum entropy method in D-Wave chips to extract information from a noise influenced signal. They observed that such technique produces closely better results than another approach called maximum likelihood method. They generalized their method to address the limitations of QA algorithm in case of Ising spin glass.

With the aim of testing the quantum nature of the D-Wave quantum annealer, Albash et al. (2015) rigorously analyzed the data obtained from a D-Wave quantum annealer with 100 qubits. They inferred that simulated quantum annealing and classical rotor model correlate well with the experimental data in predicting the ground state probability distribution. However, both the algorithms fail to explain the observed excited state spectrum. In fact there is no single algorithm which can successfully deal with the entire set of experimental data. Inoue (2015) gave a precise description of QA through the analysis of infinite range spin glass models. He emphasized on the issue of replica

ARTICLES

nature physics

PUBLISHED ONLINE: 28 FEBRUARY 2014 | DOI: 10.1038/NPHYS2900

Evidence for quantum annealing with more than one hundred qubits

Sergio Boixo[1], Troels F. Rønnow[2], Sergei V. Isakov[2], Zhihui Wang[3], David Wecker[4], Daniel A. Lidar[5], John M. Martinis[6] and Matthias Troyer[2*]

Quantum technology is maturing to the point where quantum devices, such as quantum communication systems, quantum random number generators and quantum simulators may be built with capabilities exceeding classical computers. A quantum annealer, in particular, solves optimization problems by evolving a known initial configuration at non-zero temperature towards the ground state of a Hamiltonian encoding a given problem. Here, we present results from tests on a 108 qubit D-Wave One device based on superconducting flux qubits. By studying correlations we find that the device performance is inconsistent with classical annealing or that it is governed by classical spin dynamics. In contrast, we find that the device correlates well with simulated quantum annealing. We find further evidence for quantum annealing in the form of small-gap avoided level crossings characterizing the hard problems. To assess the computational power of the device we compare it against optimized classical algorithms.

Fig. 9.13 Title, abstract and some excerpts from a paper by scientists from the University of Southern California, University of California, ETH Zurich and Microsoft Research. They argued in favour of the quantum nature of the D-Wave quantum annealer by comparing its performance with that of a classical one. In the introduction, the authors had commented "The phenomena of quantum tunneling suggests that it can be more efficient to explore state space quantum mechanically in a quantum annealer (Ray et al. 1989, Finnila et al. 1994, Kadowaki and Nishimori 1998)". Some other related references here are Brooke et al. (1999), Farhi et al. (2001) and Johnson et al. (2011). (Permission to use title and abstract from the paper is given by Nature Publishing Group)

symmetry breaking in such spin glass models. We are yet to have any clear evidence to state decidedly whether such replica symmetry breaking or restoration occurs in spin glass systems. O'Gorman et al. (2015) used the QA scheme in the problem of Bayesian network structure learning BNSL. A Bayesian network is a graphical model which represents a set of random variables and their inter correlation. To adopt QA in this problem, they mapped BNSL into QUBO, which is equivalent to a set of Ising spins with arbitrary 2-body interactions. O'Gorman et al. (2015) addressed the limitation in system size of the present quantum annealer machine for which such a device becomes unable to compete with the classical algorithm. del Campo and Sengupta (2015) discussed the lose of adiabaticity of quantum systems near the critical point which essentially does not allow the system to settle in the ground state. They also suggested a few remedies to reduce the defect density which is a quantitative measure of production of excited states. Zhu et al. (2015) have discussed their Monte Carlo studies on the models associated with a D-Wave Two quantum

Science
AAAS

Science **384,** 215 (2015)

Quantum versus classical annealing of Ising spin glasses

Bettina Heim,[1] Troels F. Rønnow,[1] Sergei V. Isakov,[2] Matthias Troyer[1*]

Quantum annealers use quantum fluctuations to escape local minima and find low-energy configurations of a physical system. Strong evidence for superiority of quantum annealing (QA) has come from comparing QA implemented through quantum Monte Carlo (QMC) simulations to classical annealing. Motivated by recent experiments, we revisit the question of when quantum speedup may be expected. Although a better scaling is seen for QA in two-dimensional Ising spin glasses, this advantage is due to time discretization artifacts and measurements that are not possible on a physical quantum annealer. Simulations in the physically relevant continuous time limit, on the other hand, do not show superiority. Our results imply that care must be taken when using QMC simulations to assess the potential for quantum speedup.

Fig. 9.14 Title and abstract from a paper where authors critically investigate the claim of the probable quantum speed-up over classical annealing. Their investigations suggest that the previously observed (Santoro et al., 2002; Martoňák et al., 2002) advantage of QA compared to SA in searching for the ground state of 2D spin glasses was due to the discretization of time in the quantum Monte Carlo algorithm. They have not found any advantage in continuous time limit. Other related cited papers are Ray et al. (1989), Finnila et al. (1994), Kadowaki and Nishimori (1998), Farhi et al. (2001) and Das and Chakrabarti (2008). (Permission to use title and abstract from the paper is given by The American Association for the Advancement of Science)

annealer. Their analysis suggests that either error correction of the implementation technique or noise reduction is required to increase the number of qubits. Zhu et al. (2015) also provided the way to develop hard instances, which are robust to noise. Gra ß et al. (2015) showed that a Mattis spin glass phase can be obtained in a trapped ions system by supervising the detuning of a spin–phonon coupling. Their studies on a trapped ions chain under the influence of transverse field speculate that such systems can be used to implement QA. The hardware limitations of the quantum annealer force such machine to approximate a large set of interactions between the qubits, which essentially introduces error in the QA technique. A remedy to this problem has been found by Vinci et al. (2015)

PRL **114,** 147203 (2015) PHYSICAL REVIEW LETTERS week ending 10 APRIL 2015

Quantum versus Classical Annealing: Insights from Scaling Theory and Results for Spin Glasses on 3-Regular Graphs

Cheng-Wei Liu, Anatoli Polkovnikov, and Anders W. Sandvik
Department of Physics, Boston University, 590 Commonwealth Avenue, Boston, Massachusetts 02215, USA
(Received 25 September 2014; published 7 April 2015)

We discuss an Ising spin glass where each $S = 1/2$ spin is coupled antiferromagnetically to three other spins (3-regular graphs). Inducing quantum fluctuations by a time-dependent transverse field, we use out-of-equilibrium quantum Monte Carlo simulations to study dynamic scaling at the quantum glass transition. Comparing the dynamic exponent and other critical exponents with those of the classical (temperature-driven) transition, we conclude that quantum annealing is less efficient than classical simulated annealing in bringing the system into the glass phase. Quantum computing based on the quantum annealing paradigm is therefore inferior to classical simulated annealing for this class of problems. We also comment on previous simulations where a parameter is changed with the simulation time, which is very different from the true Hamiltonian dynamics simulated here.

DOI: 10.1103/PhysRevLett.114.147203 PACS numbers: 75.10.Nr, 03.67.Ac, 05.30.Rt, 75.10.Jm

Fig. 9.15 Title and abstract from a paper discussing the out of equilibrium quantum Monte Carlo study of spin glasses on 3-regular graphs. Authors observe the inability of quantum algorithm regarding fastness relative to SA in bringing down the system to glassy state. Some related references here are Ray et al. (1989), Finnila et al. (1994), Kadowaki and Nishimori (1998), Brooke et al. (1999), Farhi et al. (2001), Santoro et al. (2002), Das and Chakrabarti (2008) and Boixo et al. (2014). (Permission to use title and abstract from the paper is given by American Physical Society)

and the efficiency of such a scheme is tested in the D-Wave machine. Quantum annealing correction (QAC) improves the performance of the quantum annealer by limiting the excitations over the ground state. Matsuura et al. (2015) provided the theoretical analysis of QAC. They tried to explain the success of QAC through the mean field analysis of p-body infinite range transverse Ising model and quantum Hopfield model.

Along with these keen positive discussions, there are also some gloomy forecasts about QA. Laumann et al. (2015) expressed their suspicion over the capability of QA in handling NP hardness. They debated on the rendition of QA on the basis of the possibility of Anderson localization of the system during adiabatic evolution. They also questioned the quantumness of D-Wave machine due to the presence of noise in the course of operation. Hen and Young (2015) unveiled the absence of quantum speed-up in optimization problems. In building the general algorithm, they recommended the use of any previous information regarding the optimization problem which is already solved for any instance. Katzgraber et al. (2015) proposed that proper scaling can display the poor performance of the D-Wave machine relative to classical algorithms in cases of computationally hard spin glass instances (see Fig. 9.20). They are also doubtful about the quantum character of such machines.

PHYSICAL REVIEW A **91**, 062320 (2015)

Decoherence in adiabatic quantum computation

Tameem Albash[1,2,3] and Daniel A. Lidar[2,3,4,5]
[1]*Information Sciences Institute, University of Southern California, Marina del Rey, California 90292, USA*
[2]*Department of Physics and Astronomy, University of Southern California, Los Angeles, California 90089, USA*
[3]*Center for Quantum Information Science & Technology, University of Southern California, Los Angeles, California 90089, USA*
[4]*Department of Electrical Engineering, University of Southern California, Los Angeles, California 90089, USA*
[5]*Department of Chemistry, University of Southern California, Los Angeles, California 90089, USA*
(Received 4 April 2015; published 17 June 2015)

Recent experiments with increasingly larger numbers of qubits have sparked renewed interest in adiabatic quantum computation, and in particular quantum annealing. A central question that is repeatedly asked is whether quantum features of the evolution can survive over the long time scales used for quantum annealing relative to standard measures of the decoherence time. We reconsider the role of decoherence in adiabatic quantum computation and quantum annealing using the adiabatic quantum master-equation formalism. We restrict ourselves to the weak-coupling and singular-coupling limits, which correspond to decoherence in the energy eigenbasis and in the computational basis, respectively. We demonstrate that decoherence in the instantaneous energy eigenbasis does not necessarily detrimentally affect adiabatic quantum computation, and in particular that a short single-qubit T_2 time need not imply adverse consequences for the success of the quantum adiabatic algorithm. We further demonstrate that boundary cancellation methods, designed to improve the fidelity of adiabatic quantum computing in the closed-system setting, remain beneficial in the open-system setting. To address the high computational cost of master-equation simulations, we also demonstrate that a quantum Monte Carlo algorithm that explicitly accounts for a thermal bosonic bath can be used to interpolate between classical and quantum annealing. Our study highlights and clarifies the significantly different role played by decoherence in the adiabatic and circuit models of quantum computing.

Fig. 9.16 Title and abstract from a paper where authors scrutinize the role of decoherence in quantum computation as well as quantum annealing. They indicate that decoherence does not affect adiabatic quantum computation. The related references here are Ray et al. (1989), Kadowaki and Nishimori (1998), Farhi et al. (2001), Das and Chakrabarti (2008), Boixo et al. (2013) and Boixo et al. (2014). (Permission to use title and abstract from the paper is given by American Physical Society)

9.4 Applications of QA

Perdomo-Ortiz et al. (2012) has reported the first successful application of quantum annealers in evaluating the low energy conformations of lattice protein models (see Fig. 9.21). Using a quantum annealer, Rosenberg et al. (2015) searched for an optimal way to invest money in a set of assets such that after certain time, the expected return will be maximum. To solve this problem, they produced a formalism taking account of the transition costs as well as permanent and temporary market impact costs. Adachi and Henderson (2015) employed D-Wave machines for the training of deep neural networks. They concluded that quantum sampling based training of such networks can outperform classical contrastive divergence training. These investigations are showing bright prospects of the interdisciplinary applications of QA.

PHYSICAL REVIEW E **92**, 023302 (2015)

Singular-value decomposition using quantum annealing

Yoichiro Hashizume,* Takashi Koizumi, Kento Akitaya, Takashi Nakajima,† and Soichiro Okamura‡
Department of Applied Physics, Tokyo University of Science, Tokyo 125-8585, Japan

Masuo Suzuki§
Computational Astrophysics Laboratory, RIKEN, 2-1 Hirosawa, Wako, Saitama 351-0198, Japan
(Received 21 April 2015; published 10 August 2015)

In the present study, we demonstrate how to perform, using quantum annealing, the singular value decomposition and the principal component analysis. Quantum annealing gives a way to find a ground state of a system, while the singular value decomposition requires the maximum eigenstate. The key idea is to transform the sign of the final Hamiltonian, and the maximum eigenstate is obtained by quantum annealing. Furthermore, the adiabatic time scale is obtained by the approximation focusing on the maximum eigenvalue.

DOI: 10.1103/PhysRevE.92.023302 PACS number(s): 95.75.Pq, 03.67.Ac

Fig. 9.17 Title and abstract from a paper reporting on an interesting implementation of singular-value decomposition by QA in cases of big data and image analysis. The related references are Ray et al. (1989), Finnila et al. (1994), Kadowaki and Nishimori (1998), Farhi et al. (2001), Das and Chakrabarti (2005), Das and Chakrabarti (2008) and Suzuki et al. (2013). (Permission to use title and abstract from the paper is given by American Physical Society)

9.5 Summary and Discussions

Instead of sequential search of entire the configurational space, SA shows an alternative way to find the global minima of an optimization problem. In case of NP-hard problems, the thermal escape probability from a local minima scales as $\exp(-N)$. As a result of this, the search time to find the global minima is not bounded by any polynomial of N. The quantum tunneling probability for the same N order barrier goes as $\exp(-\sqrt{N})$ (see e.g., Ray et al., 1989; Mukherjee and Chakrabarti, 2015). Such benefit indicates a probable quantum speed-up over SA. Although this kind of indication was used to suggest (e.g., Ray et al., 1989; Farhi et al., 2001) some polynomial bound solution to NP-hard problems, QA has not yet been able to bound the search time by any polynomial in N. There are two major open questions related to QA. One of which is the role of quantum tunneling across the free energy barriers. Starting from the pioneering indication by Ray et al. (1989) and the successful numerical implementation of QA by Kadowaki and Nishimori (1998), several exhaustive theoretical efforts (see e.g., Boxio et al. (2014), Heim et al. (2015) for some recent advances) have been made in the development of QA. The experimental works of Wu et al. (1991) and Brooke et al. (1999) have been able to show a reliable physical picture of QA. Building of the quantum annealer by Johnson et al. (2011) has really stepped

PHYSICAL REVIEW A **92**, 062320 (2015)

Adiabatic quantum optimization in the presence of discrete noise: Reducing the problem dimensionality

Salvatore Mandrà, Gian Giacomo Guerreschi, and Alán Aspuru-Guzik
Department of Chemistry and Chemical Biology, Harvard University, 12 Oxford Street, 02138 Cambridge, Massachusetts, USA
(Received 26 April 2015; revised manuscript received 29 July 2015; published 10 December 2015)

Adiabatic quantum optimization is a procedure to solve a vast class of optimization problems by slowly changing the Hamiltonian of a quantum system. The evolution time necessary for the algorithm to be successful scales inversely with the minimum energy gap encountered during the dynamics. Unfortunately, the direct calculation of the gap is strongly limited by the exponential growth in the dimensionality of the Hilbert space associated to the quantum system. Although many special-purpose methods have been devised to reduce the effective dimensionality, they are strongly limited to particular classes of problems with evident symmetries. Moreover, little is known about the computational power of adiabatic quantum optimizers in real-world conditions. Here we propose and implement a general purposes reduction method that does not rely on any explicit symmetry and which requires, under certain general conditions, only a polynomial amount of classical resources. Thanks to this method, we are able to analyze the performance of "nonideal" quantum adiabatic optimizers to solve the well-known Grover problem, namely the search of target entries in an unsorted database, in the presence of discrete local defects. In this case, we show that adiabatic quantum optimization, even if affected by random noise, is still potentially faster than any classical algorithm.

DOI: 10.1103/PhysRevA.92.062320 PACS number(s): 03.67.Ac, 03.67.Lx, 02.70.−c

Fig. 9.18 Title and abstract from the paper suggesting a generalized method for the reduction of the dimension of the Hilbert space of quantum systems. The related references are Ray et al. (1989), Finnila et al. (1994), Kadowaki and Nishimori (1998), Farhi et al. (2001) and Perdomo-Ortiz et al. (2012). (Permission to use title and abstract from the paper is given by American Physical Society)

up the field to a new level. The recent detailed investigations by Katzgraber et al. (2014), Albash and Lidar (2015) and Vinci et al. (2015) made functional contributions for the successful implementation of QA and thereby quantum computation. For some specific problems, Denchev et al. (2015) claimed that the D-Wave quantum annealer is almost $\sim 10^8$ times faster than the classical algorithms. Such outcomes make the present scenario very exciting. Rapidly growing publications and fundamental works on the development of QA techniques are indicating a new era of quantum computing.

nature COMMUNICATIONS

ARTICLE

Received 3 Jun 2015 | Accepted 26 Nov 2015 | Published 7 Jan 2016 DOI: 10.1038/ncomms10327 OPEN

Computational multiqubit tunnelling in programmable quantum annealers

Sergio Boixo[1], Vadim N. Smelyanskiy[1,2], Alireza Shabani[1], Sergei V. Isakov[1], Mark Dykman[3], Vasil S. Denchev[1], Mohammad H. Amin[4,5], Anatoly Yu. Smirnov[4], Masoud Mohseni[1] & Hartmut Neven[1]

Quantum tunnelling is a phenomenon in which a quantum state traverses energy barriers higher than the energy of the state itself. Quantum tunnelling has been hypothesized as an advantageous physical resource for optimization in quantum annealing. However, computational multiqubit tunnelling has not yet been observed, and a theory of co-tunnelling under high- and low-frequency noises is lacking. Here we show that 8-qubit tunnelling plays a computational role in a currently available programmable quantum annealer. We devise a probe for tunnelling, a computational primitive where classical paths are trapped in a false minimum. In support of the design of quantum annealers we develop a nonperturbative theory of open quantum dynamics under realistic noise characteristics. This theory accurately predicts the rate of many-body dissipative quantum tunnelling subject to the polaron effect. Furthermore, we experimentally demonstrate that quantum tunnelling outperforms thermal hopping along classical paths for problems with up to 200 qubits containing the computational primitive.

[1] Google, Venice, California 90291, USA. [2] NASA Ames Research Center, Moffett Field, California 94035, USA. [3] Department of Physics and Astronomy, Michigan State University, East Lansing, Michigan 48824, USA. [4] D-Wave Systems Inc., Burnaby, British Columbia, Canada V5C 6G9. [5] Department of Physics, Simon Fraser University, Burnaby, British Columbia, Canada V5A 1S6. Correspondence and requests for materials should be addressed to S.B. (email: boixo@google.com) or to V.N.S. (smelyan@google.com).

NATURE COMMUNICATIONS | 7:10327 | DOI: 10.1038/ncomms10327 | www.nature.com/naturecommunications

Fig. 9.19 Title and abstract from the paper reporting on experimental evidence of the success of QA using quantum tunneling across free energy barriers. The related references are Ray et al. (1989), Finnila et al. (1994), Kadowaki and Nishimori (1998), Brooke et al. (1999), Santoro et al. (2002), Johnson et al. (2011) and Boixo et al. (2013).

PHYSICAL REVIEW X 5, 031026 (2015)

Seeking Quantum Speedup Through Spin Glasses: The Good, the Bad, and the Ugly*

Helmut G. Katzgraber,[1,2,3] Firas Hamze,[4] Zheng Zhu,[1] Andrew J. Ochoa,[1] and H. Munoz-Bauza[1]
[1]*Department of Physics and Astronomy, Texas A&M University, College Station, Texas 77843-4242, USA*
[2]*Materials Science and Engineering Program, Texas A&M University, College Station, Texas 77843, USA*
[3]*Santa Fe Institute, 1399 Hyde Park Road, Santa Fe, New Mexico 87501, USA*
[4]*D-Wave Systems, Inc., 3033 Beta Avenue, Burnaby, British Columbia, V5G 4M9, Canada*
(Received 6 May 2015; revised manuscript received 3 August 2015; published 1 September 2015)

There has been considerable progress in the design and construction of quantum annealing devices. However, a conclusive detection of quantum speedup over traditional silicon-based machines remains elusive, despite multiple careful studies. In this work we outline strategies to design hard tunable benchmark instances based on insights from the study of spin glasses—the archetypal random benchmark problem for novel algorithms and optimization devices. We propose to complement head-to-head scaling studies that compare quantum annealing machines to state-of-the-art classical codes with an approach that compares the performance of different algorithms and/or computing architectures on different classes of computationally hard tunable spin-glass instances. The advantage of such an approach lies in having to compare only the performance hit felt by a given algorithm and/or architecture when the instance complexity is increased. Furthermore, we propose a methodology that might not directly translate into the detection of quantum speedup but might elucidate whether quantum annealing has a "quantum advantage" over corresponding classical algorithms, such as simulated annealing. Our results on a 496-qubit D-Wave Two quantum annealing device are compared to recently used state-of-the-art thermal simulated annealing codes.

DOI: 10.1103/PhysRevX.5.031026 Subject Areas: Computational Physics, Quantum Information, Statistical Physics

Fig. 9.20 Title and abstract from a paper reporting on the failure of the quantum annealer in context of speed enhancement with respect to SA. Some related references here are Finnila et al. (1994), Kadowaki and Nishimori (1998), Brooke et al. (1999), Santoro et al. (2002), Das and Chakrabarti (2008), Boixo et al. (2014) and Mukherjee and Chakrabarti (2015). (Permission to use title and abstract from the paper is given by American Physical Society)

SCIENTIFIC REPORTS

Scientific Reports **2**, Article number: 571

Received 10 May 2012

Accepted 16 July 2012

Published 13 August 2012

Finding low-energy conformations of lattice protein models by quantum annealing

Alejandro Perdomo-Ortiz[1], Neil Dickson[2], Marshall Drew-Brook[2], Geordie Rose[2] & Alán Aspuru-Guzik[1]

[1]Department of Chemistry and Chemical Biology, Harvard University, Cambridge, MA 02138, USA, D-Wave Systems, Inc., 100-4401 Still Creek Drive, Burnaby, British Columbia V5C 6G9, Canada.

Lattice protein folding models are a cornerstone of computational biophysics. Although these models are a coarse grained representation, they provide useful insight into the energy landscape of natural proteins. Finding low-energy threedimensional structures is an intractable problem even in the simplest model, the Hydrophobic-Polar (HP) model. Description of protein-like properties are more accurately described by generalized models, such as the one proposed by Miyazawa and Jernigan (MJ), which explicitly take into account the unique interactions among all 20 amino acids. There is theoretical and experimental evidence of the advantage of solving classical optimization problems using quantum annealing over its classical analogue (simulated annealing). In this report, we present a benchmark implementation of quantum annealing for lattice protein folding problems (six different experiments up to 81 superconducting quantum bits). This first implementation of a biophysical problem paves the way towards studying optimization problems in biophysics and statistical mechanics using quantum devices.

Fig. 9.21 Title, abstract and excerpts from the first major paper supporting the claim of D-Wave quantum computer used in searching for low energy conformations of the lattice protein model, reported by the Harvard University group. The authors commented in the introductory section "Harnessing quantum-mechanical effects to speed up the solving of classical optimization problems is at the heart of quantum annealing algorithms (Finnila et al., 1994; Kadowaki and Nishimori, 1998; Farhi et al., 2001; Santoro and Tosatti, 2006; Das and Chakrabarti, 2008; Ray et al., 1989)". Some other relevant references here are Amara et al. (1993), Brooke et al. (1999), Farhi et al. (2001) and Johnson et al. (2011). (Permission to use title and abstract from the paper is given by Nature Publishing Group)

Acknowledgements

I am grateful to my collaborators A. Ghosh and A. Rajak for their useful suggestions regarding this chapter.

References

Adachi, S. H., and M. P. Henderson. 2015. 'Application of quantum annealing to training of deep neural networks'. arXiv:1510.06356.

Albash, T., and D. A. Lidar. 2015. 'Decoherence in adiabatic quantum computation'. *Phys. Rev. A* 91: 062320.

Albash, T., T. F. Rønnow, M. Troyer, and D. A. Lidar. 2015. 'Reexamining classical and quantum models for the D-Wave One processor'. *Eur. Phys. J. Special Topics* 224: 111.

Altshuler, B., H. Krovi, J. Roland, and B. I. Halperin. 2010. 'Anderson localization makes adiabatic quantum optimization fail'. *Proceedings of the National Academy of Sciences of the United States of America* 107: 12446.

Amara, P., D. Hsu, and J. E. Straub. 1993. 'Global energy minimum searches using an approximate solution of the imaginary time Schrödinger equation'. *J. Phys. Chem.* 97: 6715.

Ancona-Torres, C., D. M. Silevitch, G. Aeppli, and T. F. Rosenbaum. 2008. 'Quantum and classical glass transitions in $LiHo_xY_{1-x}F_4$'. *Phys. Rev. Lett.* 101: 057201.

Bapst, V., L. Foini, F. Krzakala, G. Semerjian, and F. Zamponi. 2013. 'The quantum adiabatic algorithm applied to random optimization problems: The quantum spin glass perspective'. *Phys. Rep.* 523: 127.

Bitko, D.,T. F. Rosenbaum, and G. Aeppli. 1996. 'Quantum critical behavior for a model magnet'. *Phys. Rev. Lett.* 77: 940.

Boixo, S., T. Albash, F. Spedalieri, N. Chancellor, and D. A. Lidar. 2013. 'Experimental signature of programmable quantum annealing'. *Nat. Commun.* 4: 2067.

Boixo, S., T. F. Rønnow, S. V. Isakov, Z. Wang, D. Wecker, D. A. Lidar, J. M. Martinis, and M. Troyer. 2014. 'Evidence for quantum annealing with more than one hundred qubits'. *Nat. Phys.* 10: 218.

Boixo, S., G. Ortiz, and R. Somma. 2015. 'Fast quantum methods for optimization'. *Eur. Phys. J. Special Topics* 224: 35.

Boixo, S., V. N. Smelyanskiy, A. Shabani, S. V. Isakov, M. Dykman, V. S. Denchev, M. H. Amin, A. Y. Smirnov, M. Mohseni, and H. Neven. 2016. 'Computational multiqubit tunnelling in programmable quantum annealers'. *Nat. Commun.* 7.

Brooke, J., D. Bitko, T. F. Rosenbaum, and G. Aeppli. 1999. 'Quantum annealing of a disordered magnet'. *Science* 284: 779.

Brooke, J., T. F. Rosenbaum, and G. Aeppli. 2001. 'Tunable quantum tunnelling of magnetic domain walls'. *Nature* 413: 610.

Chakrabarti, B. K. 1981. 'Critical behavior of the Ising spin-glass models in a transverse field'. *Phys. Rev. B* 24: 4062.

Chakrabarti, B. K., A. Dutta, and P. Sen. 1996. *Quantum Ising Phases and Transitions in Transverse Ising Models*. Springer.

Chancellor, N., S. Szoke, W. Vinci, G. Aeppli, and P. A. Warburton. 2015. 'Maximum-entropy inference with a programmable annealer'. arXiv:1506.08140.

Cohen, E., and B. Tamir. 2014. 'D-Wave and predecessors: From simulated to quantum annealing'. *Int. J. Quantum Inf.* 12: 1430002.

Das, A., and B. K. Chakrabarti. 2005. *Quantum Annealing and Related Optimization Methods*. Springer Science & Business Media.

Das, A., and B. K. Chakrabarti. 2008. 'Colloquium: Quantum annealing and analog quantum computation'. *Rev. Mod. Phys.* 80: 1061.

Das, A., B. K. Chakrabarti, and R. B. Stinchcombe. 2005. 'Quantum annealing in a kinetically constrained system'. *Phys. Rev. E* 72: 026701.

del Campo, A., and K. Sengupta. 2015. 'Controlling quantum critical dynamics of isolated systems'. *Eur. Phys. J. Special Topics* 224: 189.

Denchev, V. S., S. Boixo, S. V. Isakov, N. Ding, R. Babbush, V. Smelyanskiy, J. Martinis, and H. Neven. 2015. 'What is the computational value of finite range tunneling?' arXiv:1512.02206.

Ebeling, W., A. Engel, B. Esser, and R. Feistel. 1984. 'Diffusion and reaction in random media and models of evolution processes'. *J. Stat. Phys.* 37: 369.

Farhi, E., J. Goldstone, S. Gutmann, J. Lapan, A. Lundgren, and D. Preda. 2001. 'A quantum adiabatic evolution algorithm applied to random instances of an NP-complete problem'. *Science* 292: 472.

Finnila, A. B., M. A. Gomez, C. Sebenik, C. Stenson, and J. D. Doll. 1994. 'Quantum annealing: A new method for minimizing multidimensional functions'. *Chem. Phys. Lett.* 219: 343.

Graß, T., D. Raventós, B. Juliá-Díaz, C. Gogolin, and M. Lewenstein. 2015. 'Controlled complexity in trapped ions: From quantum Mattis glasses to number partitioning'. arXiv:1507.07863.

Hashizume, Y., T. Koizumi, K. Akitaya, T. Nakajima, S. Okamura, and M. Suzuki. 2015. 'Singular-value decomposition using quantum annealing'. *Phys. Rev. E* 92: 023302.

Heim, B., T. F. Rønnow, S. V. Isakov, and M. Troyer. 2015. 'Quantum versus classical annealing of Ising spin glasses'. *Science* 348: 215.

Hen, I., and A. P. Young. 2011. 'Exponential complexity of the quantum adiabatic algorithm for certain satisfiability problems'. *Phys. Rev. E* 84: 061152.

Hen, I., and A. P. Young. 2015. 'Performance of the quantum adiabatic algorithm on constraint satisfaction and spin glass problems'. *Eur. Phys. J. Special Topics* 224: 63.

Inoue, J. 2015. 'Infinite-range transverse field Ising models and quantum computation'. *Eur. Phys. J. Special Topics* 224: 149.

Johnson, M. W., M. H. S. Amin, S. Gildert, T. Lanting, F. Hamze, N. Dickson, R. Harris, A. J. Berkley, J. Johansson, P. Bunyk, E. M. Chapple, C. Enderud, J. P. Hilton, K. Karimi, E. Ladizinsky, N. Ladizinsky, T. Oh, I. Perminov, C. Rich, M. C. Thom, E. Tolkacheva, C. J. S. Truncik, S. Uchaikin, J. Wang, B. Wilson, and G. Rose. 2011. 'Quantum annealing with manufactured spins'. *Nature* 473: 194.

Kadowaki, T., and H. Nishimori. 1998. 'Quantum annealing in the transverse Ising model'. *Phys. Rev. E* 58: 5355.

Katzgraber, H. G., F. Hamze, and R. S. Andrist. 2014. 'Glassy chimeras could be blind to quantum speedup: Designing better benchmarks for quantum annealing machines'. *Phys. Rev. X* 4: 021008.

Katzgraber, H. G., F. Hamze, Z. Zhu, A. J. Ochoa, and H. Munoz-Bauza. 2015. 'Seeking quantum speedup through spin glasses: The good, the bad, and the ugly'. *Phys. Rev. X* 5: 031026.

Kechedzhi, K., and V. N. Smelyanskiy. 2015. 'Open system quantum annealing in mean field models with exponential degeneracy'. arXiv:1505.05878.

Kirkpatrick, S., C. D. Gelatt Jr., and M. P. Vecchi. 1983. 'Optimization by simulated annealing'. *Science* 220: 671.

Laumann, C. R., R. Moessner, A. Scardicchio, and S. L. Sondhi. 2015. 'Quantum annealing: The fastest route to quantum computation?' *Eur. Phys. J. Special Topics* 224: 75.

Liu, C., A. Polkovnikov, and A. W. Sandvik. 2015. 'Quantum versus classical annealing: Insights from scaling theory and results for spin glasses on 3-regular graphs'. *Phys. Rev. Lett.* 114: 147203.

Mandrà, S., G. G. Guerreschi, and A. Aspuru-Guzik. 2015. 'Adiabatic quantum optimization in the presence of discrete noise: Reducing the problem dimensionality'. *Phys. Rev. A* 92: 062320.

Matsuura, S., H. Nishimori, T. Albash, and D. A. Lidar. 2015. 'Mean field analysis of quantum annealing correction'. arXiv:1510.07709.

Morita, S., and H. Nishimori. 2008. 'Mathematical foundation of quantum annealing'. *J. Math. Phys.* 49: 125210.

Mukherjee, S., and B. K. Chakrabarti. 2015. 'Multivariable optimization: Quantum annealing and computation'. *Eur. Phys. J. Special Topics* 224: 17.

O'Gorman, B., R. Babbush, A. Perdomo-Ortiz, A. Aspuru-Guzik, and V. Smelyanskiy. 2015. 'Bayesian network structure learning using quantum annealing'. *Eur. Phys. J. Special Topics* 224: 163.

Perdomo-Ortiz, A., N. Dickson, M. Drew-Brook, G. Rose, and A. Aspuru-Guzik. 2012. 'Finding low-energy conformations of lattice protein models by quantum annealing'. *Sci. Rep.* 2: 571.

Perdomo-Ortiz, A., J. Fluegemann, S. Narasimhan, R. Biswas, and V. N. Smelyan-skiy.. 2015. 'A quantum annealing approach for fault detection and diagnosis of graph-based systems'. *Eur. Phys. J. Special Topics* 224: 131.

Rajak, A., and B. K. Chakrabarti. 2014. 'Quantum annealing search of Ising spin glass ground state(s) with tunable transverse and longitudinal fields'. *Ind. J. Phys.* 88: 951.

Ray, P., B. K. Chakrabarti, and A. Chakrabarti. 1989. 'Sherrington-Kirkpatrick model in a transverse field: Absence of replica symmetry breaking due to quantum fluctuations'. *Phys. Rev. B* 39: 11828.

Rosenberg, G., P. Haghnegahdar, P. Goddard, P. Carr, K. Wu, and M. de Prado. 2015. 'Solving the optimal trading trajectory problem using a quantum annealer'. arXiv:1508.06182.

Rujan, P. 1988. 'Searching for optimal configurations by simulated tunneling'. *Zeitschrift für Physik B Condensed Matter* 73: 391.

Santoro, G. E., and E. Tosatti. 2006. 'Optimization using quantum mechanics: Quantum annealing through adiabatic evolution'. *J. Phys. A: Math. Gen.* 39: R393.

Santoro, G. E., R. Martoňák, E. Tosatti, and R. Car. 2002. 'Theory of quantum annealing of an Ising spin glass'. *Science* 295: 2427.

Seki, Y., and H. Nishimori. 2015. 'Quantum annealing with antiferromagnetic transverse interactions for the Hopfield model'. *J. Phys. A: Math. Gen.* 48: 335301.

Silevitch, D. M., T. F. Rosenbaum, and G. Aeppli. 2015. 'Quantum tunneling vs. thermal effects in experiments on adiabatic quantum computing'. *Eur. Phys. J. Special Topics* 224: 25.

Suzuki, S., J. Inoue, and B. K. Chakrabarti. 2013. *Quantum Ising Phases and Transitions in Transverse Ising Models*. Springer.

Thirumalai, D., Q. Li, and T. R. Kirkpatrick. 1989. 'Infinite-range Ising spin glass in a transverse field'. *J. Phys. A: Math. Gen.* 22: 3339.

Vinci, W., T. Albash, G. Paz-Silva, I. Hen, and D. A. Lidar. 2015. 'Quantum annealing correction with minor embedding'. arXiv:1507.02658.

Wu, W., B. Ellman, T. F. Rosenbaum, G. Aeppli, and D. H. Reich. 1991. 'From classical to quantum glass'. *Phys. Rev. Lett.* 67: 2076.

Wu, W., D. Bitko, T. F. Rosenbaum, and G. Aeppli. 1993. 'Quenching of the nonlinear susceptibility at a T= 0 spin glass transition'. *Phys. Rev. Lett.* 71: 1919.

Zhu, Z., A. J. Ochoa, S. Schnabel, F. Hamze, and H. G. Katzgraber. 2015. 'Best-case performance of quantum annealers on native spin-glass bench-marks: How chaos can affect success probabilities'. arXiv:1505.02278.

Index

ε-expansion 20
$\pm J$ model 7, 22, 25
K-SAT problem 71, 269
q-fold symmetry 46
S_2 symmetry 50
Z vortex 49
Z_2 symmetry 12
3-clause 71
3-SAT problem 70–72, 184, 261, 277

acceptance probability 84–87, 90, 94, 232–233, 235, 239, 242
adiabatic 3, 150, 157–158, 176, 222, 226, 260–262, 264–267, 270, 273, 276–277, 279–282, 340, 344, 346–347, 350, 362, 364, 368, 373
adiabatic approximation 205, 215–216, 221, 231, 260
adiabatic computation 255, 259, 261, 265–266, 272, 277–278
adiabatic condition 203, 204, 215, 221–223, 226, 230–231
adiabatic quantum computation 195, 203–204, 256, 259, 262, 366, 374
adiabatic theorem 175, 203–206, 211, 215, 222, 259, 261, 276
alternating current (AC) susceptibility 201–202
Anderson localization 271, 373
annealing speed 159–164, 166, 173, 181
asymptotic expansion 156–157
AT line 36–37

balance condition 76, 266
Bayesian network 371
Bethe–Peierls–Weiss approximation 77
Binder ratio 21, 44–45, 131–136, 138–142
bipartite 24, 319
bistable systems 7–8
Bogoliubov transformation 104
Boltzmann constant 11, 61, 77, 102
Boltzmann distribution 87, 94–95, 232, 239–240
Boltzmann weight 61–62
Boolean variable 70–72, 266

canonical distribution 196
chimera graph 274, 282, 365
chiral glass order 49, 51
circuit model 255–256, 262, 265, 270, 277
classical algorithm 258, 260, 362, 366–367, 371, 373, 376
classical annealing 2, 181–182, 184–185, 189–190, 267, 273–274, 282, 372
classical Ising model 100, 108, 130, 138, 142, 171, 180, 226
classical spin glass 7, 22, 115–116
classically ordered state 101
clique cover problem 66–67
clock model 48
cluster Monte Carlo simulation 43
coherence time 164–165, 259, 277, 281
combinatorial optimization problem 1–3, 7, 22, 59–60, 63–64, 66–67, 70, 150, 170, 204,

computational complexity 64–66, 257, 273, 278, 282
computational complexity theory 64–66, 269, 273, 279
configurational average 26–27, 40, 42, 120, 131, 135, 143, 163
continuous phase transition 9, 289
continuous symmetry 50
convergence theorem 59, 84, 86, 88–89, 170, 175, 177, 181, 205, 222–223, 226, 230–234, 242
cooperative phenomena 1, 7–8, 45
coordination number 11
correlation function 9, 13, 135, 162–163, 289, 334
correlation length 10, 42–43, 51, 131–132, 138, 159–164, 289, 291, 295, 316
cost function 1, 60, 63, 66–68, 72–73, 256, 261, 266, 270, 272–273, 359, 369
critical Binder ratio 132, 135
critical phenomena 9, 18, 20, 42, 44, 104, 135, 159, 290
critical point 9, 35, 131–132, 138–142, 288–292, 294–295, 298–299, 301, 303–307, 310–311, 313, 316, 318, 321, 334, 339, 347, 349, 371
crossover 136, 150, 158
crystal field 201
Curie temperature 7
cusp 22, 84

D-Wave computer 255–256, 274–281, 365
de Almeida–Thouless line 36
decision problem 64–65
decomposition of exponential operator 192
density matrix 77–78, 193–194, 196–199, 293, 295
density matrix renormalization group 170, 191, 194–195
detailed balance condition 76, 266
deterministic Turing machine 64–65
Deutsch–Josza algorithm 257
Dirac delta function 22, 143
discontinuous phase transition 9
dislocation 62
disordered phase 8, 139
divide-and-conquer method 60
DMRG 191–192, 194
domain wall 62, 82, 150, 161–162, 164, 166
double-well 100
doublet 201
driven quantum system 296
duality technique 43
dynamical critical phenomena 159
dynamical exponent 131–132, 141, 159, 164–165, 289

edge state 288, 296, 325, 335–336, 338, 340, 344, 346, 350
Edwards–Anderson model 1, 7, 22, 27, 42, 98, 116, 137
eigenvalue 17, 74–75, 113, 134, 151, 193, 194, 196–199, 205–206, 214, 223–225, 260, 262, 265, 268, 270, 291, 308, 315, 318, 323, 336–337, 366
eigenvector 74–75, 100, 130, 134, 151, 198–199, 260–265, 267, 270, 272, 274, 276–277, 280, 315, 328–329, 336
emergent phenomena 60
energy barrier 41, 80–82, 256
energy gap 81, 150, 157, 173, 220, 223–224, 230, 277, 288, 297, 305, 311, 325, 328, 335, 338–339, 366–368
energy landscape 41, 83, 116, 271, 360
entanglement 194, 276, 281, 290, 294–295, 313, 319
entanglement entropy 288, 295, 313, 316, 320–321, 334, 347
entropy effect 62
equilibration 79
equilibration time 79
equilibrium probability distribution 75–76
equilibrium state 61–63, 73, 75, 77, 79, 81, 159, 160–161, 163, 165, 171, 196
ergodic 85–86, 91, 116, 232, 271
ergodic coefficient 86, 88, 91–93, 233, 236–238, 244
ergodic phase 41
ergodic spin glass phase 116
ergodicity 85–86, 94, 232
error correction code 259
exactly solvable model 10, 104
excited state 47, 81, 135, 150–151, 173, 216, 221, 223, 230, 260, 274, 280, 288, 293, 315, 325, 366, 370–371

fidelity susceptibility 288, 291, 299–300, 347
finite-size effect 135, 167
finite-temperature phase transition 22, 42, 44, 49–50, 104, 108, 162
first-order phase transition 8–9, 46–48, 173
Floquet theory 294
Fokker–Planck equation 83
formal solution 175
freezing process 63
freezing temperature 22
frustrated Ising spin glass 362
frustration 22, 24, 64, 72, 79, 81, 369
frustration plaquette 24

gapless phase 307–311, 340, 342–343, 347–348, 350
Geman–Geman theorem 2, 84, 96
generalized magnetization 25
generation probability 84, 88, 234, 242
Gibbs–Bogoliubov–Feynman inequality 78, 197
Glauber dynamics 163
global minimum 171, 262, 270, 272–273
graph coloring problem 66–67
Green's function Monte Carlo method 170, 185–186, 188–190, 195, 222, 242, 245, 248
ground state 7, 12, 23–25, 36, 41–42, 46–47, 49–50, 59, 62, 71, 80, 84, 100–101, 150–151, 155, 158, 162, 166, 170–173, 176–177, 179, 181, 184, 195, 201, 203, 205, 215–216, 221–223, 225–226, 228–230, 242, 248, 256, 259–260, 262, 264–265, 267, 269–271, 273–274, 276–277, 280–281, 285, 288–294, 299, 304, 313, 315–316, 325, 330–332, 339, 347, 360–366, 368, 370–373
Grover algorithm 257, 267

Hamiltonian cycles and paths problem 66
heat-bath method 76–77, 85
Heisenberg model 7, 13, 50–51, 100
Hermitian 206–207, 209, 211–212, 219, 314, 325–326
heuristic method 60
Hilbert space 268, 291, 366–367
Hopf theorem 223–225
Hopfield model 173, 367, 373
Hubbard–Stratonovich transformation 28, 32, 122, 126
Husimi–Temperely–Curie–Weiss model 13, 60, 100, 110, 115–116, 146
hypergeometric function 154

identity matrix 224
imaginary time 106, 120–121, 125, 129, 186, 268, 273
infinite-range model 13–15, 40
information theoretic measure 290, 295, 347
inhomogeneous magnetic field 68, 70, 72, 77
inhomogeneous Markov chain 84–88, 96, 231–233, 242, 248
inverse temperature 11, 45, 77, 266, 294
invisible state 47
Ising model 1–3, 4, 7–13, 15, 16–18, 19–25, 45–60, 62–63, 66–73, 77, 79, 80–82, 86, 97–101, 103–105, 108–110, 115–117, 130, 138, 142, 146–150, 164, 167, 170–171, 176–182, 189–191, 194–195, 199, 201, 226, 232, 246, 275, 292, 294, 296–297, 307, 318, 348–349, 373
iterative improvement method 60, 63

Jarzynski equality 204–205
Jordan–Wigner transformation 104, 297

kagomé lattice 24
Kibble–Zurek mechanism 3, 150, 158–160, 162–166
Kitaev model 307, 310, 348
knapsack problem 66
Kronecker delta 16

laboratory magnetic field 201
Lagrange multiplier 78, 197
Landau theory 35
Landau–Zener transition 3, 151, 154, 158, 173, 182
largest eigenvalue 75, 366
local density matrix 78, 197–199
local minima 41, 82–83, 116, 170, 178, 256, 273, 298, 359, 375
logical AND 70–71
logical OR 70–71
long-range order 9, 49–50, 333
longitudinal field 100, 368
longitudinal magnetization 114–115

Loschmidt echo 288, 292, 313, 347

macroscopic state 73
macroscopically degenerated 24, 36, 173
magnetic entropy 8, 36, 40
magnetic susceptibility 9, 13, 22, 281
magnetization 8, 11–13, 17–21, 25, 31, 61–62, 78–79, 102, 112, 114–116, 196, 198–199, 333
Majorana chain 336–337, 343
many-body localization 100
many-body problem 10
many-valley structure 41
Markov process 73–74
mass gap 135
master equation 73–74, 81, 83, 176–177, 265, 270, 279, 281
Maximum cut problem 66
maximum state set 88, 233
mean-field approximation 10–13, 15, 19–20, 77, 100–103, 170, 195, 197–199
melting process 63
Metropolis method 63, 76–77, 85, 87
Minimum spanning tree 66
molecular field 101
Monte Carlo simulation 20, 43, 108, 129–130, 133–134, 181, 185, 189, 261
multicritical point 43–44, 296, 298, 348

nonequilibrium work 205
non-deterministic Turing machine 64–65
non-ergodic 41, 270
non-ergodic phase 41
non-ergodicity 41, 116
non-negative matrix 74
nonadiabatic transition 151
nonequilibrium 79, 205
nonmagnetic 201
NP-complete 60, 65–67, 80, 258, 261
NP-complete problem 261
NP-hard 65–66, 277, 359, 363–364, 375
number partitioning problem 60, 66–67
number partitioning problem satisfiability(SAT) problem Hamiltonian cycles and paths problem 66
numerical transfer matrix calculation 43

objective function 60, 64

one-step RSB 37
Onsager reaction term 40
open boundary condition 191,194, 329, 336–337, 342, 349
order-disorder phase transition 99–100
order-disorder transition 8, 98, 100
ordered phase 8, 103, 139, 333
orthogonality catastrophe 290

Parisi equation 40
Parisi solution 36, 38–40
partial trace 193
partition function 11–12, 14, 16–17, 27–28, 54, 75, 77, 87, 104–105, 110–111, 120–122, 130, 176, 180, 227, 232, 294
path integral 268
path-integral formalism 104, 130, 180
path-integral Monte Carlo method 170, 181, 189–190, 195, 222, 231–233, 242
Pauli matrices 99, 171, 308, 328, 341
periodic boundary condition 16, 62, 295, 328–331, 336, 339, 341, 343–344
Perron–Frobenius theorem 75, 223, 229
phase factor 218
phase transition 2, 7–12, 15, 18–19, 20, 22, 24, 25, 41–50, 62, 98, 99–100, 103–104, 108–109, 114–117, 119, 128, 131, 134–142, 150, 158–160, 162, 165, 173, 261, 273, 288–290, 292, 294, 296–298, 325, 333, 347
phase-transition temperature 62
polynomial time 60, 181, 255, 268
Potts model 7, 46–49, 66–67, 72–73
probability distribution 2, 22, 27, 42–45, 48–49, 51, 73–76, 84–86, 88, 117, 181, 188, 233, 242, 248, 279, 370
projection operator 205–206, 211, 216–218

quantum adiabatic evolution 170, 203, 368
quantum adiabatic theorem 204–205, 215, 222, 259
quantum algorithm 257–258, 265, 360, 365–366
quantum annealer 256, 276–277, 279–280, 360, 364–365, 369–376
quantum annealing 1–3, 98, 150, 158, 165–167, 170–185, 188–191, 194–195, 199–205, 222, 226, 231–233, 242, 248, 255–256, 267–268, 270–274, 281, 359, 364, 368, 370, 373

quantum computing 262, 360, 376
quantum critical point 140–142, 288–290, 294, 304, 307, 311, 334, 347
quantum effect 2, 129, 131–132, 170–171, 173, 280
quantum entanglement 194, 257, 276, 293–295
quantum fidelity 290
quantum fluctuation 2, 98–100 103, 116–117, 129, 136, 142, 150, 170–171, 173, 182, 184, 200, 229–230, 270–271, 288, 291, 362, 369
quantum information 1, 100, 104, 170, 174, 256, 282, 288, 290, 292, 295–296, 327, 347
quantum Jarzynski annealing 204
quantum Monte Carlo 108–109, 138, 181–185, 261, 268, 365
quantum paramagnetic state 101
quantum phase transition 98, 115, 119, 128, 134–138, 140–142, 150, 165, 288, 290, 292, 294, 296, 325, 333, 347
quantum quenching 319–321, 324, 336–351
quantum spin glass 5, 98, 115–116, 271, 359, 364
quantum tunneling process 171
quantum Turing machine 257
quantum-classical correspondence 204, 226, 230
quantum-classical mapping 222, 226
quantum-classical transition 292
qubit 257–259, 268–270, 273–278, 280–281, 292–294, 307, 348, 364, 368–370, 372

random first-order phase transition 48
random interaction 1, 22, 25–26, 42–44, 48–49, 51, 66–68, 70, 72–73, 77, 86, 98, 165, 179, 181–182, 189, 201
random Ising model 1–3, 7, 59–60, 63, 66, 79–82, 115–116, 170, 181–182
random walk 83, 273
reduced density matrix 193–194, 293, 295
reference energy 186
relaxation time 75, 159, 160, 163–164, 203, 273, 347
renormalization group 20, 164, 170, 191, 194–195
Renyi entropy 295
replica method 26–27, 30, 40, 120
replica symmetric approximation 32, 34–36, 119, 125, 128
replica symmetric solution 32, 36–37, 118–119
replica symmetric spin glass phase 98, 116–117, 142–143, 145–146
replica symmetry breaking 36–38, 71, 116, 129, 184, 371
replica symmetry restoration 271
residual energy 80–82, 166, 181, 185, 189, 194–195, 200, 363
rotating axis 206, 210–211

saddle-point condition 30, 34, 112, 124, 128
saddle-point method 15, 30–31, 111, 124
SAT formula 71
Satisfiability (SAT) problem 66
scaling function 21, 42, 132
scaling relations 10, 295, 305, 347
Schrödinger equation 152, 170, 174–177, 185, 191–192, 194, 205, 222, 260, 268, 271, 273, 344
second largest eigenvalue 75
second-order phase transition 7–10, 12, 15, 18–20, 24, 46–47, 62, 117, 131, 150, 158–160, 162, 173
self-consistent equation 12, 15, 79
self-organizing process 63
semiclassical analysis 100, 103
semiclassical expression 182
Sherrington–Kirkpatrick model (SK model) 2, 7, 26–27, 32, 34–37, 40–42, 98, 116–119, 128–137, 143–145, 362
Shor's algorithm 258
simulated annealing 2–4, 59, 63, 73, 79, 84–88, 98, 158–159, 163–164, 166–167, 170–174, 176–179, 181–185, 189–190, 200–201, 203, 226, 230–231, 256, 266–267, 270–274, 279–280, 282
single-body problem 10, 11
single-spin approximation 118–119, 128
solidity temperature 62
spatial dimension 10, 46–51, 161
specific heat 9, 63–64, 108–109, 135
spectral gap 267, 368–369
spin glass order parameter 25, 31, 36–38, 44, 48, 119, 131, 135, 138, 146
spin glass phase 2, 7, 22, 25, 30, 32, 34–36, 41–44, 48–49, 51, 98, 116–119, 136, 142–143, 145–146, 201–203, 365–366, 372

spontaneous magnetization 102, 333
spontaneous symmetry breaking 9, 12, 162
square lattice 15, 18–19, 23–24, 43, 62, 104, 109, 138–139, 142, 181, 189, 199
SQUID 277, 280
static approximation 112, 119, 125, 128–129, 136
steady distribution 86–87, 94, 232, 239, 245–247
Stirling formula 225
stochastic dynamics 73, 84, 94, 239, 245, 247, 368
strong ergodicity 85–86, 88, 94, 96, 232–233, 239, 242, 245, 247–248
substitution 201
superconducting chain 296, 313, 326–327, 333, 340, 349–350
Suzuki–Trotter decomposition 104–105, 110, 138, 143, 180

TAP equation 40–41
thermal annealing 201–203
thermal average 21, 25, 30, 40, 131, 143, 163
thermal equilibrium state 196
thermodynamic limit 14–15, 17, 20, 71, 116, 139, 143, 150, 290
Thouless–Anderson–Palmer equation 40
three-body interaction 72
time-dependent Schrödinger equation 152, 174, 271, 344
time-evolution operator 174–175, 185, 191–194, 205, 210–211
time-ordered product operator 120, 175, 185
Toffoli gate 257
topological defect 150, 162
topological insulator 296, 325, 327
topological phase 49, 296, 325–327, 330–331, 336, 340–342, 344, 349
transfer matrix 16–17, 43, 113, 181
transition probability 73, 75, 77, 84–85, 87, 89, 91, 150–151, 157, 176, 188–189, 232–234, 236, 242–243, 246, 346
transverse field 2, 98–101, 103–105, 110, 115–119, 128–132, 134–141, 143, 145, 150, 158, 162, 164–166, 171, 173, 177, 179–185, 189, 191, 194–195, 199–201, 205, 222–223, 230, 232, 246, 267, 271, 274, 276, 280, 292–294, 296–298, 305, 313–314, 316, 319, 333, 348–349, 372
transverse magnetization 114
trap states 360
travelling salesman problem 1, 66, 68–70, 182, 184, 273, 362
triangular lattice 24
Trotter axis 106, 108, 139, 145
Trotter number 105, 108–109, 139, 143, 181–184
twisted boundary condition 292
twofold symmetry 12, 15, 24

U(1) symmetry 49
unitary matrix 193, 308
universality 10, 268, 276, 297
UNSAT formula 71

variational method 78, 197
Vertex cover problem 66
vertex *k*-coloring problem 72–73

weak ergodicity 85–86, 91, 94, 236, 239 243, 245
Weber differential equation 154–155
Whittaker function 155
XY model 7, 13, 48–51, 298, 333

zero-temperature phase transition 42–43, 98, 104, 288